# Handbook of

# ENVIRONMENTAL
## and ECOLOGICAL
# MODELING

Edited by

## S.E. Jørgensen
The Royal Danish School of Pharmacy
Section of Environmental Chemistry
Copenhagen, Denmark

## B. Halling-Sørensen
Institute of Environmental Science and Engineering
The Technical University of Denmark
Copenhagen, Denmark

## S.N. Nielsen
The Royal Danish School of Pharmacy
Section of Environmental Chemistry
Copenhagen, Denmark

**CRC Press**
Taylor & Francis Group
Boca Raton London New York

CRC Press is an imprint of the
Taylor & Francis Group, an **informa** business

CRC Press
Taylor & Francis Group
6000 Broken Sound Parkway NW, Suite 300
Boca Raton, FL 33487-2742

First issued in paperback 2019

ISBN-13: 978-1-56670-202-7 (hbk)
ISBN-13: 978-0-367-40147-4 (pbk)
Library of Congress Card Number 95-20624

**Library of Congress Cataloging-in-Publication Data**

Handbook of environmental and ecological modeling /
    edited by S.E. Jørgensen, B. Halling-Sørensen, and S.N. Neilsen.
        p.   cm.
    Includes bibliographical references and index.
    ISBN 1-56670-202-X
    1. Ecology—Simulation methods. 2. Pollution—Environmental aspects—Simulation
methods.. I. Jørgensen, Sven Erik, 1934—    . II. Halling-Sørensen, B.
III. Nielsen, Søren Nors.
QH541.15.S5H36      1995
574.5′01′1—dc20

95-20624
CIP

**Visit the Taylor & Francis Web site at**
**http://www.taylorandfrancis.com**

**and the CRC Press Web site at**
**http://www.crcpress.com**

# PREFACE

This volume attempts to capture the experience gained by the development of about 1000 models during the last decade or two. The aim has been to give modelers sufficient information about the spectrum of environmental models to be able to evaluate the applicability of previously developed models. The idea is in short: why should a modeler not stand on the shoulders of the previous modelers, when he has a model of a specific case under development? So, a modeler would typically ask the question: If somebody else has developed a model of the same or a similar problem in a similar ecosystem, what was his experience? This question is answered for about 400 models, selected among the approximately 1000 models that have been developed during the last 15-20 years. The answer is given on about 1-2 pages for each of the 400 models, and a proper list of references is included to give the reader an opportunity to get additional information if needed. The tables for the models are based upon a questionnaire, which was send to the modelers. The following questions were asked for all the models:

1. Model title:
2. Model type:
Media: water (lake ocean, estuary, river or wetland), air (local, regional or global) or terrestrial (sediment, soil, forest or agricultural)
Class: biogeochemical, toxicological or hydrological
Substance:
Nutrients (carbon, sulfur, phosphorus, oxygen, nitrogen, other (which?)
Chemicals (radionuclides, pesticides, organic compounds, other (which?)
Heavy metals (Hg,Pb,Cd,Cu,Zn, other(s) (which?)
Biological components (ecosystem, society, community, population(s), organism(s) with indications which?
3. Model purpose: (max. 50 words)
4. Short description: (max. 200 words) including lists of state variables, forcing functions, important parameters and necessary input to the model.
5. Model application(s) and state of development: application area, number of case studies.
State: conceptualization, verification, calibration, validation, prognosis made, prognosis validation.
6. Model software and hardware demands: PC-based, Mac, IBM, others (which?)
Mainframes, digital, IBM, other (which?)
Workstation, (which?)
Operative system: software demands, hardware demands.
7. Model availability: can the model be purchased, if yes what is the price?
Software available as executable code or source code?
Name, address, telephone, fax number, E-mail.
8. Model documentation and major references.
9. Other relevant information about the model.

We are very grateful to the many modelers who answered our questionnaire. More than 60% of the modelers who have received the questionnaire answered, which we have taken as an expression for a great interest among modelers for the handbook. We have recieved many letters expressing the urgent need for this work. It is therefore our hope that this handbook will be found useful to modelers all over the world in their endeavors to develop more and better models, leading to better understanding of the system properties of ecosystem, the reaction of our environment to the anthropogenic impacts, and for a better environmental management in general. The addresses of the modelers were found by a review of more than 50 modeling books and journals i.e., *Ecological Modelling, Environmental Software, Chemosphere and Water Research*. We would like to thank Cathy Thomé Halling-Sørensen for typing most of the manuscript.

S. E. Jørgensen
B. Halling-Sørensen
S. N. Nielsen
Copenhagen, Denmark

TABLE OF CONTENTS

# 1. INTRODUCTION

1.1 The Development of Environmental Models

1.2 Scope of the Book

1.3 State of the Art of Modeling

1.4 How to Use the Handbook

# 1. INTRODUCTION

## 1.1 The Development of Environmental Models

An overview of the development in ecological modeling is given in Figure 1.1 (redrawn after Jørgensen, 1994). The time axis gives approximate information on the year, when the various development steps took place. The first models of the oxygen balance in a stream (the Streeter-Phelps model) and of the prey-predator relationship (the Lotka-Volterra model) were developed already in the early 1920s. In the 1950s and the 1960s further development of population dynamic models took place. More complex river models were also developed in the 1960s. These developments could be named the second generation of models.

The wide use of ecological models in environmental management started around 1970, when the first eutrophication models emerged and very complex river models were developed. These models may be named the third generation of models. They are characterized by often being too complex, because it was so easy to write programs for a computer, which was able to handle rather complex models. To a certain extent it was the revolution in computer technology that created this model generation. It became, however, clear in the mid-1970s, that the limitations in modeling are not due to the computer and the mathematics, but to the lack of data and our knowledge about ecosystems and ecological processes. Consequently, the modelers became more critical in their acceptance of models. They realized that a profound knowledge to the ecosystem, the problem, the ecological components and the reactions of the ecosystem on the system level were the necessary basis for development of sound ecological and environmental models. The recommendations resulting from this period are
- follow strictly all the steps of the procedure, i.e., conceptualization, selection of parameters, verification, calibration, examination of the sensitivity, validation and so on.
- find a complexity of the model, which considers a balance between data, problem, ecosystem and knowledge.

Parallel to this development, ecologists became more quantitative in their approach to environmental and ecological problems, probably as a consequence of the needs formulated by environmental management. The quantitative research achieved by ecology from the late 1960s until today have been of enormous importance for the quality of the ecological models. They are probably just as important as the development in computer technology. The models from this period, going from the mid-1970s to the late 1980s, could be called the fourth generation of models. The models from this period are characterized by having a relatively sound ecological basis and with emphasis on realism and simplicity. Many models were validated in this period with an acceptable result and for some (few) it was even possible to validate the prognosis.

The conclusions from this period may be summarized in the following three points:
1.     Provided that the general modeling recommendations are followed and the underlying database was of good quality, it was possible to develop models that can be used as prognosis tool.

3

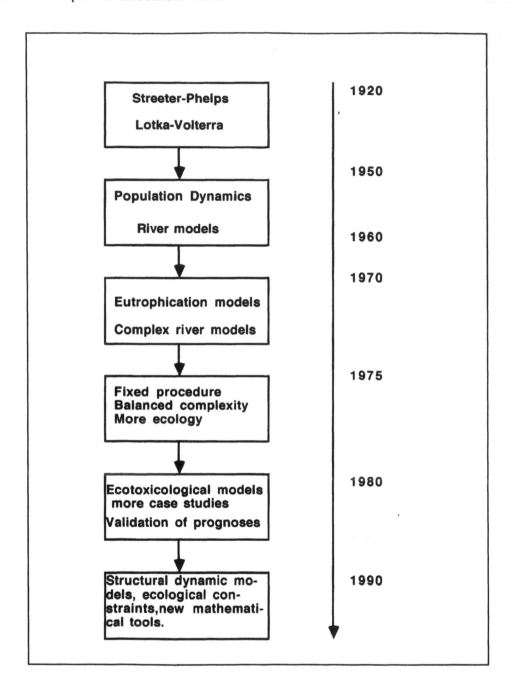

**Figure 1.1:** *The development of ecological and environmental models is shown schematically.*

2.  Models based upon a database of not completely acceptable quality should maybe not be used as prognosis tool, but they could give an insight into the mechanisms behind the environmental management problem, which could be most valuable. Simple models are often of particular value in this context.

3.  Ecological sound models, i.e., models based upon ecological knowledge, are powerful tools in understanding ecosystem behavior and as tool for setting up research priorities. The understanding may be qualitatively or semi-quantitatively, but has anyhow proved to be of importance for ecosystem theories and a better environmental management.

The shortcomings of modeling were, however, revealed in this period, too. It became clear that the models were rigid in comparison with the enormous flexibility that was characteristic for ecosystems. The hierarchy of feedback mechanisms that ecosystems possess was not accounted for in the models, which made the models incapable of predicting adaptation and structural dynamic changes. Since the mid-1980s modelers have proposed many new approaches such as 1) application of objective and individual modeling, 2) examinations of catastrophical and chaotic behavior of models and 3) application of goals functions to account for adaptation and structural changes. This development may be named the fifth generation of modeling.

## 1.2 Scope of the Book

History can teach us from gained experiences. Therefore, it is essential for modelers to be able to draw as much as possible on the previous modeling experience. The general experience gained in modeling can be found in the textbooks written about ecological and environmental modeling; see for instance *Fundamentals of Ecological Modeling*, Second Edition, 1994 by S.E. Jørgensen. Previous experience and recommendations on parameter values can be found in the wide ecological and biological literature. A good selection of parameters are also given in *Handbook of Ecological Parameters and Ecotoxicology*, 1991 by S.E. Jørgensen et al.

The experience on how to model a problem for a specific type of ecosystem is, however, also needed to provide the optimum initial conditions for a modeling task. A modeler would like to build on the previous experience on modeling a problem related to a given ecosystem, i.e., to get an answer to the following questions: which state variables and forcing functions are the most important to include in the model? Which characteristics were considered in the development of all the models, that have been focusing on the same or almost the same problem and ecosystem? Can any of the previous models be purchased? Would a particular model without modifications or only with minor modifications solve my problem? Which calibration and validation results have been achieved by application of the previously developed models, and are these achievements sufficient for my specific case or should I rather aim for a more accurate and/or a more detailed model?

This volume attempts to answer these questions on basis of a wide spectrum of models. The aim is clearly to give a modeler the previously gained experience for development of the model that is in focus.

The modeling literature has been reviewed by the editors of this handbook and a limited number of models were selected to give a reasonably good picture of the *entire* spectrum of environmental models. In total about 1000 models have been described and referred in the literature during the last decade. This handbook gives a detailed description of about 400 models, which have been selected to represent the entire modeling experience resulting from the about 1000 models. Questionnaires were sent to all the authors and users of the 400 selected models and the handbook contains and summarizes the information collected in these questionnaires. For each model all relevant references are given to enable the readers to find additional information when needed. The questionnaires were carefully developed as a trade off between giving as much information as possible for the individual models on the one hand and keeping the volume on a reasonable number of pages on the other. Some readers may therefore not find all the necessary information relevant to him or her for a model, but the tables should give sufficient information for the modeler to be able to evaluate whether a model can be used at least partially in a given problem context. The modeler would then anyhow study the relevant references before his or her final model is developed in order to build upon previous modeling experience.

## 1.3. State of the Art of Modeling

Table 1.1 reviews types of ecosystems, which have been modeled by the use of *biogeochemical* models up to 1994 (see also Jørgensen, 1994). It includes ecotoxicological models, but population dynamic models are excluded from this table, but of course not from this volume, as they focus mainly on populations. An attempt has been made to indicate the modeling effort by use of a scale from 0-5. Five means very intense modeling effort; 4, intense modeling effort; 3, some modeling effort; 2, a few models which have been fairly good studies; 1, one good study or a few not sufficiently well calibrated and validated models; and 0, almost no modeling effort at all. Population dynamic models have been subject to a very intense modeling effort, corresponding to 5 on the scale. Table 1.2 (also taken from Jørgensen, 1994) similarly reviews the environmental problems, which have been modeled up today. The same scale is applied to show the modeling effort as in Table 1.1. Besides biogeochemical models Table 1.2 also covers models used for management of population dynamics in national parks and steady state models applied as ecological indicators. It is beneficial to apply goal functions in conjunction with a steady state model to obtain a good ecological indication, as proposed by Christensen (1993). Biogeochemical, non-steady state models with goal functions are shortly reviewed in Chapter 13, Jørgensen (1994), where various goal functions and their application are presented. The few models that have been developed in this area so far are presented here by a comparative overview. It is possible to conclude 1) that we are able to develop structural dynamic models 2) they reflect better the behavior of nature in their description 3) they are also more easy to calibrate as they possess the flexibility and ariability which is characteristic for nature.

**Table 1.1.** Review of Ecosystems Modeled by Use of Biogeochemical Models (Jørgensen, 1994)

| Ecosystem | Modeling effort |
|---|---|
| Rivers | 5 |
| Lakes and reservoirs | 5 |
| Fish ponds | 5 |
| Estuaries | 5 |
| Coastal zone, lagoons, fjords | 4 |
| Open sea | 3 |
| Wetland | 4 |
| Grassland | 4 |
| Desert | 1 |
| Forests | 4 |
| Agriculture land | 5 |
| Savanna | 2 |
| Mountain lands (above timberline) | 0 |
| Arctic ecosystems | 0 |
| Contaminated soil | 3 |
| Atmosphere | 5 |

The development of models for an ecosystem or a particular pollution problem is almost always related to the environmental actuality. Eutrophication models were developed in the early 1970s, when eutrophication was widely discussed in environmental management. Development of ecotoxicological models were initiated in the late 1970s, when toxic substances in the environment came into focus. Correspondingly, the interest for wetland models increased during the early eighties, when the importance of wetlands as buffer zones in the landscape was discussed in conjunction with non-point pollution. Also, the interest for the global pollution problems, the greenhouse effect and the destruction of the ozone layer, provoked development of models dealing with these problems. Recently, regional models for toxic substances, particularly pesticides, are in focus and it may be foreseen that a major step forward in the development of this type of model will take place in the coming year. Simultaneously, the entire spectrum of environmental models will currently be improved, as increased experience is gained in the application of models in environmental context. Hopefully, this volume will be able to contribute to a rapid dissipation of this experience.

**Table 1.2** Review of Environmental Problems Modeled by Use of Models. (Jørgensen, 1994)

| Problem | Modeling effort |
| --- | --- |
| Oxygen balance | 5 |
| Eutrophication | 5 |
| Heavy metal pollution, all types of ecosystems | 4 |
| Pesticide pollution of terrestrial ecosystems | 4 |
| Protection of national parks | 3 |
| Management of populations in national parks | 3 |
| Ground water pollution | 5 |
| Carbon dioxide/greenhouse effect | 5 |
| Acid rain | 5 |
| Total or regional distribution of air pollutants | 5 |
| Change in microclimate | 3 |
| As ecological indicator | 4 |
| Decomposition of the ozone layer | 3 |
| Fate models for chemicals | 4 |

## 1.4. How to Use the Handbook

A comprehensive list of contents is given in the beginning of this volume. The model classification (see also the preface) is reflected in the chapter and section headings to facilitate the search for models of a particular ecosystem and/or problem. 100 models of aquatic ecosystems are reviewed, 74 models in ecotoxicology, 77 models of terrestrial ecosystems, 37 models of atmospheric pollution and climate, 30 landscape models and auxiliary models and 85 models of population dynamics. A comprehensive list of references is given for each model to obtain additional information.

The models are reviewed on three levels. First, there is a short review of each model (about 3-7 lines). Next the models are included in an overview table which also includes slightly older models of general interest. Finally, the models are reviewed in more detail in a presentation of the information gathered in the questionnaire (see the Preface). Particularly, the overview tables give the readers the opportunity to find the previous modeling studies that can be applied as basis for a new model development.

The index provides more detailed model information about specific models, elements, physical factors and biological components, particular populations and organisms.

It is recommended that both the list of contents and the index to be used to find the needed combination of information. Through the list of contents it might, for instance, be possible to find models, that cover a particular ecosystem. A very short review of the existing models can be obtained by application of the overview table and it may be sufficient to select a few models of interest for the focal problem. By reading the 3-7 lines briefly presenting the model, the comparison of models are facilitated. For a final selection, it will be necessary to read the 1-2 pages corresponding to the questionnaire for the few models which still may seem to offer a solution to the focal problem.

Modifications of the existing models due to the presence of other biological components, physical factors or elements than used in the selected models may be needed. Inspiration to which modifications are necessary to perform may be found in other models by use of the index.

There are only a few general models that can be used for case studies other than those for which they have been developed. It is more common that a model for a specific ecosystem and a specific problem must be tailored to the case study. It does not imply that the modeler has to start from scratch, but that different model experiences should be combined to develop the model. Often, various model components from different models **must** be combined to achieve a proper working model. The previous modeling experience from similar modeling cases provides the knowledge from which the model can be developed. The use of the handbook should be seen in this light: the user will rarely find a model that can be used unchanged, but the basic information for modifications of existing models can be found. If the model, however, can be used unmodified, the address of the modeler is available for direct contact.

# 2. BIOGEOCHEMICAL MODELS OF AQUATIC ECOSYSTEMS

## 2.1 Characteristics of Models for Aquatic Ecosystems

In accordance with the history of modeling, presented in the introduction, Chapter 1, this type of model already commenced by Streeter-Phelps simple BOD-DO model, which could be solved analytically. BOD-DO models developed further during the 1950s, 1960s and 1970s. Eutrophication models emerged in the early 1970s, as a result of our interest for eutrophication in environmental management. They represented a step forward toward incorporating more ecology in the models, but also the river models included more and more ecological components and processes, which were known to affect the oxygen balance for a stream. The models were in the first hand a further development of already existing hydrodynamical models, because the need for models that would consider not only transport processes but also chemical and biological processes and constituents in environmental management, was obvious.

During the 1970s ecologists started to be interested in the application of models as a powerful synthesizing tool in their research and in application of their view points in environmental management. The two different starting points, ecology and hydrodynamics, are still reflected today in many models that are either strong in coverage of the ecological processes and weak in the coverage of the hydrodynamics, or vice versa. It has long been acknowledged that the best-balanced models are developed by a team, that encompasses both ecological and hydrodynamical viewpoints. Such well-balanced biogeochemical models are available today, see, for instance, Biotest and Ersem. It must, however, be concluded that too little experience has been gained on models *integrating* ecological processes and hydrodynamics. We still do not have a clear, general answer to such crucial questions, such as what would be the right time steps for the two different processes and which accuracy of the various available data sets should be required to ensure the highest possible accuracy of the final model results. Some solutions to these problems are resulting from the above mentioned two models (and other similar model development), but wider experience is needed before general lines can be set up. It is, however, foreseen that the coming years will give general solutions to these problems, as more and more experience on combined ecological and hydrodynamical models will be gained.

These considerations do not imply that models without a detailed description of hydrodynamics cannot be used in research or environmental management. As pointed out in the introduction, the goals of the models and the characteristics of the model ecosystem in relation to the focal problem determine the components of the models. The lake model Glumsø in its basic version has, for instance, very simple hydrodynamics, which is sufficient for many shallow lakes with a simple morphology. The model named Ulva is a zero dimensional model, but the model serves the purpose to reveal the factors, that determine the growth of macro-algae. The model Chemsee is a third example of a model with relatively simple hydrodynamics, as the horizontal concentration differences are neglected, which is often completely allowable for deep lakes. The central question is always, which processes, independent on whether they are hydrodynamical, chemical or biological, are most crucial for the focal problem? The transport processes are of course often vital for the solution of the problem, but they are not always necessary in all ecological contexts.

Aquatic ecosystems have been modeled heavily since the late 1960s and modellers should draw upon the entire experience from the last 25 years. It implies that a modeller should use the overview Tables 2.1-2.3 as well as the more detailed description on the recently developed models as a basis for development and selection of any new model that would either focus on a new problem or just on a new case study. Hydrodynamical models considering 2-3 dimensions are today standard, and the experience presented here should make it possible to build on this experience and couple it with the relevant formulations of the focal biogeochemical processes, where wide experience is available from the last 25 years. The Danish Hydrological Institute has, for instance, developed several hydrodynamical models, called MIKE + a number, which cover the hydrodynamics at different levels of complexity and which can in a very flexible manner be coupled to biogeochemical processes which the user wishes to include.

During the last decade several specific models solving very specific problems have been developed. The specific models may be applied directly with or without modifications to solve similar problems in relation with new case studies or it may be beneficial to combine a model developed for a special problem with a more general hydrodynamical, BOD-DO or eutrophication model.

As seen from this spectrum of models, very specific needs for water quality modelling in relation to the problem, type of ecosystem, hydrodynamics and so on, can be met by the presented models.

## 2.2 An Overview of Models for Rivers and Streams

Models of rivers and streams focus almost entirely on the dissolved oxygen and the biological oxygen demand. This type of model was launched with the Streeter-Phelps model in the twenties. During the 1960s and 1970s, further development of this type of model took place, mainly by inclusion of more processes, which could cope with more complex hydrodynamics and with the components that influenced the dissolved oxygen concentration, for instance, the photosynthesis, the decomposition of the suspended organic matter and the organic matter in the sediment. The models from these two decades are not included in this handbook in details, as many of them are accessible in textbooks: see for instance Jørgensen (1994), and Thomann and Mueller (1987). They are, however, to a certain extent included in Table 2.1, which gives an overview of the processes included in the various types of river and stream models. The development of river and streams models during the last 10-15 years has been concentrated on inclusion of more processes determining the oxygen concentrations in streams and rivers, modeling particular processes and application of models on several new case studies. Table 2.1. gives a good expression of the spectrum of available river models, including the recently developed models.

The selection of the most appropriate model for a specific case study is as usually a matter of which processes are most significant in the modeled situations. More detailed information is given in Section 2.3, where the recently developed models are included.

The model PRAM is a further development of QUAL 2 (see Table 2.1 for comparison of the two models). The model named Microbial Benthos Method considers complex hydrodynamics and the development of a biofilm. The simulation of the self-purification ability of a stream is the main objective for this model, and here have previous models not considered the importance of the biofilm formation.

The model Stream is based on a moving segment in the cell-in-series (CIS) system, where streams are modeled as a series of completely mixed flow reactors. The model Stress is developed as a Monte Carlo technique, which can be linked to the Stream model. It is designed to account for randomness of the water head, point sources and non-point sources. It has particular relevance if a step and shallow stream has large variations of water quantities and qualities.

**Table 2.1.** Survey of Some Useful River and Stream Models. Processes in Addition to Decomposition of Organic Matter, Nitrification and Reaeration are Indicated.

| Model and reference | Processes |
| --- | --- |
| Eckenfelder and O'Connor (1961) | Turbulence of water, biological growth in river beds, acclimatization of microorganisms, toxicity. |
| Thomas (1961) | Settling. |
| Eckenfelder (1970) | Autocatalytic nitrification. |
| O'Connell and Thomas (1965) | Photosynthesis and respiration. |
| Fair et al. (1941) | Benthic oxygen uptake. |
| Edwards and Rolley (1965) | Benthic oxygen uptake. |
| O'Connor (1962) | Longitudinal mixing (dispersion). |
| Dobbins (1964) | Longitudinal mixing (dispersion). |
| Hansen and Frankel (1965) | Diurnal dissolved oxygen profiles. |
| O'Connor (1967) | Temporal and spatial distribution in due dimension. |
| O'Connor and Di Toro (1970) | Photosynthesis, respiration benthic respiration |

15

**Table 2.1 (continued)**

| Model and reference | Processes |
|---|---|
| DOSAG I: Texas Water Development Board (1970) | Spatial and temporal variation under various conditions of flow and temperature. |
| DOSAG M: Armstrong (1977) Improved version of DOSAG I. | Benthic oxygen demand, coliform load. |
| QUAL 1: Texas Water Development Board (1970) | Detailed hydrodynamics in one dimension. |
| QUAL 2: Water Resources (1973) | As QUAL 1 + benthic oxygen demand, photo-synthesis, engineers, respiration, 2 steps nitrification, coliforms, radioactive material, phosphorus. |
| RECEIV II: | State variables: BOD, DO, $NH_4^+$ $N_2$, $NO_2^-$, N-total. |
| Raytheon Company (1974) | Coliforms, chlorophyll. |
| PRAM<br><br>U.S. EPA | State variables: as Qual ll, but with more complex hydrodynamics |
| STREAM Dr. Seok Park | State variables: DO, BOD, suspended solid, coliform bacteria, organic nitrogen, ammonium, nitrate, phosphorus, phytoplankton, periphyton, macrophytes, and sediment oxygen demand. |
| RBM10 John R. Yearsley | Includes also benthic animals and zooplankton in addition to state variables of "Stream". |
| EXWAT and RIVER | Watercourse models applied to toxic risk estimation of a chemical spill. |

The model Mixzon focuses on modeling of the mixing zone characteristics of natural streams. The output of the model is the 3-D distribution of the water quality in the mixing zone. The model developed by Taskinen is particularly focusing on the transport of phosphorus by a river or stream.

The model RBM10 is a comprehensive river model similar to the comprehensive models developed in the seventies, but it considers also ecological risk analysis and environmental impact assessment. As state variables are applied all forms of nutrients, phytoplankton, zooplankton, benthic animals, benthic plants, coliform bacteria and macrophytes.

Hearne has developed a stream model, which looks into the nutrient dynamics in a stream. The model is foreseen coupled to a lake model, as it is able to consider the inputs of nutrients coming from non-point sources and the removal processes by riparian vegetation.

The model named Asram is not included in the table, because it is a model for predictions of the potential extent of recovery of Atlantic salmon populations as stream pH increases due to reduced emission or liming. The model could also be considered a model of population dynamics, but is included here as it is closely related to models of streams and rivers.

Finally, the model denoted Laxbreak should be mentioned. It is 1-D hydrodynamical model which enable numerical calculations of waves caused by instantaneous or partial and gradual dambreak.

## 2.3 Models of Rivers and Streams

---

**Model identification**
Model name/title: PRAM (Pigeon River Allocation Model)

---

**Model type**
Media: river
Class: biogeochemical
Substance(s)/compounds: nutrients, phosphorus, nitrogen, oxygen
Biological components: ecosystem - riverine community - water column

---

**Model purpose**
To assess the effects of added carbonaceous materials and enhanced DO to stream dynamics with regard to DO and nutrient concentrations spatially and temporally.

---

**Short description**
A conceptual overview of PRAM shows a wide range of processes determining dissolved oxygen dynamics. The processes in conjunction with point discharges control DO concentrations in the river. Physically, the river is primarily controlled by advective flow, instantaneous mixing, point source discharges and tributary flows. The model contains 13 biological/chemical state variables (DO, BOD, Chlorophyll a, N, TON, NH3, NO3, NO2, TOP, DIP, TP, Cl, and temperature) and 6 physical state variables (streamflow, travel time, velocity, depth, stream width and volume). These variables combine to represent the carbon, oxygen, nutrient, and water storages of the model.

---

**Model application(s) and State of development**
Application area(s): Pigeon River, NC.
Number of cases studied: 1
State: prognosis made, validation of prognosis

---

**Model software and hardware demands**
Platform: PC-based, Mainframe
OS: DOS or VMS
Software demands: QUAL2E-UNCAS
Hardware demands: 386-level machine

---

**Model availability**
Purchase: no, executable code
Contact person(s) Dr. Kevin Summers
        U.S. EPA
        1 Sabine Island Drive
        Gulf Breeze, FL 32561 USA
Phone: 904-934-9244
Fax: 904-934-9201

**Model documentation and major references**
Brown, L.C. and Barnwell, T.O., 1987. The enhanced stream water quality model QUAL2E and QUAL2E- UNCAS. Documentation and user model. EPA-600-3-87-007.

**Other relevant information**

## Model identification
Model name/title: Microbial Benthos Method

## Model type
Media: water, river, sediment
Class: biogeochemical
Substance(s)/compounds: carbon, organic compounds
Biological components: ecosystem, lotic ecosystem, community, microbial community

## Model purpose
The purpose of the model was twofold: (i) to simulate the self-purification mechanisms in a lotic ecosystem and more generally, organic matter decomposition processes; (ii) to couple the two approaches which are commonly used in aquatic modeling, the transport oriented and the ecology oriented approaches.

## Short description
The model was developed to simulate self-purification mechanisms that occur in a small stream heavily polluted by an organic load. As the main detritical processes that occur in small streams take place in the sediment, the model was applied to simulate the dynamics of both dissolved organic matter and benthic microorganisms. The physical part of the model included a hydrodynamical component derived from Saint-Venant's equations, coupled to a transport model based on a convection-dispersion equation under uniform, unsteady flow conditions. The simulation of the biodegradation mechanisms was based on biofilm kinetics. The linkage between hydrophysical and benthic mechanisms affords a dynamic description of the system. Transport modeling provides a picture in space and time of the evolution of organic carbon concentration in water. This evolution makes it possible to predict the fate of the benthic biological component. The main characteristics of the model were
STATE VARIABLES: cross sectional area, discharge, organic substrate in the flow, biofilm, organic substrate, microbial biofilm biomass.
FORCING FUNCTION: temperature, tributary inputs.
MAIN PARAMETERS: dispersion coefficient, transfer velocity at the interface, biofilm shearing rate, maximum growth rate, bacterial decay rate, half saturation coefficient, biofilm cellular density.
INPUTS: Stream length, geometric description of the stream, time series of the discharge at the point x=0, time series of the organic load at x=0.

19

## Model application(s) and State of development
Application area(s): the Albenche river (Savoie, France)
Number of cases studied:
State: conceptualization, verification, calibration, partial validation

## Model software and hardware demands
Platform: Bull mainframe
OS: Fortran 77
Software demands: Benson plotters
Hardware demands:

## Model availability
Purchase: no, as sourcecode
Contact person(s) Dr. Bernard Cazelles
                  Centre de Bioinformatique and INSERM U263
                  T53, 2 Place Jussieu
                  F-75251 Paris, France
Phone: +33 1-4427-7722
Fax: +33 1-4326-3830
E-mail: cazelles@urbb.jussieu.fr

## Model documentation and major references
Cazelles, B., 1987. Modelisation d'un Ecosystem
Cazelles, B., Fontvieille and Chau, N.P., 1991. Self-purification in a lotic ecosystem: a model of dissolved organic carbon and benthic microorganisms dynamic. *Ecol. Modelling*, 58, 91-117.

## Other relevant information
This original organic carbon based model designed for the Albenche river could be applied to other heterotrophic streams where decomposition processes occur mainly in the benthic compartment, in addition to/or replacing the usual oxygen deficit models. The biological part of the model has been used to optimize fixed biomass reactors used in waste water treatment from pilot scale to industrial scale.

## Model identification
Model name/title: STREAM (Segment Travel River Ecosystem Autograph Model)

## Model type
Media: water, river
Class: biogeochemical
Substance(s)/compounds: nutrients, carbon, nitrogen, oxygen, suspended solid
Biological components: organism, bacteria

**Model purpose**
To simulate water qualities of steep and shallow streams and display the result in graphic format.

**Short description**
The formulation of STREAM was based on moving segment approach in the Cell-In-Series (CIS) system where stream is assumed as a series of completely mixed flow reactors. Major constituents included in STREAM are Dissolved Oxygen (DO), Biochemical Oxygen Demand (BOD), Suspended Solid (SS), Coliform Bacteria, Nitrogen Series, Phosphorus Series and Phytoplankton. Dissolved oxygen change caused by Sediment Oxygen Demand (SOD) and respiration and photosynthesis of periphyton and macrophyte are included in the model structure. STREAM was programmed in a BASIC language and can be executed on the personal computer. STREAM was designed for implementation with EGA/VGA color graphic. The model results and the measured data are displayed in the computer screen and the output device in graphic format. A demonstration application is presented to validate the STREAM model. The results indicated that the STREAM model could effectively simulate water qualities of steep and sow streams where longitudinal dispersive transport is negligible.

**Model application(s) and State of development**
Application area(s): Waste Load allocation Analysis
Number of cases studied: 1
State: conceptualization, verification, calibration, validation

**Model software and hardware demands**
Platform: PC-based, IBM
OS: DOS
Software demands: NO, BASIC
Hardware demands: VGA/EGA color, 640 40 Mb hard drive

**Model availability**
Purchase: yes, $100 executable code
Contact person(s) Dr. Seok Park
              Dept. of Environmental Science
              Kangwon National University
              Chuncheon
              200-701 Korea

Phone: 82-361-50-8575
Fax: 82-361-56-4043
E-mail: sspark@cc.kangwon.ac.kr

**Model documentation and major references**
STREAM user Manual and Documentation, Kangwon National University, 1993.

## Other relevant informations
Development and Application of a Multi Constituent Moving Segment Model for Water Quality Predictions in Steep and Shallow Streams. *Water Resource Bulletin*, 1993 in review.

## Model identification
Model name/title: STRESS (Segment Travel River Ecosystem Stochastic Simulation)

## Model type
Media: water, river
Class: biogeochemical
Substance(s)/compounds: nutrients, carbon, phosphorus, nitrogen, oxygen, suspended solid
Biological components: organism, bacteria

## Model purpose
To predict water quality with random variations and present the results in graphic format.

## Short description
The model was developed by incorporating the Monte Carlo technique with the previously developed and validated deterministic model called STREAM (Segment Travel River Ecosystem Autograph Model). STRESS was designed to be more representative of steep and shallow streams with large variations of water quantity and quality which are typical in Korean peninsula. The model considers randomness in the head water, point source, non-point source conditions, and the travel time of water segment, and computes the frequency distributions, the mean values, and the standard deviations of the predicted water qualities. Major constituents included in STRESS are Dissolved Oxygen (DO), Biochemical Oxygen Demand (BOD), Suspended Solid (SS), Coliform Bacteria, Nitrogen Series, Phosphorus Series, and Phytoplankton. Stress is executed on personal computer and designed for implementation with EGA/VGA color graphic. A demonstration application for the Tancheon, a tributary of the Han River, is presented to validate the stochastic nature of the model. The result indicated that STRESS could effectively simulate random variations of water qualities in steep and shallow streams.

## Model application(s) and State of development
Application area(s): Waste load Allocation Analysis
Number of cases studied: 1
State: conceptualization, verification, calibration, validation

## Model software and hardware demands
Platform: PC-based, IBM
OS: DOS
Software demands: no, BASIC
Hardware demands: 640, 40 Mb hard drive VGA/EGA color

**Model availability**
Purchase: yes, $100
Contact person(s) Dr. Seok Park
        Dept. of Environmental Science
        Kangwon National University
        Chuncheon
        200-701 Korea

Phone: 82-361-50-8575
Fax: 82-361-56-4043
E-mail: sspark@cc.kangwon.ac.kr

**Model documentation and major references**
STRESS User Manual and Documentation, Kangwon National University, 1993.

**Other relevant information**
Monte Carlo simulation of random phenomena in natural streams submitted to the International Association of Water Quality, 1993.

**Model identification**
Model name/title: MIXZON (Mixing Zone Model)

**Model type**
Media: water, river
Class: biogeochemical
Substance(s)/compounds: chemicals, organic compounds, any type, first order independent decay constituent
Biological components: organism, bacteria

**Model purpose**
To predict the spatial distribution of water quality in the mixing zone of natural streams.

**Short description**
The program accepts the discharge and river flow conditions, such as effluent discharge flow rate, effluent concentration, river flow rate, river background concentration, and a specific criterion for the input, along with the hydraulic conditions of the mixing zone reach. The output of the program is the transverse distributions of in-zone quality according to the downstream distance in matrix form and the mixing zone width and mixing zone length at a given specific criterion.

## Model application(s) and State of development
Application area(s): Mixing zone in stream: water body contiguous to wastewater disposal sites.
Number of cases studied: 1
State: conceptualization, verification, calibration, validation

## Model software and hardware demands
Platform: PC-based, IBM
OS: DOS
Software demands: no, FORTRAN
Hardware demands: 640, 40 Mb hard drive

## Model availability
Purchase: yes, $100. Executable code
Contact person(s) Dr. Seok Park
                 Dept. of Environmental Science
                 Kangwon National University
                 Chuncheon
                 200-701 Korea

Phone: 82-361-50-8575
Fax: 82-361-56-4043
E-mail: sspark@cc.kangwon.ac.kr

## Model documentation and major references
Mathematical Modeling of Mixing Zone characteristics in Natural Streams, Ph.D. Dissertation, Rutgers University, New Braunswick, NJ 08903 (1985).

## Other relevant information
Park, S.S. and Uchrim, C.G., A numerical mixing zone model for water quality assessment in natural streams: conceptual development. *Ecol. Modelling*, Vol. 42. pp. 233-244 (1988).

## Model identification
Model name/title: Probabilistic uncertainty assessment of phosphorus balance calculation

## Model type
Media: water, river, local scale
Class: biogeochemical
Substance(s)/compounds: nutrients, phosphorus
Biological components: ecosystem-river ecosystem

## Model purpose
The model purpose is to asses the uncertainties included in mass balance calculation of a river and to find the components that cause most uncertainty in balance calculation.

## Short description
The phosphorus balance calculations are divided into different components which are upstream and downstream discharges and phosphorus concentrations and loads from point sources. The residual term includes, e.g., non-point load and errors. The uncertainty, standard deviation of each component is assessed and using measurement as a mean value the components are perturbated using Latin Hypercube simulation. The overall uncertainty is assumed to accumulate into residual term, which variance is studied. The most uncertainty term, which needs more measurements, causes the biggest variance. The spreadsheets and a risk analysis add-in are used in calculations.

## Model application(s) and State of development
Application area(s): Assessing uncertainties in balance calculations
Number of cases studied: 1
State:

## Model software and hardware demands
Platform: PC-based
OS:
Software demands: spreadsheet and risk analysis add-in
Hardware demands:

## Model availability
Purchase: no
Contact person(s) Dr. Antti Taskinen
                  Vesitalonden Laboratorieo
                  Konemiehentie 2
                  SF-02150 Espoo
                  Finland

Phone: +358-0-4513830
Fax: +358-0-4513836
E-mail: ataskine@leka.hut.fi

## Model documentation and major references


## Other relevant information
It is very difficult to classify this model by terms used in this paper. The model is not a "classical model" but a case study of simulation. If someone wants to try a similar kind of study, he/she should read the article and use spreadsheets and a risk analysis add-in which are very easy to "code".

## Model identification
Model name/title: RBM10, A dynamic river basin model

## Model type
Media: water, river, lake, reservoir
Class: biogeochemical, hydrology
Substance(s)/compounds: carbon, phosphorus, nitrogen, dissolved oxygen, temperature, suspended solids
Biological components: ecosystem - phytoplankton, zooplankton, macrophytes, benthic animals

## Model purpose
The model was developed to perform planning and ecosystem risk analysis in river basins where point source and nonpoint sources can have an impact on temperature, dissolved oxygen, suspended solids and eutrophication.

## Short description
State variables: Phytoplankton, Zooplankton, Benthic Animals, Benthic Plants, Macrophytes, Detritus, $PO_4$-P, $NO_2$+$NO_3$-N, $NH_4$-N, Dissolved Oxygen, Temperature, Suspended Solids, Coliform Bacteria.
Forcing functions: Solar radiation, river inflow, groundwater inflow, nonpoint source inflow, point source inflow.

## Model application(s) and State of development
Application area(s): Environmental regulation applications such as water discharge permits environmental impact statements, ecological risk analysis.
Number of cases studied: several
State: conceptualization, parameter estimation, prognosis

## Model software and hardware demands
Platform: DEC Microvax, DG Aviion Workstation
OS: DEC VMS, Unix
Software demands: FORTRAN 77 compiler, FORTRAN 77
Hardware demands: 4 Mb RAM, 200 MB storage

## Model availability
Purchase: none, software available as FORTRAN 77 source code
Contact person(s) Dr. John R. Yearsley
        EPA Region 10 ES097
        1200 Sixth Avenue
        Seattle, WA 98101
        USA

Phone: 206-553-1532
Fax: 206-553-0119
E-mail: jyearsle@r0dg09.r10.epa.gov
     yjd@sequoia.nesc.epa.gov

**Model documentation and major references**
Bowler, P.A., Watson, C.A., Yearsley, J.R. and Cirone, P.A., 1992. Assessment of Ecosystem Quality and its Impact on Resource Allocation in the Middle Snake River Sub-Basin. Proceedings of the 1992 Annual Symposium, Volume XXIV, ISSN 1068-0381, Desert Fishes Council, Bishop, CA.
Yearsley, J.R., 1991. A dynamic river basin water quality model. EPA Publ. #910/9-91-019. EPA Region 10, Seattle, WA.

**Other relevant information**

**Model identification**
Model name/title: Nitrate removal by riparian vegetation in a springfed stream

**Model type**
Media: water, river, local scale
Class: biogeochemical
Substance(s)/compounds: nutrients, nitrogen, nitrate
Biological components: ecosystem - macrophyte/nitrate

**Model purpose**
To investigate the role of macrophytes in removing nutrients from a stream. To determine whether grazing of riparian vegetation by sheep would diminish or enhance this role. To determine the timing and extent of grazing should be allowed in riparian strips.

**Short description**
The state variables are macrophyte biomass and instream nitrate concentration. These are functions of both time and distance downstream. The growth rate of the macrophyte at any point in the stream is dependent upon the nitrate concentration at that point. Seasonal temperature changes also play a role in macrophyte production, mortality and decay. Production and decay affect the nitrate levels. The discharge was constant for the springfed stream in the case but can be otherwise specified. The nitrate concentration at the head of the stream must be specified as a function of time. To simulate grazing, a rate of biomass removal can be specified. The model consists of a coupled system of partial differential equations which are solved numerically.

**Model application(s) and State of development**
Application area(s): Whangamata Stream, Lake Taupo, New Zealand
Number of cases studied: 1
State: conceptualization, validation, calibration, prognosis made - YES, verification, validation of prognosis - NO

## Model software and hardware demands
I no longer have the code for this model. It was run a mainframe in FORTRAN.
If re-written it could be run on a PC.

Platform:
OS:
Software demands:
Hardware demands:

## Model availability
Purchase: A new version of the model could be written in Pascal for use on a PC if
demand justified. A fee of $150 would be charged. Both executable and source code could
be provided.
Contact person(s) Professor John Hearne
                Department of Mathematics
                University of Natal
                P.O. Box 375
                Pietermaritzburg
                3200 South Africa

Phone: +27 331-260-5626
Fax: +27 331-260-5599
E-mail: hearne@math.unp.ac.za

## Model documentation and major references
Hearne, J.W. and Howard-Williams, C., Modelling nitrate removal by riparian vegetation
in a springfed stream: the influence of land-use practices, 1988. *Ecol. Modelling*, 42: 179-198.
Hearne, J.W., and Wake, G.C., A numerical method for solving a stream nutrient model,
1989. *Applied Mathematical Modelling*, 13: 155-159.

## Other relevant information


## Model identification
Model name/title: ASRAM - Atlantic Salmon Regional Acidification Model

## Model type
Media: river, regional scale
Class: toxicology (also population)
Substance(s)/compounds: sulfur (leading to low pH in streams)
Biological components: population - Atlantic salmon

## Model purpose
ASRAM is a computer program designed to predict the potential extent of recovery of Atlantic salmon populations as stream pH increases (due to reduced emissions or liming), and to assess the effectiveness of alternative stocking and harvesting strategies in streams of different acidity.

## Short description
ASRAM is a habitat- and pH-sensitive age-structured salmonid life history model. It combines toxicity relationships, habitat preference, density-dependent processes, and natural and fishing mortality estimates to estimate annual salmon production at site and river system levels. Weekly pH observations (or estimates) are used to predict mortality at different juvenile stages using empirically-derived toxicity functions. These mortality rates are fed into an age-structured population model to predict smolt numbers (2-3 years later) and adult returns (3-6 years later) after fishing. An empirical relationship between growth and density is used to determine the age and size of smolts produced from each reach.

To apply the model at the river system level, each system is divided into tributaries consisting of multiple reaches separated at 5 m contour intervals. The area and water surface grade of each reach, used as input data for the model, are based on information obtained from aerial photographs and orthophotographic maps. Reach grade is used to distribute juveniles and returning adults among reaches based on a gradient-driven preference model. Annual pH time series collected from monitoring stations distributed throughout a river system are assigned to a set of spatially linked reaches to predict tributary-specific pH-driven mortality rates.

## Model application(s) and State of development
Application area(s): The model has been applied to the LaHave, Westfield, and Stewiacke Rivers, Nova Scotia, Canada. The model was calibrated and validated in the LaHave River, but has also been partially validated in the Westfield and Stewiacke Rivers.
Number of cases studied:
State:

## Model software and hardware demands
Platform: ASRAM is a DOS-based application running on IBM-PCs. It requires at least 2 Mb free disk space and 1 Mb RAM. A 486 DX is recommended to run the model at the river-system level. The model is written in MS QuickBasic version 4.5.
OS: Dos
Software demands:
Hardware demands:

## Model availability
Purchase: The executable and source code is available on request. Source code for a proprietary set of libraries for the user-interface of the model will not be supplied. A small fee will be charged to cover materials (user's guide and distribution diskettes) and shipping costs.

Contact person(s) Dr. J. Korman     or     Dr. David R. Marmorek
                3320 W 5th Ave                    ESSA Technologies LtD.
                Vancouver, British Columbia      1765 W. 8th Ave, 3rd Floor
                V6R 1R7                        Vancouver, British Columbia
                Canada                      V6J 5C6
                                          Canada

Phone: +1 604-734-1148     or     +1 604-733-2996
Fax:   no fax                     +1 604-733-4657
E-mail: korman@sfu.ca          dmarmorek@essa.com

## Model documentation and major references

Korman, J., Marmorek, D.R., Lacroix, G.L., Amiro, P.G., Ritter, J.A., Watt, W.D., Cutting, R.E., and Robinson, D.C.E., 1994. Developments and evaluation of a biological model to assess regional scale effects of acidification on Atlantic Salmon (Salmon salar). *Can. J. Fish. Aquat. Sci.*, 51: 662-680.

ESSA LtD, 1992. Atlantic salmon regional acidification model, version 2.0: User's guide. Prepared by Environmental and Social Systems Analysts Ltd. Vancouver, B.C. for Department of Fisheries and Oceans, Scotia Fundy Region. 19p.

## Other relevant information

## Model identification
Model name/title: LAXBREAK

## Model type
Media: water, river
Class: hydrology/hydraulics
Substance(s)/compounds: none
Biological components: none

## Model purpose
Numerical simulation of waves caused by instantaneous or partial and gradual dambreak.

## Short description
Fully non-linear unsteady 1-D equation of motion (with advective term included) together with the continuity equation (St. Venant's Equation) are solved by the explicit finite-difference method Lax-Wendroff. The method was completed with a term expressing local energy losses at channel expansions which also improves the numerical stability so that also flows in highly non-prismatic channels can be simulated. Supercritical flow as well as transitions from supercritical to subcritical and vice-versa can be simulated.

    The simulation is executed over the whole domain from the upstream end of the reservoir, over the dam site, to the wave front at the downstream end. Propagation over initial flow as well as over dry bottom can be simulated, the difference being in the simulation of the wave front. Total and instantaneous as well as partial and gradual dambreak flow can be simulated, the manner of collapse being prescribed by the input data.

Measurements from four physical models were used for the verification and calibration of the mathematical model. Usually a stage-discharge relationship at the downstream end of the computational domain is taken as the boundary condition but several other BCs and combinations with the 2-D models can be used.

## Model application(s) and State of development
Application area(s):Rivers and narrow valleys downstream from dams
Number of cases studied: 60
State: verification, calibration, validation, prognosis made

## Model software and hardware demands
Platform: PC-based - IBM
OS: DOS
Software demands: language FORTRAN
Hardware demands: IBM PC 386 compatible

## Model availability
Purchase: yes, DEM 5,000. Software available both as executable and sourcecode.
Contact person(s)  Prof. Rudi Rajar and Asst. Prof. Matjaz Cetina
    FAGG-Hydraulic div.
    Univerza v. Ljubljani
    Hajdrihova 28
    Sl-61000 Ljubljana
    Slovenia

Phone: +386-31-12-54-333
Fax: +386-61-219-897
E-mail: rudi.rajar@uni-lj.sl

## Model documentation and major references
Short provisional manual (four pages only).
Rajar, R., 1972. Recherche theorique et experimentale sur la propagation des ondes de rupture de barrage dans une vallee naturelle. (Theoretical and experimental study of dam - break flow in a natural valley). Doctoral Thesis, University Paul Sabatier, Toulouse, France, 1972.
Rajar, R., 1973. Modele Mathematique et abaques sans dimensions pour le determination de l'ecoulement qui suit la rupture d'un barrage. (Mathematical Model and Dimensionless Diagrams for the Detremination of Flow after a Dam Collapse). XI Congress of the International Commission on Large Dams. Madrid, 1973, Q40, R34, pp. 503- 521 (in French).
Rajar, R., 1975. Mathematical and Physical Model for the Dam-Break Wave Propagation in a Natural Valley, Chapter 5.11. (Vol. 2) in the book: Unsteady Flow in Open Channels, Eds. Mahmood and Yevyevich, Ft. Collins, USA, 1975, pp.662-665.
Rajar, R., 1978. Mathematical Simulation of Dam-Break Flow. Journal of the Hydraulics Division, Proceedings of the American Society of Civil Engineers (PASCE), HY7, 1978, pp. 1011- 1026.
Rajar, R., Zakrajsek, M., 1993. Modelling a Real Case of Landslide- Created Dam Collapse, 25th IAHR Congress, Tokyo, Japan, 30.8.-3.9. 1993. Proceedings, B-4-1, pp. 109-116.

Rajar, R., Kryzanowski, A., 1994. Self-Induced Opening of Spillway Gates on the Mavcice Dam - Slovenia, 18th Congress of the International Commision on Large Dams, Durban, Nov. 1994, Proceedings (Q71).

## Other relevant information
The program is not recommented in a form to be used by an user not experienced in this field.

## Model identification
Model name/title: EXWAT and RIVER

## Model type
Media: water, lake, river
Class: biogeochemical
Substance(s)/compounds: chemicals, organic compounds
Biological components: organism - fish

## Model purpose
Watercourse models applied to toxic risk estimation of a chemical spill.

## Short description
Watercourse models EXWAT and RIVER were validated at two pulp mill recipients using known emissions and concentrations in water, solids and fish of pulp chlorobleaching - originated chlorophenolics. Then the models were used in estimation of the fate of accidentally spilled tetraethyl led and tetrasin to these watercourses. The models consists of boxes with known length, width, depth, volume flows, amounts of suspended solids, porosity, density and fraction of organic carbon in solids, deposition rates, pH of water and temperature.

## Model application(s) and State of development
Application area(s): Two Finnish watercourses (pulp mill recipients)
Number of cases studied: 2
State: validation, prognosis made

## Model software and hardware demands
Platform: PC-based
OS: DOS
Software demands: 4 Mb disk, language - English
Hardware demands: 386-PC

## Model availability
Purchase: Software avilable as executable code.

Contact person(s) For application:

    Professor Jaakko Paasivirta

    UNiversity of jJyväskylå

    P.O. Box 35

    SF-40351 Jyväskylä

    Finland

Phone: 358-41-602652

Fax: 358-41-602501

E-mail: IN%"PAASIVIRTA@JYLK.JYU.FI"

For programs:

Professor, Dr. Michael Matthies

FB Mathe/Informatik

University of Osnabrück

Box 4469

D-4500 Osnabrück, Germany

## Model documentation and major references
Paasivirta, J., 1994. Environmental fate models in toxic risk estimation of a chemical spill. Research Centre of the Defence Forces (Finland) Publications A/4 (1994), 11-21.

## Other relevant information
The models were obtained from professor Michael Matthies, University of Osnabrück, Germany, and used in the application by professor Jaakko Paasivirta, University of Jyväskylä, Finland.

## 2.4. An Overview of Models for Lakes and Reservoirs

This section contains mainly eutrophication models. Some of the models have been applied for lakes as well as coastal areas, but they are included in this section if they have mainly been applied on inland waters. Models in this section are represented by a wide spectrum of complexity in the hydrodynamics as well as in the ecology.

Several other lake management problems have, however, also been covered by the use of models. Acidification of lake water has been covered by four models, including a model that is able to predict the concentrations of calcium and carbonate species in lake water. Some models are particularly fitted for management of lake restoration methods. Some of these models are able to predict the vertical oxygen profile. Others, for instance, the Lake Orta model, focus on specific problems - in this case copper contamination. The Glumsø model includes the entire food chain, and consider the intracellular pool of nutrients in the algae, which is essential for shallow, very eutrophic lakes.

The model named Aselm is a simple version of the Glumsø model. It has only 6-8 state variables, and does not consider the intracellular concentrations of nutrients. It should be used in stead of the Lake Glumsø model in case studies, where the available observations are more limited.

The model Aster does not consider the entire food chain, but models the primary and secondary production in more details. Phytoplankton is represented by two state variables: large diatoms, needing silica for growth and not consumed by zooplankton, and several small species, consumed by filtering zooplankton. Zooplankton is represented by three state variables: herbivorous Cladocerans and Copepods and a carnivorous group. This model is coupled to the thermocline model named Eole, which is developed to describe the thermal stratification. It is a one-dimensional, vertical model based on the energy budget. The model is particularly fitted to management of reservoirs. The model named Melodia is a result of coupling Aster with Eole. It is developed particularly for water management purposes.

Ophelie is another model, which simulates the evolution of the seasonal thermal stratification. It is a two-dimensional laterally averaged hydrodynamical and thermal model of a reservoir.

The Lake Baikal Model is a very comprehensive model, which includes the impacts by toxic chemical. It is, however, also a eutrophication model and is therefore included in this chapter. As Lake Baikal is a very large lake, so it is not surprising that the model considers a complex hydrodynamics by the use of 10 boxes and 3 layers. The ecological description is also comprehensive by inclusion of the following biotic components: bacteria, phytoplankton, zooplankton, zoobenthos, phytobenthos, fishes, and seals.

Chemsee, a 1-D vertical model is particularly fitted for modeling of chemical processes in lakes. The user selects which chemicals, elements and constituents he would like to include in the model including nutrients, radionuclides, heavy metals, pesticides and other organic compounds. Both the vertical water column and the sediment are considered.

The model, Biotest, represents the application of a very fine grid, but with a limited number of state variables (8) on a limited area of sufficient importance from a hydrodynamical point of view to require a three-dimensional description. The trophic-diffusive macromodel is to be considered an expanded application of Biotest on the Lagoon of Venice with 358400 cells (compartments), see Sections 2.6 and 2.7 on Marine Models.

The model 3DWFGAS is also three dimensional with air and soil modules for optional forcing functions determining inputs and outputs from the these spheres. The model encompasses a great number of state variables: 5 nutrients, oxygen, ligno-sulphonate, oil, organic chlorine compounds, pesticides, pH and 12 metals. Also the number of biological components is high: 5 species of fish, 3 groups of zooplankton and zoobenthos and 4 types of phytoplankton.

The model Reward is a comprehensive water quality and hydrodynamical model, which computes the surface profiles for steady state and for gradually varied flows. The water quality module of the model includes BOD, COD, temperature, algae, nitrite, nitrate, ammonium, organic N and P, dissolved phosphorus, coliforms, oxygen, alkalinity, mercury and cadmium. This model encompasses also a model bank with well-known models such as QUAL 2, Streeter-Phelps and IAWPRC. This model can be used for several types of aquatic ecosystems, but is mainly applied for lakes and reservoirs. The nutrient/biomass model for Liberty Lake, Washington, is a typical eutrophication model. The state variables are phosphorus, nitrogen, carbon and biomass measured as chlorophyll a.

Several models focus on very specific problems. The model named Lake Ecosystem is developed as a research tool to assess the trade off between increasing model complexity and collecting data of higher quality or quantity. The model is used in conjunction with the extended Kalman filter to test hypotheses regarding modeling performance. The model itself has 11 state variables, including temperature, suspended matter, benthic animals and benthic plants.

The model called Lavsoe provides a seasonal description of the algal growth, water transparency and the cycling of the two nutrients, phosphorus and nitrogen. It has the following state variables: algal biomass, algal nitrogen, algal phosphorus, dissolved inorganic nitrogen and phosphorus and non-available nitrogen and phosphorus, in total seven state variables.

A model, named Hydro, is developed for long-term predictions of eutrophication. The model has a 3-D description of the hydrodynamics, while only phosphorus is considered in inorganic form as detritus.

The model denoted Minlake is particularly developed to examine the efficiency of lake treatment methods (restoration methods, ecotechnology). The model has the following state variables: inorganic phosphorus, inorganic nitrogen in three oxidation states, BOD, phytoplankton, zooplankton and dissolved oxygen.

The model named Girl is a generalized, simple lake model consisting of three equations, describing phosphorus as limiting element, and phytoplankton and zooplankton is uniformly mixed layers of water with a specified depth.

The model called Salmo is a general model for lakes and reservoirs. It considers two layers: epilimnion and hypolimnion. Three groups of phytoplankton and two groups of zooplankton are included in the model description. The model has unfortunately 138 parameters, which make it difficult to calibrate the model.

GARDA 3-D is a three dimensional hydrodynamical model of Lake Garda. The model exemplifies a very advanced model for hydrodynamics, and it would be interesting to couple it with an advanced ecological model, which has not yet been done.

Limnod is a physical/biochemical model for long term predictions of water quality in lakes. It is particularly fitted for studies of restoration methods, such as oxygenation, deep water drainage and artificial mixing. The model is a one-dimensional vertical model that is able to consider various layers of the lake water.

Griffin and Ferrara have developed a model for long-term prediction of phosphorus dynamics in impoundments. It distinguishes between hypolimnion and epilimnion phosphorus and accounts for phosphorus throughout the entire food chain.

The model, Simple, is first of all a fishery model. It considers five piscivore fish species and three planktivore species, which are all represented as dynamic, age-structured populations. Invertebrates are grouped into four categories: large zooplankton, small zooplankton, macro-invertebrates and terrestrial insects.

Livansky has developed a model for simulation of carbon dioxide and oxygen in algae cultures and resorption into the atmosphere. The model Eyesocks is able to predict the oxygen uptake and the reaeration process for a lake, including the effect of ice cover.

The model Calcite can be used to calculate the concentrations of calcium and carbonate species in natural waters. It is included in this section, because it has been used for two lake case studies.

The model named Lake Liming Model examine the need for calcium carbonate for re-acidification to obtain a higher (neutral) pH.

Ryanzhin has developed a model which is able to predict annual mean and extreme surface temperatures in a year cycle of the Northern hemisphere freshwater lakes. It is also able to calculate ice covering time and time for convective overturning as function of latitude and altitude.

The model Orta has been applied on Lake Orta in Northern Italy. The model has been applied as a tool by restoration of Lake Orta which was contaminated by copper. The model has the following state variables: copper concentration in water, totally, copper concentration in epilimnion, copper concentration in hypolimnion and copper concentration in the sediment. The copper concentration in water is changed by addition of calcium carbonate which changes the pH and thereby the concentration of dissolved copper ions.

Rdom is a model which is able to estimate the potential impacts on projected global climate changes on dissolved oxygen in lakes. The model combined with a water temperature stratification model simulates dissolved oxygen profiles in stratified lakes.

Bobba has developed a watershed model for prediction of pH and sulphate in lakes, rivers and streams.

The model denoted Etd is a generalized water acidification model. It considers the snow-smelt, inter-flow, overland flow, groundwater flow, frozen ground processes, seepage and evapotranspiration. The model describes the hydrologic and geochemical response of surface watersheds to freshwater acidification.

The presented collection of biogeochemical models of aquatic ecosystems covers the model published during the last decade. A wide spectrum of aquatic models were, however, already developed during the 1960s and particularly during the 1970s.

Table 2.2. gives an overview of eutrophication models from the this period. The table shows clearly that a wide spectrum of complexity is represented in the models developed at this early stage. Many of models are still in use, although mathematical and computational amendments have been made. The expressions used for the various processes in these models are also still reflected in the latest models of the same type of problem in conjunction with the same type of aquatic ecosystems.

The spectrum of models listed in the table illustrates furthermore that different needs and quantities and qualities of data always will require different complexities of models. So, it is important to be able to select the right model for a given problem, related to a given case study. It is therefore recommended to draw upon the previous experience as well as upon the recently developed models, when a new model is being developed for a new case study.

This brings us to the question, how general is an aquatic ecosystem model? This is illustrated for the Glumsø eutrophication model in Table 2.3, where the modifications from case to case are listed. As seen from this table, modifications were needed for almost every new case study, although the core components and processes could be applied from case to case.

**Table 2.2** Various Eutrophication Models

| Model name | Number of st. var. per layer or segment | Nutri- ents | Seg- ments | Dimen- sion (D) or layer (L) | CS or NC* studies | C and/ or V** | Num. of case |
|---|---|---|---|---|---|---|---|
| Vollenweider | 1 | P (N) | 1 | 1L | CS | C+V | many |
| Imboden | 2 | P | 1 | 2L,ID | CS | C+V | 3 |
| O'Melia | 2 | P | 1 | 1D | CS | C | 1 |
| Girl | 3 | P | 1 | 1-2 D | CS | C+V | many |
| Larsen | 3 | P | 1 | 1L | CS | C | 1 |
| Lorenzen | 2 | P | 1 | 1L | CS | C+V | 1 |
| Thomann 1 | 8 | P,N,C | 1 | 2L | CS | C+V | 1 |
| Thomann 2 | 10 | P,N,C | 1 | 2L | CS | C | 1 |
| Thomann 3 | 1 | P,N,C | 67 | 2L | CS | - | 1 |
| Chen&Orlob | 15 | P,N,C | sev. | 2L | CS | C | min. 2 |
| Patten | 33 | P,N,C | 1 | 1L | CS | C | 1 |
| Di Toro | 7 | P,N | 7 | 1L | CS | C+V | 1 |
| Biermann | 14 | P,N,Si | 1 | 1L | NC | C | 1 |
| Canale | 25 | P,N,Si | 1 | 2L | CS | C | 1 |
| Jørgensen | 17 | P,N,C | 1 | 1-2L | NC | C+V | 21 |
| Cleaner | 40 | P,N,C,Si | sev. | sev. L | CS | C | many |
| Lavsoe | 7 | P,N | 3 | 1-2L | NC | C+V | 25 |
| Aster/ Melodia | 10 | P,N,Si | 1 | 2L | CS | C+V | 1 |
| Baikal | >16 | P,N | 10 | 3L | CS | C+V | 1 |
| Chemsee | >14 | P,N,C,S | 1 | profile | CS | C+V | many |
| 3DWFGAS | >70 | P,N,C,S | many | many | CS | C+V | 135 |
| Reward | 16 | P,N | 1 | profile | CS | C+V | 5 |
| Mahamah Lake | 8 | C,P,N | 1 | 1-2L | CS | C+V | 2 |
| Ecosystem | 11 | C,P,N | 1 | 1 | CS | C+V | 1 |
| Minlake | 9 | P,N | 1 | 1 | CS | C+V | > 10 |
| Salmo | 17 | P,N | 1 | 2L | CS | C+V | 16 |

\* CS means constant stoichiometric and NC independent nutrient cycle.
\** C means calibrated and V validated.

**Table 2.3** Survey of Eutrophication Studies Based upon the Application of a Modified Glumsø Model

| Ecosystem | Modification | Level* |
|---|---|---|
| Glumsø, version A | Basis version | 6 |
| Glumsø, version B | Nonexchangeable nitrogen | 6 |
| Ringkøbing Firth | Boxes, nitrogen fixation | 5 |
| Lake Victoria | Boxes, thermocline,other food chain | 4 |
| Lake Kyoga | Other food chain | 4 |
| Lake Mobuto Sese Seko | Boxes, thermocline, other food chain | 4 |
| Lake Fure | Boxes, nitrogen fixation, thermocline | 3 |
| Lake Esrom | Boxes, Si-cycle, thermocline | 4 |
| Lake Gyrstinge | Level fluctuations, sediment exposed to air | 4-5 |
| Lake Lyngby | Basis version | 6 |
| Lake Bergunda | Nitrogen fixation | 2 |
| Broia Reservoir | Macrophytes, 2 boxes | 2-3 |
| Lake Great Kattinge | Resuspension | 5 |
| Lake Svogerslev | Resuspension | 5 |
| Lake Bue | Resuspension | 5 |
| Lake Kornerup | Resuspension | 5 |
| Lake Balaton | Adsorption to suspended matter | 2 |
| Roskilde Fjord | Complex hydrodynamics | 4 |
| Stadsgraven, Copenhagen | 4-6 interconnected basins | 6 |
| Internal lakes of Copenhagen | 5-6 interconnected basins | 5 |

**\* Explanation to the indication of level:**
Level 1: Conceptual diagram selected.
Level 2: Verification carried out.
Level 3: Calibration using intensive measurements.
Level 4: Calibration of entire model.
Level 5: Validation. Object function and regression coefficient are found.
Level 6: Validation of a prognosis for significant changed loading.

## 2.5 Models for Lakes and Reservoirs

**Model identification**
Model name/title: The Glumsø-model

**Model type**
Media: water, lake, estuary
Class: biogeochemical
Substance(s)/compounds: nutrients, carbon, phosphorus, nitrogen, silica
Biological components: ecosystem

**Model purpose**
Management of eutrophication.

**Short description**
State variables: The model has 17 state variables -

| | | |
|---|---|---|
| CC | - | Carbon in algal cells [g/m³] |
| FNF | - | Proportion of nitrogen in fish |
| FNZ | - | Proportion of nitrogen in zooplankton |
| FPF | - | Porportion of phosphorus in fish |
| FPZ | - | Proportion of phosphorus in zooplankton |
| NC | - | Nitrogen in algal cells [g/m³] |
| ND | - | Nitrogen in detritus [g/m³] |
| NSED | - | Nitrogen in sediment [g/m³] |
| NS | - | Soluble nitrogen [g/m³] |
| PB | - | Phosphorus released biologically from sediment [g/m³] |
| PC | - | Phosphorus in algal cells [g/m³] |
| PD | - | Phosphorus in detritus [g/m³] |
| PE | - | Exchangeable phosphorus in sediment [g/m³] |
| PHYT | - | Phytoplankton biomass [g/m³] |
| PI | - | Phosphorus in interstitial water [g/m³] |
| PS | - | Soluble phosphorus [g/m³] |
| ZOO | - | Zooplankton biomass [g/m³] |

Forcing functions:

| | | |
|---|---|---|
| QTRI | - | [m³/24 h] tributaries |
| NTOTRI | - | Total N [mg/L] tributaries |
| $NH_4TRI$ | - | [mg $NH_4^+$ - N/L] tributaries |
| $NO_3TRI$ | - | [mg $NO_3^-$ - N/L] tributaries |
| PTOTRI | - | Total P [mg/L] tributaries |
| PTRI | - | [mg $PO_4^{3-}$ -P/L] tributaries |
| T | - | Temperature of lake water |
| QWAS | - | [m³/24 h] waste water |
| NTOWAS | - | Total N [mg/L] waste water |
| $NH_4WAS$ | - | [mg $NH_4^+$ - N/L] waste water |
| $NO_3WAS$ | - | [mg $NO_3^-$ - N/L] waste water |
| PWAS | - | Total P [mg/l] waste water |

rad        -      Global irradiance [kcal/m$^2$/24 h]
Q        -      Outflow [m$^3$/24 h]
QPREC     -      Precipitation [m$^3$/24 h]

Important parameters:

| | |
|---|---|
| Max. growth rate phytoplankton | 2.3000- 2.500 1/24 h |
| Max. growth rate zooplankton | 0.1900- 0.200 1/24 h |
| Settling rate phytoplankton | 0.0600- 0.300 m/24 h |
| Uptake rate phosphorus | 0.5500- 0.650 1/24 h |
| Uptake rate nitrogen | 0.0150- 0.030 1/24 h |
| Uptake rate carbon | 0.0014- 0.003 1/24 h |

Characteristic feature: independent N,C and P cycles, which is important for eutrofied and/or shallow lakes. Algae growth described as a two- steps process: uptake of nutrient followed by growth.

---

## Model application(s) and State of development

Application area(s):
Number of cases studie: 22
State: conceptualization - 1, verification - 1, calibration - 5, validation - 13, prognosis made, validation of prognosis - 2 case studies.

---

## Model software and hardware demands

Platform: PC-based, Mainframe
OS:
Software demands: FORTRAN compiler, language- SYSL
Hardware demands:

---

## Model availability

Purchase: no, available as sourcecode
Contact person(s) Dr. S.E. Jørgensen
                      Sektion for Miljøkemi
                      Danmarks Farmaceutiske Højskole
                      Universitetsparken 2
                      DK-2100 København Ø
                      Denmark

Phone: +45 35 37 08 50
Fax: +45 35 37 57 44
E-mail: sej@pommes.dfh.dk

---

## Model documentation and major references

Jørgensen, S.E., Jørgensen, L.A., Kamp Nielsen, L., and Mejer, H.F., 1981. Parameter estimation in eutrophication modelling. *Ecol. Modelling*, 13: 111-129.
Jørgensen, S.E., Kamp Nielsen, L., Christensen, T., Windolf- Nielsen, J., and Westergaard, B., 1986. Validation of a prognosis based upon a eutrophication model. *Ecol. Modelling*, 72: 165-182.
Jørgensen, S.E., Mejer, H., and Friis, M., 1978. Examination of a lake model. *Ecol. Modelling*, 4: 253-279.
Jørgensen, S.E., 1976. A eutrophication model for a lake. *Ecol. Modelling*, 2: 147-165.

**Other relevant information**

---

**Model identification**
Model name/title: A Simple Eutrophication Lake Model

---

**Model type**
Media: water, lake
Class: biogeochemical
Substance(s)/compounds: nutrients, phosphorus, nitrogen
Biological components: ecosystem

---

**Model purpose**
Management of eutrophication.

---

**Short description**
State variables: soluble P and N, phytoplankton, zooplankton, detritus, sediment exchangeable P and N, fish, totally 6-8 state variables.
Forcing functions: inputs of N and P, climate.
Important state variables: growth rate phytoplankton and zooplankton, settling rates of phytoplankton and detritus, mineralization rate.
Characteristic feature: first order reactions and Michaelis-Menten's equation.

---

**Model application(s) and State of development**
Application area(s):
Number of cases studied: 8
State: 8 validation

---

**Model software and hardware demands**
Platform: PC-based MAC
OS:
Software demands: language SYSL and STELLA
Hardware demands:

---

**Model availability**
Purchase:
Contact person(s) Dr. S.E. Jørgensen
                Sektion for Miljøkemi
                Danmarks Farmaceutiske Højskole
                Universitetsparken 2
                DK-2100 København Ø
                Denmark

Phone: +45 35 37 08 50
Fax: +45 35 37 57 44
E-mail: sej@pomme.dfh.dk

## Model documentation and major references
Jørgensen, S.E., 1994. *Fundamentals of Ecological Modelling*, 2nd edition, 1994, Elsevier, Amsterdam.

## Other relevant information

## Model identification
Model name/title: ASTER: biological model of Pareloup Lake

## Model type
Media: water, lake
Class: biogeochemical
Substance(s)/compounds: nutrients, phosphorus, silicates
Biological components: ecosystem, planktonic

## Model purpose
The ASTER biological model represents the first phase in development of the MELODIA reservoir-management model. ASTER simulates the dynamics of the principal biological and chemical variables. It has a simple physical structure (two-layer simulation of seasonal thermal stratification) and aims at improving understanding of the dominant processes involved in the dynamics of the planktonic ecosystem. The considerable information base available enabled calibration of ASTER, thereby ensuring correct simulation of the seasonal evolution of each of the variables over a five-year period.

## Short description
Pareloup Lake is a dammed mesotrophic reservoir. The ASTER biological model was developed on the basis of data for the period between 1983 and 1987. The water mass is divided into two layers representing seasonal thermal stratification. The depth of the thermocline, determined on the basis of profiles obtained from the results of EOLE stratification model, is introduced by forcing.
Forcing variables:
- incident solar irridiance,
- extinction coefficient,
- temperature of each layer,
- depth of each layer,
- volume of each layer,
- volume of the sediment layer.
State variables: - Phytoplankton, divided into two groups: large diatoms, needing silica for growth and not consumed by the zooplankton; and a group comprising several small species, consumed by the filtering zooplankton.

- Zooplankton, divided into three groups; herbivorous Cladocerans and Copepods and a carnivorous group.
- Dissolved nutrients: $PO_4$ and $SiO_2$
- Detritus permitting storing of dead organic matter.

Light, temperature and P and Si concentrations are phytoplankton growth-limiting factors. The four factors represented as governing loss are respiration, sedimentation, predation and mortality. The rate of ingestion of zooplankton depends on temperature and food concentration. Zooplankton mortality is primarily due to predation by fish. The complex dynamics of fish populations are represented with a forcing function. Mineralization of detritus and dissolution of silica frustules are temperature-dependent. Diatom sinking is one of the fundamental mechanisms involved in the evolution of the ecosystem. It appears that the behaviour of the ASTER model is extremely sensitive to forcing conditions, and most particularly to fluctuations in the level of the thermocline. It also appears that the values for the winter extinction coefficient have a major impact on collapse of algae at the end of autumn, and bloom in early spring.

---

**Model application(s) and State of development**
Application area(s): Lakes and reservoirs
Number of cases studied: one, during five years
State: validation

---

**Model software and hardware demands**
Platform: Mainframe, IBM
OS:
Software demands: SP422, FORTRAN 77 (2.5.0)
Hardware demands: ES9000/831

---

**Model availability**
Purchase: no
Contact person(s) Mme Marie-José Salencon
        Département Environnement
        Electricité de France - Direction des Etudes et Recherches AEE/ENV
        6, Quai Watier BP49
        F-78401 Chatou Cedex
        France
Phone: +33 1 30 87 77 87
Fax: +33 1 30 87 81 09
E-mail:

---

**Model documentation and major references**
Thébault, J.M. and Salencon, M.J., 1983. Simulation model of a mestrophic reservoir (Lac de Pareloup): biological model. *Ecol. Modelling*, 65: 1-30.

## Other relevant information

A vertical thermal model of the reservoir, EOLE, which takes through flow, bathometry and hydraulic management into account, was also developed and calibrated for Pareloup Lake. The extreme sensitivity of the ecosystem to hydrodynamics fully justifies coupling the two models, so as to permit their potential use as a water management tool. The complete model MELODIA is presented in another model form.

## Model identification

Model name/title: EOLE, thermal stratification model of reservoir

## Model type

Media: water, lake
Class: hydrology
Substance(s)/compounds: temperature, water density
Biological components:

## Model purpose

EOLE is a hydrodynamic and thermal model (one-dimensional vertical model) which simulates the evolution in the seasonal thermal stratification of a reservoir. It is a well-mixed layer model which takes into account bathometry, energy exchanges at the air-water interface and inflow and outflow of water (rivers, pumping, turbining, residual stream flow).

## Short description

EOLE is a one-dimensional, vertical, energy budget model. On the basis of energy exchanges through the air-water interface, an energy budget is calculated in the water column:
- internal energy budget, on the basis of heat fluxes at the water surface,
- turbulent kinetic energy (TKE): the TKE at the water surface (wind shear, penetrative convection) creates a well-mixed surface layer whose depth is determined by the balance between TKE and the variation in potential energy of the well-mixed layer. Movement of the water masses due to inflow and outflow of water in the reservoir is represented by means of a simple transport model coupled to the preceding model. Natural inflow into the reservoir is located in all cases at the level of its density. Forced inflow creates a convective plume. Outflow come from a horizontal layer of the reservoir which is dependent on outflow rate and thermal stratification. Inflow and outflow rates are vertically distributed around inflow or outflow levels according to a gaussian profile representing the entrainment velocity of nearby water masses.
State variables: temperature, density
Necessary input: meteorological parameters (3-hour), rate of flow and temperature of inflow and outflow (daily), bathometry, extinction coefficient (weekly).
EOLE was calibrated for the Pareloup reservoir with extremely accurate simulations of the thermal dynamics over a consecutive 12-year period.

---

**Model application(s) and State of development**
Application area(s): Lakes and reservoirs
Number of cases studied: two (one during five years, another during twelve years)
State: validation, prognosis made?

---

**Model software and hardware demands**
Platform: Mainframe, IBM CRAY, Workstation HP
OS:
Software demands: IBM SP422 , FORTRAN 2.5.0
   CRAY Unicos 7C3  FORTRAN CFT77 5.03
   HPUX 8.07 , FORTRAN A.B8.07.1B
Hardware demands: IBM ES9000/831
   CRAY C98/5256
   HP 9000/730

---

**Model availability**
Purchase: no
Contact person(s) Mme Marie-José Salencon
   Département Environnement
   Electricité de France
   Direction des Etudes et Recherches AEE/ENV
   6, Quai Watier
   F-78401
   France
Phone: +33 1 30 87 77 87
Fax: +33 1 30 87 81 09
E-mail:

---

**Model documentation and major references**
Enderlé M.J., 1982. Impact of a pumped storage station on temperature and water quality of the two reservoirs. Third International Conference on State of the Art in Ecological Modelling-May 24-28 1982, Fort Collins, Colorado. *ISEM Journal*, 4 (3-4), 87-95.
Salencon, M.J., 1994a. Stratification thermique d'un réservoir: le modéle à bilan d'énergie, EOLE. Rapport HE31/94-1, Electricité de France, Paris.
Salencon, M.J., 1994b. Study of the thermal dynamics of two dammed lakes (Pareloup and Rochebut) using the EOLE model. *Ecol. Modelling*, (in prep).

---

**Other relevant information**
The EOLE model was applied to two reservoirs with quite different residence times (1 year for Pareloup and 3 months for Rochebut). The close agreement in both cases between calculated and measured profiles over a long period shows that the model enables reproducing the thermal dynamics in response to highly variable hydrometeorological conditions. If input data is adequately representative of the site under study, the model does not require calibration as it is based on a reprensentation of the processes.

## Model identification
Model name/title: MELODIA: an ecosystem management reservoir model

## Model type
Media: water, lake
Class: biogeochemical, hydrology
Substance(s)/compounds: nutrients, phosphorus, silicates, temperature, water density
Biological components: ecosystem, planktonic

## Model purpose
The MELODIA model is a result of the coupling of the ASTER biological model with the EOLE vertical hydrodynamic and thermal model, applied to Pareloup Lake.

Its purpose is to help in improving understanding of the dynamics of this ecosystem in both biological and hydrodynamic terms, and to provide a tool for water management.

## Short description
MELODIA is a one-dimensional vertical model representing a coupling of two models: EOLE (vertical hydrodynamic) and ASTER (biological).
State variables in water and sediment:
- temperature
- phytoplankton: diatoms needing silica for growth, not consumed by the zooplankton non-siliceous algae consumed by the zooplankton
- zooplankton: 2 groups of herbivores, 1 group of carnivores
- dissolved nutrients, $PO_4$ and $SiO_2$
- detritus
Necessary inputs:
- meteorological parameters (3-hour)
- inflow and outflow rates (daily)
- water quality of inflow (daily)
- bathometry
- non-chlorophyllous extinction coefficient (weekly)
MELODIA enabled simulation on a daily scale of the vertical dynamics of the ecosystem over the five-year period of measurements in the Pareloup data base. Comparison of calculated and measured profiles for each variable shows that it represents the ecosystem well in terms of vertical dynamics, seasonal evolution and year-to-year fluctuations.

This study improved our understanding of the role played by the *Asterionella formsa* diatom which is dominant in spring, and the sensitivity of the ecosystem to the hydrodynamic structure. This diatom, which as it sinks in the spring traps part of the phosphorus in the lake, acts as a purifier. Spring stratification, resulting from meteorological conditions and hydraulic management, determines diatom population dynamics. If any given phenomenon disturbs diatom development, the quantity of $PO_4$ available in summer is greater, fostering rapid summer growth of algae. It therefore plays a key role in the seasonal cycle of the ecosystem.

**Model application(s) and State of development**
Application area(s): Lakes and reservoirs
Number of cases studied: one during five years
State: validation, prognosis made?

**Model software and hardware demands**
Platform: Mainframe, IBM, CRAY. Work station, HP
OS:
Software demands: IBM SP422, FORTRAN 77, 2.5.0.
        CRAY UNICOS 7C3, FORTRAN 77, CFT77 5.03
        HP HPUX 8.07, FORTRAN 77, A.B.8.07.1B
Hardware demands: IBM ES9000/831
        CRAY C98/5256
        HP 9000/730

**Model availability**
Purchase: no
Contact person(s) Mme Marie-José Salencon
        Département Environnement
        Electricité de France
        Direction des Etudes et Recherches AEE/ENV
        6, Quai Watier
        F-78401 Chatou Cedex
        France

Phone: +33 1 30 87 77 87
Fax: +33 1 30 87 81 09
E-mail:

**Model documentation and major references**
Salencon, M.J. and Thébault, J.M., 1994. Simulation model of a mesotrophic reservoir (Lac Pareloup): MELODIA, an ecosystem management model. *Ecol. Modelling* (submitted).

**Other relevant information**
Because the model simulated well both the thermal dynamics of the reservoir and the growth and sinking of diatoms, it was used as a water management tool for evaluating the repercussions on the Pareloup ecosystem of 8 reservoir management scenarios.

**Model identification**
Model name/title: OPHELIE: a two-dimensional (x-z) laterally-averaged reservoir model

**Model type**
Media: water, lake
Class: hydrology
Substance(s)/compounds: temperature, water density
Biological components:

**Model purpose**
OPHELIE is a two-dimensional (x-z) laterally-averaged hydrodynamic and thermal model of a reservoir, which simulates the evolution in seasonal thermal stratification of the reservoir, as well as circulation of water masses. Its objective is to represent the effect of the longitudinal dimension (the axis of the river in general) on the spatial distribution of the components in the ecosystem.

**Short description**
OPHELIE is a two-dimensional (x-z) laterally-averaged reservoir model. Resolution is made with a finite volume scheme. Topographical effects due to the transversal dimension are taken into account by averaging punctual equations in the lateral direction (perpendicular to the average flow in the reservoir). Boussinesq and hydrostatic pressure assumptions are made. Horizontal dispersion coefficients are treated as constants but modeling of vertical thermal stratification is similar to the principle of the EOLE one-dimensional integral model: the turbulent kinetic energy (TKE) brought to the surface (wind friction, penetrative convection) creates a well-mixed surface layer whose level is determined by the balance between this TKE and the variation in potential energy of the well-mixed layer.

The vertical sections are coupled by resolving, for the entire reservoir, the dynamic equation governing all evolution in the average horizontal current (convection, vertical diffusion, horizontal dispersion, density current). The free surface is treated as a horizontal rigid lid which evolves vertically depending on the water budget. Unstable stratification is handled in a convective manner.
States variables: temperature, density
Necessary input: meteorological parameters (3-hour), rate of flow and temperature of inflow and outflow (daily or hourly), bathometry, extinction coefficient (weekly)
OPHELIE was developed on the Sainte-Croix reservoir, which has good longitudinal heterogeneity. Simulation of this reservoir over one year, under real management conditions, showed a highly satisfactory agreement between calculated and measured thermal profiles and a significant improvement in the representation of the hypolimnion.

**Model application(s) and State of development**
Application area(s): Lakes and reservoirs
Number of cases studied: one during one year
State: verification

---

**Model software and hardware demands**
Platform: Mainframe, CRAY
OS:
Software demands: CRAY: UNICOS 7C3, FORTRAN 77 (CFT5.03)
Hardware demands: CRAY: C98/5256

---

**Model availability**
Purchase: no
Contact person(s) Mme Marie-José Salencon
         Département Environnement
         Electricité de France
         Direction des Etudes et Recherches AEE/ENV
         6, Quai Watier
         F-78401 Chatou Cedex
         France

Phone: +33 1 30 87 77 87
Fax: + 33 1 30 87 81 09
E-mail:

---

**Model documentation and major references**
Salencon, M.J., and Simonot, J.Y., 1989. Proposition de modèle bidimensionnel (x-z) dynamique et thermique de retenue. Rapport HE31/89-21, Electricité de France, Paris, 34 pp.
Salencon, M.J., 1994a. Stratification thermique d'un réservoir: le modèle à bilan d'énergie, EOLE. Rapport HE31/94-1, Electricité de France, Paris.
Salencon, M.J., 1994b. Study of the thermal dynamics of two damned lakes (Pareloup and Rochebut) using the EOLE model. *Ecol. Modelling*, (in preperation)

---

**Other relevant information**
Application of the model to the Sainte-Croix reservoir provided an opportunity to test numerical resolution methods under model boundary conditions showing sharp variations (alternate pumping and turbining during the day, resurgence of warm water at a certain depth, extremely variable river inflow temperature due to turbining at an upstream dam). Despite these difficult conditions, the model remained stable over an entire annual cycle.

---

**Model identification**
Model name/title: Model of Lake Baikal Ecosystem Disturbance

---

**Model type**
Media: water, lake, regional scale
Class: biogeochemical, toxicology, hydrology
Substance(s)/compounds: nutrients, phosphorus, nitrogen, chemicals, organic compounds, oil products, phenols, heavy metals, Cd
Biological components: ecosystem

## Model purpose
The basic object of the model is optimization of interaction of the anthropogenic factors with ecosystem of Baikal, therefore the model was based on the method of disturbances. The model describes effects of anthropogenic influence on the state of the lake ecosystem components and destruction of pollutants under the action of biotic and abiotic factors.

## Short description
The lake water surface is divided into 10 regions, significantly differing by their conditions. In each region the water body is divided into three layers (0-50, 50-250), the layer 250 m bottom was supposed to be homogenous. 21 boxes are thus obtained and ecosystem dynamics within each box are described by an equation
Biotic components described by the model bacterio-, phyto-, zooplankton, fishes, seal, zoo- and phytobenthos, hydrochemical parameters - water mineralization, nitrogen and phosphorus content, dissolved and particulated organic matter, content of pollutants - phenols, oil products, heavy metals, other anthropogenic influences - catchment of fishes and seal. The inputs of the model are the input of pollutants with tributaries and precipitation, other economic activity. Model describes both seasonal and annual changes. Information provision is based on the series of *in situ* field experiments made in different seasons for five years.

## Model application(s) and State of development
Application area(s): ecological prognostics
Number of cases studied:
State: validation of prognosis

## Model software and hardware demands
Platform: PC-based, IBM
OS: MS-DOS
Software demands: FORTRAN
Hardware demands:

## Model availability
Purchase: yes, executable code and source code
Contact person(s) Dr. Eugene A. Silow
    Scientific Research Institute of Biology
    P.O. Box 24
    Irkutsk-3
    664003 Russia

Phone: (3952) 33-44-79
Fax: (3952) 24-30-79
E-mail:

## Model documentation and major references
Silow, E.A. and Stom, D.J., 1992. Model ecosystems and models of ecosystem in hydrobiology. Irkutsk. University Press. (In Russian).

**Other relevant information**

---

**Model identification**
Model name/title: CHEMSEE

---

**Model type**
Media: lake
Class: biogeochemical, toxicology
Substance(s)/compounds: nutrients, carbon, phosphorus, nitrogen, sulphur, oxygen, other - user-defined, chemicals, radionuclides, pesticides, organic compound, heavy metals, Hg, Pb, Cd, Cu, Zn, other - user-defined
Biological components:

---

**Model purpose**
CHEMSEE is a flexible "model construction kit" for the modeling of chemical processes in lakes. The program with a user interface consisting of menus standard dialog boxes and graphic windows, was developed on the Apple Macintosh personal computer, and is currently in use both in teaching and research.

---

**Short description**
Underlying the program is the mathematical description of the behaviour of a substance in terms of a one-dimensional vertical model. Such a model, in which horizontal concentration differences are neglected, is suitable for the description of reactive substances in deep lakes.

Physical processes occurring in a lake are not influenced by the presence of trace substances and are therefore represented in the model by corresponding input parameters (e.g., epilimnion depth, temperature, etc.).

Any number of variables may be defined by the user to represent the concentrations of substances and particles in the water column, and of substances in the sediments. Transport and transformation processes (loading, adsorption, sedimentation, air-water and sediment-water exchange, chemical reactions) can be defined interactively. The results of simulation runs are output both graphically and on file, and the interactively defined models can be saved for future use.

The program can be used to solve problems related to the following problems:
- setup of a dynamic models using information on a system of interest,
- discrimination between different hypotheses which explain measured concentrations of substances of interest,
- identification of relevant transport of transformation processes,
- development of teaching models.

**Model application(s) and State of development**
Application area(s): Modelling of dynamic processes in lakes
Number of cases studied: numerous
State: validation

**Model software and hardware demands**
Platform: MAC
OS: Macintosh OS
Software demands: Macintosh, English
Hardware demands: 500 Kb RAM, two floppies or hard disk

**Model availability**
Purchase: yes, US $150.00 non-commercial, US$ 750.00 commercial. Executable code.
Contact person(s) Dr. Markus Ulrich
    Swiss Federal Institute for Environmental Science and Technology
    (EAWAG)
    CH-8600 Duebendorf
    Switzerland
Phone: +41 1 823 5464
Fax: +41 1 823 5210
E-mail: ulrich@eawag.ch

**Model documentation and major references**
Ulrich, M., 1991. Modeling of Chemicals in Lakes - Development and Application of User-Friendly Simulation Software (MASAS & CHEMSEE) on Personal Computers. Dissertation ETH No.9632.
Johnson, C.A, Ulrich, M., Sigg, L., and Imboden, D.M., 1991. A mathematical model of the manganese cycle in a seasonally anoxic lake. *Limnol. Oceanogr.*, 36/7, 1415-1426.

**Other relevant information**

**Model identification**
Model name/title: BIOTEST

**Model type**
Media: water, lake, estuary
Class: biogeochemical
Substance(s)/compounds: nutrients, phosphorous, nitrogen, oxygen
Biological components: populations, phytoplankton, zooplankton

## Model purpose
The simulation of the yearly dynamic of a two trophic level aquatic ecosystem under continuous flow of nutrients, dispersed by turbulent diffusion through the open boundaries, is achieved by using a very small grid size. This makes the model particularly suitable for testing possible improvements for larger models of the same type, at very low computational costs.

## Short description
This finite-difference model considers a small three dimensional grid (min 4 x 4 x 2 cells) where macronutrients and heat are introduced at constant rate in the upper corner cell and dispersed by means of a turbulent diffusive process. Eight state variables are followed: Phyto- and zooplankton densities, reduced and oxidized nitrogen concentrations, reactive phosphorus concentration, dissolved organic detritus, dissolved oxygen and sedimented organic detritus. The forcing functions are the input rates, the water temperature and the sunlight intensity. The last two can be calculated as indicated in the 1-D vertical model proposed by the same authors, or introduced by the user. The system can reach a steady state condition if the forcing functions are kept constant because the two horizontal walls opposite the input corner are open to the outward fluxes. A smooth behaviour of the numerical solution of the reaction diffusion equation which connect the state variables is assured by purposely studied boundary conditions. According with this approach, external values are extrapolated by assuming that the three last grid points follow a gaussian profile, whose asymptotic behaviour can be set by the user or computed using the 1-D model. The model can represent with sufficient approximation a small lake or pond not completely mixed or a shallow water basin strongly influenced by the tide, where the advection does not, on average, contribute significantly to the dispersion process. It can also been used as a quick tool for testing some improvements for an already existing combined transport-water quality model, as the computational request of the program can be conveniently satisfied by a 486 PC.

## Model application(s) and State of development
Application area(s): Venice Lagoon
Number of cases studied:
State: verification, calibration

## Model software and hardware demands
Platform: PC-based, IBM
OS: UNIX, DOS
Software demands: FORTRAN
Hardware demands:

## Model availability
Purchase: no, executable code and sourcecode
Contact person(s) Dejak Camillo, Pasters Roberto, Pecenik Giovanni
        Universit' di Venezia
        Dip. Chimica Fisica
        Dorsoduro 2137
        I-30123 Venezia, Italy

Phone:
Fax:
E-mail: Pasters@unive.it

## Model documentation and major references
Dejak, C., Franco D., Pasters R., Pecenik G, 1989. A steady state achieving 3D eutrophication-diffusion submodel. *Environmental software*, 4(2): 94-101.

## Other relevant information

## Model identification
Model name/title: 3DWFGAS: Three-dimensional Water Flow and Quality Model with Air and Soil Modules for optional Forcing Functions (FINNWIND - FINNSHED - FINNFLOW - FIINQAU - FIINWAVE- FINNBIO - FINNFISH - FINNESOC)

## Model type
Media: water, lake, ocean, estuary, river, swamp, coastal waters, reservoir, air, local scale, terrestrial, sediment, soil, forest, agricultural
Class: biogeochemical (55%), toxicology (5%), hydrology (40%)
Substance(s)/compounds: nutrients, carbon, phosphorus, nitrogen, sulphur, oxygen, BOD, COD, color, Lignosulphonate, chemicals, radionuclides, pesticides, organic compounds, oils and organic chlorines, alkalinity upto 12 ions for pH, heavy metals, Hg, Pb, Cd, Cu, Zn, Ti, V, Fe, Cr, Ni, Mn, Na, etc. (most of them not here).
Biological components: populations - 5 species of fish, 3 groups of zooplankton, zoobenthos, 4 types of phytoplankton, bacteria organism - Perch and Pike, Daphnia, Algae

## Model purpose
Real-time: I. Operational use: to support ongoing oil of chemical combating or rescue.
Prospective: II. Management support: advice to harm mitigation. III. Decision support: to predict and compare the effects of planned alternatives.
Retrospective: IV. Research: to show contributions of specified factors, test hypotheses and understand the causal dynamics. V. Education: to make nature's responses understood.

## Short description
Hydrodynamics:
Three-dimensional time-dependent calculation of water currents and volumes (flow components u, v, w + surface level elevation) is forced by surface shear (winds), river flows and density effects. Selected freely-varying Cartesian mesh and the depths therein are the main inputs. The surface drag, bottom drag, latitude (Coriolis effect), and the components of viscosity and diffusivities are coefficients needed.
Transport and water quality: The transport of each substance and bio-component with the varying (x,y,z,t) flow components (u,v,w) is calculated in the 3-D flow grid. Each component is simultaneously mixed with eddies, and settled, decayed or elsewhere changed either independently or in interaction with other components.

The forcing discharges are from point sources, bottom, rivers, external areas and air. For each component, settling, sedimentation, decay, and reaction rates, and their temperature dependencies are needed as model coefficients.

Calculation has been applied to about 70 substances and bio-components, most frequently to BOD, oxygen and phosphorus. In one run - for periods from one season to 10 years - not more than 2-5 components at 80.000 points can fluently be solved. Only in food web applications there have been more than 20 components in one run (6 bio-components and several fractions of nutrients or impurities within them, in sediment and in water).

---

**Model application(s) and State of development**
Application area(s): Natural and artificial lakes in North and Central Europe and in Africa; Coastal and off-shore areas around the Baltic, the Mediterranean and the Red Sea; River networks and channels in Finland, Sweden and a few areas of the former Soviet Union.
Number of cases studied: 135
State: validation, prognosis made, validation of prognosis

---

**Model software and hardware demands**
Platform: PC-based, IBM or IBM compatible. Mainframe Digital IBM, Cyber, Eclipse, Univac, Workstation, CRAY X-MP, Convex
OS: DOS, Windows, OS/2, Unix/Motif, UNIX, VMS, NOS, AOS/VS, etc.
Software demands: No other software needed, no compilers needed for run; program written in FORTRAN
Hardware demands: PCs 16 Mb memory, 386 processor and VGA graphics needed (Math/Coprocessor, 66 MHz and about 200 Mb disc recommended).
In other computers there are no specific demands.

---

**Model availability**
Purchase: yes, delivery with applications, see below. Executable code.
Contact person(s) Dr. Jarma Koponen or Dr. Markku Virtanen
        EIA Ltd. (Otaniemi Science Park)
        Tekniikantie 17
        SF-02150 Espoo
        Finland

Phone: +358-0-7001-8680
Fax: +358-0-7001-8682
E-mail: koponen@convex.csc.fi.

---

**Model documentation and major references**
User instructions, detailed description of the needed files, and partial documentation are written and duplicated as the internal reports of EIA Ltd. and the National Board of Waters and the Environment, Finland; more comprehensive documentation is inset within the source code; comprehensive external documentation is not produced; help system and easy selections are installed within the user interface.
Koponen, J., and Alasaarela, E:, 1993. Use of mathematical models for solving environmental management problems. CSC News, 5, No. 3, August (Center for Scientific Computation, P.O. Box 40, 02101 Espoo, Finland), pp.3-6.

Koponen, J., Virtanen, M., Vepsä, H., and Alasaarela, E., 1993. Operational model and its validation with drift tests in water areas around the Baltic Sea. *Water Pollution Research Journal of Canada*, 16 pages (in print).

Virtanen, M., Koponen, J., Hellsten, S., Nenonen, O., and Kinnunen, K., 1993. Principles for calculation of transport and water quality in strongly regulated reservoirs. *Ecol. Modelling*, 21 p. (in print).

Koponen, J., Alasaarela, E., Lehtinen, K., Sarkkula, J., Simbierowcz, P., Vepsä, H., and Virtanen, M., 1992. Modelling the Dynamics of a Large Sea Area. Publications of the Water and Environment Research Institute, 7, National Board of Waters and the Environment, Finland, Helsinki, 91 p.

Virtanen, M., Koponen, J., Dahlbo, K., and Sarkkula, J., 1986. Three-dimensional water-quality-transport model compared with field observations. *Ecol. Modelling*, 31, pp. 185-199.

---

### Other relevant information

A training period of three to eight months is typically needed for new users of the system. That is why the model is not recommended to be directly purchased, but should to be delivered along with an application.

For validation, in every case study the model results have been compared with the available field observations of flow velocities, *in situ* detected and sampled concentrations components, tracer studies, and remote sensing. No serious disagreements have ever been found. In addition to the most severe practical tests of 25 emergency cases, the validity of published prognosis is tested in about 15 cases which have been completed according to plans. Outstanding agreement is found in all of these cases.

Recently, during the last two years, the three-dimensional Water Flow and Quality Model has been supplemented with proper models of local atmospheric wind forcing and loading from remote transport, discharges from the water shed, wave effects on the bottom interactions (as included within the water quality model), food web interactions (including two groups of algae, three groups of zooplankton, and bacteria, as extensions to the water quality model), bioenergetic accumulation of impurities throughout the web until fish, effects on fish stocks, and (as quite preliminary schedules) consequences on private and political economy, health and social welfare.

---

### Model identification

Model name/title: REWARD: Regional Water Quality and Resources Decision Model

---

### Model type

Media: water, lake, river, wetland, local scale, regional scale, sediment
Class: biogeochemical
Substance(s)/compounds: carbon, phosphorus, nitrogen, oxygen, chloride, Hg, Cd
Biological components: ecosystem, organism - 4 algal groups, heterotrophic bacteria, autotrophic bacteria, Coliforms

---

### Model purpose

REWARD is a decision support system created for the purpose of
- river basin modeling,
- lake management and controlling waste water treatment processes.

It can be used as a software tool for the purpose of
- process description (e.g., eutrophication, river water quality deterioration),
- planning (e.g., wastewater treatment plants),
- evaluation of water quality management policies.

---

**Short description**
REWARD is a user-friendly and comprehensive water quality and hydrodynamical model. It computes surface profiles (and discharge, flow depth, velocity etc.) for steady state and for gradually varied flows (including 1-D Saint-Venant Eq. or 2-D finite difference models). The constituents which can be simulated in the water quality model (under steady state or dynamical conditions) are: BOD, COD, Temperature, Algae, $NO_2$-N, $NO_3$-N, $NH_4$-N, Organic Nitrogen, Organic Phosphorus, Dissolved Phosphorus, Coliforms, readily (and slowly) biodegradable substrate, Heterotrophic biomass, Autotrophic biomass, Oxygen, Alkalinity, Arbitrary conservative (and non-conservative) constituents Hg, Cd.
Inputs includes: geometrical and load data, temperature, measured data and hydrological information.
The main features are as follows:
- The MODEL-BANK of REWARD includes well-known models (e.g., QUAL2E, IAWPRC, Streeter-Phelps, etc.), or the user is allowed to define new model structures.
- The boundary conditions can be flexibly defined with the help of a graphical program (e.g., lake, river reach or wastewater treatment plant).
- Graphical user interface for the construction of river basins and data bank.
- Number of river reaches is unlimited.
- Mode selection for planning or simulation.
- Automatic calibration (or sensitivity analysis) of the model parameters.
- Graphical user interface (icon menus, push-button menus, mouse, etc.)
- Advanced results presentation (2-D, 3-D plots), output HPGL files.

---

**Model application(s) and State of development**
Application area(s): River Ruhr (Germany), Lake Kemnade (Germany), Lake Balaton and Lake Balaton (Hungary), River S. Bartolomeu (Brazil)
Number of cases studied: 5
State: validation

---

**Model software and hardware demands**
Platform: PC-based (IBM PC and compatible)
OS: MS-DOS 4.0 or higher
Software demands: FORTRAN, C, HPGL
Hardware demands: IBM PC (or compatible), Mouse, >640 KB RAM, >8 MB storage capacity (on hard disk)

**Model availability**
Purchase: yes, 290- 2800 US$ (depending on the system configuration) demo version: available (price: - ) executable code

Contact person(s) Dr. L. Koncsos      or      Dr. L. Somlyody
         Ruhr- Universität Bochum            Water Resource Research Center
         Lehrstuhl für Hydrologie            VITUKI
         und Wasserwirtschaft            Pf. 27
         Postfach 102148            H-1095 Budapest
         D-44721 Bochum            Hungary
         Germany

Phone:                                       36 11 33 81 60
Fax: 0049 234 7094 153           telex: 22-4959h
E-mail:

**Model documentation and major references**
Koncsos, L., and Somlyody, L., An Interactive Model System for Operating the Kis-Balaton reservoir. XVI General Assembly of European Geophysical Society. Wiesbaden, 1991.

Koncsos, L., Schumann, A., and Schultz, G.A., Water Quality Model for the River Ruhr. Research Report, Institute for Hydrology and Water Resources Management and Environmental Techniques, Ruhr University Bochum (Germany), January 1994.

Koncsos, L., Computer Program Documentation for the Regional Water Quality and Resources Decision Model (REWARD). Institute for Hydrology and Water Resources Management and Environmental Techniques, Ruhr University Bochum (Germany), February 1994.

Luettich, R.A., Harleman, D.R.F., Somlyody, L., and Koncsos, L., Wind-induced Sediment Resuspension and its Impact on Algal Growth for Lake Balaton. IIASA RR-93-3, February 1993.

Schumann, A., Koncsos, L., and Schultz, G.A.. Estimation on dissolved pollutant transport to rivers from urban areas: A modelling approach. Sediment and Stream Water Quality in a Changing Environment: Trends and Explanation (proceedings of the Vienna Symposium, August 1991) IAHS Publ. No. 203, 1991, p 267-275.

Somlyody, L., and Koncsos, L., Influence of sediment resuspension on the light conditions and algal growth in Lake Balaton. *Ecol.Modelling*, 57: p173-192.

Szilagyi, F., Somlyody, L., and Koncsos, L., The Kis-Balaton reservoir system as means of controlling eutrophication of Lake Balaton. XXIV Congress of the International Assoc. of Limnology. München, 1989.

**Other relevant information**

**Model identification**
Model name/title: Nutrient/Biomass Model for Liberty Lake, Washington

## Model type
Media: water, lake, local scale
Class: biogeochemical, hydrology
Substance(s)/compounds: nutrients, carbon, phosphorus, nitrogen, biomass, organic compounds
Biological components: ecosystem - aquatic

## Model purpose
To simulate the effect of nutrient discharges on an ecosystem. Long- and short-term effects are investigated.

## Short description
State variables of biomass (as measured by chlorophyll a), phosphorus, nitrogen, and carbon are simulated in a lake ecosystem. Both internal (sediment) and external (runoff, septic discharges) are modeled as input. Taking into account the complex interchanges between sediments, overlying water, growth, model equations are developed as differential equations and solved by finite difference methods. The effects of various reductions in inputs are also simulated.

## Model application(s) and State of development
Application area(s): Eutrophication in ecosystems
Number of cases studied: two
State: conceptualization, verification, calibration, validation, prognosis made, validation of prognosis

## Model Software and hardware demands
Platform: Mainframe IBM
OS: IBM-OS
Software demands: Fortran compiler, FORTRAN
Hardware demands: Mainframe Computer

## Model availability
Purchase: no
Contact person(s) Dr. D.S. Mahamah
    Engineering Division
    Saint Martin's College
    5300 Pacific Ave SE
    Lacey, WA 98503, USA

Phone: 206-438-4321
Fax: 206-438-4124
E-mail: DMAHAMAH@STMARTIN.WA.COM

## Model documentation and major references
Published by NTIS, USA

**Other relevant information**

---

## Model identification
Model name/title: Lake Ecosystem

---

## Model type
Media: water, lake
Class: biogeochemical, hydrology
Substance(s)/compounds: carbon, phosphorus, nitrogen, dissolved oxygen, temperature, suspended solids
Biological components: ecosystem - phytoplankton, zooplankton

---

## Model purpose
The model was developed as a research tool to assess the tradeoff between increasing model complexity and collecting data of higher quality or quantity. The lake ecosystem model is used in conjunction with the extended Kalman filter to test hypotheses regarding modeling performance.

---

## Short description
State variables: Phytoplankton, Zooplankton, Benthic Animals, Benthic Plants, Detritus, $PO_4$-P, $NO_2$+$NO_3$-N, $NH_4$-N, Dissolved Oxygen, Temperature, Suspended Solids.
Forcing functions: Solar radiation, river inflow, point source inflow.

---

## Model application(s) and State of development
Application area(s): Evaluation of value of complexity in lake ecosystem models compared to value of data collection programs.
Number of cases studied: one case study with 18 scenarios
State: conceptualization, simple parameter estimation

---

## Model software and hardware demands
Platform: DEC Microvax, DG Aviion Workstation
OS: DEC VMS, Unix
Software demands: FORTRAN 77 compiler, FORTRAN
Hardware demands: 4 Mb RAM, 200 Mb storage

---

## Model availability
Purchase: none
Contact person(s) Dr. John R. Yearsley
        EPA Region 10  ES097
        1200 Sixth Avenue
        Seattle, WA 98101
        USA

Phone: 206-553-1532
Fax: 206-553-0119
E-mail: jyearsle@r0dg09.r10.epa.gov
        yjd@sequoia.nesc.epa.gov

## Model documentation and major references
Yearsley, J.R., 1989. State Estimation and hypothesis testing: A framework for the assessment of model complexity and data worth in environmental systems. Technical Report No. 116, University of Washington, Seattle, WA, 232 pp.

## Other relevant information


## Model identification
Model name/title: Hydrodynamic model: Hydro.Mass balance model: Ph-ala

## Model type
Media: lake
Class: biogeochemical, hydrology
Substance(s)/compounds: nutrients, phosphorus, organic compounds
Biological components:

## Model purpose
To develop a model able to analyze over many years the eutrophication trend in a lake taking in account the three-dimensional characteristics of the hydrodynamic field.

## Short description
The system is composed by two different models, a hydrodynamic model and a mass balance model:
Hydrodynamic model parameters: velocity components, temperature, baroclinic component of the pressure, density anomaly.
Mass balance model parameters: phytoplankton concentration, phosphorus concentration, rate of phosphorus release from the bottom sediment, phosphorus concentration in the bottom sediment, residual components of the velocity, phytoplankton growth rate, phosphorus fraction in the algal biomass, sedimentation velocity.
Forcing functions: wind action
Necessary input: wind velocity and direction over long periods, initial temperature distribution in the lake, pollutant injection location and mass injection, daily hours of sunshine.

## Model application(s) and State of development
Application area(s): lakes
Number of cases studied: 2
State: validated

**Model software and hardware demands**
Platform: The hardware demand depend on the dimensions of the domain discretization (PC, Work Station).
OS:
Software demands: language - FORTRAN 77
Hardware demands:

**Model availability**
Purchase: yes, model is available as sourcecode
Contact person(s) Dr. Paolo Viotti
    Dept. Hydraulic, Transportation and Roads
    Faculty of Engineering
    University of Rome "La Sapienza"
    Via Eudossiana 18
    I-00184 Roma, Italy

Phone: +39 6 44585638
Fax: +39 6 44585217
E-mail: Viotti@cenedese2.ing.uniroma1.it

**Model documentation and major references**

**Other relevant information**

---

**Model identification**
Model name/title: LAVSOE

**Model type**
Media: water, lake
Class: biogeochemical
Substance(s)/compounds: nutrients, carbon, phosphorus, nitrogen
Biological components: community - phytoplankton

**Model purpose**
To provide a dynamic (seasonal) description of the algal growth, water transparency and the phosphorus and nitrogen loads and internal cycling in shallow eutrophic lakes.

**Short description**
The model describes average conditions on a time scale resolution of about 1 week to 1 month and includes the following state variables: algal biomass, algal nitrogen, algal phosphorus, dissolved inorganic nitrogen and phosphorus, and non-available nitrogen and phosphorus.

The forcing functions (mostly in the form of tabulated weekly to monthly average values) are: Solar radiation, water temperature, loads of total and dissolved nitrogen and phosphorus, and water throughflow.

Important site specific input parameters are: lake surface area and mean depth and background water extinction coefficient. Certain rate parameters must be fine tuned by calibration. These are settling rates for algae and non-available nitrogen and phosphorus, release rates of dissolved nitrogen and phosphorus from the sediment surface, algal growth rate and grazing rate.

---

### Model application(s) and State of development
Application area(s):
Number of cases studied: ca. 25
State: verification, calibration, validation, prognosis made

---

### Model software and hardware demands
Platform:PC-based, Mainframe
OS:DOS (PC version)
Software demands: language: FORTRAN
Hardware demands: minimum: 386 PC with co-processor

---

### Model availability
Purchase: yes, special arrangements have to be made with current version.
Software available both as executable and sourcecode.
Contact person(s) Dr. Niels Nyholm
           Institute of Environmental Science and Engineering
           Bld.113
           Technical University of Denmark
           DK-2800 Lyngby
           Denmark

Phone: +45 45 93 39 08
Fax: +45 45 93 28 50
E-mail:

---

### Model documentation and major references
Nyholm, N., 1978. A simulation model for phytoplankton growth and nutrient cycling in eutrophic shallow lakes. *Ecol. Modelling*, (1978), Vol. 4, pp. 279-310.

---

### Other relevant information
An improved PC version is under development. The mainframe version is documented in Danish.

## Model identification
Model name/title: MINLAKE

## Model type
Media: water, lake
Class: hydrology
Substance(s)/compounds: phosphorus, nitrogen, oxygen, suspended sediment
Biological components: community - phytoplankton, zooplankton

## Model purpose
To evaluate the effectiveness of lake treatment methods, MINLAKE models the principal physical and kinetic relationships among water temperature, suspended sediment, nutrients, dissolved oxygen and phytoplankton in a lake. Treatment methods can be modeled by modifying those components of the model which are directly affected by the treatment.

## Short description
MINLAKE is a process-oriented deterministic, one-dimensional water quality model for lakes. The model can simulate water temperatures, biologically available phosphorus, nitrogen (ammonium, nitrate, nitrite), BOD (biochemical oxygen demand), phytoplankton (represented as chlorophyll concentration), zooplankton, and dissolved oxygen in a daily timestep.
  Weather data, which include daily air temperature, dew point temperature, wind speed, solar radiation and precipitation, are forcing parameters. Lake morphometry, inflow and outflow, biological characteristics of the lake, and proposed lake treatment methods are model input. The model can be run in different combinations depending on what kind of information is available for input and needed for output. Lake treatment methods, e.g., in-flow controls through watershed management, and in-lake controls, e.g., through sediment/ water column nutrient inactivation, can be tested by MINLAKE. MINLAKE is for an individual lake and some field measurements are needed to calibrate the model before simulating treatment methods.

## Model application(s) and State of development
Application area(s): Lakes in temperate climates, especially North- Central USA
Number of cases studied: more than 10
State: verification, calibration, validation

## Model software and hardware demands
Platform: PC-based - IBM
OS: MS-DOS
Software demands: PLOT88, language FORTRAN 77
Hardware demands: 286 or higher PC

## Model availability
Purchase: yes, price on request. Software available both as executable and sourcecode
Contact person(s) Dr. M. Riley

> St. Anthony Falls Hydraulic Laboratory
> University of Minnesota
> Mississippi River at 3rd Avenue S.E.
> Minneapolis, MN 55414 USA

Phone:
Fax: +1 612-627-4609
E-mail:

## Model documentation and major references
Riley, M., and Stefan, H.G., 1987. Dynamic lake water quality simulation model "MINLAKE." Project Report No. 263, St. Anthony Falls Hydraulic Laboratory, University of Minnesota, Minneapolis, MN 55414, USA.
Riley, M.J., 1998. User's Manual for the dynamics lake water quality simulation program "MINLAKE". External Memorandum No. 213 (including program in disk), St. Anthony Falls Hydraulic Laboratory, University of Minnesota, Minneapolis, MN 55414, USA.

## Other relevant information

## Model identification
Model name/title: GIRL: General simulation of Reservoirs and Lakes

## Model type
Media: water, lake
Class: biogeochemical
Substance(s)/compounds: nutrients, phosphorus
Biological components: ecosystem - aquatic

## Model purpose
To investigate some general features of lake ecosystems, particular for management purposes.

## Short description
A generalized simple lake model consisting of three equations describing phosphorus as limiting element, and phytoplankton and zooplankton in uniformly mixed layers of water with specified depth (MIX) is developed. The reservoir/lake due to variable throughflow (Q/V), extinction coefficient of rates for, light and annually varying temperature. Annual changes of sedimentation are expressed by means of the function PIDI. Geographical location is given by annual changes of temperature, radiation and photoperiod. Inflow of phosphorus, algae and zooplankton are inputs.

## Model application(s) and State of development
Application area(s): eutrophication management
Number of cases studies: 3
State: verification

## Model software and hardware demands
Platform: PC-based - IBM
OS: DOS
Software demands: language FORTRAN
Hardware demands:

## Model availability
Purchase: no, software available as executable code
Contact person(s) Dr. Milan Straskraba
        Biomathematical Laboratory
        Acad. Sci. Czech Rep.
        Branisouska 31
        CZ-37005 Ceske Budejovice
        Czech Republic

Phone: 042 38 41 928
Fax: 042 38 45 985
E-mail: stras@entu.cas.cz

## Model documentation and major references
Kmet, T., and Straskraba, M., 1189. Global behaviour of a generalized aquatic ecosystem model. *Ecol. Modelling*, 45: 95-110.

## Other relevant information
Extension for management- optimization model see Kalime et al., 1982. An optimization model for the economic control of reservoirs eutrophication. *Ecol. Modelling*, 17: 121- 128. Straskraba, M., 1985. Managing of eutrophication by means of ecotechnology and mathematical modelling. International Congress "Lakes Pollution and Recovery", Rome, 15th- 18th April, 1985 : 17-28.

## Model identification
Model name/title: SALMO: Simulation by an Analytical Lake Model

## Model type
Media: water, lake
Class: biogeochemical
Substance(s)/compounds: nutrients, soluble reactive phosphorus (SRP), inorganic nitrogen, oxygen, phytoplankton biovolume, zooplankton biovolume, allochthonous particulate organic matter (POM)
Biological components: population - phytoplankton (2 out of 3 functional groups)

## Model purpose
1. Tool to find weak points in limnological research.
2. Water quality management.

## Short description
SALMO (Salmulation by an Analytical Lake Model) was developed as a "general" model for lakes and reservoirs with different trophic state. The model is a two layer system with mixed hypolimnion and epilimnion. The boundary between epilimnion and hypolimnion is variable in time. For both layers the state variables SRP, nitrogen, phytoplankton (2 functional groups out from 3 parameter sets) zooplankton, POM and oxygen are calculated. The phytoplankton groups are:
1. microplanktonic diatoms (*Asterionella formosa*-type)
2. green algae and nanoplanktonic diatoms (*Scenedesmus quadricauda* and *Cyclotella meneghiniana*-type)
3. blue-green algae of the *Oscillatoria redekei*-type.

Sediment water interactions are considered by empirical relationships for nutrient release and oxygen consumption.

The egg development of zooplankton is involved with time delay.

The input variables are given as discrete values (10 day intervals) for a particular water body.

They are: depth and volume of the two layers, water temperature, photosynthetic active radiation (PAR) at the water surface, water influx and outflow, import of phosphorus, nitrogen, POM and living phytoplankton (if necessary).

Despite of the low number of state variables in SALMO, values for 138 parameters are required as a consequence of numerous control mechanisms.

The parameters are assumed to be "natural constants" except for 7 lake specific parameters (e.g., light extinction) and 7 initial values.

Fish is not described as a state variable. But there is a variable mortality rate of zooplankton. The parameters of which allow to simulate a certain zooplanktivorous fish density.

## Model application(s) and State of development
Application area(s): Lakes and reservoirs in Germany, Netherlands, Great Britain, Finland
Number of cases studied: 16
State: All states from conceptualization to prognosis, including verification, validation, sensitivity analysis. propagation of input errors, prognosis, validation of prognosis, NO CALIBRATION. The model should be a general model valid for oligo to eutrophic stratified lakes (mean depth >3m).

## Model software and hardware demands
Platform: PC-based - IBM
OS: MS- DOS 3.3 or higher
Software demands: MS-WINDOWS 3.x with MS-EXCEL 4.0 (as optional user front- end)
Language: Microsoft FORTRAN 77, vers. 5.x
Hardware demands: 8086 + 1 Mb RAM (without front-end) 80386 (or better) + 4 Mb RAM (with front-end) Coprocessor recommended.

## Model availability
Purchase: The model is available without fee only for scientific use within scientific cooperation. Depending on the task, executable code or source may be used. Commercial and governmental applications are performed in cooperation with the following consulting laboratory:
Contact person(s) Drs. Th. Guderitz and R. Kruspe
IDUS Biologisch Analytisches Umweltlabor GmbH
Dresdner Str. 43
D-01458 Ottendorf-Okrilla
Germany

Phone:
Fax:
E-mail:

## Model documentation and major references
Benndorf, J., and Recknagel, F., (1982). Problems of application of the ecological model {SALMO} to lakes and reservoirs having various trophic states. *Ecol. Modelling*, Vol. 17, p. 139-145.
Recknagel, F., and Benndorf, J. 1982, Validation of the ecological simulation model {SALMO}. *Int. Rev*, Vol. 67, p.113-125

## Other relevant information

## Model identification
Model name/title: GARDA 3-D

## Model type
Media: water, lake
Class: hydrology
Substance(s)/compounds: nutrients, temperature
Biological components:

## Model purpose
The model represents the first known attempt to simulate water circulation in the Garda Lake.

## Short description
The model represents the first known attempt to simulate water circulation in the Garda Lake.
THREE-DIMENSIONAL HYDRODYNAMIC MODEL OF THE GARDA LAKE.
The lake of Garda is the widest Italian fresh water resource. Its major physiographic features are: area 379 km²; perimeter 185 km; height 65 msl; maximum depth 346 m; major axis 51.6 km; minimum axis 17.2 km. Garda's lake circulation is predominantly wind driven with periodic convective overturn.

The hydrodynamic model is based on the three-dimensional Reynold's equation. The numerical method is based on a semi-implicit finite-difference formulation allowing to reproduce the circulation with a very fine mesh adopting a relatively high time step. This horizontal grid covers the whole surface with 350 x 210 nodes with a spatial step of 125 m. Along the vertical a space step of 1 m is included up to the maximum depth, thus yielding 6176100 grid's point of which 553379 are active. The model has been seen to be computationally stable and simulation performed adheres to hydrodynamic situations under constant wind conditions. The model, which is the first and, so far, the only one describing Garda water circulation, is optimized for the supercomputer CRAY Y-MP8/432 and it is easily extendible to include biochemical and ecological relevant variables.

**Model application(s) and State of development**
Application area(s):
Number of cases studied:
State: calibration

**Model software and hardware demands**
Platform: PC-based, Mainframe, Workstation, Supercomputer
OS:
Software demands:
Hardware demands:

**Model availability**
Purchase: yes, price to be discussed. Software available as executable code.
Contact person(s) Dr. Giovanni Pecenik   and   Dr. Vincenzo Casulli
Dept. Env. Physical-Chemistry
Dorsoduro 2137
I-30123 Venezia
Italy

Phone: +49 41 5298587
Fax: +49 41 5298594
E-mail: pcnk@pcnk.dcf.unive.it

**Model documentation and major references**

**Other relevant information**
Casulli, V., and Pecenik, G., 1994. Modello Idrodinamico Tridimensionale dei Lago di Garda. In: Cimeca, Ed. *Scienza e supercalcolo al Cimeca*, 1994, pp. 255-258.

## Model identification
Model name/title: LIMNOD

## Model type
Media: lake, estuary, local scale
Class: biogeochemical, hydrology
Substance(s)/compounds: carbon, phosphorus, oxygen
Biological components:

## Model purpose
LIMNOD is a physical-biochemical model for long-term prediction of water quality in lakes. The effects of restoration measures such as oxygenation, deep water drainage and artificial mixing can also be studied. The model is adaptable to most lakes by adjusting some lake specific parameters or by adding new state variables.

## Short description
LIMNOD is a one-dimensional vertical lake model which considers coupled physical, biochemical and sedimentation processes. For the physical description (state variables are temperatures and conductivity), the lake is divided into fully mixed epilimnion and a hypolimnion with a sharp thermocline between these layers. The thermocline depth is calculated daily by means of an energy-balance considering heat and radiation exchange with the atmosphere and energy input by the wind. In the hypolimnion the turbulent mixing processes are expressed by the concept of a time and depth dependent eddy diffusion.

Based on the physical processes a cycle of nutrients is calculated with phosphorus as limiting nutrient (state variables are the concentrations of particular organic carbon (biomass), dissolved oxygen and dissolved and particular phosphorus). In the sediment two types of organic phosphorus, inorganic phosphorus and organic carbon are considered.

In this coupled system, transport of chemical species is governed by the physical model. Inversely, biochemical processes influence the stability of the water column due to light extinction by biomass reducing the penetration depth of the incoming short wave radiation and due to dissolution of settling particles from the epilimnion, enhancing the concentration of dissolved species in the hypolimnion.

## Model application(s) and State of development
Application area(s): Lake Lugano
Number of cases studied: 1
State: prognosis made

## Model software and hardware demands
Platform: PC-based, Work-station
OS:
Software demands: language - FORTRAN 77
Hardware demands:

## Model availability
Purchase: yes, 200 $. Available both as executable and source code.
Contact person(s) Dr. J. Trösch
           Lab. of Hydraulics, Hydrology and Glaciology
           Eldgenössische Technische Hochschule
           CH-8092 Zürich
           Switzerland
Phone:
Fax:
E-mail: troesch@VAW.ethz.ch

## Model documentation and major references
Karagounis, I., 1992. Ein physikalisch- biochemisches Seemodell; Anwendung auf das Nordbecken des Luganersees. Communications Nr. 116 of the Laboratory of Hydraulics, Hydrology and Glaciology, Federal Institute of Technology, Zürich, Switzerland.
Karagounis, I., Trösch, J., and Zamboni, F., 1993. A coupled physical-biochemical lake model for forecasting water quality. *Aquatic Sciences*, Vol. 2: pp. 87-102.

## Other relevant information

## Model identification
Model name/title: Multi-Component Model of Phosphorus Dynamics in Reservoirs

## Model type
Media: lake
Class: biogeochemical
Substance(s)/compounds: phosphorus
Biological components:

## Model purpose
Long-term (i.e., years) simulation of phosphorus and the trophic state in impoundments.

## Short description
The models available for simulating phosphorus dynamics and trophic states in impoundments vary widely. The simpler empirically derived phosphorus models tend to be appropriate for long-term, steady or near steady state analyses. The more complex ecosystem models, because of computational expense and the importance of input parameter uncertainty, are impractical for very long-term simulation and most applicable for time-variable water quality simulations generally of short to intermediate time frames. An improved model for time variable, long-term simulation of trophic state in reservoirs with fluctuating inflow and outflow rates and volume is needed. Such a model is developed in this paper representing the phosphorus cycle in two-layer (i.e., epilimnion and hypolimnion) reservoirs.

The model is designed to simulate seasonally varying reservoir water quality and eutrophication potential by using the phosphorus state variables as the water quality indicator. Long-term simulations with fluctuating volumes and variable influent an effluent flow rates are feasible and practical. The model utility is demonstrated through application to a pumped storage reservoir characteristic of these conditions. (KEY TERMS: phosphorus, eutrophication, reservoirs, impoundments, water quality modeling.)

## Model application(s) and State of development
Application area(s): lakes and reservoirs
Number of cases studied:
State: calibration, prognosis made

## Model software and hardware demands
Platform:
OS:
Software demands: language - FORTRAN
Hardware demands:

## Model availability
Purchase: yes, free. Software available as sourcecode.
Contact person(s) Dr. Raymond A. Ferrara
        Omni Environmental Corporation
        211 College Road East
        Princeton, NJ 08540-6623 USA

Phone: +1 609-243-9393
Fax: +1 609-243-9297
E-mail:

## Model documentation and major references
Griffin, T.T., and Ferrara, R.A., 1984. A Multicomponent Model of Phosphorus Dynamics in Reservoirs. Paper No. 83124 of the Water Resources Bulletin, Vol. 20, No. 5, October 1984.
Gray, W.G.. Physics-based Modelling of Lakes, Reservoirs, and Impoundments, page 229. Edited by author. (Documented Models, Round Valley Reservoir Multi-Component Phosphorus Model Documentation by Raymond A. Ferrara and Thomas Griffin).

## Other relevant information

## Model identification
Model name/title: SIMPLE: Sustainability of Intensively Managed Populations in Lake Ecosystems

## Model type
Media: water, lake
Class: population, simulation
Substance(s)/compounds: none

Biological components: ecosystem - pelagic lake community - pelagic lake pisicvores and planktivores population - five piscivore species, three planktivores also non-dynamic representation of invertebrate groups.

## Model purpose
To allow investigation of the consequence for fish community dynamics of various fisheries management strategies, most particularly changes in stocking rates and harvest levels of the five salmonine piscivore species abundant in the Laurentian Great Lakes. Model development has focused on Lake Ontario but a working version has also been developed for Lake Michigan.

## Short description
The SIMPLE model is an integrated age-structured population and bioenergetics simulation model that includes a dynamic representation of the predator-prey interactions between piscivores, planktivores and invertebrates. The model includes five piscivore species: lake trout, chinook salmon, coho salmon, rainbow trout and brown trout; and three plantivore species (for Lake Ontario): alewife, rainbow smelt and slimy sculpin. Invertebrates are grouped into four categories: large zooplankton (>1 mm), small zooplankton, macroinvertebrates (Mysis, Diporeia), and terrestrial insects. All fish species are represented as dynamic, age-structured populations. Recruitment of piscivores is driven by stocking, which essentially acts as a forcing function.

Recruitment of planktivores is computed from stock-recruitment relationships. The invertebrates are simply represented as fixed pools of available prey. Predation is modeled using a Holling type-II multiple-species functional response, and growth is computed from a simple conversion efficiency ration size relationship.

## Model application(s) and State of development
Application area(s): Laurentian Great Lakes, potentially relevant to other areas with hatchery dependent salmonines fisheries
Number of cases studied: 2 so far, Lake Ontario, Lake Michigan
State: partial verification/calibration, validation impractical, prognosis has been made

## Model software and hardware demands
Platform: PC-based, current version requires Microsoft basic PDS V 7.1 Windows version using Visual Basic is under development expected completion was fall 1994.
OS:
Software demands:
Hardware demands:

## Model availability
Purchase: no software available
Contact person(s) Dr. Michael L. Jones
        Glenora Fisheries Station
        Rural Route 4,
        Picton, Ontario
        Canada K0K 2T0

Phone: 613-476-2400
Fax: 613-476-7131
E-mail: jonesmi@gov.on.ca

## Model documentation and major references
Jones, M.L., et al., 1993. Sustainability of hatchery-dependent salmonine fisheries in Lake Ontario: the conflict between predator demand and prey supply. *Trans. Am. Fish. Soc.* 122: 1002-1018.

## Other relevant information
User documentation was not available as of fall 1994.

## Model identification
Model name/title: $pCO_2$ and $pO_2$ profiles along the flow of algal suspension in open solar culture units

## Model type
Media: water
Class:
Substance(s)/compounds: carbon, oxygen
Biological components: organism - microscopic algae

## Model purpose
Measured values of partial pressures of $CO_2$ and $O_2$ in flowing algae suspension on culture surface are used in the model to compute rates of $CO_2$ consumption and $O_2$ evolution in algae culture and for estimation of mass transfer coefficients of $CO_2$ and $O_2$ desorption into atmosphere.

## Short description
Model consists from two parts: (a) Analytical expression of the course of $pCO_2$ (partial pressure of dissolved $CO_2$ in algae culture flowing down culture surface; (b) Analytical expression of the course of $pO_2$ (partial pressure of dissolved $O_2$ in algae culture). Input variables are: velocity of culture flow, thickness of culture layer, distance from the beginning of culture surface and $pCO_2$, $pO_2$ values. Using least squares nonlinear regression analysis (procedure) the following parameters are estimated: Rates of photosynthesis (in terms of $CO_2$ consumption and $O_2$ evolution) and mass transfer coefficient Kl for $CO_2$ and $O_2$ desorption from algae culture into atmosphere. Model can be used to the description of the state of culture of algae during cultivation and to evaluate $CO_2$ economy during growth of microscopic algae in a commercial scale.

## Model application(s) and State of development
Application area(s): commercial growth of microscopic algae in open culture system
Number of cases studied:
State: verification

## Model software and hardware demands
Platform: PC-based
OS:
Software demands: nonlinear regression, Basic
Hardware demands:

## Model availability
Purchase: no
Contact person(s) Dr. K. Livansky
           Institut of Microbiology CAV
           Department of Autotrophic Microorganism
           37901 Trebon, Opoitovieky Miyn
           Tjekiet
Phone:
Fax:
E-mail:

## Model documentation and major references
Model is not documented well enough to be distributed. It is in a preliminary form which served for verification of the models.

## Other relevant informations
Part of model describing $pO_2$ profiles was verified using Leuvenberg-Marquardt nonlinear regression. The served part, describing $pCO_2$ profiles was verified by a published (other author) procedure for a TI-59 pocket calculation. Today, in the age of the PC, this procedure is of course obsolete.

## Model identification
Model name/title: EYESOCKS

## Model type
Media: lake
Class: oxygen
Substance(s)/compounds:
Biological components:

## Model purpose
Prediction of the winter oxygen uptake of a lake, including the effect of ice cover.

## Short description

A one-box exponential model with constant coefficients generally suffices to describe the winter oxygen uptake of a holomictic lake with no ice cover. A simple modification of the model - settling the gas transfer velocity to zero during periods of ice cover - allows the mean oxygen concentration in the lake to be predicted irrespective of whether the lake is frozen over or not. Required input variables:

Cs       =       oxygen saturation concentration $[M*L^{-3}]$
Co       =       homogeneous $O_2$ concentration at time t=0 (early winter) $[M* L^{-3}]$
beta     =       $O_2$ transfer velocity $[L*T-1]$
sigma    =       mean lake depth [L]
J        =       net $O_2$ consumption rate $[M*L^{-3}*T^{-1}]$
ta       =       time of freeze-up [T]
tb       =       time of break-up [T]

Other Variables:

C        =       Cs - J*sigma/beta = equilibrium $O_2$ concentration $[M*L^{-3}]$
Ca       =       C  - (C  - C0) exp[(-beta/2)ta] = $O_2$ concentration at t = ta $[M*L^{-3}]$
Cb       =       Ca - J(tb - ta) = $O_2$ Concentration at t = tb $[M*L^{-3}]$
Cm       =       C  - (C  - Cb)exp[(-beta/2) (tm - tb)] = $O_2$ concentration at t = tm
                 After timebreak) $[M*L^{-3}]$

## Model application(s) and State of development

Application area(s): Lake Oxygenation
Number of cases studied: 1
State: validation, validation of prognosis

## Model software and hardware demands

Platform: MAC
OS:
Software demands: EXCEL
Hardware demands:

## Model availability

Purchase:
Contact person(s) Dr. David M. Livingstone
                 Umweltphysik,
                 EAWAG/ETH
                 Überlandstrasse 133
                 CH-8600 Dübendorf
                 Switzerland

Phone: +41 1 823 5540
Fax: +41 1 823 5210
E-mail: living@eawag.ch

## Model documentation and major references
Livingstone, D.M., 1993. Lake oxygenation: Application of a one-box model with ice cover, *Int. Rev. Ges. Hydrobiol.*, 78 (4) 465-480.

## Other relevant information

## Model identification
Model name/title: Calcite precipitation model for natural waters

## Model type
Media: water, lake
Class: biogeochemical
Substance(s)/compounds: nutrients, carbon
Biological components:

## Model purpose
Computation of $CO_2$, $HCO_3^-$, $CO_3^{2-}$ and $CaCO_3$ concentration in natural waters (after Rossknecht 1977).

## Short description
Input: Average concentration of all single- and double-valence ions or the electric conductivity of the water, water- temperature, alkalinity, Ca-concentration, pH, result of the marble test, which determines the actual super- or undersaturation with $CaCO_3$.
Output: Concentrations of $CO_2$, $HCO_3^-$, $CO_3^{2-}$ and $CaCO_3$ in equilibrium and in the natural water.
Validation: by comparison of the computed pH in equilibrium and the pH measured after the marble test.

## Model application(s) and State of development
Application area(s): Lake Constance, Lake Belau (Northern Germany)
Number of cases studied: 2
State: validation

## Model software and hardware demands
Platform: PC-based
OS: DOS
Software demands: SYSL, language - FORTRAN
Hardware demands:

## Model availability
Purchase: no, available as executable code.
Contact person(s) Dr. Gerald Schernewski   and   Dr. Lars Theesen
             Projektzentrum Ökosystemforschung
             Schauenberger Str. 112
             D-24118 Kiel
             Germany

Phone: + 49 431 880 4002
Fax: + 49 431 880 4083
E-mail:

## Model documentation and major references
Rossknecht, H., 1977. Zur automatiche Calcite fällung im Bodensee-Obersee. *Arch. Hydrobiol.*, 81: 35-64.
Schernewski, G., Theesen, L., and Kerger, K.E., 1994. Modelling thermal stratification and calcite precipitation of Lake Belau (Northern Germany). *Ecol. Modelling*, 75/76: 421-433.
Schernewski, G., (in press). Kohlenstofflimitierung, Selbstreinigung and Auswirkungen des Global Change im Belauer See: Dokumentation und Anwendung des Modells für den anorgishe C-Haushalt in Seen. EcoSys bd.3.

## Other relevant information

## Model identification
Model name/title: Lake Liming Model

## Model type
Media: lake, local scale
Class: biogeochemical
Substance(s)/compounds: nutrients, Ca
Biological components:

## Model purpose
The purpose of the model is to predict the reacidification of limed lakes as a function of catchment characteristics (hydrology, hydrochemistry, morphology) and design parameters (post- liming ANC, bottom coverage of undissolved calcite).

## Short description
The model is dynamic, deterministic and has low structural complexity.
Mixing model: Continuous stirred tank reactor representing lake volume.
Key reactions are: Dissolution of calcite from lake bottom, acid-base relationships.
State variables are: Ca-concentration (as a surrogate for ANC).
Important parameters are: Lake water titration curve, lake depth and volume, residual calcite on lake bottom and bottom coverage.
Forcing functions include: Runoff and temporal variations, annual average Ca and pH in inflow.

## Model application(s) and State of development
Application area(s):
Number of cases studied: ca. 50
State: validation of prognosis

## Model software and hardware demands
Platform: Mainframe IBM
OS:
Software demands:
Hardware demands:

## Model availability
Purchase: yes, price is low. Software available as executable code.
Contact person(s) Dr. Harald Sverdrup
             Dept. Chemical Engineering II
             Lund Institute of Technology
             P.O. BOX 124
             S-22100 Lund
             Sweden

Phone: +46 46 10 82 74
Fax: +46 46 10 82 74
E-mail: harald.sverdrup@chemeng.lth.se

## Model documentation and major references
Sverdrup, H., and Warfvinge, P., 1985. A reacidification model for acidified lakes neutralized with calcite. *Water Resources Research* 21, 1374-1380.

## Other relevant information

## Model identification
Model name/title: PROFILE

## Model type
Media: local scale, regional scale, terrestrial, soil
Class: biogeochemical
Substance(s)/compounds: nutrients, nitrogen, sulphur, Ca, Mg, K, Al , $H^+$, OH., $(HCO_3)^-$, $(CO_3)$-, DOC
Biological components:

## Model purpose
The purpose of the model is to predict the effects of atmospheric deposition and nutrient uptake on the chemistry of soil systems as a function of soil properties (mineralogy, morphology) and boundary conditions (nutrient uptake, water balance, temperature). Regional application for calculation of critical loads of acid deposition.

## Short description
The model is steady-state, deterministic and has moderate structural complexity. Large emphasis on weathering reactions.

Mixing model: a series of continuous stirred tank reactors each representing a soil horizon.

Key reactions are: chemical weathering, solution acid-base reactions.

State variables are: soil solution ANC (pH, Al, organic acids), base cation concentration.

Important parameters are: mineralogy, dissolution rate constants, mineral surface area, water content, Al-solubility, DOC, soil solution equilibrium constants.

Forcing functions includes: atmospheric input to soil profile, nutrient uptake, temperature, water content and water flux.

## Model application(s) and State of development
Application area(s):

Number of cases studied: >50

State: validation

## Model software and hardware demands
Platform: PC-based MAC, IBM

OS: System 7 (MAC)

Software demands: hypercard 2.1 (MAC), language: FORTRAN, HyperScript (MAC)

Hardware demands: Math co-processor

## Model availability
Purchase: yes, free. Software available as executable code

Contact person(s) Dr. Per Warfvinge
> Dept. Chemical Engineering II
> Lund Institute of Technology
> P.O. BOX 124
> S-22100 Lund
> Sweden

Phone: +46 46 10 36 26

Fax: +46 46 10 82 74

E-mail: per. warfvinge@chemeng.lth.se

## Model documentation and major references
Warfvinge, P., and Sverdrup, H., 1992. Calculating critical loads of acid deposition with PROFILE - a steady-state soil chemistry model. *Water, Air, Soil Pollut.*, 63, 119-143.

## Other relevant information
A very user-friendly, self-explanatory and powerful model. Unique for calculation of weathering rates. Material for use in teaching is developed.

## Model identification
Model name/title: SAFE: Soil Acidification in Forest Ecosystems

## Model type
Media: local scale, terrestrial, soil
Class: biogeochemical
Substance(s)/compounds: Ca, Mg, K, Al, $H^+$, OH-, $HCO_3^-$, $CO_3^{2-}$-, DOC
Biological components:

## Model purpose
The purpose of the model is to predict the effects of atmospheric deposition and nutrient uptake on the chemistry of soil systems as a function of soil properties (mineralogy, morphology) and boundary conditions (nutrient uptake, water balance, temperature). Used for analysis of temporal aspects of impact of acid deposition, viz., scenario analysis.

## Short description
The model is a dynamic, deterministic and has moderate structural complexity. SAFE is the dynamic counterpart of PROFILE.
Mixing model: a series of continuous stirred tank reactors each representing a soil horizon.
Key reactions are: chemical weathering, solution acid-base reactions.
State variables are: soil solution ANC (pH, Al, organic acids), base cation concentration.
Important parameters are: mineralogy, dissolution rate constants, mineral surface area, water content, Al-solubility, DOC, soil solution equilibrium constants.
Forcing functions include: atmospheric input to soil profile, nutrient uptake, temperature, water content and water flux.

## Model application(s) and State of development
Application area(s):
Number of cases studied: x10
State: validation, validation of prognosis

## Model software and hardware demands
Platform: MAC
OS: System 7
Software demands:
Hardware demands: Math co-processor

## Model availability
Purchase: yes, free. Software available as executable code.
Contact person(s) Dr. Per Warfvinge
        Dept. Chemical Engineering II
        Lund Institute of Technology
        P.O.Box 124
        S-22100 Lund
        Sweden

Phone: +46 46 10 36 26
Fax: +46 46 10 82 74
E-mail: per.warfvinge@chemeng.lth.se

**Model documentation and major references**
Warfvinge, P., Falkengren-Grerup, U., and Sverdrup, H., 1993. Modeling long-term base cation supply to acidified forest stands. *Environmental Pollution*, 80, 209-220.

**Other relevant information**

**Model identification**
Model name/title: Latitudinal-altitudinal regression- and spline-models for the surface temperatures of the Northern hemisphere freshwater lakes.

**Model type**
Media: water, lake, global scale
Class: hydrology
Substance(s)/compounds: annual mean and extreme temperatures, time of ice covering
Biological components:

**Model purpose**
To predict annual mean and extreme surface temperatures in a year cycle of the Northern hemisphere freshwater lakes as well as ice covering time and time period between fall and spring convective overturning depending on latitude and altitude.

**Short description**
Listed above variables are computed as some universal nondimensional functions of nondimensional latitude and altitude. The important variables to input are latitude and altitude. The main governing parameters are: average freezing latitude (53°N), the average northern boundary of cold monomictic lakes distribution (80°N), annual mean surface temperature of the equatorial lake (30°C), snow line altitude at given latitude.

**Model application(s) and State of development**
Application area(s): The Northern hemisphere
Number of cases studied:
State: conceptualization, verification, prognosis made

**Model software and hardware demands**
Platform: PC-based - IBM
OS: MS DOS
Software demands:
Hardware demands:

---

**Model availability**
Purchase: yes
Contact person(s) Dr. S.V. Ryanzhin
                 Limnology Institute of the Russian Academy of Science
                 Sevastyanova 9
                 196199 St. Petersburg
                 Russia

Phone: +7 812 294 8050
Fax: +7 812 298 7327
E-mail:

---

**Model documentation and major references**
Ryanzhin, S.V.,1991. *Docl. Acad. Sci.* (USSR), 1991, 317: 3, 628-634.
Ryanzhin, S.V., 1990. *Docl. Acad. Sci.* (RUSSIA), 1990, 312: 1, 209-214.
Ryanzhin, S.V., 1994, *Ecol. Modelling*, 1994, (to be published).

---

**Other relevant information**

---

**Model identification**
Model name/title: ORTA

---

**Model type**
Media: lake
Class: toxicology
Substance(s)/compounds: heavy metals - Cu
Biological components:

---

**Model purpose**
To calculate time evolution of total concentration in lake (epilimnion + hypolimnion) and lake sediments.

---

**Short description**
The mathematical model consists of a system of two linear differential equations with constant coefficients, which can be analytically solved, complemented by a computer program to solve the equations numerically in the case when the time dependence of the coefficients must be taken into account.
State variables: Total metal concentration in lake $C_t$; metal concentration in solid sediment $C_{ads}$ for the lake-sediment version; total metal concentration in epilimnion $C_{te}$ and hypolimnion $C_{th}$ for the epilimnion-hypolimnion version
Input parameters: external metal load; mean water residence time; vertical mixing coefficient; metal distribution coefficient; settling velocity of particulate matter; sediment parameters.

## Model application(s) and State of development
Application area(s): lakes (shallow + deep), lake bays
Number of cases studied: Lake Orta; Monvalle Bay of Lake Maggiore
State: validation

## Model software and hardware demands
Platform: PC-based
OS: MS-DOS
Software demands:
Hardware demands:

## Model availability
Purchase: no, software available as sourcecode.
Contact person(s) Dr. Giovanni Rossi
                Via Bazaggia 43
                I-21100 Varese
                Italy

Phone: +39 332 222838
Fax: + 39 332 222838
E-mail:

## Model documentation and major references
*Ecol. Modelling*, 64: 23-45. For the main model.
*Hydrol. Sci. Bull.*, XX: 4-12. For the water submodel.

## Other relevant information
The asymptotic concentrations and time constants can be analytically determined if the model coefficients are time-constant (e.g., long term evolution of the lake-sediment system with constant metal loading).

## Model identification
Model name/title: Regional Dissolved Oxygen Model (RDOM)

## Model type
Media: water, lake, regional scale
Class: biogeochemical
Substance(s)/compounds: oxygen, temperature
Biological components:

## Model purpose
To simulate dissolved oxygen and temperature profiles in several classes of lakes of the central North American temperate zone in order to make comparisons among lakes with a wide range of morphometries and trophic levels and their responses to different climate scenarios.

## Short description
The model is to estimate the potential impacts of projected global climate change on dissolved oxygen in lakes. The model combined with a water temperature stratification model simulates dissolved oxygen profiles in stratified lakes in daily timesteps throughout the open water season. The dissolved oxygen transport equation includes photosynthesis as a source term; biochemical oxygen demand and plant respiration are sink terms. Oxygen exchange through the air water interface and sedimentary oxygen demand through the water- sediment interface are physical boundary conditions for the dissolved oxygen transport equation. The model relates biological variables to specified trophic levels. For example, biochemical oxygen demand and sedimentary oxygen demand are specified in that way. Rate coefficients for a region were determined by sequential literature search, model calibration, validation and sensitivity analysis for an array of lakes. The model simulates water temperatures first and dissolved oxygen afterwards. The forcing parameters are weather parameters including daily air temperature, dew point temperature (relative humidity), solar radiation, wind speed, sunshine percentage, and precipitation. Lake morphometries and trophic status are model input.

## Model application(s) and State of development
Application area(s): lakes in temperate climates, especially north-central USA
Number of cases studied: one region (27 lake classes)
State: verification, calibration, validation

## Model software and hardware demands
Platform: PC-based - IBM
OS: MS-DOS
Software demands: PLOTIT88, language - FORTRAN 77
Hardware demands: 386 or higher

## Model availability
Purchase: no
Contact person(s) Dr. H. G. Stefan
                University of Minnesota
                Minneapolis, MN 55414 USA
Phone:
Fax:
E-mail:

## Model documentation and major references
Stefan, H.G., Hondzo, M., and Fang, X., 1993. Lake water quality modeling for projected future climate scenarios. *Journal of Environmental Quality*, Vol. 22 (3), pp. 417-431.

Fang, X., and Stefan, H.G., 1994. Modeling of dissolved oxygen stratification dynamics in Minnesota lakes under different climate scenarios. Project Report 339, St. Anthony Falls Hydraulic Laboratory, University of Minnesota.

## Other relevant information
This regional dissolved oxygen and water temperature model for the open water season has been extended to a year-round simulation model including ice cover. The model development and results will be reported.

## Model identification
Model name/title: WATERSHED ACIDIFICATION MODEL

## Model type
Media: water, lake, river, soil
Class: biogeochemical, hydrology
Substance(s)/compounds: sulphur, nitrogen
Biological components: ecosystem

## Model purpose
To predict pH and sulphate in lakes, rivers and streams.

## Short description

## Model application(s) and State of development
Application area(s): In Canada
Number of cases studied: five to seven watersheds
State: conceptualization, calibration, validation.

## Model software and hardware demands
Platform: Mainframe
OS:
Software demands: FORTRAN   language - FORTRAN
Hardware demands:

## Model availability
Purchase:yes, N/A. Software available as executable code
Contact person(s) Dr. A. Ghosh Bobba
            National Water Research Institute
            P.O. Box 5050, Burlington
            Ontario, Canada L7R4A6

Phone:
Fax: +1 905 336 4972
E-mail:Ghosh.Bobba@cciw.ca

**Model documentation and major references**
Bobba, A.G., and Lam, C.L., 1990. Hydrological modelling of acidified Canadian Watersheds. *Ecol. Modelling*, Vol. 50, pp. 5-32, 1990.

**Other relevant informations**
Bobba, A.G., et al., 1990. Stochastic analysis of acid shocks generated by mixed hydrological processes. *Water, Air, Soil Pollution*, Vol. 53, pp. 239-250, 1990.

**Model identification**
Model name/title: Enhanced Trickle Down (ETD) Model

**Model type**
Media: water, lake, river, local scale, terrestrial, sediment, soil
Class: biogeochemical, hydrology
Substance(s)/compounds: sulphur, alkanity
Biological components:

**Model purpose**
The Enhanced Trickle- Down model can be used to model the hydrologic and geochemical response of surface water watersheds to freshwater acidification.

**Short description**
A generalized soft water acidification model has been developed. The enhanced trickle-down (ETD) model is driven by precipitation, evaporation, acidity, sulphate, and chloride loading time series daily input data. The hydrologic component simulates snowmelt, interflow, overland flow, groundwater flow, frozen ground processes, seepage, and evapotranspiration. Physicochemical and biological processes that affect the alkalinity or sulphate balance and are included in the formulation are cation exchange, chemical weathering, sulphate sorption, and sulphate reduction. The system of 20 ordinary differential equations is solved by using a variable timestep fourth-order predictor-corrector numerical scheme. Shown here is calibration of the ETD model for two lakes in the Adirondack Park of New York. ETD is relatively simple and requires limited input data, and yet it accounts for the predominant hydrologic and biochemical processes of the ecosystem.

**Model application(s) and State of development**
Application area(s): Assessment of the effects of acid precipitation on lake and river watersheds
Number of cases studie: 8 watersheds
State: conceptualization, verification, calibration, validation, prognosis made

**Model software and hardware demands**
Platform: PC-based - IBM, Workstation - SUN
OS:
Software demands: language: FORTRAN 77
Hardware demands:

**Model availability**
Purchase: software available as source code
Contact person(s) Dr. Nikolaos P. Nikolaidis
                    Dept. of Civil Engineering
                    University of Connecticut
                    BOX U-37
                    Storrs, CT 06269 USA

Phone: +1 203-486-5648
Fax: +1 203-486-2298
E-mail: nikos@engz.uconn.edu

**Model documentation and major references**
Nikolaidis, N.P, Hu, H., and Ecsedy, C., 1994. Effects of climatic variability on freshwater watershed: case studies. *Aquatic Sciences*, Vol. 56, No. 2, pp. 161-178.
Nikolaidis, N.P., Hu, H., Ecsedy, C., and Lin, J.D., 1993. Hydrologic response of freshwater watersheds to climatic variability; model development. *Water Resources Research*, Vol. 29, No. 10, pp. 3317-3328.
Nikolaidis, N.P., Muller, P.K., Schnoor, J.L., and Hu, H.L., 1991. Modeling the hydrogeochemical response of a stream to acid deposition using the enhanced trickle-down model. *Research Journal of Water Pollution Control Federation*, Vol. 63, No. 3, pp. 220-227.
Georgakakos, K.P., Valle-Folho, G.M., Nikolaidis, N.P., and Schnoor, J.L., 1989. Lake acidification studies: the role of input uncertainty in long-term predictions. *Water Resources Research*, Vol. 25, No. 7, pp. 1511-1518.
Nikolaidis, N.P., Schnoor, J.L., and Georgkakaos, K.P., 1989. Modeling of Long-Term Lake Alkalinity Responses to Acid Deposition. *Journal of Water Pollution Control Federation*, Vol. 61, No. 2, pp. 188-199.
Nikolaidis, N.P., Rajaram, H., Schnoor, J.L., and Georgakakos, K.P., 1988. Generalized softwater acidification model. *Water Resources Research*, Vol. 24, No. 12, pp. 1983-1996.
Nikolaidis, N.P., 1987. Modeling the Direct Versus Delayed response of Surface Waters to Acid Deposition in the Northeastern United States. Ph.D. Thesis, Civil and Environmental Engineering, University of Iowa, pp. 288.
Schnoor, J.L., Georgakakos, K.P., Lee, S., Nikolaidis, N.P., and Rajaram, H., 1990. Lakes Resources at Risk to Acidification in the Northeastern United States: Direct-Delayed Response Project. Final Report, Cooperative Agreement CR-812329, U.S., EPA, ERL, Corvallis, Oregon.

**Other relevant information**

## Model identification
Model name/title: LEEDS

## Model type
Media: water, lake
Class: biogeochemical
Substance(s)/compounds: nutrients, phosphorus
Biological components: ecosystem

## Model purpose
A management model accounting for seasonal variations, all important rates and including effect variables.

## Short description
Catchment area: transport of P from land.
Fish farms: input of P from fish farm.
Precipitations: input of P from atmosphere.
Other sources: input of P from industries and purification plants.
Lake variables: area, volume, latitude, longitude, altitude, size of watershed, etc.

## Model application(s) and State of development
Application area(s): LEEDS-model (Lake Eutrophication- Effect- Dose- Sensitivity - Model)
Number of cases studied:
State: verification

## Model software and hardware demands
Platform: MAC
OS:
Software demands:
Hardware demands:

## Model availability
Purchase: No
Contact person(s) Dr. Lars Håkanson
　　　　　　　　Institute of Earth Sciences
　　　　　　　　Uppsala
　　　　　　　　Sweden
Phone: 018 183897
Fax: 018 555920
E-mail: lars.hakanson@natgeo.uu.se

## Model documentation and major references
Håkanson and Peters, 1995. *Predictive Limnology*. SPB Academic Publishers, Amsterdam (in press).

**Other relevant information**
A large lake ecosystem model.
Dissolved and particulate P.
Seasonal variations; sedimentation; resuspension; diffusion; stabilization.
Predictions of phytoplankton biomass.

**Model identification**
Model name/title: AQUAMOD1, AQUAMOD2

**Model type**
Media: water, lake
Class: biogeochemical
Substance(s)/compounds: nutrients, phosphorus
Biological components: ecosystem - aquatic

**Model purpose**
Simulation of solutions in temperate lakes of different size. Nutrient load, extinction has light and fishpressure on zooplankton, as well as retention time.

**Short description**
One- layer and two-layer model of lake and reservoir eutrophication. The one-layer model represents the mixing zone, the two-layer one also the hypolimnion, the exchange through metalimnion being included into the dynamic equations for both layers. A generalized morphometry and exchange rates for lakes of different size are covered according to equations developed in Straskraba 1980: The effect of physical variables on freshwater production: analyses based on models. In: E.D. LeCren and R.H. Lowe-McConnel (Eds.) *The Functioning of Freshwater Ecosystems*, Cambridge University, Press pp. 13-84. Equations for phosphorus, phytoplankton and zooplankton. The forcing functions are annual changes of temperature and radiation, and a sedimentation function. Parameters are derived from observations and parameter estimation by an optimization procedure. Inputs quantify extinction coefficient, inflow (=outflow) rate and zooplankton mortality, by means of which the predation pressure by fish on zooplankton is simulated.
For extension to cover sediment see AQUAMOD3.

**Model application(s) and State of development**
Application area(s): eutrophication management
Number of cases studied: 5
State: verification, calibration, validation

**Model software and hardware demands**
Platform: PC-based - IBM, Workstation - HP
OS: DOS and UNIX
Software demands: language - FORTRAN
Hardware demands:

## Model availability
Purchase: no, software available as executable code.
Contact person(s) Dr. Milan Strakraba
        Biomathematical Laboratory
        Acad. Sci. Czech Rep.
        Branisovska 31
        37005 C. Budejovice
        Czech Republic

Phone: +42 38 41928
Fax: +42 38 45985
E-mail: stras@entu.cas.cz

## Model documentation and major references
Straskraba M. et al., 1985. *Aquatic Ecosystems. Modelling and Simulation*, Elsevier.

## Other relevant information
Straskraba, M., 1980. Periodicity, sensitivity of periodic solutions and parameters optimization of a theoretical ecosystem model. *J. Int. Soc. , Ecol. Modelling*, 2, pp. 42-62.
A., Matthäus, E., Straskraba, M., Affa, I., 1990. The use of SONCHES has aquatic ecosystem modelling. *Syst. Anal. Model. Sim.*, 17, pp. 438-458.

## Model identification
Model name/title: AQUAMOD 3

## Model type
Media: water, lake
Class: biogeochemical
Substance(s)/compounds: nutrients, phosphorus
Biological components: ecosystem - aquatic

## Model purpose
To investigate the relationship between water in a stratified lake and sediments in respect to phosphorus and eutrophication.

## Short description
The model represents an extension of the two- layer eutrophication model AQUAMOD2. Included are processes of phosphorus changes in the sediment and exchange of phosphorus at the water- sediment interface. The distribution of phosphorus in the hypolimnion and mixing layers are modelled by the simulated exchange through the metalimnion as well as P-uptake and liberation by phyto- and zooplankton. In addition to phosphorus, phytoplankton and zooplankton the equations cover organic matter in the sediment and the liberation of phosphorus, the dissolved phosphorus originating from the exchange with water, autolysis, decomposition of organic matter, adsorption and desorption from the adsorbed phosphorus pool. Organic matter and adsorbed phosphorus are burried into deeper, inactive sediment strata.

**Model application(s) and State of development**
Application area(s):
Number of cases studied:
State: conceptualization

**Model software and hardware demands**
Platform: PC-based - HP 45
OS:
Software demands: language - HP BASIC
Hardware demands:

**Model availability**
Purchase: no.
Contact person(s) Dr. Milan Straskraba
            Biomathematical Laboratory
            Acad. Sci. Czech. Rep
            Branisovska 31
            37005 e. Budejovice
            Czech Republic

Phone: +42 38 41928
Fax: +42 38 45985
E-mail: stras@entu.cas.cz

**Model documentation and major references**
Straskraba, M. et al., 1985. *Aquatic Ecosystems. Modelling and Simulation*, Elsevier.

**Other relevant informations**
Dvorakova, M., and H.-P. Kozerski, 1980. Three-Layer Model of an Aquatic Ecosystem. *J. Int. Sci., Ecol. Modelling*, 2, pp.63-70.
Kozerski, H.-P., and M. Dvorakova, 1980. Simulation Experiments With a Three-Layer Model. In: *Proceedings Simulation of Systems in Biology and Medicine*, Prague, Vol. 2, pp. 144-151.

## 2.6 Overview of Marine Models

Marine models have a wide spectrum of complexity in hydrodynamics as well as in ecology. The model may be zero dimensional (examples: the Ulva model and the Lake Glumsø model used as eutrophication model in a coastal ecosystem), one dimensional (Cams, 1-D vertical model; see Section 2.5 ), two dimensional (as many models with a box division, for instance, Mases) or 2.5 D (as Ersem) or even three dimensional (Biotest, Trophic-diffusive macromodel, SOR3D, PCFLOW3D and Trim-3D). Marine models are generally more complex in the description of the hydrodynamics than lake and reservoir models due to the much shorter retention time and higher extent of openness. Only a few marine models, on the other hand have the complex descriptions of the ecological processes which characterize many lake models. It must therefore be foreseen that the hydrodynamics of many marine models in the future will be applied in lake models to account for a more complex pattern of water flows, while many marine models probably will adopt the descriptions of ecological processes that are used in lake models.

The models which are focusing on general marine pollution problems are summarized in Table 2.4.

Ersem is a model that includes several major nutrients and their dynamics and covers a huge geographical area, namely the North Sea. It represents the attempt to construct a model applicable for environmental management in large scales and not surprisingly the model is developed by a multidisciplinary group and has cost several man years to develop. The model is as indicated above 2.5 dimensional, but it has also a high ecological complexity, as it has 40 state variables, including the nutrient dynamics of carbon, phosphorus and nitrogen.

The trophic-diffusive macromodel is a 3-D model developed to cope with the complex hydrodynamics of the Venice Lagoon, which has been discretized in 358400 cells, but it could be implemented also for other aquatic ecosystems, including lakes. The model has several biogeochemical variables, including various forms of carbon, nitrogen and phosphorus and phytoplankton and zooplankton.

The model SOR3D is a 3-D model particularly adapted to deal with complex hydrodynamics due to interactions between advection and diffusion. The biogeochemical variables are dissolved oxygen, temperature and suspended solid.

PCFLOW2D and PCFLOW3D simulates free surface flows in coastal areas (or lakes) in a two-dimensional or three-dimensional description. It uses conservative pollutants and constituents, heat and salinity as state variables.

A number of models focus entirely on primary production in its various forms. The 1-D model developed by Moll simulates phytoplankton standing stock and phosphate in the water column, while zooplankton is considered a forcing function. The model considers the following processes affecting the primary production and the phytoplankton concentration: turbulence, sinking, photosynthesis, respiration, mortality and grazing.

The Ulva-model simulates the growth of ulva determined by nutrient concentrations, light penetration and temperature. Intracellular nutrient concentrations are applied, which implies that the growth is described as a two-step process.

The Estuarine Phytoplankton Model is a 1-D model focusing on all the processes and factor, that determine the changes in phytoplankton biomass within a tidal estuary: photosynthesis, light penetration. zooplankton grazing, sinking, turbulent mixing and grazing by benthic macrofauna.

The model Fjord Ecosystem Model is developed particularly for modeling the flow of nutrient within a pelagic fjord and to and from the external sources and sinks. The model has a medium complexity with 14 state variables.

Several marine models have been develop for simulating nutrient and water flows in marine ecosystem of particular interest. The model MKP02 for instance describe the nutrient flow and the primary and secondary production in an upwelling system.

The model named Matsalu is model for determination of the impact of non-point agricultural pollution by nutrients on a shallow water bay. Emphasis is given to the macrophyte-type of eutrophication.

The model or expert system Recon assists in developing an optimum environmental management plan of a polluted coastal ecosystem by assessment of the current and concentration field.

A model of a coral reef is developed by McClanahan in his Coral-reef Ecosystem-Fisheries Model. The model has the following state variables; calcium carbonate, corals, algae and four groups of fish. The model is particularly fitted to optimize the fishery, but can also be used to assess the eutrophication.

Balyev and Sovga have developed a eutrophication model for a marine shelf ecosystem, describing the ecosystem with many details. The biogeochemistry is described by seven state variables for each compartment: phytoplankton, zooplankton, macrophytes, fish, nitrogen, phosphorus and detritus.

A model developed by Fong explores the importance of resource competition to algal community structure of shallow coastal lagoons. Three groups of algae are included in the model which may be considered a eutrophication model for a lagoon.

The model Ssem is intended to be used for prediction of the impact on fisheries caused by coastal development activities, including discharge of waste water. It can account for many different species of fish with different values as a fishery resource and different behaviour for the same impact.

**Table 2.4.** An Overview of Marine Models

| Model name/ main component | Characteristics of the model Hydrodynamics | Ecological Processes | References |
|---|---|---|---|
| Ersem, North Sea | 2.5 D | 40 state variables | Radford |
| Venice Lagoon | 3 D | C, N, P, phytoplankton and zooplankton | Dejak et al., 1990 |
| SOR3D | 3 D | Oxygen, temperature, susp. matter, biology | Yearsley , 1993 |
| PCFLOW2D/3D | 2-3 D | Only hydrodynamics | Rajar, 1990 |
| North Sea | 1 D | P + phyt as f(depth,time) detritus as f(time) | Radack and Moll, 1993 |
| Ulva | 0 D | Complex ulva model intracellular nut. conc. | Bendoricchio et al. , 1995 |
| Estuarine model | 1 D | Complex optional biology | Cloern, 1991 |
| Fjord | Simple | 14 state variables | Ross et al., 1993 |
| MKP02 | 1 D | nut., phyt., zoopl., | Valentin |
| Matsalu | Simple | Complete N and P cycles | Krysanova et al., 1989 |
| Coral reef | 0 D | 9 state variables incl. entire food chain | McClanahan, 1995 |
| Shelf ecosystem | Detailed (boxes) | 11 state variables: C, N, P + entire foodchain | Belyaev et al., 1992 |
| Algae competition | 0 D | 3 algae groups | Fong |
| Shallow sea (SSEM) | 2 D | Impact on fisheries, has entire food chain | Sekine et al., 1991 |

## 2.7 Marine Models

**Model identification**
Model name/title: ERSEM: European Regional Seas Ecosystem Model.

**Model type**
Media: water, ocean/sea, local and regional scale, sediment
Class: biogeochemical
Substance(s)/compounds: carbon, phosphorus, nitrogen, oxygen, silica, organic compounds
Biological components: ecosystem, marine ecosystem

**Model purpose**
Simulation of the seasonal cycle in temporate shelf seas of C, N, P, $O_2$ and Si as forced by light, temperatures and advective and diffusive transport.

**Short description**
The ERSEM model is a dynamical simulation model which allows the simulation of the time-dependent development of the 40 state variables, in from one to three space dimensions. It consists of a set of 40 coupled ode's with a vertically integrated differential equation for irradiance.

The model simulates the carbon dynamics and the associated dynamics of the major nutrients (N, P, Si), using functional groups of organisms as biological state variables. It simulates the dynamics of the biological system in the water column (which may be vertically resolved) as well as in the benthic system (which has three layers: oxygenated, denitrification and anaerobic), thus allowing the investigation of benthic pelagic coupling. Hydrodynamical forcing of this model is done indirectly with an advection-diffusion transport model which in turn is forced with the output from a 2.5 D circulation model of the North sea, aggregated into daily transport values through all spatial boxes.

The whole ERSEM model is written in FORTRAN-77 and runs under a modeling software package developed at NIOZ for use on UNIX machines. At present it runs on SUNS-parcstations. The model is set up as a series of submodels that communicate through COMMON, allowing (in theory, if not always in practice) independent development and testing of the submodels in the framework of a standard model.

**Model application(s) and State of development**
Application area(s): North Sea
Number of cases studied: 4
State: verification, prognosis made

**Model software and hardware demands**
Platform: Work-station SUN-microsystems
OS: Solaris 1.1 (UNIX)
Software demands: GKS FORTRAN 77
Hardware demands: dependent on spatial resolution. Min Spars.station 1+

## Model availability
Purchase: yes, media price. Sourcecode : copyrighted.
Contact person(s) For distribution
>P.J. Radford
>Plymouth Marine Lab
>The Hoe
>Plymouth
>UK

Phone: +44 752 299 772
Fax: + 44 752 670 637
E-mail: pjr(tegn)ibma.nerc-pml.ac.uk

## Model documentation and major references
ERSEM- The Modules compiled by P.J. Radford, PML Special issue of NJSR, in prep.

## Other relevant information
The model runs in the software environment of Sesanc, which is distributed together with the model.

## Model identification
Model name/title: Trophic-diffusive macromodel

## Model type

Media: water, estuary, lagoon
Class: biogeochemical
Substance(s)/compounds: nutrients, phosphorus, nitrogen, oxygen, temperature
Biological components: populations, phytoplankton, zooplankton

## Model purpose
The aim of this interdisciplinary model is to provide a description of the eutrophication phenomenon in a complex coastal environment as the Venice Lagoon, a water basin affected by macronutrient input linked to urban, industrial and agricultural activities and whose hydrodynamic regime was also heavily altered during the 1960s, by opening a wide and deep artificial channel.

## Short description
The finite difference three-dimensional macromodel covers the central part of the Venice Lagoon, which has been discretized in 358400 cells of the size Dx=Dy=100m, Dz=1m. The evolution of the state vector is followed by solving numerically a reaction-diffusion equation. A smooth behaviour of the solution at the open boundaries is assured by a purposely designed set of conditions, adopted also for a small test version of the model presented by the same authors (Biotest). The advective term has been dropped, because the hydrodynamic regime of the basin is governed by the tide and the importance of the residual currents has not yet been clearly demonstrated: in this case the tidally averaged

dispersion can be modeled by using the diffusion term only, with a properly tuned eddy diffusivity matrix. The components of the state vector are: phyto- and zooplankton densities, oxidized and reduced nitrogen concentration, reactive phosphorus concentration, BOD, dissolved oxygen (DO) and sedimented organic detritus. The forcing functions are represented by a set of metereological data which allow to estimate the water temperature by solving a nineth reaction-diffusion equation, by the incident solar radiation and by solving the continuous input of macronutrients.

The program is fairly time consuming, as an yearly simulation takes about 90 minutes on a supercomputer mainframe, after a proper coding of the program, aimed at exploiting the vector and parallel architectures. Nevertheless, the new generation of alpha processors can cope with its computational request, as the same simulation takes about 14 hours.

## Model application(s) and State of development
Application area(s): Venice Lagoon
Number of cases studied:
State: calibration

## Model software and hardware demands
Platform: Mainframe: Cray YMP, Digital Axp 3300, work station: Silicon graphics
OS: UNIX
Software demands: FORTRAN
Hardware demands:

## Model availability
Purchase: no, executable code and sourcecode
Contact person(s) Dejak Camillo, Pasters Roberto, Pecenik Giovanni
        Universit' di Venezia
        Dip. Chimica Fisica
        Dorsoduro 2137
        I-30123 Venezia, Italy
Phone:
Fax:
E-mail: pasters@unive.it

## Model documentation and major references
Dejak, C., and Pecenik G., 1987. Special issue: Venice Lagoon. *Ecol. Modelling*, 37 pp.1-101.
Dejak C., Franco D., Pastres R., Pecenik G., and Solidoro C., 1992. Thermal exchanges at air-water interfacies and reproduction of temperature vertical profiles in water columns. *J. Marine Systems*, 3: 465-467, Elsevier, Amsterdam.
Dejak C., Franco, D., Pastres R., and Pecenik, G., 1990. A 3-D eutrophication-diffusion model of the Venice Lagoon: Some application. In: R.T. Cheng (Ed.) Residual Corrents and Long-Term Transport. *Coastal and Estuarine Studies*, 38: 526-538. Springer -Verlag, NY.
Dejak C., Franco D., Pastres, R., and Pecenik, G., 1990. Application of parallel computers in environmental modelling. In: *Supercomputing Tools in Science and Engineering*, Angeli, Milan, pp. 101-107.

Dejak C., Franco D., Pastres R., and Pecenik G., 1990. #D Modelling of the Venice Lagoon: estimating through variable diffusitivities the impacts of frequent inlet closures due to greenhouse effect. 5th Int. Biennal Conference Physics of Estuaries and Coastal Seas. University of Wales Gregynog, UK, July 9-13. pp. 33-36.

Dejak C., and Pecenik G., 1991. A Physico-Chemical approah to modelling transport processes: an application to perturbed water bodies. In: *Ecological Physical Chemistry*. Tiezzi E. and Rossi C., Eds., Elsevier.

Pastres R., Franco D., Pecenik G., Solidoro C., and Dejak C., 1993. Using Parallel computers in environmental modelling: a working example. *Ecol. Modelling* (accepted for publication and in press).

### Other relevant information

Model output can be visualized on a work Silicon Graphic station by using a purposely coded program. The program displays at the same time the average evolution of the state variables at 13 points, as reconstructed by analyzing the time series of experimental data. This tool will be used in a forthcoming recalibration and validation of the model.

### Model identification

Model name/title: SOR3D: A Three-Dimensional Model of Diffusion and Avection in an Estuarine Environment

### Model type

Media: water, estuary
Class: biogeochemical, hydrology
Substance(s)/compounds: dissolved oxygen, temperature, suspended solids
Biological components:

### Model purpose

The model was developed to perform planning and ecosystem risk analysis in estuaries or coastal embayments where point source may have an impact on temperature, dissolved oxygen, and suspended solids.

### Short description

State variables:   Dissolved oxygen, temperature, suspended solids
Forcing Functions: Oceanic inflow, river inflow, point source inflow.

### Model application(s) and State of development

Application area(s): Development of water discharge permits, ecological risk analysis, and environmental impact statements.
Number of cases studied: 4
State: conceptualization, parameter estimation, prognosis

### Model software and hardware demands

Platform: DEC Microvax, DG Aviion Workstation
OS: DEC VMS, Unix

Software demands: FORTRAN 77 compiler, FORTRAN 77
Hardware demands: 4 Mb RAM, 200 Mb storage

## Model availability
Purchase: none, software available as FORTRAN 77 source code
Contact person(s) Dr. John R. Yearsley
>                 EPA Region 10        ES097
>                 1200 Sixth Avenue
>                 Seattle, WA 98101
>                 USA

Phone: +1 206-553-1532
Fax: +1 206-553-0119
E-mail: jyearsl@r0dg09.r10.epa.gov
>        yjd@sequoia.nesc.epa.gov

## Model documentation and major references
Yearsley, J.R., 1993. Cumulative impacts of seafood processing on dissolved oxygen in Akutan Harbor, Alaska. In: *Environmental Assessment of Deep Sea Fisheries Shore Plant and Cumulative Effects of Seafood Processing Activities in Akutan Harbor, Alaska.* EPA Region 10, Seattle, WA.
Yearsley, J.R., Estimating impacts of discharges from Ketchikan Pulp Co. on the surface waters of Ward Cove, Alaska. EPA 910/R-93-004. EPA Region 10, Seattle, WA.
Cirone, P.A. and Yearsley J.R., 1993. Comparative analysis of mining tailing disposal for the Quartz Hill molybdenum mining project. In a review of ecological assessment case studies from a risk assessment perspective. EPA/630/R-92/005. EPA Risk Assessment Forum, Washington D.C.

## Other relevant information

## Model identification
Model name/title: PCFLOW2D

## Model type
Media: water, lake, ocean, estuary, river
Class: hydrology/hydraulics
Substance(s)/compounds: please see "other relevant information about the model" conservative pollutants
Biological components:

## Model purpose
Numerical simulation of HD and AD phenomena in surface waters.

## Short description
The basic 2-D depth averaged-nonlinear with advective terms and time-dependent equations of motion and the continuity equation are solved by the finite volume method of Patankar and Spalding. The upwind/central differencing is used depending on the Peclet number. The Cartesian orthogonal numerical grid is used in which the topography of the terrain is taken into account by the data input. The two-equation (k-epsilon) turbulence model is included to compute the time and space dependent turbulent (eddy) viscosity. One version of the program was also transformed to curvilinear orthogonal coordinate system to compute dam-break flow in a very steep and curved channel.
Input: bottom levels matrix, Manning's friction coefficient matrix, wind velocity over the whole area, different boundary conditions.

## Model application(s) and State of development
Application area(s): Free surface flows in rivers, lakes and seas
Number of cases studied: 8
State: verification, calibration, validation, prognosis made

## Model software and hardware demands
Platform: PC-based IBM
OS: DOS
Software demands: Autodesk AutoCAD ver. 10 (only for graphics)  language FORTRAN
Hardware demands: IBM PC 386 (486 Pentium preferred)
NOTE: With minor modifications it can run on any platform using FORTRAN (VAX, SGOUNIX, AIX- IBM, UNIX)

## Model availability
Purchase: yes, DEM 10,000
Contact person(s) Prof. Rudi Rajar    and    Asst. Prof. Matjaz Cetina
                FAGG-Hydraulic Div.
                Univerza v. Ljubljani
                Hajdrihova 28
                Sl-61000 Ljubljana
                Slovenia

Phone: +386 61 1254 333
Fax: +386 61 219 897
E-mail: rudi.rajar@uni-lj.sl

## Model documentation and major references
Cetina, M., and Rajar, R., 1993. Mathematical Simulation of Flow in a Kyak Racing Channel. 5th Int. Symp. on Refined Flow Modelling and Turbulence Measurements, Paris, 7.-10. 9. 1993, pp. 673-644.
Rajar, R., and Cetina, M., 1994. Two-dimensional Modelling of Flow in Sava River at Tacen. Int. Conf. Modelling of 2D Flow over Initially Dry Regions, Milan, 29.6.-2.7. 1994, Proceedings, pp. 309-323.

Cetina, M., and Rajar, R., 1994. Two-dimensional Dam-break Flow in a Sudden Enlargement. Int. Conf. Modelling of 2D Flow over Initially Dry Regions, Milan, 29.6.-2.7. 1994, Proceedings, pp. 268-282.

## Other relevant information

Additional modifications of the source term in advection - dispersion equation are presently being bulit into the model. On the basis of this step, including of the sediment transport (erosion-deposition) and Hg (Mercury) dynamics submodels is planned for the year 1995. An user-friendly interface for data input and presentation of the results is planned in the near future. The program has not been generally tested yet and is not yet in a user-friendly form to be utilized by a user not experienced in this field.

## Model identification

Model name/title: PCFLOW3D

## Model type

Media: water, lake, ocean, estuary
Class: hydrology/hydraulics
Substance(s)/compounds: conservative pollutants, heat, salinity
Biological components: none

## Model purpose

3-D numerical simulation of HD and AD in lakes and coastal seas.

## Short description

The basic 3-D (nonlinear and time-dependent) equations of motion and the continuity equation are solved by the finite volume method of Patankar and Spalding. The hydrostatic approximation is used for the distribution of pressure in the vertical direction. After determining the velocity field the 3-D convection- diffusion equations are used to calculate the transport and dispersion of heat, salinity or any conservative pollutant. Their influence on the density field is accounted for by the equation of state. The variable density influences the velocity field in the next time step (baroclinic model). The "vertical" turbulent (eddy) viscosity is computed by a simplified one-equation turbulence model proposed by Koutitas (equation of turbulence kinetic energy). Constant "horizontal" turbulent viscosity is supposed.
The following boundary conditions can be taken into account:
-- fixed boundaries,
-- inflow of rivers (with different temperature, salinity or density),
-- tidal water-level changes at the open boundaries,
-- wind stress on the surface (possibly time-dependent),
-- radiation or continuity BC at the open boundaries.
Input: bottom topography matrix, Manning's coefficient of bottom friction matrix, wind speed, inflow of rivers and pollutants, open boundary condition (tide, radiation or continuity BC).

---

**Model application(s) and State of development**
Application area(s): Free surface flows in lakes and coastal seas
Number of cases studied: 6
State: verification, calibration

---

**Model software and hardware demands**
Platform: PC-based - IBM
OS: DOS
Software demands: AutoCAD (only for graphics), language FORTRAN
Hardware demands: IBM PC 386 (486/Pentinum preferred)
NOTE: With minor changes it can run on any platform using FORTRAN (VAX, VMS, SCO UNIX,AIX- IBM UNIX)

---

**Model availability**
Purchase: yes, DEM 15,000
Contact person(s) Prof. Rudi Rajar and Asst. Prof. Matjaz Cetina
             FAGG-Hydraulic div
             Univerza v. Ljubljani
             Hajdrihova 28
             Sl-61000 Ljubljana
             Slovenia

Phone: +386 61 12 54 333
Fax: +386 61 129 897
E-mail: rudi.rajar@uni-lj.sl

---

**Model documentation and major references**
Rajar, R., 1989. Three-Dimensional Modelling of Currents in the Northern Adriatic Sea. XXIII Congres of the International Association for Hydraulic Research, Ottawa, Canada, August 1989, Proceedings, pp. C335-C342.
Rajar, R., 1990. Mathematical Modelling of Currents and Dispersion in the Northern Adriatic Sea, XXXLLth Congress CIESM, Perpignan, Oct. 1990, Proceedings.
Rajar, R., 1990. Three-dimensional mathematical model of currents and of transport of pollutants in the Adriatic, 8th International Conference on Computational Methods in Water Resources, Venice, June 1990, Proceedings, Computational Methods in Surface Hydrology, pp. 57-62.
Rajar, R., and Cetina, M., 1991. Modelling Wind- Induced Circulation and Dispersion in the Northern Adriatic. XXIV. Congress of the International Association for Hydraulic Research, Madrid, Sept. 1991, Proceedings, 8 pp.
Rajar, R., and Cetine, M., 1991. Modelling of Wind-Induced Currents and Dispersion in the Northern Adriatic, Int. Seminar Protection of Adriatic Sea, Urbino, Italy, Proceedings.
Rajar, R., 1992. Modelling of Tidal and Wind-Induced Currents and Dispersion of Pollutants in the Northern Adriatic, Acta Adriatica, Split, Croatia, 1992, 32(2): 785-812.
Rajar, R., 1992. Application of the Three-Dimensional Model to Slovenian Coastal Sea. International Conference on Computer Modelling of Seas and Coastal Regions. Southampton, April 1992, Proceedings.

**Other relevant information**
An user-friendly interface for data input and results presentation is planned for the near future. NOTE: The program is not yet generally tested and not prepared in an user-friendly form to be utilized by a user not experienced in this field.

**Model identification**
Model name/title: Modelling the primary production in the central North Sea.

**Model type**
Media: water, ocean, regional scale
Class: biogeochemical
Substance(s)/compounds: carbon, phosphorus
Biological components: ecosystem - coastal, North Sea

**Model purpose**
A physical and a biological one-dimensional upper layer model for the simulation of the animal cycles of both the physical and phytoplankton dynamics, and used to estimate the primary production in the central North Sea.

**Short description**
Phytoplankton standing stock and phosphate in the water column serve as time and depth-dependent pools. Detritus is a time-dependent pool at the bottom. All three pools are prognostic state variables. Underwater light is calculated as a diagnostic variable. Zooplankton standing stock (copepods) is prescribed and serves as a forcing function variable. Bacteria activity only appears implicitly in the parameterizations of the remineralization terms. Benthic detritus accumulates out of the water column. Light penetration is determined by the extinction by the water itself and by the phytoplankton in the water (self-shading). Phytoplankton standing stock is affected by turbulent, sinking, primary production, respiration, mortality, and grazing. The nutrient, phosphate, is taken up by phytoplankton and is regenerated through remineralization by bacteria. This occurs via several pathways.

**Model application(s) and State of development**
Application area(s): North Sea. Flex'76 position, Ocean Weather Ship Farmetor, LV Elbe1
Number of case studied: 4
State: validation

**Model software and hardware demands**
Platform: Mainframe
OS: MVS
Software demands: Language FORTRAN
Hardware demands:

**Model availability**
Purchase: available as source code
Contact person(s) Dr. Andreas Moll
University of Hamburg
Troplowitz Str. 7
D-22529 Hamburg
Germany

Phone:
Fax:
E-mail: Moll@ifm.uni-hamburg.de

**Model documentation and major references**
Radack, G., and Moll, A., 1993. Estimation of the variability of production by simulating annual cycles of phytoplankton in the central North Sea. *Prag. Oceanografi*. Vol 31(4): 339-419.

**Other relevant information**

**Model identification**
Model name/title: ULVA

**Model type**
Media: water
Class: biogeochemical
Substance(s)/compounds: nutrients, phosphorus, nitrogen, oxygen
Biological components: population - algae: *Ulva rigida*

**Model purpose**
To allow a deep knowledge of *U. rigida* life cycle in lagoon and to point out relative importance of processes such as: nutrient limitation and light and temperature effects.

**Short description**
The state variables included in the model are: biomass of *Ulva rigida*, and its intermol quotes of nitrogen and phosphorus. The model is zero-dimensional thus the forcing functions concern: light, temperature, oxygen and nutrients dissolved in the water column. The latter three forcing functions should be introduced as input to the model. Sensitivity analysis shows that parameters related to temperature influence are the most important to fitt better the life cycle.

**Model application(s) and State of development**
Application area(s): large growth in shallow waters
Number of cases studied: 4 (all in lagoon of Venice)
State: Italy, calibration, no prognosis made

**Model software and hardware demands**
Platform: PC-based
OS: MS-DOS
Software demands: none, language C++
Hardware demands: mathematic coprocessor

**Model availability**
Purchase: no, available without charges and as executable code
Contact person(s) Dr. Giovanni Coffaro
        Facoltà di Ingegneria, Istituto di Chimica Industriale
        Università di Padova
        Via Marzolo 9
        I-35131 Padova, Italy
Phone: 49 39 831682
Fax: 4939 831683
E-mail: Gbendo@ipduniux.unipd.it

**Model documentation and major references**
Bendoricchio, G. et al., (1995) A Trophic model for Ulva rigida in the lagoon of Venice. *Ecol. Modelling* (in press).

**Other relevant information**

The model is under improvement and we are increasing the case studies.

**Model identification**

Model name/title: Estuarine Phytoplankton Model

**Model type**

Media: estuary
Class: biogeochemical
Substance(s)/compounds:
Biological components: population - phytoplankton

**Model purpose**
A research tool to pose and test hypotheses about coupled physical-biological processes that give rise to phytoplankton blooms in estuarines.

**Short description**
This is a 1-D vertical model that began simply and is evolving into a more complex representation of all processes throughout to cause changes in phytoplankton biomass within tidal estuaries. The original version (Cloern, 1991) considered simple representations

of light-limited growth; zooplankton grazing; sinking; turbulent mixing; and grazing by benthic macrofauna. The next step in the evolution (Koseff et al., 1993) examined nondimensional formulation of the model to identify critical parameter ratios; to identify combinations of those parameters that would allow phytoplakton biomass to increase; and incorporated time-variable and depth-variable eddy diffusivities. Own current work is devoted to: (1) expansion of the model to include 2 phytoplankton components (diatoms and nondiatoms, dissolved nitrogen and dissolved silica; (2) Mellor-Yamada 2.5 turbulence closure to calculate eddy diffusivities as functions of wind and tidal currents; (3) external sources and regeneration rates of N and Si; and (4) simultaneous light- and nutrient-(N,Si) limitation of phytoplankton growth. Over the next two year we intend to expand the model to two dimensions.

---

**Model application(s) and State of development**
Application area(s): San Francisco Bay - the model is used as an explorators tool, not as a predictive tool
Number of cases studied:
State: validation of prognosis

---

**Model software and hardware demands**
Platform: PC-based, Workstation
OS:
Software demands: language FORTRAN
Hardware demands:

---

**Model availability**
Purchase: We intend to publish a model description with source code listing in 1995 or 1996
Contact person(s) Dr. J. Cloern
U.S. Geological Survey
Water Resource Division MS496
345 Middlefield Rd.
Menlo Park, CA 94025, USA

Phone:
Fax:
E-mail:

---

**Model documentation and major references**
Cloern, J.E., 1991. Tidal stirring and phytoplankton bloom dynamics in a esturary. *Journal of Marine Research*, Vol. 49, pp. 203-221.
Koseff, J.R., et al., 1993. Coupled effects of vertical mixing and benthric grazing on phytoplankton populations in shallow, turbid estuaries. *Journal of Marine Research*, Vol. 51, pp. 843-868.

---

**Other relevant information**

## Model identification
Model name/title: Fjord ecosystem model

## Model type
Media: water, estuary, local scale
Class: biogeochemical, hydrology
Substance(s)/compounds: nutrients, carbon, nitrogen
Biological components: ecosystem - pelagic ecosystem

## Model purpose
To model the flow of nutrients within a fjord pelagic ecosystem, and between the system and external sinks and sources; to understand what are the critical factors driving productivity; to asses what impact, if any, marine farming and other activities may have.

## Short description
A strategic simulation model of a fjord ecosystem is developed and tested. This model is sufficiently detailed to allow a general agreement with data from a number of pelagic ecosystems; yet is sufficiently simple to allow us to determine unambiguously what the key processes are which govern its behaviour. The full version contains fourteen state variables, reflecting the major pools of nutrients and the trophic levels present in these systems. Simpler versions have been developed to demonstrate which variables and processes are the critical ones for describing the dynamics of the system.

In the conditions prevailing in the western British Isles, the flow of nutrients between the fjord and the external coastal water is found to be the critical flux supplying nutrient for primary production. This flux is sufficiently large that growth is almost limited by the availability of light. Although the persistance of the system is determined by this flux, and although primary productivity is determined by the availability of light; for much of the year primary production is controlled by zooplankton grazing.

We conclude that the system is most sensitive to disturbance which impact directly on the zooplankton grazing rate.

## Model application(s) and State of development
Application area(s):
Number of cases studied: 4
State: verified, calibrated, validated and the prognosis has been made

## Model software and hardware demands
Platform: PC DOS, and Sun UNIX workstation versions
OS: UNIX, MS-DOS
Software demands: SOLVER, SUN Pascal and OpenWindows or Borland Turbo Pascal language PASCAL
Hardware demands: Sun workstation or IBM PC compatible

---

**Model availability**
Purchase: The model itself is fully specified in the literature. Solver, the software package used for numerical integration and display of the model, is available from:
Contact person(s) Dr. W.S.C. Gurney (contact Ms. A. Mitchell)
                Department of Statistics and Modeling Science
                University of Strathclyde
                Livingstone Tower
                Richmond Street
                Glasgow G1 1XH, Scotland
                UK
Phone: 041 552 400
Fax: 041 552 2079
E-mail: bill@stams.strath.ac.uk

| Prices (USD) | PC/DOS | PC/Windows | SUN/UNIX |
|---|---|---|---|
| Single license | 100 | 400 | 600 |
| Site license | 400 | 800 | 1000 |
| Site license/ lifetime support | NA | 1250 | 1500 |

Appropriate Pascal compiler reqired: Borland Turbo Pascal, Turbo Pascal for Windows or SUN Pascal. Prices include manuals and installation media.

---

**Model documentation and major references**
Ross, A.H., Gurney, W.S.C., Heath, M.R., Hay, S.J., and Henderson, E.W., 1993. A strategic simulation model of a fjord ecosystem. *Limnology and Oceanography*, 38: 128-153.
Ross, A.H., Gurney, W.S., and C., Heath, M.R., 1994. A Comparative study of the ecosystem dynamics of four fjords. *Limnology and Oceanography*, 39: 319-343.
Ross, A.H., 1994. Fjord ecosystem models. Ph.D. Thesis, University of Strathclyde, Glasgow.

---

**Other relevant information**

---

**Model identification**
Model name/title: MKPØ2: Modelling of the Vertical Distribution of Marine Primary Biomass in the Cabo Frio Upwelling Region.

---

**Model type**
Media: water, ocean, local scale
Class: hydrology
Substance(s)/compounds: nitrogen
Biological components: ecosystem - upwelling

## Model purpose
For simulating the nutrient (nitrogen) flow through the lower trophic levels in a water column of the Cabo Frio upwelling area.

## Short description
Model developed in FORTRAN 77 - compartment model with water column divided into N vertical blocks.

State variables are: nutrients, phytoplankton, herbivorous, zooplankton, all expressed in myM nitrogen.

Forcing functions: Light varying according to hour, depth, absorption by water and chlorophyll, diffusion and/or advection.

Parameters: Phytoplankton assimilation rate, phytoplankton sinking rate, zooplankton excretion rate, grazing rate, zooplankton mortality rate, light extinction coefficients, optimum photosynthesis rate.

Inputs: Nutrient, phytoplankton and zooplankton initial values in each block, dimensions of blocks, initial and final time of simulation, time step of simulation.

## Model application(s) and State of development
Application area(s): Marine production - upwelling zones.
Number of cases studied: 1
State: conceptualization, verification

## Model software and hardware demands
Platform: PC-based
OS: DOS
Software demands: 1 FD 5 1/4, FORTRAN 77
Hardware demands: Basic configuration (1 driver)

## Model availability
Purchase: yes, free. Source code.
Contact person(s) Dr. Jean Louis Valentin
             Rua Maria Amalia 628/509, 20510.130
             28910 Rio de Janeiro
             Brazil

Phone: 0055 21 2807893
Fax: 0055 21 2807993
E-mail:

## Model documentation and major references


## Other relevant information

## Model identification
Model name/title: MATSALU: the system of simulation models for agricultural watershed and shallow sea bay

## Model type
Media: water, estuary, river, wetland, local scale, regional scale, terrestrial, soil, agricultural
Class: biogeochemical, hydrology
Substance(s)/compounds: nutrients, phosphorus, nitrogen
Biological components: ecosystem - agricultural watershed and shallow sea bay

## Model purpose
The purpose is two-fold: first, modelling transport of nutrients from nonpoint (main focus) and point sources of pollution at the territory of meso-scale watershed, and second, modeling the effect of nutrient input on the macrophyte-type eutrophication in the shallow sea bay ecosystem. The model allows to evaluate different management scenarios aimed for eutrophication control.

## Short description
The model consists of four coupled submodels. They are: submodel of terrestrial water balance, submodel of nutrient transformations in soil, submodel of nutrient transfer by stream flow, and submodel of the sea bay ecosystem. The first two submodels are developed in the point and watershed-distributed versions. The main processes simulated are: evapotranspiration, dynamics of soil water moisture, infiltration, surface and subsurface runoff, organic nitrogen (N) and phosphorus (P) mineralization, nitrification, P adsorption in soil, denitrification, N and P uptake by plants, erosion, N percolation, dissolved N and P run-off, sediment-bound N and P run-off, N and P retention, water discharge by stream flow, nutrients transfer by stream flow, and dynamics of nitrogen, phosphorus, macrophytes, phytoplankton, and other components of biota in the sea bay ecosystem. Input data: standard meteodata, hydrochemical monitoring in rivers, groundwater level, soil properties, and human activities.

The modeling system is based on difference balance equations with semi-empirical coefficient. Several layers of spatial information were used for spatial disaggregation of the watersheds (which have no tributaries), land use map, soil map (only for arable land). First of all the watershed submodels were verified for a small representative subbasin, where more detailed hydrochemical monitoring was provided over two years. After that the modeling system was applied for the Matsalu Bay drainage basin (3500 km²) and the bay ecosystem (90 km², 1.5 m depth). After verification a number of scenarios were calculated aiming in the eutrophication control.

## Model application(s) and State of development
Application area(s): Matsula Bay and its drainage area, West Estonia
Number of cases studied: There was one main case study, and several more applications evaluation of nonpoint source pollution in Estonia.
State: validation

**Model software and hardware demands**
Platform: PC-based IBM
OS: MS-DOS 3.1 or higher
Software demands: LOTUS 1-2-3 or similar, FORTRAN 77
Hardware demands: IBM compatible PC

**Model availability**
Purchase: yes, practically free, only the cost of mailing and documentation
(if necessary) for submodels, executable code
Contact person(s) Dr. Valentina Krysanova
Potsdam- Institut für Klimafolgenforschung
Postfach 601203, Telegrafenberg
14412 Postdam, Germany

Phone: +49 331 288 2515
Fax: +49 331 288 2600
E-mail: valen@pik-potsdam.de

**Model documentation and major references**
At present, no documentation is available, but just some short comments in the source code.
Krysanova, V., and Luik, H. (Eds.), 1989. Simulation modeling of the system Watershed-River- Sea Bay". Tallinn, Valgus, 428 p. (in Russian).
Krysanova, V., Meiner, A., Roosaare, J., and Vasilyev, A.. Simulation modelling of the coastal waters pollution from agricultural watershed. *Ecol. Modelling*, Vol. 49, 1989, p. 7-29.

**Other relevant information**
Since 1989 the new version of the model has been developed together with different databases, and some related software (in dBASE and FoxPro) for calculations and simple visualization are available. The submodel of nutrient transformations in soil has been transfered into UNIX environment and coupled with GIS software ARC/INFO, Release 6.0.

**Model identification**
Model name/title: Expert System RECON

**Model type**
Media: water, lake, coastal sea, estuary, wetland, local scale, regional scale
Class: biogeochemical, hydrology
Substance(s)/compounds: any substance released to water, e.g., nutrients, carbon, sulfur, nitrogen, heavy metals
Biological components: a complete aquatic ecosystem is taken into account

## Model purpose
Assists in developing an optimum environmental management plan of polluted coastal seas and lakes by first RECON-structing the existing current field and concentration field by use of optimized models and then by predicting the concentration field of any potential environmental management plan.

## Short description
Based on geometry of a bay, coastal sea, lake or an estuary (with islands, peninsulas, pears etc.) and boundary conditions, the expert system RECON first constructs movement of water, i.e., the current field. If data on current measurements exist, they are used. Then, by using measurements of the substance in water (nutrient, pollutant or any other non-conservative substance) the package automatically constructs an optimized model of transport. In the process it estimates possible unknown point and diffuse sources and parameters that scientists have no other way of estimating, such as the extinction rate of substances in water. When models are constructed, RECON gives the present concentration field and the total mass balance of the substance in the region (where does the substance originates from and where does it leave the region?). The decomposition of the concentration field due to each source is also given, i.e., it answers the question: What is the contribution of each source to the concentration at any location in the region? RECON is now ready for simulation of scenarios corresponding to any pollution management plan including relocation, merging or decreasing existing and/or installing new sources. Since to each plan one can assign the cost and the resulting concentration field, the optimum may be found. RECON has been used for predictions and they have been validated in the most stringent conditions that the package can be used profitably for the most sensitive aquatic environments. The package uses a series of new algoritms for solving a combination of direct and inverse problems.

## Model application(s) and State of development
Application area(s): Adratic Sea, Gulf of Lyons, Cannes Bay, Marmara Sea
Number of cases studied: 8
State: First running version 1982. Predictions validated 1991.

## Model software and hardware demands
Platform: PC-based
OS: DOS 3.0 and higher
Software demands:
Hardware demands: RAM: 640 Kb or higher, EGA/VGA, 5 Mb on disk, printer/plotter

## Model availability
Purchase: yes, US $10.000
Contact person(s) Dr. T. Legovic
    The Rudjer Boskoviv Institute
    P.O. Box 1016
    Bijenicka 54
    41001 Zagreb, Croatia

Phone: +385 41 425 498  or  +385 41 461 111 switchboard
Fax: +385 41 425 497
E-mail:

---

**Model documentation and major references**
User's Manual, 4 diskettes 3.5" HD, examples on diskettes

---

**Other relevant information**
Legovic, T., Limic, N., and Sekulic, B., Reconstruction of a concentration field in a coastal sea. *Estuarine Coastal and Shelf Science*, 29 (1989), 217-231.
Legovic, L., Limic, N., and Valkovic, V., Estimation of diffuse inputs to a coastal sea: Solution to an inverse modelling problem. *Estuarine Coastal and Shelf Science*, 30 (1990), 619-634.

---

**Model identification**
Model name/title: Coral-reef Ecosystem - Fisheries Model

---

**Model type**
Media: ocean, coral reef, local scale, benthic
Class:
Substance(s)/compounds: carbon, organic compounds
Biological components: ecosystem - coral reef community - grazers population - sea urchins and herbivorous fishes

---

**Model purpose**
Determine the energetic properties of coral reef organisms and ecosystems and impacts of fishing on energetics and predator-prey relationships.

---

**Short description**
The variables used in the model include:
(1) sunlight
(2) calcium carbonate
(3) sea urchins
(4) herbivorous fish
(5) piscivores
(6) invertivores
(7) fishermen
(8) algae
(9) coral

---

**Model application(s) and State of development**
Application area(s): coral reefs - Kenya
Number of cases studied:
State: conceptualization, verification, calibration, validation

**Model software and hardware demands**
Platform: MAC
OS:
Software demands: < 200 K, language BASIC
Hardware demands: math coprocessor

**Model availability**
Purchase: no
Contact person(s) Dr. Timothy R. McClanahan
          P.O. Box 99470
          Mombasa, Kenya

Phone:
Fax:
E-mail:

**Model documentation and major references**
McClanahan, T.R., 1995. Harvesting in an uncertain world: impact of resource competition on harvesting dynamics. *Ecol. Modelling.*
McClanahan, T.R. (in press). A coral reef ecosystem.

**Other relevant information**
Fisheries model: Impacts of fishing intensity and catch selection on reef structure and processes. *Ecol. Modelling.*

**Model identification**
Model name/title: "Model of the shelf ecosystem".

**Model type**
Media: ocean, local scale, regional scale
Class: biogechemical
Substance(s)/compounds: nutrients, carbon, phosphorus, nitrogen, oxygen, organic compounds
Biological components: ecosystem - water, marine

**Model purpose**
The purpose of the model is to study the response of the shelf ecosystem to external natural and antropogenic influences. The natural influences are meteorological conditions, water exchange on outer boundary, river discharge. The antropogenic ones are fish catch and pollutions, which are estimated by changes of organism mortalities.

116

**Short description**
The method of aggregation and averaging of the system components with the consequent hierarchical decomposition of the model is used. The shelf ecosystem is modeled by constructing a succession of models describing the ecosystem with an increasing extent of detail. The first in this series of models is that of basic components which determine the general view of the system.

The seven basic components of the biogeochemical marine system are state variables of the model: biomass concentration of phytoplankton, zooplankton, macrophytes, fish, compounds of nitrogen, phosphorus, dead organic matter (both suspended and dissolved).

The block structure of the model allows the application of different versions of submodels and, if necessary, substitution for more accurate ones.

The inputs for the model are meteorological data, intensity of solar radiation at the sea surface, conditions on the liquid boundary with an open sea, river discharge, fish catch and other. Some of these variables may considered as forcing functions. The outputs were fields of all basic components of the ecosystem and fields of optical characteristics.

**Model application(s) and State of development**
Application area(s): 1. Black Sea North-Western Shelf. 2. Gdansk Gulf of the Baltic Sea
Number of cases studied: 2
State: Verification, prognosis made

**Model software and hardware demands**
Platform: IBM PC-based
OS: MS DOS v. 3.0 and more
Software demands: no
Hardware demands: IBM PC 80286 with math coprocessor

**Model availability**
Purchase: Software available as executable code
Contact person(s)  Prof. V. I. Belyaev
            Marine Hydrophysical Institute
            National Academy of Sciences of the Ukraine
            2, Kapitanskaya St., Sevastopol, 335000
            Crimea, Ukraine

Phone:(0690) 520382
Fax:  (0690) 411325
E-mail:

**Model documentation and major references**
Belyaev, V.I., and Konduforova, N.V., 1992. Modelling of the shelf ecosystem. *Ecol. Modelling*, 60 pp. 95-118, 1992.
Belyaev, V.I., and Konduforova, N.V., 1990. The mathematical modelling of shelf ecological system (in Russian). Naukova Dumka, Kiev, p. 240.
Belyaev, V.I., and Sovga, E.E., (Editors). 1993. Modelling the geochemical processes in marin coastal ecoton. Kiev, Naukova Dumka, p. 240 (in Russian).

**Other relevant information**
The model is very complicated and is not now adopted for use by arbitrary specialists. It might be that its adaptation will be possible in the future. Varying the block structure of the model we have the possibility to use it for modelling of any water ecosystem influenced by antropogenic factors. Authors have the possibility to carry out the modeling of any water ecosystem for diagnosis and prognosis of its states in the framework of the work made to order.

**Model identification**
Model name/title: Prediction of the effects of nutrient loading from the watershed on algal productivity and biomass accumulation.

**Model type**
Media: water, ocean, estuary
Class: biogeochemical
Substance(s)/compounds: nutrients, nitrogen
Biological components: ecosytem - shallow coastal lagoons; ocean embayments community - mixed algal; primary producers

**Model purpose**
This model explores the importance of resource competition to algal community structure of shallow coastal lagoons. The shorter-term objective was to test alternate mechanisms of resource competition among functional-forms of algae. Ultimately, we want to predict the effects of nutrient loading from the watershed on algal productivity and biomass accumulation.

**Short description**
We used modified Michaelis-Menten equations to describe the kinetics of nitrogen uptake for 3 co-occuring algal functional forms: phytoplankton, foliose algae, and cyanobacterial mats. Uptake rates and subsequent growth were modified by differential internal storage and recycling of nitrogen from decaying tissues. Analysis of model behavior indicated that a two-step model uncoupling nutrient uptake and growth was capable of producing a wide variety of dynamic behaviour not possible with simple uptake kinetics. A nutrient storage term allowed for time lags between nutrient input and growth, and resulted in periodic changes in dominance between foliose algae and mats. Analyses of the most important parameters of the model, nutrient uptake and growth rates, indicate that our model can be adapted for algal communities composed of different dominant algal forms. The model demonstrates a dynamic response in both magnitude and timing of algal growth when the parameter values for nutrient uptake are manipulated. From this we hypothesize that the dynamic relationships between the algal groups are similar in different communities, but that the parameter values in the Michaelis-Menten equations for each algal form must be tailored to each system.

**Model application(s) and State of development**
Application area(s):
Number of cases studied: composed to 3 experiments
State: conceptualization, verification, calibration, validation

**Model software and hardware demands**
Platform: PC-based MAC
OS: presently or a DOS machine in "Pro-dynamo" - am currently producing it in "STELLA" on MAC
Software demands:
Hardware demands:

**Model availability**
Purchase: no, copies sent to interested parties upon request. Software available as sourcecode
Contact person(s) Dr. Peggy Fong
                Pacific Estaurine Research Laboratory
                Department of Biology
                San Diego State University
                San Diego, CA 92182-0057 USA

Phone: 619-594-8714
Fax: 619-594-2035
E-mail: pfong@pcrl.sdsu.edu

**Model documentation and major references**

**Other relevant information**
I am currently producing a MacIntosh version with nutrient dynamics added, including watershed loading + sediment sources/sinks. This will be in STELLA.

**Model identification**
Model name/title: SSEM (a Shallow Sea Ecological Model)

**Model type**
Media: water, lake, ocean, estuary, river
Class: biogeochemical, toxicology
Substance(s)/compounds: nutrients - nitrogen, chemical pesticides, others - you can employ anything you want easily
Biological components: ecosystem - biomass based

**Model purpose**
SSEM is intended to be a modeling tool to predict the impact on fisheries caused by coastal development activities. It can handle many species of fish and their swimming, because each type of fish has a different value as a fishery resource and a different behaviour for the same impact.

---

**Short description**

SSEM is programmed using an object-oriented programming language. SSEM has three major classes, BOX, COMPONENTS and ECOMODEL. COMPONENTS represents the model component such as fish, plankton, nutrient, etc. BOX represents a water area that contains COMPONENTS in it. ECOMODEL represents a whole model that controls the time schedule of swimming, diffusion, and advection. SSEM describes an object water area as a connected set of BOXes. Physiological parameters for living COMPONENTS are maximum growth rate, half-saturation constant for food, faecal ratio, excretion ratio, respiration ratio, and death ratio. When you use the feature of movement by swimming, a set of preference parameters are also required. Forcing functions are water temperature, water current and nutrient load from land area. Not only those basic parameters but SSEM is intended to be able to handle special events in specific problems easily. For example, you may define pesticide as a member of COMPONENTS. In this case you just need to focus on the effect of pesticide. Other features such as growth or swimming are already defined and you do not need to think about them. In this sense, SSEM is not a completed model but a set of tools to model various areas of water.

---

**Model application(s) and State of development**

Application area(s): Shijiki Bay, Nagasaki, JAPAN (Change of behaviour of sea bream population caused by its growth) Oumi Bay, Yamaguchi, JAPAN (Change of catch of little neck clam caused by reclamation). Ajisu Channel, Yamaguchi, JAPAN (Mullet mortality caused by a pesticide). Indoor experiment on physiological parameters using goldfish and Chroococcus. Indoor experiment on preference parameters using shiner.

Number of cases studied: 5

State: verification, calibration

---

**Model software and hardware demands**

Platform: PC-based MAC, IBM, NEC PC9800 Series. Work-station SUN, SONY NEWS, etc.

OS: UNIX, PC-DOS, MS-DOS, Windows (Any system which can execute Objectworks/Smalltalk or Smalltalk-80 or Smalltalk/V)

Software demands: Objectworks/Smalltalk or Smalltalk-80 or Smalltalk/V, Smalltalk

Hardware demands: Any system which can execute Objectworks/smalltalk or Smalltalk-80 or Smalltalk/V

---

**Model availability**

Purchase: no, available model through anonymous FTP from env.civil.yamaguchi u.ac.jp(1333.62.144.2)

Contact person(s)  Dr. Masahiko Sekine
              Department of Civil Engineering
              Yamaguchi University
              Ube Yamaguchi 755
              Japan

Phone: 81-836-31-5100 (ext. 3613)
Fax: 81-836-35-9429
E-mail: ms@env.civil.yamaguchi-u.ac.jp

## Model documentation and major references

Sekine, M., Nakanishi, H., Ukita, M. and S. Murakami. A shallow-sea ecological model using an object-oriented programming language. *Ecol. Modelling*, 57 (1991), 221-236.

Sekine, M., Nakanishi, H., and Ukita, M., A shallow sea ecological model to assess the impact of coastal development. Extended Abstracts of 4th International Conference on Computing in Civil and Building Engineering, (1991) 189.

Sekine, M., Nakanishi, H. and Ukita, M.. Study on fish mortality caused by the combined effects of pesticide and changes in environmental conditions. International Congress on Modelling and Simulation Proceedings, Vol. 4 (1993), pp. 1693-1698.

## Other relevant information

SSEM is intended to be applicable to various situations from very simple to very complicated. However, while SSEM has already been tested in rather simple situations, it is very difficult to test it in complicated ecosystems because it requires a wide range of knowledge of physics, biology, chemistry, and/or ecology. Developers of the model hope that SSEM would be tried in various situations and the results be feed-backed. It would be pleasurable for the developers to cooperate with researchers who want to try SSEM.

## Model identification

Model name/title: TRIM-3D

## Model type

Media: water, lake, ocean, estuary, local scale, regional scale, global scale
Class: hydrology
Substance(s)/compounds: nutrients, temperature
Biological components:

## Model purpose

3-D Model for Accurate Flow Simulation in the Lagoon of Venice.

## Short description

The Lagoon of Venice (50 km²) consists of several inter-connected narrow channels with a maximum width of 1 km, and up to 50 m deep encircling large and flat shallow areas. A considerable portion of the water body consists of tidal flats, and proper numerical treatment of flooding and drying of these areas are essential. The model uses a semi-implicit finite difference formulation for the numerical solution of the three-dimensional Reynolds equations in which the pressure is assumed to by hydrostatic. A minimal degree of implicitness has been introduced in the finite-difference formula so that the resulting algorithm permits the use of large time steps at a minimal computational cost. This formulation includes the simulation of flooding and drying of tidal flats, and is fully vectorizable for an efficient implementation on modern vector computers. The Lagoon has been covered with a 642 by 827 by 200 finite-difference mesh of dx=dy=50 m and with the maximum dz being 0.25 m. Thus, the total number of grid points is 108,754,800, but only 1,177,729 of these are active.

This fine computational mesh allows for a very accurate description of the tree-like structure of the main channels. The high computational efficiency of this method has made it possible to provide the fine details of circulation structure that previous studies were unable to obtain.

## Model application(s) and State of development
Application area(s):
Number of cases studied:
State: validation

## Model software and hardware demands
Platform: PC-based, Mainframe, Workstation, Supercomputers
OS: DOS- UNIX
Software demands: language - English
Hardware demands:

## Model availability
Purchase: yes, price to be discussed. Software available as executable code
Contact person(s) Dr. Vincenzo Casulli and Dr. Enrico Bertolazzi
        Dept. Civil and Env. Engineering
        Mesiano
        Trento, Italy

Phone: +49 461 882676
Fax: +49 461 882672
E-mail: casulli@itnvax.science.unitn.it

## Model documentation and major references
Casulli, V., 1990. Semi-implicit finite difference methods for the two-dimensional shallow water equations. *J. of Computational Physics*, Vol. 86, No. 1, pp. 56-74, 1990.
Casulli, V., and Cattani, E., 1994. Stability, accuracy and efficiency of a semi-implicit method for three-dimensional shallow water flow, *Computers & Mathematics with Applications*, Vol. 27, No. 4, pp. 99-112, 1994.
Casulli, V., and Cheng, R.T., 1992. Semi-implicit finite difference methods for three-dimensional shallow water flow. *Int. J. Numerical Methods in Fluids*, Vol. 15, pp. 629-648, 1992.
Cheng, R.T., Casulli, V., and Gartner, J.W., Tidal, Residual, Inter-tidal Mud-flat (TRIM) model with applications to San Francisco Bay. *Estuarine, Coastal Shelf Science*, Vol.36, pp. 235-280.

## Other relevant information

---

## Model identification
Model name/title: A biomass-based model for the sand lance in Seto Inland Sea, Japan

## Model type
Media: ocean, regional scale
Class: biogeochemical
Substance(s)/compounds: biomass, organic components
Biological components: ecosystem - biomass-based, prey-predator

## Model purpose
To study important biological parameters for stock fluctuation and the role of young sand lance, zooplankton, aestivation of sand lance.

## Short description
State variables: egg, larvae, young, one-year-old adult, more than one-year-old adult of sand lance and copepoda.
Forcing functions: catch data of sand lance.
Important parameters: initial state of zooplankton, grazin rate.
Necessary inputs: selection constants for sand lance.

## Model application(s) and State of development
Application area(s): Seto Inland Sea, Japan
Number of cases studied: 30
State: verification

## Model software and hardware demands
Platform: Mainframe IBM, Workstation IBM
OS:
Software demands: GDDM, FORTRAN 77
Hardware demands:

## Model availability
Purchase: case by case (can be distributed), as source code
Contact person(s) Dr. Michio J. Kishi
                Ocean Research Institute
                University of Tokyo
                Minamidai 1-15-1, Nakano-ku
                Tokyo, 113 Japan

Phone: 81-3-5351-6507
Fax: 81-3-5351-6506
E-mail: kishi@ori.u-tokyo.ac.jp

## Model documentation and major references
*Ecol. Modelling*, 54 (1991): 247-263

## Other relevant information

## Model identification
Model name/title: Numerical simulation model for quantitative management of marine culture

## Model type
Media: ocean, regional scale
Class: biogeochemical
Substance(s)/compounds: biomass, organic compound
Biological components: ecosystem - (carbon cycle, prey-predator)

## Model purpose
To assess the influence of the location or the area of mariculture rafts on the ecological and/or environmental system. To obtain a better distribution of rafts in bay area.

## Short description
State variables: COD, DO, accumulated matter
Forcing function: precipitation, biomass of fish, shellfish in mariculture rafts, amount of food for fish in mariculture, three-dimensional ocean current.
Important parameters: Consumption of oxygen by fish
Necessary input: three-dimensionally calculated ocean currents

## Model application(s) and State of development
Application area(s): Mikame Bay, Kusu-ura Bay, Nanao-sei Bay, (Japan)
Number of cases studied: 8
State: verification

## Model software and hardware demands
Platform: Mainframe - HITACHI, Workstation - SUN
OS:
Software demands: 10 Mb of memory, language: FORTRAN 77
Hardware demands:

## Model availability
Purchase: No (if you can get admittance. Software available as sourcecode.
Contact person(s) Dr. Michio J. Kishi
                  Ocean Research Institute, University of Tokyo,
                  Minamidai 1-15-1, Nakano-ku
                  Tokyo, 113 Japan

Phone: +81 3 5351 6507
Fax: + 81 3 5351 6506
E-mail: kishi@ori.u-tokyo.ac.jp

## Model documentation and major references
*Ecol. Modelling*, 66, (1993).

**Other relevant information**
The copyright of the model belongs to the Fisheries Agency of Japan, if you can get permission of the Fisheries Agency of Japan, the program is available to be obtained for free.

---

**Model identification**
Model name/title: POPCYCLE: acronym for Population Cycle.

---

**Model type**
Media: water, developed for open ocean, but suitable for other pelagic, aquatic populations as well.
Class:
Substance(s)/compounds: used carbon as currency
Biological components: population, *Metridia pacifica* (a pelagic copepod) organism, *Metridia pacifica*

---

**Model purpose**
The model was developed to describe the numerical abundance and phenology (life history timing) of a pelagic copepod, *Metridia pacifica*, in the Subarctic Pacific Ocean.

---

**Short description**
POPCYCLE is an individual based model (IBM) in which growth, development, reproduction and mortality of individual organisms are tracked. The model as developed operates in one dimension (depth in the vertical), but can be readily extended to two or three dimensions (by embedding it within a model of three-dimensional transports and adding additional state variables representing position). Equations describing growth, mortality, and reproduction of individuals are formulated and used to describe and generate a population dynamics history of the population for year-long model runs. Time step is one day. Growth is described using carbon-based, input-output model. Mortality is implemented as a constant daily probability of predation, but is allowed to vary with development. Reproductive parameters of clutch size, clutch frequency, and total number are specified based on literature or experimentally determined values. Growth in the published version of the model is assumed to be food-limited. Subsequently, an influence of temperature has been added to the model. The principal forcing functions are food resources available (each day and by depth) and temperature. In the published version, state variables tracked for each individual are its birth date, current weight, life stage, and number of days as reproducing adult. Subsequently, other state variables, such as degree of recent hunger/satiation have been added to keep track of feeding history. Other state variables can be easily added, as appropiate, to the model. Initial conditions that need to be specified for the one-dimensional version are the numbers and weights of individuals present on day one of the model run. Because individual organisms are tracked and potentially the number of organisms can exceed 100 thousand with some parameter selections, the model uses a factor of five scalling to reduce population size whenever it gets so large that run times exceed several hours (on an IBM-PC).

---

**Model application(s) and State of development**
Application area(s):
Number of cases studied: 2. The model was initially developed for the copepod, *Metridia pacifica*, in the Pacific (as published -- see reference below). Subsequently, the model has been applied to *Metridia lucens* (sibling species) in the North Atlantic with the goal of evaluating the spatio-temporal distribution of zooplankton-derived bioluminescence (unpublished).
State:

---

**Model software and hardware demands**
Platform: The model operates on an IBM-PC with 640 K and at least 5 Mb of free space on a hard disk. A custom written disk paging routine is used to implement "virtual memory" using hard disk space (can also use a ram disk if available).
OS: PC-DOS 4.0 or later. The code is written in Borland Turbo Pascal (vers. 5.0 or later). Results are saved to files which are imported into spreadsheets for display/graphing/analysis.
Software demands:
Hardware demands:

---

**Model availability**
Purchase: Model is not available commercially. Sourcecode can be made available but is not very well documented.
Contact person(s) Dr. Harold P. Batchelder
                    Division of Environmental Studies
                    University of California
                    Davis, CA 95616-8576
                    USA
Phone: 916-752-8576
Fax: 916-752-3350
E-mail: hpbatchelder(tegn)ucdavis.edu

---

**Model documentation and major references**
Batchelder, H.P. and Miller, C.B., 1989. Life history and population dynamics of Metridia pacifica: results from simulation modelling. *Ecol. Modelling*, 48: 113-136.

---

**Other relevant information**

## 2.8 Overview of Models of Wetlands

Models of wetlands have increased in importance since the early 1980s, where discussion on the role of non-point pollution started. Since then a wide spectrum of wetlands models have been developed.

An overview of the models presented in Section 2.8 is given in Table 2.5. The model Methane.sim., focuses on the methane production from wetlands. The state variables encompass three types of microorganisms and three chemical components in addition to oxygen, which, of course, determines the shift between aerobic and anaerobic conditions.

Several models have been developed to account for the nutrient retention of wetlands: Old Woman Creek Phosphorus Model and Silow's model for purification of waste water by means of plants. The model developed by Logofet and Alexandrov, named Bog, is able to simulate the nitrogen retention as function of the development of a bog.

Jørgensen et al. (1988) have developed a very complex model accounting for four layers, water, nitrogen and phosphorus, but the model is difficult to use in practice due to its demand for many and detailed data. Therefore, it can be recommended in most situations, where the amount of data is more limited, to use the model named wetlands for the assessment of the nitrogen removal capacity of wetlands.

Duever has developed a hydrological model for wetlands. As the hydrology is very important also for the nutrient removal capacity, this model is not only significant for modeling the hydrology of a wetland, but will probably be used widely to assess the nutrient removal in a wetland with a non-simple hydrology.

Feuwa is another hydrological model based upon the combined water balance of saturated and unsaturated soil, for the water exchange with the atmosphere and the ground water.

A model named Chinese Fish Pond is able to assess the importance of the various components of an aquaculture system and quantify the performance possibilities of such systems. As many as six different carp species are included in this model, which can be used mainly to manage fishponds from an aquaculture point of view.

Reyes et al. have developed a model for simulation of aquatic primary production and fish migration in a lagoon. The model is a combination of two models, Lappter, describing the spatial hydrodynamics and Roeter, aimed for a description of the regional organisms exchange. The model is able to give the spatial distribution of primary producers, zooplankton and fish.

**Table 2.5** Overview of Wetland Models

| Model name/ Model comp. | Type of wetland | Characteristics | Reference |
|---|---|---|---|
| Methane.sim | All | Methane, sulfate, oxygen, 3 bacteria | Thomas, 1993 |
| Old Woman Creek | All | Nutrient retention: phosphorus resuspension important | Mitsch and Reeder, 1991 |
| Waste water | All | Aquatic plant / waste water | Silow et al., 1992 |
| Bog | Bog | St. var. N, C, trees, shrubs, mosses and peat | Logofet and Alexandrov, 1988 |
| Hydrology | All | Water balances | Duever, 1988 |
| Feuwa | All | Water exchange processes | See model, 2.9 |
| Chinese Fish Pond | Fish pond | P, Si, O, 2 phyt., zoopl., 5 fish species | Hiroyuki and Mitsch, 1993 |
| Laguna de Terminos, Mexico | Lagoons | 16 cells, 9 st. variables. Focus: primary prod. + fish migration | Reyes, 1994 |
| SMP | All | spatial dynamics | Sklar, 1994 |
| Space/Time | Fens | 3 levels, 15 state variables, hydrology, spatial vegetation | Bekker, 1994 |
| Aquatic bed marsh | Swamps | 23 st. variables, use of unfolding methodology | Whippler and Patten, 1994 |
| Aqua culture | Fish ponds | Fish growth, ammonia, oxygen | Jørgensen, 1976 |

Sklar et al., have developed a model, SMP, for simulation of the spatial distribution of various wetland processes to evaluate the the impacts on coastal land-use changes. The model results are the changes in the landscape due to change in salinity and hydrology, caused by a change in precipitation and as indicated above in land use.

Bekker (1994) has developed a spatial dynamic model, Space/Time model, for a fen ecosystem. The model combines dynamic modeling formalism, the automata theorem and GIS; (the latter to give the spatial distribution). The model is built into three levels. The first level is a traditional dynamic model for nutrient and carbon cycles. The second level describes the hydrology of the fen system as a whole. Interactions among cells are expressed by simple and deterministic cellular automata. This level sends the information on nutrient status and accumulated organic matter to the third level, which determines the vegetation type.

Whipple and Patten have developed a model, which consider the primary production-based and the detritus-based food pathways in a complex energy flow network. The model is static, but various static situation can be modeled in accordance with change in forcing functions. The model uses Patten's unfolding energy-partitioning methodology. The Aquaculture model is useful for management of artificial fish ponds. It focuses on the oxygen concentration, the ammonia, pH, the growth of fish, the number of fish and the required and optimal feeding of fish.

## 2.9 Models of Wetlands

### Model identification
Model name/title: Methane.sim

### Model type
Media: water, wetland, regional scale, sediment
Class: biogeochemical
Substance(s)/compounds: nutrients, carbon, sulfur, nitrogen, oxygen, organic compounds
Biological components: community - microbial population - methanogens, sulfur bacteria, methane oxidizers

### Model purpose
Methane production from wetlands is highly variable. To account for this variability, this model was used to determine factors having the greatest impact on methane flux from the Florida Everglades.

### Short description
State variables: acetate, sulfate, methane bacteria, sulfate bacteria, methane, methane oxidizers, oxygen.
Forcing Functions: temperature, algal mat photosynthesis, acetate production, sulfate input.
Important Parameters: Diffusion rates of: acetate, sulfate, methane.
Non-Comsumptive Losses: acetate, sulfate, methanogens, sulfate bacteria, methane, methane oxidizing bacteria.
Maximum Uptake: acetate by methanogens, acetate by sulfate bacteria, sulfate by sulfate bacteria, methane by methane oxidizers
Half-Saturation Constants: acetate by methanogens, acetate by sulfate bacteria, sulfate by sulfate bacteria, methane by plants, methane by methane oxidizers, oxygen tolerance for methanogens, oxygen tolerance for sulfate bacteria
Yield Constants: methanogens, sulfate bacteria, methane oxidizers, methane flux, flux from plants
Liquid phase exchange coefficients: methane, oxygen
Tolerances for methane oxidizers: oxygen maximum (total inhibition), oxygen minimum (total inhibition), oxygen optimum

## Model application(s) and State of development
Application area(s): specific model for methane production in wetlands
Number of cases studie: 1
State: conceptualization, verification

## Model software and hardware demands
Platform: PC-based IBM, UNIX-based SUN Workstation
OS:
Software demands: none, PASCAL. Hardware demands: at least a 286 PC with a VGB monitor, (unix version no graphics capability)

## Model availability
Purchase: Source code from author
Contact person(s) Dr. R. Thomas James
        Department of Research
        South Florida Water Management District
        P.O. BOX 24680
        3301 Gun Club Road
        West Palm Beach, FL 33416-4680 USA

Phone: +1 407-687-6356
Fax: +1 407-687-6442
E-mail: (internet) tom.james@sfwmd.gov

## Model documentation and major references
James, R.T., 1993. Sensitivity analysis of a simulation model of methane flux from the Florida Everglades. *Ecol. Modelling*, 68: 119-146.

## Other relevant information


## Model identification
Model name/title: Old Woman Creek Phosphorus Model: nutrient retention of a freshwater coastal wetland

## Model type
Media: water, lake, wetland
Class: biogeochemical, hydrology
Substance(s)/compounds: phosphorus
Biological components: ecosystem - wetland and lake

## Model purpose
Model was developed to determine fate and retention of phosphorus in a coastal freshwater wetland receiving agricultural runoff.

## Short description
Development of a simulation model for a wetland along Lake Erie. To determine fate and retention of phosphorus. Submodels for hydrology, productivity, and phosphorous and a simulated barrier beach were incorporated. Re-suspension was incorporated to help predict p-concentrations in the wetland water-column. Model predicted high retention with high in-flow and lake levels.
State variables: phosphorous retention/fate in wetland to lake
Forcing functions: runoff (flow), lake levels
Parameter and necessary inputs: re-suspension of sediments.

## Model application(s) and State of development
Application area(s): wetland, pretention
Number of cases studied: 6
State: calibration

## Model software and hardware demands
Platform: MAC
OS:
Software demands: STELLA, language STELLA
Hardware demands:

## Model availability
Purchase: yes, $ 50 donation to OSW wetland project, available as source code
Contact person(s) Dr. William J. Mitsch
        School of Natural Resources
        Ohio State University
        2021 Coffey Road
        Columbus, OH 43210 USA

Phone: +1 614-292-9774
Fax: +1 614-292-9773
E-mail:

## Model documentation and major references
Mitsch, W., and Reeder, B., Modelling nutrient retention of a freshwater coastal wetland: estimating the roles of primary productivity, sedimentation, re-suspension and hydrology. *Ecol. Modelling*, 55: 151-187.

## Other relevant information

## Model identification
Model name/title: The Model of Waste Waters Purification with the use of Aquatic Plants.

## Model type
Media: water, local scale, agricultural
Class: biogeochemical
Substance(s)/compounds: nutrients, phosphorus, nitrogen, chemicals, organic compounds
Biological components: populations of aquatic plants

## Model purpose
The model was created to optimize the processes of waste waters (of cattle- and fish-breeding complexes) purification with the use of aquatic plants.

## Short description
The following model describing waste purification and the growth of phytomass:

$$dB_i/dt = B_i * \{ D_{ij} * C_i + Q_i * f \}$$

$$dC_j/dt = C_j * \{ K_j * f + H_{ij} * B_i * f + A_{ij} * B_i * p \}$$

where:
   B = the vector of macrophyte biomasses (i species),
   C = the vector of pollutants concentration (j compounds),
   Q = the vector of hydrophytes growth rate dependence from the temperature,
   K = the vector of abiotic elimination of pollutants,
   D = the matrix of dependence of plants growth rate from the concentration of pollutants,
   H = the matrix of dependence of pollutants elimination from the density of plants and temperature,
   A = the natrix of dependence of pollutants elimination from the density of plants and pH,
   f = temperature,  p - pH.
The model is created on the basis of laboratory and semi- field experiments.

## Model application(s) and State of development
Application area(s): The optimization of waste water purification.
Number of cases studied:
State:

## Model software and hardware demands
Platform: PC-based, IBM
OS: MS-DOS
Software demands: C++
Hardware demands:

**Model availability**
Purchase: yes, executable code or source code
Contact person(s) Dr. Eugene A. Silow
        Scientific Research Institute of Biology
        P.O. Box 24
        Irkutsk-3
        664003 Russia

Phone: (3952) 33-44-79
Fax: (3952) 24-30-79
E-mail:

**Model documentation and major references**
Silow, E.A. and Stom, D.J., 1992. Model ecosystems and models of ecosystems in hydrobiology. Irkutsk. University press. (In Russian).

**Other relevant information**

**Model identification**
Model name/title: BOG

**Model type**
Media: water, wetland
Class: biogeochemical
Substance(s)/compounds: nutrients, nitrogen, water
Biological components:

**Model purpose**
Development of a bog and its ability to remove nutrients, mainly nitrogen.

**Short description**
State variables: N and C of trees, shrubs mosses and peat.

**Model application(s) and State of development**
Application area(s):
Number of cases studied: 1
State: conceptualization, verification, calibration

**Model software and hardware demands**
Platform: PC-based - IBM, Mainframe
OS:
Software demands:
Hardware demands:

**Model availability**
Purchase:
Contact person(s) Dr. D. O. Logofet
          Computation Centre
          Russian Academy of Sciences
          40 Vavilov Street
          Moscow 117 333 Russia

**Model documentation and major references**
Logofet, D.O., and Alexandrov, G.A., Interferences between mosses and trees in the framework of a dynamic model of carbon and nitrogen cycling in a mesotrophic bog ecosystem. In: Mitsh, W.J., Swashraba, M., and Jørgensen, S.E. (Eds): *Wetland Modelling*, Elsevier, Amsterdam, 1988.

**Other relevant information**

---

**Model identification**
Model name/title: INLET

**Model type**
Media: water, wetland
Class: biogeochemical
Substance(s)/compounds: water only
Biological components: ecosystem

**Model purpose**
Model the hydrology of wetlands: output a hydrographs as function of time.

**Short description**
State variables: atmospheric moisture, water in vegetation, surface water, water in unsaturated zone, water in saturated zone and groundwater.
Forcing functions: temperature, precipitation, water, in- and outflows, evapotranspiration.

**Model application(s) and State of development**
Application area(s):
Number of cases studied: several
State: conceptualization, verification, calibration, validation

**Model software and hardware demands**
Platform: PC-based - IBM, MAC
OS:
Software demands:
Hardware demands:

**Model availability**
Purchase:
Contact person(s) Dr. M. J. Duever
                Ecosystem Research Unit
                National Audubon Society
                Box 1877, Route 6
                Naples, FL 33999 USA

**Model documentation and major references**
Duever, M.J., 1988. Hydrological processes in models of freshwater wetlands. In: W.J. Mitsch, M. Shrashuba, and S.E. Jørgensen (Eds.), *Wetland Modelling*. Elsevier, Amsterdam, 1988.

**Other relevant information**

**Model identification**
Model name/title: Chinese Fish Pond: Black carp - common carp pond

**Model type**
Media: water
Class: biogeochemical
Substance(s)/compounds: phosphorus, oxygen, silicon
Biological components: ecosystem - fish pond

**Model purpose**
The purpose of this model is to assess the importance of the various components of a chinese fish aquaculture system and quantify the performance possibilities of such systems.

**Short description**
The model includes sunlight and temperature forcing functions and phosphorus, silicon, and oxygen chemistry. These abiotic components affect producers (diatom and non-diatom phytoplankton) and consumers (zooplankton and fish). The aquacultural inputs of fish stocking and feeding with grass and fine feeds is incorporated. The fish species are those used in a Suzhou pond:   black carp (*Mylopharyngodon piceus*)
                            common carp (*Cyprinus carpio*)
                            Wuchang fish (*Megalobrama amblycephala*)

silver carp (*Hypophtalmichtys molitrix*)
grass carp (*Ctenopharyngodon idella*)
bighead carp (*Aristichthys nobilis*).

## Model application(s) and State of development
Application area(s): Fish aquaculture
Number of cases studied:
State: calibration

## Model software and hardware demands
Platform: MAC
OS:
Software demands: language STELLA II v.1.0.2

## Model availability
Purchase: no
Contact person(s) Dr. William J. Mitsch
School of Natural Resorces
Ohio State University
2021 Coffey Road
Columbus, OH 43210 USA

Phone: 614-292-9774
Fax: 614-292-9773
E-mail:

## Model documentation and major references
Hiroyuki, H. and Mitsch, W., Ecosystem modelling of a multi-species integrated aquaculture pond in South China. *Ecol. Modelling*.

## Other relevant information

## Model identification
Model name/title: FEUWA: FEUchtgebiet-WAsserstand or Wetland-Waterlevel

## Model type
Media: water, local scale, wetland
Class: hydrology
Substance(s)/compounds:
Biological components:

## Model purpose
Simulation of the water level, of the water balance and of the ground water exchange for mostly unflooded wetlands between terrestrial and aquatic ecosystems. Completion of measured data sets by simulation of short-term water level dynamics. Submodel for subsurface mass exchange across riparian ecotones.

## Short description
The box model FEUWA is based on the combined water balance for the saturated and the unsaturated soil zones, for the water exchange with the atmosphere and for the lateral water exchange between ground water and surface waters (lakes, streams, etc.). The lateral water flux is controlled by hydraulic resistances. The water volume stored in the balanced system consists of the water content in the saturated and the unsaturated soil zones, of the interception storage and of the snow storage. With the relation between the matrix potential and the water content, the volume of water in the unsaturated zone depends only on the water level (H(t)) in the box as a quasi-stationary equilibrium between the saturated and the unsaturated soil zones.
State variables: water level H(t).
Forcing functions: groundwater level of the adjacent aquifer, water level of the adjacent lake or stream, precipitation, evapotranspiration, air temperature.
Important model parameters: soil water characteristics, elevation head of the soil surface, lateral hydraulic resistance to the adjacent aquifer, lateral hydraulic resistance to the adjacent surface water system, head of flooding, leaf area index of vegetation.

## Model application(s) and State of development
Application area(s): ripirian zone management, research tool for groundwater/surface ecotones
Number of cases studied: 2
State: verification, calibration

## Model software and hardware demands
Platform: PC-based - IBM, Digital
OS: MS- DOS, VMS
Software demands: language: SYSL/FORTRAN
Hardware demands:

## Model availability
Purchase: yes, software available as source code.
Contact person(s) Dr. Winfrid Kluge or Dr. Lars Theesen
  Ecosystem Research Center
  University of Kiel
  Schauenburgerstr. 112
  D-24118 Kiel
  Germany

Phone: + 49 0 431 880 4034
Fax: +49 0 431 880 4083
E-mail: winfrid@pz.oekosys.uni-kiel.d.400.de   or
  lars@pz-oekosys.uni-kiel.d.400.de

## Model documentation and major references
Kluge, W., Mueller-Buschbaum, P., and Theesen, L., 1994. Parameter aquisition for modelling exchange processes between terrestrial and aquatic ecosystems. *Ecol. Modelling*, 75/76, 399-408.
Kluge, W., and Theseen, L., (in preparation, 1994). Wasseraustausch- und Wasserhaushaltmodelle für Uferoekotone. Ecosystem, Kiel.

## Other relevant information
E2 consulting, 1985. SYSL. System Simulating Language. User's Guide. E2 consulting Poway.

## Model identification
Model name/title: Lappter and Roeter

## Model type
Media: water, wetland
Class:
Substance(s)/compounds: nutrients
Biological components: ecosystem, organism

## Model purpose
To assess the spatial distribution of primary production and fish migration.

## Short description
State variables: mangrove plants, detritus, nitrogen in water, nitrogen in soil, phytoplankton, seagrass, benthos, zooplankton and fish. 16 equally-sized interacting cells represent the wetland. Light, secchi disk transparency, temperature and sensitivity are forcing functions for each cell. The results are the spatial distribution of the state variables.

## Model application(s) and State of development
Application area(s):
Number of cases studied: 1
State: conceptualization, verification, calibration, validation

## Model software and hardware demands
Platform: PC-based - IBM, MAC
OS:
Software demands:
Hardware demands:

## Model availability
Purchase: no
Contact person(s) Dr. E. Reyes
               Department of Oceanography and Coastal Sciences
               Louisiana State University
               Baton Rouge, LA, USA

Phone:
Fax:
E-mail:

## Model documentation and major references
Ecosystem models of aquatic primary production and fish migration in Laguna de Terminos, Mexico. In: *Global Wetlands, Old World and New*, W.J. Mitsh (Ed.), pp. 519-536. Elsevier, Amsterdam, 1994.

## Other relevant information

## Model identification
Model name/title: SMP

## Model type
Media: water, wetland
Class:
Substance(s)/compounds: nutrients, nitrogen
Biological components: ecosystem, organism - primary producers

## Model purpose
Evaluate environmental regulations in relation to site-specific management options, global change or natural processes.

## Short description
State variable: inorganic nitrogen, detrital community, detritus, various primary producers, water level, salinity.
Forcing functions: temperature, water flows (including tides), precipitation, dominage. The model gives the spatial distribution of the habitat types, water level, salinity and other state variables. The model outputs are maps with colours indicating various types of habitats, water levels, etc.

## Model application(s) and State of development
Application area(s):
Number of cases studied: 1
State: conceptualization, verification, calibration, validation

**Model software and hardware demands**
Platform: PC-based - MAC, IBM
OS:
Software demands:
Hardware demands:

---

**Model availability**
Purchase: no.
Contact person(s) Dr. F.H. Sklan
              Everglades Systems Research Division
              South Florida Water Management District
              3301 Gun Club Road
              West Palm Beach, FL 33416, USA
Phone:
Fax:
E-mail:

---

**Model documentation and major references**
Spatial explicit and implicit dynamic simulations of wetland processes. In: *Global Wetlands, Old and New*, W.J. Mitsh (Ed.), pp. 537-554. Elsevier, Amsterdam, 1994.

---

**Other relevant information**

---

**Model identification**
Model name/title: Space/Time Model  (Fen Model)

---

**Model type**
Media: water, wetland
Class:
Substance(s)/compounds: nutrients, carbon, phosphorus, nitrogen
Biological components: ecosystem, organism - primary producers

---

**Model purpose**
To describe heterogeneous landscape development.

---

**Short description**
Level 1: 15 state variables accounting for carbon, phosphorus and nitrogen cycles, detritus and primary producers.
Level 2: hydrology of the fen system. Interactions between cells are described by means of cellular automata.
Level 3: determines the vegetation type based upon nutrients and organic matter accumulation.

## Model application(s) and State of development
Application area(s):
Number of cases studied: 1
State: conceptualization, verification, calibration, validation, prognosis made, validation of prognosis

## Model software and hardware demands
Platform: PC-based - MAC, IBM
OS:
Software demands:
Hardware demands:

## Model availability
Purchase: no
Contact person(s) Dr. S.A. Bekker
  Department of Plant Ecology and Evolutionary Biology
  Utrecht University
  Sorbonnelaan 16
  NL-3584 CA Utrecht
  The Netherlands

Phone:
Fax:
E-mail:

## Model documentation and major references
Spatial and dynamic modelling: describing the terrastrialization of fen ecosystems. In: *Global Wetland, Old World and New*, W.J. Mitsh (Ed.), pp. 555-562. Elsevier, Amsterdam, 1994.

## Other relevant information

## Model identification
Model name/title: Aquatic bed marsh

## Model type
Media: water, wetland
Class:
Substance(s)/compounds: nutrients
Biological components: ecosystem, organism - decomposers, producers, microinvertebrates, macroinvertebrates and vertebrates.

## Model purpose
Model the dual food web structure in a wetland.

**Short description**
23 state variables: 2 organic matter (peat and non-peat detritus), nutrients, 3 decomposers, 3 producers (peat-forming macrophytes, non-peat forming macrophytes, algae), 3 microinvertebrates, 6 macroinvertebrates, 6 vertebrates.
The model uses unfolding methodology with flow partitioning in steady state.
Forcing functions: (inflows + precipitation - outflows) determine the resulting steady state.

**Model application(s) and State of development**
Application area(s):
Number of cases studied: 1
State: conceptualization, verification, calibration, validation

**Model software and hardware demands**
Platform: PC-based - MAC, IBM
OS:
Software demands:
Hardware demands:

**Model availability**
Purchase: no.
Contact person(s) Dr. S.J. Whippler and Dr. B.C. Patten
                 Institute of Ecology
                 Georgia University
                 Athens, GA 30602, USA

Phone:
Fax:
E-mail:

**Model documentation and major references**
The complex trophic structure of an aquatic bed marsh ecosystem in Okefenokee Swarp, USA. In: *Global Wetlands, Old World and New*, W.J. Mitsh (Ed.), pp. 593-611. Elsevier, Amsterdam, 1994.

**Other relevant information**

**Model identification**
Model name/title: Aquaculture Model

**Model type**
Media: water
Class: biogeochemical
Substance(s)/compounds:
Biological components:

142

## Model purpose
Management of a fishpond (fish growth, oxygen and ammonia contamination).

## Short description
Forcing functions: feeding, addition of oxygen, recirculation and treatment of water.
State variables: weight of one fish, number of fish, ammonia, BOD and oxygen concentrations in water.
Important parameters: excretion rates of ammonia by fish, growth rate of fish as f(weight).

## Model application(s) and State of development
Application area(s):
Number of cases studied: 2
State: 1 validation, 1 validation of prognosis

## Model software and hardware demands
Platform: PC-based
OS:
Software demands: language SYSL
Hardware demands:

## Model availability
Purchase: no
Contact person(s) Dr. S.E. Jørgensen
　　　　　　　　Sektion for Miljøkemi
　　　　　　　　Danmarks Farmaceutiske Højskole
　　　　　　　　Universitetsparken 2
　　　　　　　　DK-2100 København Ø
　　　　　　　　Denmark

Phone: +45 35 37 08 50
Fax: +45 35 37 57 44
E-mail: sej@pommes.dfh.dk

## Model documentation and major references
Jørgensen, S.E., 1976. A model of fish growth. *J. Ecol. Model.*, 2: 303-313.

## Other relevant information

## Model identification
Model name/title: Wetland Model

## Model type
Media: water, wetland
Class: biogeochemical
Substance(s)/compounds: nutrients, phosphorus, nitrogen
Biological components: ecosystem

## Model purpose
Determination of a wetland's capacity for removal of N and P.

## Short description
State variables: N (+ in some cases also P) as 1) inorganic compounds in surface water and soil water, both ammonium and nitrate, 2) detritus, 3) plant. Water amount [$m^3/m^2$] as surface water and soil water. Totally 8 (only N) or 12 (also P).
Forcing functions: inputs of N (and P), climate.
Important variables: rate of nitrification and denitrification, permeability of soil, mineralization rate, uptake rate of N (and P) by plants.

## Model application(s) and State of development
Application area(s):
Number of cases studied: 4
State: 4 validation

## Model software and hardware demands
Platform: PC-based MAC
OS:
Software demands: language SYSL and STELLA
Hardware demands:

## Model availability
Purchase: no
Contact person(s) Dr. S.E. Jørgensen
                       Sektion for Miljøkemi
                       Danmarks Farmaceutiske Højskole
                       Universitetsparken 2
                       DK-2100 København Ø
                       Denmark

Phone: +45 35 37 08 50
Fax: +45 35 37 57 44
E-mail: sej@pommes.dfh.dk

## Model documentation and major references
Jørgensen, S.E., Hoffmann, C.C., and Mitsch, W.J., 1988. Modelling nutrient retention by a reed-swamp and wet meadow in Denmark. In: W.J. Mitsch, M. Straskraba and S.E. Jørgensen (Eds.), *Wetland Modelling*, Elsevier, Amsterdam, pp. 133-151.

## Other relevant information

## Model identification
Model name/title: SIMO-NEW: Simulation MOdel for Nutrients in European Wetlands

## Model type
Media: wetland, local scales
Class: biogeochemcial
Substance(s)/compounds: nutrients, carbon, phosphorus, nitrogen
Biological components: ecosystem - river marginal wetland

## Model purpose
With the model the consequences of certain human impacts for wetland functioning can be investigated. These functions include wetlands acting as sink, source of buffer for nutrients and the wetland's food chain support function.

## Short description
The model consists of submodels of C, N and P dynamics, that are connected. Spatial aspects are implemented by defining separate models for the various hydro-geomorphic units within the wetland site and by connecting these through nutrient and carbon flows. In this way the 'unit models' together form a 'site model'.

A unit model consists of 27 state variables, of which 7 are in the Carbon submodel, 10 are in the Nitrogen submodel and 10 are in the Phosphorus submodel. State variables in the Carbon submodel are expressed as $gC/m^2$, state variables in the Nitrogen submodel as $gN/m^2$ and in the Phosphorus submodel as $gP/m^2$.

State variables can be divided into three categories: namely, plants, soil and herbivores. The plant category consists of shoot, root, nutrient store, LITTER, dead shoot and dead root. State variables related to soil are Carbon, Nitrogen and Phosphorus content of soil organic matter, available nitrogen ($NH_4^+$ and $NO_3^-$), available phosphorus and 'unavailable Phosphorus'. Herbivores are also included in the model. The two main controlling factors are temperature (air temperature and calculated from that soil temperature) and a factor called 'mode'. 'Mode' contains information related to oxygen in the soil, redox potential and soil moisture or water level.

## Model application(s) and State of development
Application area(s): Wetland impact assessment, analysis of eutrophication and productivity
Number of cases studie: 2
State: conceptualization, verification, calibration

**Model software and hardware demands**
Platform: PC-based - MAC
OS:
Software demands: language - STELLA
Hardware demands:

**Model availability**
Purchase: no
Contact person(s) Dr. J.T.A. Verhoeven
Utrecht University
P.O. Box 80084
NL-3508 TB Utrecht
The Netherlands

Phone: +31 30 536851
Fax: +31 30 518366
E-mail: josv@boev.biol.ruu.nl

**Model documentation and major references**
Van der Peijl, M.J., Huismann, G., and Verhoeven, J.T.A., 1994. Modelling of spatial patterns and dynamic processes in the river corridor ecosystem; model description. In: FAEWE, Final report.
Van der Peijl, M.J., Huismann, G., and Verhoeven, J.T.A., 1994. Modelling of spatial patterns and dynamics processes in the river corridor ecosystem; developments in modelling. In: FAEWE, Final report.

**Other relevant information**
Model was developed by M.J. Van der Peijl, Utrecht University, Department of Plant Ecologie and Evolutionary Biology.

## 2.10 Overview of Groundwater Models

Groundwater models have come into focus due to the contamination by nitrate, oil spills and pesticides of this important freshwater resource. It is therefore not surprising that the use of models in this area is increasing at an accelerated rate. It is, however, extremely difficult to model the contamination of groundwater at a specific site due to the heterogeneity of soil in terrestrial ecosystems. It would require detailed knowledge to the spatial distribution of soil composition and porosity, which would demand a huge data-set for the focal system. It is, on the other hand, possible to indicate the relative hazard for groundwater contamination by comparison of different fertilizer practice and application of different pesticides, knowing the average soil composition and porosity of the modeled area.

Groundwater models encompass the following hydrological processes: advection, diffusion and dispersion, and the following physical-chemical processes: radioactive decay, sorption, dissolution/precipitation, acid/base reactions, complexation, hydrolysis, chemical redox reactions, biodegradation and biological transformations. It is important in an early phase of the model development to select which processes the modeler should include in a model when the relevant processes for a specific problem and contaminant are considered.

Table 2.7 gives an overview of the groundwater models presented in Section 2.10. Some useful information for development of models of groundwater contamination may also be found in Chapter 3, where ecotoxicological models are reviewed.

DHI's model for the unsaturated zone describes the transport and fate of solutes in the soil, including optional chemical reactions, plant uptake of ions and transport and consumption of oxygen. The model development has aimed at describing the accumulation in soil and the groundwater contamination of the following cations and anions: calcium, magnesium, sodium, potassium, sulfate, chloride, nitrate, hydrogen-carbonate and carbonate. The interactions of these ions are considered in the model.

Albanis et al. have developed a model of pesticide movement in soil. The model is able to account for adsorption/desorption processes, degradation of pesticides by chemical, biochemical and photolytic processes and evaporation. The movement in soil is considered by the use of a series of continuously stirred tank reactors.

OS&MODFLOW is a groundwater model, which studies the flow of nitrates and water soluble compounds. A finite-difference approach is applied to describe the transport processes, where adsorption/desorption and the water flow are the major processes.

The model Magic is developed to simulate the effects of acidic deposition on soils and surface waters, operating at the catchment scale and using seasonal or annual time steps. The effects of carbon dioxide on pH and the speciation of inorganic carbon is computed from known equilibrium reactions. Organic acids are included as mono-, di- and tri-protic acids. Uptake of nitrate is covered by a first order reaction. Weathering rates are either a constant or function of pH.

**Table 2.7** Overview of Groundwater Models

| Model name/<br>component(s) | Model characteristics<br>(processes) | References |
|---|---|---|
| SHE-groundwater | Convection, dispersion, chem. processes, adsorption, decay, oxygen consumption, uptake by plants | Ammentorp and Refsgaard, 1991 |
| Pesticides | Convection, diffusion, decay and adsorption | Albanis et al., 1991 |
| OS&MODFLOW | Convection, diffusion, 3-D description | Donald and Harbough, 1984 |
| Magic acidification | Ion exchange, adsorption, chem. reactions, complexation | Wright, see section 2.11 |

## 2.11 Groundwater Models

**Model identification**
Model name/title: SHE - Groundwater Model (developed by DHI)

**Model type**
Media: water
Class: biogeochemical, hydrology
Substance(s)/compounds: nutrients, phosphorus, nitrogen, chemicals, radionuclides, pesticides, organic compounds, heavy metals - Hg, Pb, Cd, Cu, Zn, salts.
Biological components: none

**Model purpose**
Predict groundwater contamination.

**Short description**
The describes the transport and fate of solutes in the soil, including optional chemical reactions, plant uptake of ions, and transport and consumption of oxygen. The model has a wide range of applications. It provides the possibility of describing the transport and fate of pollutants, or species used as water quality indicator parameters in the unsaturated zone. Thus, estimates can be given of the consequences for the groundwater resources following an imaginary or actual pollution. Further, the effect of alternative remedial measures can be analysed. With only minor modification the chemical submodel may also be used in the groundwater zone.

Solute transport simulations, including the chemical equilibrium model and the submodel for plant uptake of ions, can be applied for design and evaluation of irrigation schemes in regions of the Earth where salt accumulation in the root zone occurs. Similarly, the effect of a planned reclamation of saline or sodic soils, e.g., by adding chemical amendments (as gypsum) or by adding a surplus amount of water and thus leaching the salts, can be evaluated.

The submodel describing oxygen transport and oxygen consumption, in connection with degradation of organic matter, can be used as a tool in optimization of waste water infiltration of plants. As the oxygen model further provides estimates of the volumetric anaerobic soil fraction, it constitutes a useful tool in denitrification modeling.

The model has been designed to suit the overall structure of the European Hydrologic System (SHE). Here, it has been assumed that the water flow in the unsaturated zone can be described by independent one-dimensional flow columns. Hence, the solute transport model has been developed from the assumption that the water flow is vertical. Further, it has been assumed that the solute concentartions are small, so that density differences are not affecting the water flow processes. This implies that the water flow can be calculated independently of the solute transport calculations.

The following components are included in the model:

a. Convection and dispersion.   As the water flow is assumed to be one-dimensional, this restriction also applies to the convective transport of the solutes. On the other hand, the model allows both longitudinal and transversal dispersion. The model can thus simulate the dispersion process from an area and a line source. In addition, a cone-shaped spreading of pollutants from a point source can be calculated by solving the transport equation in polar coordinates.

b. A chemical equilibrium model.   For the simulation of a ionic solution an optional description of the chemical equilibrium of the system is provided, comprising the following reactions: ion-exchange, complexation, precipitation/dissolution, a carbonate system.

c. Immobile water.   Several experimental solute transport studies in unsaturated soils have indicated asymmetrical concentration distribution, either as concentration versus depth or as effluent curves, under conditions where the distribution should be symmetrical according to the theory. This phenomenon has been explained by the existence of immobile water in dead-end or blind-pores, where the exchange of mass with the displacing solution rely on diffusion processes. The presence of immobile water can be accounted for in the model. The chemical processes included in the transport calculations will be applied for both mobile and immobile water.

d. Adsorption. For a single-species transport simulation, adsorption can be described by various equilibrium isotherms such as the Freundlich and Langmuir isotherms. Further, some selected non-equilibrium isotherms are included. These describe a kinetic approach towards equilibrium, when the rate of sorption is slow compared to the rate at which the chemical moves through the soil.

e. Decay. The model can describe the transport of individual species which undergo decay to a first order reaction.

f. Oxygen transport and consumption.   This sub-model describes the convective transport of oxygen in water, the convective and diffusive transport of oxygen in soil air, the exchange of oxygen between air and water, the diffusion into water saturated soil crumbs, and the consumption of oxygen in free water as well as in soil crumbs. Further, estimates of the aerobic and anaerobic fractions of the soil are given.

g. Plant uptake. Two models of different complexity for describing the uptake of cations by plants are proposed. As the amount of ions removed in this process is often insignificant the simpler model has been chosen. In this, a constant rate uptake of ions during the growing season is assumed.

## Model application(s) and State of development
Application area(s):
Number of cases studied: several
State: conceptualization, verification, calibration, validation

## Model software and hardware demands
Platform: PC-based, Mainframe
OS:
Software demands:
Hardware demands:

## Model availability
Purchase: yes, price
Contact person(s) Dr. H. C. Ammentorp
            DHI
            Agern Alle 5
            DK-2970 Hørsholm, Denmark

Phone:
Fax:
E-mail:

## Model documentation and major references
Ammentorp, H.C., and Refsgaard, J.C., A model for the unsaturated zone, chapter 9. In: Modelling in Environmental Chemistry", by S.E. Jørgensen (Ed.), pp. 277-374. Elsevier, Amsterdam, 1991.

## Other relevant information

## Model identification
Model name/title: Pesticide Movement in Soil

## Model type
Media: water
Class: biogeochemical, toxicology
Substance(s)/compounds: pesticides
Biological components:

## Model purpose
To predict groundwater contamination by pesticides.

**Short description**
Processes: advection, diffusion and adsorption. The latter is described as a two-steps process: the entry of the compound into solution and adsorption on soil particles. movement is simulated by a series of continously stirred tank reactors.

**Model application(s) and State of development**
Application area(s):
Number of cases studied: several
State: conceptualization, verification, calibration, validation

**Model software and hardware demands**
Platform: PC-based, Mainframe
OS:

**Model availability**
Purchase: no.
Contact person(s) Dr. T.A. Albanis
University of Ioannina
Department of Chemistry
Section of indsutrial and food chemistry
Ioannina 451 10, Greece

Phone: +30-651-98-348
Fax: +30-651-44-836
E-mail:

**Model documentation and major references**
Albanis, T.A., Pomonis, P.J., and A.T. Sdoukas. Model of pesticide movement in soil. In: *Modelling in Environmental Chemistry*, S.E. Jørgensen (Ed.),, Elsevier, Amsterdam, 1991.

**Other relevant information**

**Model identification**
Model name/title: groundwater model "OS&MODFLOW": A transport program is added to the groundwater-flow model "MODFLOW", described in ref. (2).

**Model type**
Media: water, lake, river, regional scale, global scale, terrestrial, agricultural
Class: hydrology
Substance(s)/compounds: nitrogen
Biological components: ecosystem - agroecosystem

## Model purpose

The model can be used to study the flow of nitrates and water soluble compounds with similar properties in groundwater bodies with a porous substrate matrix. It may help to manage water quality and to predict concentrations changes of nitrates an agriculturally used landscapes.

## Short description

The three-dimensional groundwater model OS&MODFLOW describes the transport of nitrate in the saturated zone of groundwater. The transport routine considers the two main transport mechanisms, convection and hydromechanical dispersion. The groundwater-flows, calculated by MODFLOW, result from the gradients of piezometric heads. So properties of the aquifer must be known: transmissivities/hydraulic conductivities, porosity, depth. Input from the unsaturated zone are the rates of seepage and nitrate. The numerical solution uses the block-centered finite-difference approach. The transport program may be adapted to other groundwater-flow models, which have the same numerical approach.

## Model application(s) and State of development

Application area(s): landscapes with porous substrate and identifiable groundwater properties, especially agriculturally used areas.
Number of cases studied: 1
State: conceptualization, verification

## Model software and hardware demands

Platform: PC-based, Workstation SUN spark; large area handling may require additional storage space on the use of mainframe.
OS:
Software demands: OS&MODFLOW
Hardware demands: see above, depending on project area

## Model availability

Purchase: yes, public domain
Contact person(s) Prof Dr. Helmut Lieth
                FB-5- Biologie
                Universität Postfach 44 69
                D-4500 Osnabrück, Germany

Phone: +49 541 969 2535
Fax: +49 541 969 2570
E-mail:    Internet: 131.173.17.10
            Login: Anonymous
            Password: e-mail address
            Subdirectory: pub/local/ASF

## Model documentation and major references
Wuttke, G., Thober, B., and Lieth, H., Simulation of nitrate transport in groundwater with a 3-dimensional groundwater model run as a subroutine in an agroecosystem. *Ecol. Modelling*, 57 (1991), pp. 263-276. Elsevier, Amsterdam.

## Other relevant information
McDonald, G.M., and Harbaugh, A.W., 1984. *A Modular Three-dimensional Finite-difference Groundwater Model*, Scientific Publications Co., Washington DC, pp. 528.
Bear, J., 1979. *Hydraulics of Groundwater*, McGraw-Hill, New York, 569 pp. System Research Group of University of Osnabrück, 1990. Intensivlandwirtschaft and Nitratbelastung des Grundwassers im Kreis Vechta, Final Report, Osnabrück, 294 pp.

The running time of our model for nearly 5000 matrix elements for an area of about 200 km² with 5 aquifer layers over the time span of 20 years was 1 hour CPU/IO on a SUN SPARK.

## Model identification
Model name/title: MAGIC, Model of Acidification of Groundwater In Catchments

## Model type
Media: water, lake, river, local scale, regional scale, soil
Class: biogeochemical
Substance(s)/compounds: sulfur, nitrogen, base cations, pH, alkalinity, aluminum
Biological components: ecosystem, shallow to moderate soil depths; principally upland catchments, forest, moorlands, etc.

## Model purpose
MAGIC (Model of Acidification of Groundwater In Catchments; Cosby et al., 1985a,b,c) was developed to simulate the effects of acidic deposition on soils and surface waters. The model operates at the catchment scale. Simulation typically involve seasonal or annual time steps and are implemented on decadal or centennial time scales.

## Short description
MAGIC is based on simultaneous reactions describing sulfate adsorption, cation exchange, dissolution-precipitation-speciation of aluminum, dissociation of organic acids and dissolution-speciation of inorganic carbon. Mass balances are controlled by atmospheric inputs, chemical weathering inputs, net uptake in biomass and losses to runoff. At the heart of MAGIC is the size of the pool of exchangeable base cations in the soil. As the fluxes to and from this pool change over time in response to changes in atmospheric deposition, the chemical equilibria between soil and soil solution shift producing changes in surface water chemistry. Cation exchange is modeled using equilibrium (Gaines-Thomas) equations for each base cation and aluminum. Sulfate adsorption is represented by a Langmuir isotherm. Aluminum dissolution-precipitation is assumed to be controlled by equilibrium with a solid phase of aluminum trihydroxide. Aluminum speciation is calculated by considering hydrolysis reactions as well as complexation with sulfate and fluoride.

153

The effects of dissolved $CO_2$ on pH and the speciation of inorganic carbon is computed from known equilibrium reactions. Organic acids are represented in the model by mono, di- and tri- protic acid analogs. First order reactions are used for uptake of nitrate and ammonium. Weathering rates may be constant or a function of pH. Mass and charge balance are imposed in a numerical solution of the combined equations.

## Model application(s) and State of development

Application area(s): MAGIC has been used to reconstruct the history of acidification and to simulate the future trends on a regional basis and in a large number of individual catchments in both North America and Europe (see e.g., Cosby et al., 1986b, 1989, 1990; Hornberger et al., 1989; Jenkins et al., 1990a,b, 1992; Lepisto et al., 1988; Norton et al., 1992; Whitehead et al., 1988a,b; Wright et al., 1986, 1990; Wright and Cosby, 1987).

Number of cases studied: (NOTE: The exact number of case studies at this time is not known. The model has been used extensively by a large number of investigators who are not collaborating with the model author).

State:

## Model software and hardware demands

Platform:

OS:

Software demands:

Hardware demands:

## Model availability

Purchase: Copies of the MAGIC model are available from the Norwegian Institute for Water Research (NIVA), in Oslo, Norway. The current version of MAGIC was developed at NIVA and is distributed by them. Contact Dick Wright at the following address:

Contact person(s)  Dr. R.F. Wright

NIVA

BOX 173 Kjelsas

N-0411 Oslo, Norway

Phone:

Fax:

E-mail:

## Model documentation and major references

## Other relevant information

## Model identification

Model name/title: MAGIC - "Model of Acidification of Groundwater in Catchments"

**Model type**
Media: water, lake, river, local scale, regional scale, soil, forest
Class: biogeochemical
Substance(s)/compunds: nutrients, carbon, sulfur, nitrogen, (Ca, Mg, K, Na, Al, Cl, F), organic compounds
Biological components: ecosystem, Forest-ecosystems (Forest-soil/soil-water-river)

**Model purpose**
Estimating long-term effects of acid deposition on soil and freshwater systems.

**Short description**
Model description.

| FORCING FUNCTIONS | VARIABLES |
|---|---|
| Air pollution: | |
| Dry deposition flux | $SO_4$, $NO_3$, $NH_4$, Ca, Mg, Na, K, Cl |
| Wet deposition flux | $SO_4$, $NO_3$, $NH_4$, Ca, Mg, Na, K, Cl |
| Vertical solute transport: | |
| Convection | H, Ca, Mg, K, Na, Al, Cl, $SO_4$, $NH_4$, $NO_3$, F, $HCO_3$, RCOO, Al-species |
| Soil chemical processes: | |
| (See table I) | |
| Ion exchange | H, Ca, Mg, K, Na, Al. |
| Complexation | Al, F, $SO_4$, RCOO. |
| Equilibrium equations | $H/Ca/Mg/K/Na/Al/Cl/SO_4/NH_4/NO_3$ $F/HCO_3/RCOO/$Al-species. |
| Hydroxide weathering | Al. |
| Adsorption | $SO_4$. |
| Volatilization | $CO_2$, $HCO_3$ |
| Nutrient cycling: | |
| Nutrient uptake by roots | N, K, S, Ca, Mg, P, Na (Net uptake only) |

MODEL INPUT:
1) Boundary conditions:
   Rainfall characteristics (Rainfall depth).
   Evapotranspiration (Potential evapotranspiration, air temperature).
   Deposition fluxes (Pollutants, Ca, Mg, K, Na).
   Ambient concentrations (Pollutants).
2) System properties:
   Water balance (Yearly precipitation, yearly runoff). Soil chemistry (CEC, exchangeable Na, exchangeable K, exchangeable Ca, exchangeable Mg, bulk density, layer thickness, porosity, $SO_4$ adsorption, isotherm coefficients, Al-hydroxide solubility, releasing rate of Ca, Mg, K, Na by weathering. These parameters are lumped to obtain characteristic values for 1 or two soil boxes arranged either vertically or horizontally, representing the entire soil profile). Others (Mean soil temperature; stream/lake: $pCO_2$, temperature).
3) Initial contents: Soil chemistry (Exchangeable Na, exchangeable K, exchangeable Ca, exchangeable Mg).

4) Calibration requirements: Water chemistry: (soil solution or leachate from lowest soil horizon or runoff). Yearly annual volume and volume-weighted concentrations of H, Ca, Mg, Na, K, Al, $NH_4$, $NO_3$, Cl, $SO_4$, F, RCOO. Soil chemistry: Yearly exchangeable Na, exchangeable K, exchangeable Ca, exchangeable Mg.

## Model application(s) and State of development
Application area(s): Calibration and validation by use of catchment data. Calibration and validation by use of soil-lysimeter (Column) data. Calibration to field-data (soil probes) on local scale, and then prognosis made on regional scales in order to set "critical loads" for soil-and freshwater acidification.
Number of cases studied: (several) >50 sites in >10 countries.
State: conceptualization, verification, calibration, validation, prognosis made

## Model software and hardware demands
Platform:PC-based, IBM, JBM-compatible
OS: DOS
Software demands: FORTRAN
Hardware demands: 286, Math co-processor, RAM 600 Kb

## Model availability
Purchase: yes, software available as executable code
Contact person(s)DEVELOPER:

| | |
|---|---|
| Dr. Richard F. Wright | Dr. Bill J. Cosby |
| Norwegian Institute for Water Research | Dept. of Environmental Sciences |
| NIVA, P.O. Box 69, Korsvoll | University of Virginia, |
| N-0808 Oslo, Norway | Charlottesville, VA 22903, USA |

Phone: +47 22 18 51 00      +1 804-924-7787
Fax: + 47 22 18 52 00      +1 804-982-2300
E-mail:

## Model documentation and major references
Cosby, B.J., Hornberger, G.M., Galloway, J.N., Wright, R.F., 1985. Modelling the effects of acid deposition: assessment of a lumped-parameter model of soil water and streamwater chemistry. *Water Resour. Res.*, 21, 51.
Cosby, B.J., Wright, R.F., Hornberger, G.M., Galloway, J.N., 1985. Modelling the effects of acid deposition: estimation of long-term water quality responses in a small forested catchment. *Water Resour. Res.*, 21, 1591.

## Other relevant information

## 2.12 Models of Waste Water Treatment Plants, An Overview

This section is confined to biological waste water treatment plants, as physical-chemical plants are not considered, at least not in this context, an environmental system. Biological treatment plants may be considered artificial ecosystems, and it seems therefore appropriate to include models of these systems in this volume. The processes characterizing waste water treatment plants are furthermore very similar to the processes known from natural ecosystems, which is only partially true for physical-chemical treatment plants. For those interested in models of physical-chemical treatment plants can be referred to S.E. Jørgensen "Models of physical-chemical treatment processes" in *Modelling in Environmental Chemistry* by S.E. Jørgensen (1990).

Most models in this area are developed for specific types of waste water treatment plants or specific processes: trickling filters, activated sludge, lagoon systems, oxidation ditch, nitrogen removal processes or denitrification. The models may be applied for design of the various types of waste water treatment plants.

Table 2.8 gives an survey of the models presented in Section 2.13 plus a few models which are also mentioned below.

Aquasim is developed for the identification and simulation of aquatic systems in general and could therefore also be used to model rivers, lakes, etc., but has found the widest application for development of models of waste water treatment plants. The model allows the user to define spatial system configuration of the system to be investigated as a set of compartment. The available compartments are mixed reactors, biofilm reactors (consisting of a biofilm, a liquid boundary layer and a bulk fluid phase) and river sections. The compartment can be connected by advective and diffusive links.

SND is a model for simulation of simultaneous nitrification and denitrification. The processes take place in a bio-bed, which is able also to bind ammonium by ion exchange. The model allows to design a column for simultaneous nitrification and denitrification.

The model "Activated Sludge" is developed to design the activated sludge process based upon material balances and the concept of active mass. The model distinguishes between biodegradable and non-biodegradable organic matter, consider the oxygen and nutrient requirements for biological growth and includes nitrification processes.

DPMC is a model for aerated lagoons. The model is rather simple, as it considers BOD-removal, oxygen concentration and suspended matter in a multicellular configuration.

**Table 2.8** Models of Biological Waste Water Treatment Systems

| Model name/ model focus | Model characteristics (processes) | References |
|---|---|---|
| Aquasim | Biological processes + hydrodynamics | Reichert, 1993 |
| SND | Nitrification and denitrification on biofilter | Halling-Sørensen and Nielsen, 1995 |
| Activated sludge | Mass balances on biodegradation, settling, oxygen and nutrients for designal., | Eckenfelder et al., 1985 |
| DPMC | BOD, oxygen susp. matter: aerated lagoons | Rich, 1985 |
| WSP | Algae and bacteria (growth, sedimentation, decay), decomp. of organic matter | Fritz, 1985 |
| Trickling filter | BOD, susp., nutrients/recycling and retention time | Roberts, 1985 |
| Rotating biological contactors | N in various oxidation states: nitrification + denitrification | Watanabe, 1985 |
| Fluidized bed | BOD, nutrients, biofilm + transfer processes, advection and diffusion | Elmaleh and Grasmick, 1985 |
| Anerobic treatment | Degradation of organic matter in several steps, gas production | Rozzi and Passino, 1985 |

A model for waste stabilization ponds is developed by Fritz. It considers the following processes: bacterial growth, sedimentation of bacteria and algae, algal growth determined by the temperature and the concentration of nutrients, photosynthesis, bacterial decay, aeration, decomposition of organic nitrogen compounds, nitrification, benthic regeneration, detritus, upwelling from bottom sediment and benthic decay. The model is based upon mass balances of the nutrient, oxygen and organic matter, expressed as concentration of BOD and detritus.

Roberts has developed a model for trickling filters. The forcing functions are the usual water quality parameters, the temperature of the water, recycling and the filter performance (surface, retention time and volume). The state variables are the water quality parameters for the recycled water and for the final effluent.

Watanabe has developed a special model for nitrogen removal by a rotating biological contactor, based upon nitrogen balance for the various forms of nitrogen: organic nitrogen, ammonium, nitrite and nitrate. Section 2.11 includes finally a model of a fluidized bed reactor and a model for anaerobic treatment processes, aimed for simulation of the gas-production and the degradation of organic matter by these processes.

158

## 2.13 Waste Water Models

---

### Model identification
Model name/title: AQUASIM: Computer Program for the Identification and Simulation of Aquatic Systems

---

### Model type
Media: water
Class: biogeochemical, toxicology
Substance(s)/compounds: nutrients, user-defined, chemicals, heavy metals user-defined
Biological components: user-defined components

---

### Model purpose
The program AQUASIM was developed for the identification and simulation of aquatic systems in nature, in technical plants and in the laboratory. It lets the user define a model using a set of predefined compartments and links and arbitrary transformation processes and perform simulations, sensitivity analyses and parameter estimations with this model.

---

### Short description
AQUASIM allows its user to define the spatial configuration of the system to be investigated as a set of compartments, which can be connected to each other by links. Currently, the available compartment types include mixed reactors, biofilm reactors (consisting a biofilm, a liquid boundary layer and a bulk fluid phase) and river sections (describing water flow and substance transport and transformation in open channels). It is planned to extend this set of compartment types in future versions of the program. Compartments can be connected by advective or diffusive links. The user of the program is free in specifying any set of state variables and transformation processes to be active within the compartments. For the model as defined by the user, the program is able to perform simulations, sensitivity analyses and parameter estimations using measured data. These features make the program to a very useful research tool. Due to the possibility of starting with a simple model and gradually increasing model complexity later by considering additional variables and processes, AQUASIM is also well suited for teaching. Three versions of the program are available. The window interface version uses the graphical user interface of a machine, the character interface version can be run on a primitive teletype terminal, and the batch version is designed for long calculations to be submitted as batch jobs.

---

### Model application(s) and State of development
Application area(s): Identification and simulation of aquatic systems in nature, in technical plants and in the laboratory
Number of cases studied:
State:

**Model software and hardware demands**
Platform: PC-based, MAC, IBM. Workstation SUN, HP, IBM
OS: Windows NT, MacOS, UNIX
Software demands:
Hardware demands: 10 Mb disk space

**Model availability**
Purchase: yes, SFR. 1.000,00 executable code
Contact person(s) Dr. Peter Reichert
 Swiss Federal Institute for Environmental Science and Technology (EAWAG)
 CH-8600 Dübendorf
 Switzerland

Phone: +41 1 823 52 81
Fax: +41 1 823 53 98
E-mail: reichert@eawag.ch

**Model documentation and major references**
Reichert, P., AQUASIM - A Tool for Simulation and Data ANalysis of Aquatic Systems, submitted to *Water Science and Technology*, July 1993.
Reichert, P., Object-Oriented Design of a Program for the Identification and Simulation of Aquatic Systems, to be published.
Reichert, P., and Ruchti J., AQUASIM - Computer Program for the Identification an simulation of Aquatic Systems: User Manual, Swiss Federal Institute for Environmental Science and Technology (EAWAG), Ch-8600 Dübendorf, Switzerland, 1994.
Reichert, P., Concepts of a Computer Program for the Identification and Simulation of Aquatic Systems, to be published.

**Other relevant information**

**Model identification**
Model name/title: Activated Sludge

**Model type**
Media: water
Class:
Substance(s)/compounds: nutrients, nitrogen
Biological components: community - waste water plant

**Model purpose**
Facilitate design of activated sludge.

## Short description
The design parameters to be balanced to the forcing functions are
volume of a.s., recording ration, volume of clarifier, BOD of treated water, sludge removal.
The forcing functions are: flow and BOD5 of untreated water.
State variables are: BOD5 after a.s., BOD5 after clarifier, suspended matter (MLVSS and MLSS) in a.s. and clarifier.
Equations are all leaned on mass balances. Nutrient requirement is also calculated, as well as adsorption/desorption to suspended matter. Nitrification may be included, if required.

## Model application(s) and State of development
Application area(s):
Number of cases studied: several
State: conceptualization, verification, calibration, validation, used for design

## Model software and hardware demands
Platform: Mainframe
OS:
Software demands:
Hardware demands:

## Model availability
Purchase: no.
Contact person(s) Dr. W. W. Eckenfelder
                 Austin Engineering
                 Texas University
                 Texas, USA

## Model documentation and major references
Eckenfelder, W.W., Goronszy, H.C., and Watkin, A.T., Comprehensive activated sludge process design. In: *Mathematical Models in Waste Water Treatment*, S.E. Jørgensen and M.J. Gromiec (Eds.), Elsevier, Amsterdam, 1985.

## Other relevant information

## Model identification
Model name/title: Model of Waste Stabilization Ponds (WSP)

## Model type
Media: water
Class: biogeochemical
Substance(s)/compounds: nutrients, carbon, phosphorus, nitrogen
Biological components: ecosystem - (WSP)

161

## Model purpose
Design of WSP.

## Short description
Mass balances are used to set up differential equations.
State variables: Algae, Detritius, Bacteria, Oxygen, $CO_2$, Nitrogen,
The model considers: hydrodynamics BOD-removal, algal growth, bacteria growth, adsorption and settling.

## Model application(s) and State of development
Application area(s):
Number of cases studied: several
State: conceptualization, verification, calibration, validation

## Model software and hardware demands
Platform: PC-based
OS:
Software demands:
Hardware demands:

## Model availability
Purchase: no
Contact person(s) Dr. J.C. Fritz
               National Research Council
               Washington D.C.
               USA

Phone:
Fax:
E-mail:

## Model documentation and major references
Fritz, J.C., Mathematical Models for Wates Stabilization Ponds. In: *Mathematical Models in B.W.W.*, pp. 169-242.

## Other relevant information

# 3. BIOGEOCHEMICAL MODELS IN ECOTOXICOLOGY

3.1 Introduction

3.2 Characteristic Features and Classification of Ecotoxicological Models

3.3 The Application of Models in Ecotoxicology

3.4 An Overview: Pesticide Models

3.5 Models of Pesticides

3.6 An Overview: Models of Other Toxic Organic Compounds

3.7 Models of Other Organic Compounds

3.8 An Overview: Models of Heavy Metals, Radionuclides and Other Inorganic Compounds in the Environment

3.9 Models of Heavy Metals, Radionuclides and Other Inorganic Compounds in the Environment

# 3.1 Introduction

An increasing interest in management of toxic substance pollution has emerged during the last decade and this has caused an equally large interest in toxic substance modeling.

Toxic substance models attemp to model the fate and effect of toxic substances in ecosystems. They are most often biogeochemical models, because they attempt to describe the mass flows of the considered toxic substances, although there are models of the population dynamics, which include the influence of toxic substances on the birth rate and/or the mortality, and therefore should be considered a toxic substance model. Ecotoxicological models are still treated under the label of biogeochemical models because their use is very dominant in ecotoxicology.

Toxic substance models differ from other ecological models by:
1.    The need for parameters to cover all possible toxic substance models is great, and general estimation methods are therefore used widely. Some of the models included in this chapter encompass suitable estimation methods as an integrated part of the model application.
2.    The safety margin should be high, for instance, expressed as the ratio between the actual concentration and the concentration that gives undesired effects. This implies that a higher uncertainty can be accepted in most cases.
3.    The possible inclusion of an effect component, which relates the output concentration to its effect. It is easy to include an effect component in the model; it is, however, often a problem to find a well examined relationship in the literature to base it on.
4.    The possibility and need of simple models due to point 1 and 2, and our limited knowledge of process details, parameters, sublethal effects, antagonistic and synergistic effects.

It may be an advantage to clarify several questions before developing a toxic substance model:
1.    Obtain the best possible knowledge about the possible processes of the considered toxic substance in the ecosystem. All available knowledge about the quantitative role of the processes should be obtained.
2.    Attempt to get parameters from the literature and/or from own experiments (*in situ* or in the laboratory)
3.    Estimate all parameters by the available methods.
4.    Compare the results from 2) and 3) and attempt to explain discrepancies.
5.    Estimate which processes and state variables it would be feasible and relevant to include into the model. If there is the slightest doubt, then it is better at this stage to include too many processes and state variables rather than too few.
6.    Use a sensitivity analysis to evaluate the significance of the individual processes and state variables. This often may lead to further simplification.

## 3.2 Characteristic Features and Classification of Ecotoxicological Models

Ecotoxicological models differ from ecological models in general by:

1. Being more simple in general.
2. Require more parameters.
3. Use of parameter estimation methods more widely applied.
4. Possible inclusion of an effect component.

Ecotoxicological models may be divided into six different classes. The classification presented here is based on differences in the modeling structure. The decision on which model class to apply is based upon the ecotoxicological problem that the model aims to solve. The definitions of the model classes are given below. It is indicated where it is most appropriate to use the six model types.

### A. Food Chain or Food Web Dynamic Models

This class of models considers the flow of toxic substance through the food chain or food web. Such models will be relatively complex and contain many state variables. The models will furthermore contain many parameters, which often have to be estimated. This type of models will typically be used when many organisms are affected by the toxic substance, or the entire structure of the ecosystem is threatened by the presence of a toxic substance. Because of the complexity of these models, they have not been used widely. They are similar to the more complex eutrophication models that consider the flow of nutrients through the food chain or even through the food web. Sometimes they are even constructed as submodels of a eutrophication model, see, for instance, Thomann et al., 1974. Figure 3.1 shows a conceptual diagram of an ecotoxicological food chain model for lead. The flow of lead from atmospheric fall out and water to an aquatic ecosystem, where it is concentrated through the food chain - the so-called "bioaccumulation". A simplification is hardly possible for this model type because it is the aim of the model to describe and quantify the bioaccumulation through the food chain.

### B. Static Models of the Mass Flows of Toxic Substances

If the seasonal changes are minor, or of minor importance, a static model of the mass flows will often be sufficient to describe the situation and even to show the expected changes if the input of toxic substance is reduced or enlarged. This type of model is based upon a mass balance as clearly seen from the example in Figure 3.2. It will often, but not necessarily, contain more trophic levels, but the modeler is often concerned with the flow of the toxic substance through the food chain. The example in Figure 3.2 considers only one trophic level. If there are some seasonal changes, this type, which usually is simpler than type A, can still be an advantage to use, for instance, if the modeler is concerned with the worst case and not the changes.

## C. A Dynamic Model of a Toxic Substance in a Trophic Level

It is often only the toxic substance concentration in one trophic level that is of concern. This includes the zero trophic level, which is understood as the medium - either soil, water or air. Figure 3.3 gives an example. It is a model of copper contamination in an aquatic ecosystem. It is included as one of the models reviewed in Section 3.9. The main concern is the copper concentration in the water, as it may reach a toxic level for the phytoplankton. Zooplankton and fish are much less sensitive to copper contamination, so the alarm clock rings first at the concentration level that is harmful to phytoplankton. But, only the ionic form is toxic and is therefore necessary to model the partition of copper in ionic form, complex bound form and adsorbed form. The exchange between copper in the water phase and in the sediment is also included, because the sediment can accumulate relatively large amounts of heavy metals. The amount released from the sediment may be significant under certain circumstances - for instance low pH.

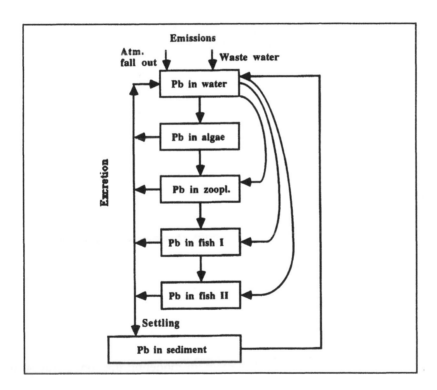

**Figure 3.1:** *Conceptual diagram of the bioaccumulation of lead through a food chain in an aquatic ecosystem.*

Figure 3.4 gives another example, which is included in Section 3.5. Here the main concern is the DDT concentration in fish, where these may be such high concentration of DDT that, according to WHO´s standards, they are unfit for human consumption. The model can therefore be simplified by not including the entire food chain but only the fish. Some physico-chemical reactions in the water phase are still of importance and they are included as shown on the conceptual diagram outlined in Figure 3.4. As seen from these examples, simplifications are often feasible when the problem is well defined, including which component is most sensitive to toxic matter, and which processes are most important for concentration changes.

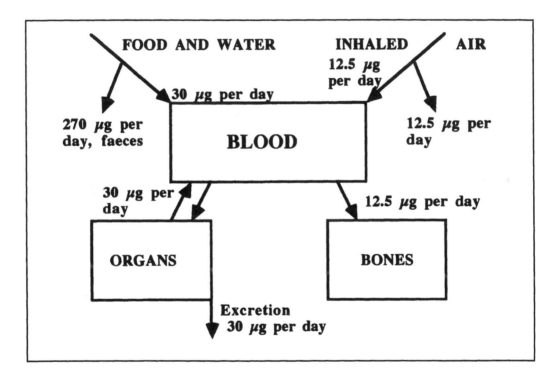

**Figure 3.2**: *A static model of the lead uptake by an average Dane before introduction of unleaded gasoline*

Figure 3.5 shows the processes of interest for modeling the concentration of a toxic component at one trophic level. The inputs are uptake from the medium (water and air) and from digested food equals total food minus non-digested food. The outputs are mortality (transfer to detritus), excretion and predation from next level in the food chain. Many toxic substance models are based upon these processes.

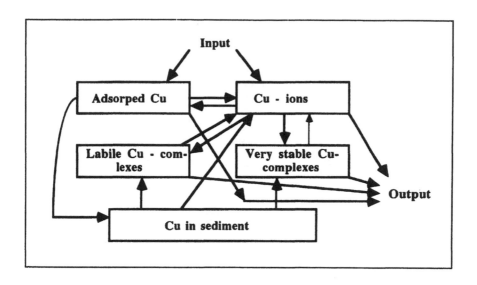

**Figure 3.3:** *Conceptual diagram of a simple copper model.*

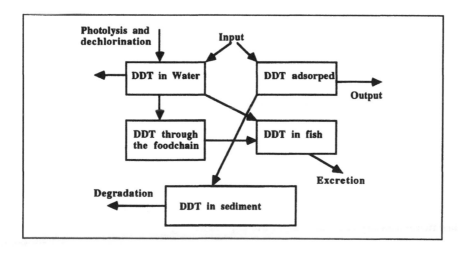

**Figure 3.4**: *Conceptual diagram of a simple DDT model.*

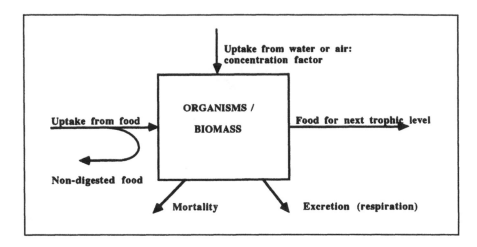

**Figure 3.5:** *Processes of interest for modeling the concentration of a toxic substance at one trophic level (Jørgensen, 1994).*

### D. Ecotoxicological Models in Population Dynamics

Population models are biodemographic models and therefore have numbers of individuals or species as state variables. The simple population models considered only one population. The growth of the population is a result of the differences between natality and mortality:

$$dN/dt = B \cdot N \cdot M \cdot N = r \cdot N \tag{3.1}$$

where N is the number of individuals, B is the natality, i.e., the number of new individuals per unit of time and per unit of population, M is the mortality, i.e., the number of organisms that died per unit of time and per unit of population; and, r is the increase in the number of organisms per unit of time and per unit of population, and = B - M. B, N and r are not necessary constants as in the exponential growth equation, but are dependent on N, the carrying capacity and other factors. The concentration of toxic substance in the environment or in the organisms also may influence the natality and the mortality, and if the relation between a toxic substance concentration and these population dynamics parameters is included in the model, it becomes an ecotoxicological model of population dynamics.

170

Population dynamic models may include two or more trophic levels and ecotoxicological models will include the influence of the toxic substance concentration on natality, mortality and interactions between these populations. In other words, an ecotoxicological model of population dynamics is a general model of population dynamics with the inclusion of relations between toxic substance concentrations and some important model parameters.

## E. Ecotoxicological Models with Effect Components

Though class E models already include relations between concentration of toxic substance and their effects, these are limited to population dynamic parameters. In comparison class E models include more comprehensive relations between toxic substance concentrations and effects. These models may include not only lethal and/or sublethal effects, but also effects on biochemical reactions or on the enzyme system. Figure 3.6 gives a conceptualization of this type of model. As shown on this figure, the effects may be considered on various levels of the biological hierarchy from the cells to the ecosystems.

In many problems it may be necessary to go into more detail on the effect to answer the following relevant questions:

1. Does the toxic substance accumulate in the organism?
2. What will be the long term concentration in the organism when uptake rate, extraction rate and biochemical decomposition rate are considered?
3. What is the chronic effect of this concentration?
4. Does the toxic substance accumulate in one or more organisms
5. What is the transfer between various parts of the organism?
6. Will decomposition products eventually cause additional effects?

A detailed answer to these questions may require a model of the processes that take place in the organism, and a translation of concentrations in various parts of the organisms into effects. This implies, of cource, that the intake = (uptake by the organisms) * (efficiency of uptake) is known. Intake may either be from water or air, which also may be expressed by concentration factors, which are the ratios between the concentration in the organism and in the air or water.

But, if all the above mentioned processes should be taken into consideration for just a few organisms, the model will easily become too complex, contain too many parameters to calibrate, and require more detailed knowledge than it is possible to provide. Often we even do not have all the relations needed for a detailed model, as toxicology and ecotoxicology still are in their infancy. Therefore, most models in this class will not consider too many details of the partition of the toxic substances in organisms and their corresponding effects. Usually, accumulation is rather easy to model and the following simple equation is often sufficiently accurate:

$$C/dt = (ef \cdot Cf \cdot F + em \cdot Cm \cdot V) / W - Ex \cdot C = (INT) / W - Ex \cdot C \tag{3.2}$$

where C is the concentration of the toxic substance in the organism; ef and em are the efficiencies for the uptake from the food and medium respectively (water or air); Cf and Cm are the concentration of the toxic substance in the food and medium respectively; F is the amount of food uptake per day; V is the volume of water or air taken up per day; W is the body weight as dry or wet matter; and Ex is the excretion coefficient (1/day). As seen from the equation INT covers the total intake of toxic substance per day.

The equation has a numerical solution:

$$C/C(max) = (INT \ (1 - exp \ (Ex \cdot t))) / (W \cdot Ex) \tag{3..3}$$

where C(max) is the steady state value of C:

$$C(max) = INT / ( W \cdot Ex) \tag{3.4}$$

Synergistic and antagonistic effects have not been touched so far. They are rarely considered in this type of models for the simple reason, that we do not have much knowledge about these effects. If we have to model combined effects of two or more toxic substances, we can only assume effects, unless we can provide empirical relationships for the combined effect.

### F. Fate or Multimedia Models with or without a Risk Assessment Component

The last type of ecotoxicological models focuses on the fate of the toxic substance - where in the ecosystem will the toxic substance be found? In which concentration?

The complete solution of an ecotoxicological problem requires in principle four (sub)models, of which the fate model may be considered the first model in the chain; see Figure 3.7. As seen on the figure the four compartments are (see Morgan 1984):

1. A fate or exposure model.
2. An effect model, translating the concentration into an effect; see type E above.
3. A model for human perception processes.
4. A model for human evaluation processes.

The first two submodels are in principle "objective", predictive models, corresponding to the types E and F, while the latter two are value oriented. The development of submodels (1) and (2) are based upon physico-chemical and biological processes.

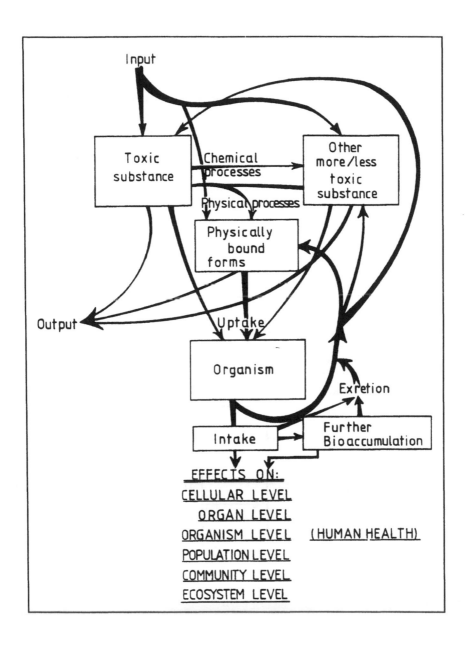

**Figure 3.6**: *Conceptual diagram of the principle in ecotoxicological models with the effect components.*

The second submodels requires a good knowledge of the effects of the toxic components. The submodels (3) and (4) are different from the generally applied environmental management models and are presented in some details below, covering risk assessment.

Fugacity models are a special type of fate or multimedia models. Mackay (1991) gives a comprehensive treatment of this type of toxic substance models. They attempt to answer the following questions: In which of these six compartments, corresponding to the spheres, can we expect the greatest problem of the chemicals emitted to the environment? Which concentration may we expect in each compartment? What are the implications of this concentration? The fugacity models are to a large extent based on physico-chemical parameters and the use of these parameters to estimate the other required parameters. The availability of the fugacity model is therefore very dependent on the use of the estimation methods.

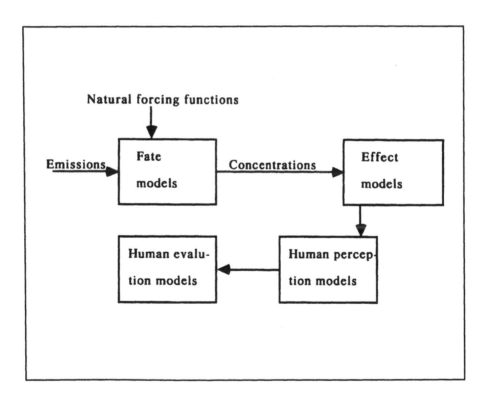

**Figure 3.7**: *The four submodels of a total ecotoxicological model are shown.*

The risk assessment component, associated with the fate model, comprises human perception and evaluation processes, (see Figure 3.7). These submodels are explicitly value laden, but must of course build on objective information on concentrations effects.

The acknowledgement of the uncertainty is of great importance and may be taken into consideration, either qualitatively or quantitatively.

Until 10-15 years ago the researchers had developed very little understanding of the processes by which people actually perceived the exposure and effects of toxic chemicals, but these processes are just as important for the risk assessment as the exposure and effects processes. The characteristics of risks and effects are the perceptions of people.

Several risk management systems are available, but no attempt will be made here to evaluate them.

The presentation of the six classes of ecotoxicological models above, clearly show the advantages and limitations of these models. The simplifications used in classes B, C and F (at least without risk assesment components) often offer great advantages. They are sufficient accurate to give a very applicable picture of the concentrations of toxic substance in the environment, due to the application of great safety factors.

## 3.3 The Application of Models in Ecotoxicology

A number of toxic substance models are reviewed in this chapter. They are divided into models of pesticides, models of other organic substances and models of heavy metals and radionuclides. Overviews of the ecotoxicological models available today are given in Tables 3.1.-3.3. The tables include not only the models from this chapter, but also a wide spectrum of slightly older toxic substance models to give a broad impression of all the models developed in this area.

Model characteristics shown in the tables are the most important state variables and/ or processes considered in the model. The model class according to the classification given in Section 3.2. is shown in brackets after the toxic substance modeled. The references are indicated for all the models.

The most difficult part of modeling the effect and distribution of toxic substances is to obtain the relevant knowledge about the behavior of the toxic substance in the environment, and to use this knowledge to make feasible simplifications. It gives the modeler of ecotoxicological problems a particular challenge in the selection of the right and balanced complexity, and there are many examples of rather simple ecotoxicological models, which can solve the focal problem. It is feasible in many case studies to apply simple models, because the use of safety factors render it possible sometimes to accept a high uncertainty, provided that the applied safety factor is significantly larger than the uncertainty. It may also occur that the uncertainty in the assessment of parameters renders it impertinent to include too many details in the model.

It can be seen from the three tables that most ecotoxicological models have been developed during the last decade. Before the mid-1970-s, toxic substance models were hardly associated with environmental modeling, as the problems seemed straight forward. The solution was simply to eliminate the source of toxic substance.

Later in the seventies, it was acknowledged that the environmental problems of toxic substances are very complex due to the many possible interacting processes and components. Several accidental releases of toxic substances into the environment, together with the detection of small, but still harmful, concentrations of very toxic substances such as PCB, mercury compounds, DDT, dioxin, and freon have reinforced the need for models. They were developed to answer pertinent questions such as which sources should be eliminated and which sources should be reduced to obtain a sufficient and acceptable reduction of the various potential risks, which concentration level would under these circumstances be acceptable in the environment, and many more questions related to the specific case.

## 3.4 An Overview of Pesticides Models

The application of pesticides implies that toxic substances are spread in nature. It is therefore not surprising that the application of pesticides have provoked development of many models to assess which harm the pesticides could involve.

Table 3.1. gives an overview of pesticides models, including the models presented below and most of the other models developed during the last about 15 years. The spectrum of models is wide. There are models focusing on the regional distribution of pesticides and there are models dealing with the details of one process only, for instance photodegradation and microbiological decomposition.

The model Espelor estimates the amount of pesticides released through surface waters from agricultural areas. It is a collective model for estimation of the total amount of pesticide run-off from draining streams, pumping stations and rivers.

The CSTR-model is included in Section 2.9, which is dealing with groundwater. The model describes the movement of pesticides (or other organic or inorganic tracers) in successive soil layers, considered as a cascade of continuously stirred tank reactors connected in series.

Liu and Zhang developed a model in 1987 with only two parameters for microbial degradation of pesticides. The state variables are the pesticide concentration and the concentration of pesticide degrading microorganisms. An improved version of this model with three parameters is developed slightly later. Liu and Zhang have furthermore developed a model, which takes into account the influence of the temperature on the degradation rate of pesticides.

Zhong and Overcash have developed a model for photodegradation of TCDD in soil. The concentration of TCDD is in the model determined by five processes: 1) convection of TCDD together with the organic solution in which TCDD is dissolved, 2) diffusion of the organic solution within the top-layer, 3) photoreaction of TCDD within the top-layer, 4) photoreaction of TCDD in organic solution within the top-layer and 5) photoreaction of TCDD adsorbed on soil at the surface.

The model Leachp is one module of a software package that includes various submodels dealing with inorganic ions (Leachc), transport and transformation of nitrogen (Leachn) and water flow (Leachw). Leachp is a able to predict displacement and degradation of pesticides in the root zone and the underlying unsaturated zone. The simulation includes the soil water profile. The model considers the solute movement, the adsorption using Freundlich's adsorption isotherm and degradation processes of pesticides.

The model Polmod.Pest. consists of five submodels according to the units of an ecosystem: atmosphere, soil, vegetation, underground and surface water. The forcing functions are the meteorological data (temperature, precipitation, wind speed, etc.). The state variables are the concentrations of the considered pesticides in the five compartments.

The model Gleams has had a wide application of simulations on regional distribution of pesticides and nutrients. The forcing functions are daily precipitation, air temperature, wind speed, dew point, topography and physical soil characteristics. State variables are concentrations of nutrients and pesticides in the various compartments (the root zone, the surface water, the sediment, etc.).

The DDT model was developed to focus on the concentration of DDT in fish, which is the most pertinent state variables. As this concentration is determined by the concentrations throughout the food chain, it is necessary to include at least the concentration of DDT in phytoplankton. The adsorption, microbiological decomposition and photolytic degradation must also be included to determine the concentration in water, which determines the concentrations in all the biological components of the considered aquatic ecosystem.

**Table 3.1.** Overview of Pesticides Models

| Toxic substance | Model class | Model characteristics | Reference |
|---|---|---|---|
| **Pesticides in fish (DDT, methoxy-chlor)** | E | CF, adsorption on body, defecation, ingestion, decom-position, excretion | Leung, 1978 |
| **Methylparathion** | A | Adsorption, microbiological decomposition, 2-4 trophic levels, chemical processes | Lassiter, 1978 |
| **Pesticides** | D | Effects on insect populations | Schallje et al., 1989 |
| **Insecticides** | B | Resistance of var. insecticides | Longstaff, 1988 |
| **Pesticides** | C | Degradation in soil 2 or 3 parameters | Liu and Zhang, 1987 and 1988 |
| **Pesticides** | C | Leaching to groundwater | Carsel et al., 1985 |
| **Pesticides** | C | Run-off losses | Albanis, 1992 |
| **Pesticides** | C | Influence of temperature on loss rates | Zhang et , al., 1993 |
| **TCDD** | C | Photodegradation | Zhong and Overcash, 1993 |

**Table 3.1(Continued)** Overview of Pesticides Models

| Toxic substance | Model class | Model characteristics | Reference |
|---|---|---|---|
| **Pesticides** | C | Displacement and degrada-<br>in the root zonetion in<br>the root zone | Hutson and<br>Wagenet,<br>1992 |
| **Pesticides in<br>ecosystems** | C | Atmosphere, soil, vegetation<br>ground- and surface water | Pykh<br>et al., 1992 |
| **Pesticides** | C | Distribution in a watershed/<br>interaction with nutrients | Leonard et<br>al., 1987 |
| **DDT** | | Bioaccumulation, CF, degra-<br>dation processes | Jørgensen<br>et al., 1982 |
| **Methomyl** | C | Groundwater contamination:<br>adsorption, decay, flow<br>pattern | Dowling et<br>al., 1994 |
| **Pesticides** | C | Degradation and movement | Jones, 1994 |

# 3.5 Models of Pesticides

**Model identification**
Model name/title: ESPELOR- EStimation of PEsticide LOsses in Runoff from agricultural areas in surface waters.

**Model type**
Media: water, estuary, river, wetland, local scale, regional scale, agricultural
Class: hydrology
Substance(s)/compounds: nutrients, chemicals, pesticides, organic compounds
Biological components:

**Model purpose**
ESPELOR is a model which can calculates the amounts of pesticides released through surface waters from agricultural areas. This collective model depends on pesticide concentrations in surface waters, water flow rates and on their changes with time.

**Short description**
Movement beyond the point of application of the pesticides used in tillage systems can be determined by water losses from field. Significant amounts of water can exit the fields as surface runoff, carrying the pesticides dissolved, or absorbed to particulate soil material. ESPELOR is a collective model for estimation of pesticide runoff losses from agricultural areas through draining streams, pumping stations and rivers in their final destination. The amounts of pesticides that are transported in each sub-period (time between sampling dates) i, within agricultural waste waters through the j draining point (sampling) can be calculated.

The total amount of each pesticide which are transported through the surface waters can be determined by the collective equation. The water amounts which flow out from fields in each subarea in different seasons is one important factor which affects the concentration differences in sampling points. The concentration changes according to the total water flow through each draining-sampling point.

**Model application(s) and State of development**
Application area(s): Pesticide transportation in surface water/Pesticide losses in runoff
Number of cases studied: 4
State: conceptualization, validation, prognosis made?

**Model software and hardware demands**
Platform: PC-based, MAC, digital
OS:
Software demands:
Hardware demands:

## Model availability
Purchase: no
Contact person(s) Dr. Triandafillos A. Albanis
     Department of Chemistry
     University of Ioannina
     Ioannina 45110
     Greece
Phone: + 30-651-98-348
Fax: + 30-651-44-836
E-mail:

## Model documentation and major references
Albanis, T.A., Runoff losses of EPTC, molinate, simazine. propanil and metolachlor in Thermaikos Gulf, N. Greece, *Chemosphere*, 22, 645-653 (1991).
Albanis, T.A., "Herbicide losses in runoff from agricultural area of Thessaloniki in Thermaikos Gulf, N. Greece, *The Science of Total Environment*, 114, 59-71 (1992).
Albanis, T.A., "Transportation of herbicides in estuaries of .Axios, Loudias, Aliakmon, Louros and Arachthos rivers (Greece), Report of the FAO/IAEA/UNEP, Review Meeting, Athens, 7-9 April 1992.
Albanis, T.A., Danis, T.G., and M.G. Kourgia, Transportation of pesticides in estuaries of Axios, Loudias and Aliakmon rivers (Thermaikos Gulf), *The Science of the Total Environment* (1994).

## Other relevant information


## Model identification
Model name/title: The model with two parameters for microbial degradation of pesticides

## Model type
Media: soil
Class: biogeochemical
Substance(s)/compounds: pesticides, organic compounds
Biological components: population, pesticide, degrading microorganisms

## Model purpose
Describing the variation with time of the concentration of organic compounds (e.g. pesticides) in soil.

## Short description
The concentration of pesticide at time t, is related to the growth of pesticide-degrading microorganisms, at the initial concentration of pesticide, and the concentration of pesticide when -dx/dt reaches its maximum, tf the time when -dx/dt reaches its maximum, and t1/2 the half-time for the loss of pesticide.

## Model application(s) and State of development
Application area(s): The degradation linetics of organic compounds which can be degraded by microorganisms
Number of cases studied: 17
State: conceptualization, verification, validation, validation of prognosis

## Model software and hardware demands
Platform: PC-based, IBM
OS: MS-DOS
Software demands: BASIC
Hardware demands:

## Model availability
Purchase: no
Contact person(s)

Phone:
Fax:
E-mail:

## Model documentation and major references
Liu, D.S. and Zhang S.M., 1987. Kinetic model for degradative processes of pesticides in soil. *Ecol. Modelling*, 37: 131-138.

## Other relevant information
The rate of degradation is proportional to the concentration of pesticide (x) and the number of pesticide-degrading microorganisms (m), i.e., $-dx/dt=kxm$, where k is the rate constant. This concept is the basic premise from which the model with two parameters is derived.

## Model identification
Model name/title: The model with three parameters for microbial degradation of pesticides

## Model type
Media: soil
Class: biogeochemical
Substance(s)/compounds: pesticides, organic compounds
Biological components: population: pesticide-degrading microorganisms

## Model purpose
Describing the variation with time of the concentration of organic compound (e.g., pesticide) in soil.

## Short description
The concentration of pesticide at time t, related to the growth of pesticide-degrading microorganism, at the initial concentration of pesticide, two roots of the polynomial $k1+k2x+k3x^2$, xf the concentration of pesticide when -dx/dt reaches its maximum, tf the time when x=xf, and t1/2 the half-life for the loss of pesticide.

## Model application(s) and State of development
Application area(s): The degradation kinetics of organic compounds which can be degraded by microorganisms
Number of cases studied: 2
State: conceptualization, verification, validation, validation of prognosis

## Model software and hardware demands
Platform: PC-based, IBM
OS: MS-DOS
Software demands: BASIC
Hardware demands:

## Model availability
Purchase: no
Contact person(s)

Phone:
Fax:
E-mail:

## Model documentation and major references
Liu, D.S., Zhang, S.M. and Li, Z.G., 1988. Study on rate model of microbial degradation of pesticides in soil. *Ecol. Modelling*, 41: 75-84.

## Other relevant information
The rate of degradation is proportional to the concentration of pesticide and the number of pesticide-degrading microorganisms. This concept is the basic premise from which the model with three parameters is derived.

## Model identification
Model name/title: The kinetic model describing the effect of temperature on pesticidal loss rate

## Model type
Media: soil
Class: biogeochemical
Substance(s)/compounds: pesticides, organic compounds
Biological components: population: pesticide- degrading microorganisms

## Model purpose
Describing the relation between the pesticde loss rate constants and temperature in soil

## Short description
The concentration of pesticide at time t, with a first-order rate constant K1 at absolute temperature T, and k2 a second-order rate constant at T, c the minimum growth temperature for pesticide-degrading microorganisms, h the maximum growth temperature for the microorganisms, A1* and A2* the constants relating to the pre-exponential factor of growth  growth process of the microorganisms, E* the activation energy for growth process of the microorganism, R the gas constant, A1 and A2 the constants relating to the pre-exponential factor of nonbiological loss process of the pesticide, and E the activation energy for nonbiological loss process of the pesticide.

## Model application(s) and State of development
Application area(s): The temperature effects on the loss rate of organic compounds which can be degraded by microorganisms
Number of cases studied: 1
State: conceptualization, verification, validation, validation of prognosis

## Model software and hardware demands
Platform: PC-based, IBM
OS: MS-DOS
Software demands: BASIC
Hardware demands:

## Model availability
Purchase: no
Contact person(s)

Phone:
Fax:
E-mail:

## Model documentation and major references
Zhang, S.M., Liu, D.S., Wang, Z.S. and Ma, X.F., 1993. A kinetic model describing the effect of temperature on the loss rate pesticides in soil. *Ecol. Modelling*, 70: 115-125.

## Other relevant information

## Model identification
Model name/title: Near sunlight zone model for photodegradation of TCDD in soils containing organic solvents

## Model type
Media: soil
Class: toxicology
Substance(s)/compunds: chemicals (dioxins)
Biological components: none

## Model purpose
The near sunlight zone model was developed to identify and quantify the controlling factors governing the processes of transport and photolysis of TCDD in soil.

## Short description
The model is based on the physical and chemical processes affecting TCDD concentrations near the soil surface. The TCDD concentration is determined by the following four processes: (1) convection of TCDD together with the movement of the organic solution in which TCDD is dissolved; (2) diffusion in the organic solution along a TCDD concentration gradient; (3) photoreaction of TCDD in organic solution within the top layer; (4) photoreaction of TCDD absorbed on soil at the surface. The model needs the following information: porosity of the soil, saturation of solvent in the soil matrix, photoreaction rate constant. Based on these inputs, the model can tell which process is the controlling step, and draw a clear picture of the TCDD concentration variation versus time in the top layer.

## Model application(s) and State of development
Application area(s): soil decontamination
Number of cases studied: two
State: verification

## Model software and hardware demands
Platform: PC-based or Digit VAX
OS:
Software demands: FORTRAN 77
Hardware demands:

## Model availability
Purchase: no
Contact person(s) Dr. Yaping Zhong  or    M.R. Overcash
Department of Chemical Engineering
North Carolina State University
Raleigh, NC 27695, USA

---

**Model documentation and major references**
*Chemosphere*, Vol. 26, No. 7, pp. 1263-1272, 1993.

---

**Other relevant information**

---

**Model identification**
Model name/title: LEACHP: Leaching Estimation and Chemistry Model; Module Pesticide

---

**Model type**

Media: local scale, terrestrial, soil, agricultural
Class: hydrology
Substance(s)/compounds: chemicals, pesticides
Biological components:

---

**Model purpose**
Simulation model for predicting pesticide displacement and degradation in the root zone of agricultural crops, and in the underlying unsaturated zone. Including simulation of the soil water profile.

---

**Short description**
LEACHP is one module of the process based Leaching Estimation And Chemistry Model (LEACHM) software package which includes additional sub-models: LEACHC (transient movement of several inorganic ions), LEACHN (transport and transformation of nitrogen), LEACHB (microbial population dynamics) and LEACHW (water flow).

A fixed depth water table, a free draining profile, zero flux or a lysimeter tank can be chosen to represent the lower boundary in order to evaluate field and laboratory (column) experiments. One-dimensional (vertical) water flow is calculated using Richard's equation. Fitted parameters of a modified Campbell's equation fitted to the soil moisture release curve are required as input data.

The convective dispersive equation (CDE, apparent diffusion coefficient as input) is utilized to calculate the solute movement. Either the linear or the Freundlich adsorption coefficient is necessary to characterize the adsorption behaviour of the considered chemical. Sequential degradation of pesticides can be calculated if first-order degradation constants are available.

---

**Model application(s) and State of development**
Application area(s): Field (ha) and laboratory column studies
Number of cases studied: several
State: validation

---

**Model software and hardware demands**
Platform: PC-based IBM
OS: DOS
Software demands: FORTRAN
Hardware demands: 286 Math coproceesor

---

**Model availability**
Purchase: yes, US $100, source code
Contact person(s) Dr. J.L. Hutson
               Department of Agronomy
               235 Emerson Hall
               Cornell University
               Ithaca, NY 14853
               USA
Phone:
Fax:
E-mail:

---

**Model documentation and major references**
Hutson, J.L. and Wagenet, R.J., 1992. LEACHM: A process-based model of water and solute movement, transformations, plant uptake and chemical reactions in the unsaturated zone. Version 3.0, N.Y. State College of Agriculture and Life Science, Cornell University, Ithaca, NY, Department of Soil, Crop and Atmosphere Sciences, Research Series No.92-3.

---

**Other relevant information**

---

**Model identification**
Model name/title: POLMOD.PEST: model for simulation of pesticides' dynamics in the elementary ecosystem

---

**Model type**
Media: local scale
Class: toxicology
Substance(s)/compounds: pesticides
Biological components: ecosystem - elementary ecosystem

---

**Model purpose**
Model is intended to describe the flow of pesticides in the units of ecosystem -- atmosphere, soil, vegetation and surface and underground water. The model also calculates the level of pesticide or radioactive pollution accumulated in each unit of the elementary ecosystem.

## Short description
POLMOD.PEST is a software product for simulation the pesticides in the units of elementary ecosystem. The main term of POLMOD.PEST model is the resistance index. We define the resistance index of ecosystem's compartments as the index of ability to resist the pollutions' flow either due to self-purification ability or due to the decrease of the accumulation rate.

POLMOD.PEST consists of five submodels in accordance with the main ecosystems' units -atmosphere, soil, vegetation, underground and surface water submodels.

Major state variables are the concentrations of pesticide in the units of ecosystem enumerated above. Time step equals one year.

Model input includes the date on meteorological, physical and chemical properties of each unit of the ecosystem: precipitations, probability of temperature inversions in the atmosphere, wind speed, pH (soil acidty), soil humus content, cation exchange capacity of soils, calcium cation content, as well as the area of drainage basin, illumination, water pH (acidity) and water temperature.

## Model application(s) and State of development
Application area(s): Scientific investigations of pollutants' dynamics in the ecosystems; education and training decisions support system for various kinds of environmental problems.
Number of cases studied: several hundred
State: validation, validation of predictions

## Model software and hardware demands
Platform: PC-based IBM PC AT 286 (386)
OS: MS DOS, version 3.2 or higher
Software demands: Turbo Pascal compiler, Turbo Pascal, version 5.0.
Hardware demands: 640 k RAM, EGA graphic adapter, math coprocessor

## Model availability
Purchase: yes, available as source code
Contact person(s) Dr. Irina G. Malkina-Pykh
    Center for International Environmental Cooperation (INENCO)
    St. Petersburg, 190000
    Chernomorsky per., 4
    Russia
Phone: 812 311-7622
Fax: 812 311-8523
E-mail: pykh@inenco.spb.su

## Model documentation and major references
Pykh, Yu.A., and Malkina-Pykh, I.G., 1992. POLMOD.PEST - the model of pesticides' dynamics in the elementary ecosystem. Preprint, Moscow-St. Petersburg, Center INENCO, 1992.

Iouri Pykh
International Institute for Applied System Analysis, A-2361 Laxenburg, Austria

Email: pykh@iiasa.ac.at
phone: +43 2236-71521 0
fax: +43 2236-71313

## Other relevant information

## Model identification
Model name/title: GLEAMS

## Model type
Media: water, local scale, sediment, soil, forest, agricultural
Class: hydrology
Substance(s)/compounds: nutrients, phosphorus, nitrogen, radionuclides, pesticides
Biological components:

## Model purpose
To assess soil-climate-pesticides-nutrients interactions.

## Short description
Daily climatic data input (daily precipitation and temperature, monthly wind and dew point temperature), soil characteristics (water retention, sat. conductivity, texture, chemical characteristics), Topography. Pesticide characteristics and applications Nutrient input, State variables include soil water content nitrogen and phosphorus pools. GLEAMS is a daily simulation model to compare response from alternate management practices. Response includes runoff, percolation below the root zone, sediment yield, water and sediment fractions of pesticides and nutrients.

## Model application(s) and State of development
Application area(s): USA + 32 foreign countries
Number of cases studied: numerous
State: completed

## Model software and hardware demands
Platform: PC-based IBM compatible, Mainframe digital/IBM, Wotkstation SUN
OS: DOS, UNIX
Software demands: 640 kB RAM; 25 files language FORTRAN/C+
Hardware demands: 5 Mb hard drive

## Model availability
Purchase: no cost for software, available both as executable and source code
Contact person(s) Dr. Kevin King
808 E. Blackland Road
Temple, TX 76502
USA

Phone: +1 817-770-6500
Fax: +1 817-770-6561
E-mail:

## Model documentation and major references
Knisel (Ed.) 1993. Gleams: Groundwater Loading Effects of Agricultural Management Systems. UGA-CPES, Biological and Agricultural Engineering Department, Publ. No. 5, 260 pp.

## Other relevant information
Leonard, R.A., Knisel, W.G., and Still, D.A., 1987. GLEAMS: Groundwater Loading Effects of Agricultural Management Systems. *Trans. Am. Soc. Agric. Engrs.*, 30(5): 1403-1418.

## Model identification
Model name/title: DDT-model

## Model type
Media: water, lake
Class:
Substance(s)/compounds: chemicals, pesticides DDT
Biological components: ecosystem

## Model purpose
Control of DDT pollution.

## Short description
State variables: DDT in water, on suspended matter, in sediment, in phytoplankton and in fish. Forcing functions: inputs of DDT, climate. Important parameters: adsorption isotherm for DDT on suspended matter, biodegradation rate of DDT, uptake of DDT by phytoplankton and fish, bioaccumulation phytoplankton to fish.

## Model application(s) and State of development
Application area(s):
Number of cases studied: 1
State: calibration

## Model software and hardware demands
Platform: PC-based
OS:
Software demands: language SYSL
Hardware demands:

## Model availability
Purchase: no
Contact person(s) Dr. S.E. Jørgensen
> Section of Environmental Chemistry
> Royal Danish School of Pharmacy
> Danmarks Farmaceutiske Højskole
> Universitetsparken 2
> DK-2100 København Ø
> Denmark

Phone: +45 35 37 08 50
Fax: +45 35 37 57 44
E-mail: sej@pommes.dfh.dk

## Model documentation and major references
Jørgensen, S.E., Kamp Nielsen, L., Jørgensen, L.A., and Mejer, H., 1982. An environmental management model of the Upper Nile lake system. *ISEM Journal*, 4: 5-72.

## Other relevant information

## Model identification
Model name/title: Plant Uptake Fugacity Model

## Model type
Media:
Class:
Substance(s)/compounds: organic compounds
Biological components: organism - plants

## Model purpose
To calculate the dynamic uptake of organic chemicals by plants from soil and the atmosphere.

## Short description
This is a five compartment model consisting of air, soil, roots, stem (including fruit) and foliage. Chemical concentrations in air and soil are defined as forcing functions. The transport of chemical between soil and root, root and stem, stem and foliage, foliage and air, and air and soil is computed. Metabolism or degradation can be included.

The dynamic behaviour of the chemicals is computed over a time period of hundreds of hours. Plant growth is not involved. The model and its application to Bromacil uptake from hydroponic solution is described in a paper by Palerson et al.,*Environ. Sci. Technol.*, 1994 28 p. 2259.

## Model application(s) and State of development
Application area(s): Uptake of Bromacil by the Soybean plant.
Number of cases studied: 1
State: calibration

## Model software and hardware demands
Platform: IBM
OS:
Software demands: Minimal, language - BASIC
Hardware demands:

## Model availability
Purchase: no. Software available as source code.
Contact person(s) Dr. D. Mackay
　　　　　　　　Institute for Environmental Studies
　　　　　　　　University of Toronto
　　　　　　　　Toronto, Ontario M5S1A4
　　　　　　　　Canada
Phone:
Fax:
E-mail:

## Model documentation and major references
Palerson et al., *Environ. Sci. Technol.*, 1994, 28, p. 2259.

## Other relevant information
The model is still under development.

## Model identification
Model name/title: Rate Constant Model of Chemical Fate in Lakes

## Model type
Media: lake
Class: hydrology
Substance(s)/compounds: chemicals
Biological components: community - aquatic food chains

## Model purpose
To deduce the fate of chemical discharges to a lake system, consisting of air, water, suspended matter, bottom sediments and an aquatic food chain using a rate constant approach.

## Short description
This model is intended for use in describing the fate of organics in a large lake such as Lake Ontario. The processes treated include discharge, evaporation, wet and dry deposition from the atmosphere, sediment-water exchange, transformation in the water column and sediment and outflow. Each process is expressed in terms of rate constant to facilitate interpretation. A simple bioaccumulation food chain is included. The input data include the dimensions of the lake, physical-chemical properties of the substance and discharge rates and atmospheric concentration which may vary with time.

## Model application(s) and State of development
Application area(s): Lake Ontario
Number of cases studied: 1
State: calibration

## Model software and hardware demands
Platform: PC-based - IBM
OS:
Software demands: language - BASIC
Hardware demands:

## Model availability
Purchase: no, available gratis. Available as source code
Contact person(s) Dr. D. Mackay
                  Institute for Environmental Studies
                  University of Toronto
                  Toronto, Ontario M5S1A4
                  Canada
Phone:
Fax:
E-mail:

## Model documentation and major references
Mackay, D., et al., 1994. A rate constant model of organic chemical behavior in a large lake. *J. Great Lakes Res.*, Dec. 1994.

## Other relevant information

## 3.6 An Overview: Models of Other Toxic Organic Compounds

Pesticides are treated separately from other organic compounds; see Section 3.4, because pesticides are applied purposely and directly in nature (agriculture and forestry), while all other organic toxic substances are more or less accidentally finding their ways into nature, as a consequence of anthropogenic activity. This difference implies that the models of pesticides are concerned with the risks, that are consequences of an accepted and quantitatively known application, while the models developed for other organic compounds are concerned with the risks of leakage to nature as contaminant of air, water and/or soil. The modeling difference is therefore that the focal forcing function - the amount of toxic substance - is known for pesticides, but not - at least only to a certain extent - for other toxic organic compounds.

Table 3.2 gives an overview of the models presented here and other slightly older models, focusing on organic compounds other than pesticides. As for Table 3.1., the model class, the characteristics of the model and the main reference are all given in the table.

The model Ecofate is an integrated fate and food-chain model for organic chemicals in aquatic ecosystems. The model is based upon 1) information about the ecosystem: the configuration of the aquatic ecosystem, 2) information about the organisms/the biological components: the weight, the lipid content and the feeding rate and preference of the aquatic organisms and 3) information about the toxic organic compounds: the octanol-water partition coefficient, the dissociation constant and the Henry's Law constant. The state variables are the concentration of the organic compound in water, in sediment, on suspended matter, in all the included aquatic organisms.

The model Masas is a one-dimensional vertical model, which is developed to investigate the dynamic behavior of organic micropollutants. The model goes into detail about the processes that are crucial for the dissolved concentration in water. It implies that the model considers the transport processes, adsorption, sedimentation, air-water and sediment-water exchange processes and chemical reactions, including hydrolysis and photolysis.

MacKay has developed several fate models aimed for answering the question: which concentrations are to be expected in the various media. The models are excellent tools to compare various chemicals and to indicate where measures should be taken to reduce the risk of harm.

Cemos is another fate model, concerned with exposure concentrations after point and diffuse releases to the environment of organic compounds. The model consists of the following compartments: water (a box model for discharge into an aquatic ecosystem), plume, air, soil, plant and chain (bioaccumulation through a food-chain). The model has two levels: a steady-state, equilibrium level and a non-steady-state, equilibrium level. The implementation of the model allows an easy integration of additional model modules, developed by the user.

194

"Seabird" is a model developed to deal with the impact of oil spills on seabird populations. The model synthesizes the most well-known factors determining the susceptibility of seabird to oil spills. It is a two-dimensional model describing the two-dimensional spreading of an oil spill toward land and of the two-dimensional distribution of the seabird population on the coast. The model is developed on basis of meteorological data, information about a specific oil spill, and information about the initial distribution of seabirds. An assessment of the effect of oil on population parameters is needed.

The Ohio River Oil Spill Model is developed to deal with a particular accident, but its simplicity could inspire modelers to develop their own models for similar cases. It has many of the characteristics of a simple river model concerned with the transport processes. In addition it considers the sedimentation and resuspension.

The Unifac model is a model aimed for prediction of physical-chemical properties of chemicals based upon the group contribution method.

Endicott and Cook have developed a model of the partitioning and bioaccumulation of hydrophobic chemicals in Lake Ontario. The model is based upon the use of an Excel spreadsheet, which solves the steady state mass balance equations for a completely mixed water column and underlying sediment layer. The steady state bioaccumulation in a four level food chain is included in the model. The approach could easily be applied for other lakes than Lake Ontario.

Toxfate (version 3.6) is a model developed to simulate and predict the fate of toxic organic contaminants in large lakes. It integrates information on the properties of a chemical with the environment. The model simulates the time varying concentrations of a toxic contaminant in the water column and in the bottom sediment by the use of differential equations.

**Table 3.2** Models of Organic Compounds, other than Pesticides

| Toxic substance | Model class | Model characteristics | Reference |
|---|---|---|---|
| Vinyl chloride | C | Chemical processes in water | Gillett et al., 1974 |
| Toxic compounds | F | Hazard ranking and assessment of chemicals in general | Bro-Rasmussen Christiansen, 1984 |
| PAH | C | Transport, degradation and bioaccumulation | O'Neill, 1984 |
| Persistent toxic organic substances | C | Groundwater movement, transport and accumulation | Uchrin, 1984 |
| PCB | B | Hydraulic overflow rate (settling), sediment interactions, steady state food chain submodel | Thomann, 1984 |
| Hydrophobic organic | C | Gas-exchange, sorption/desorp., hydrolysis, photolysis, hydrodynamics | Schwarzenbach and Imboden, 1984 |
| Mirex | C | Water-sediment exchange, adsorption, volatilization, bioaccumulation | Halfon, 1984 |
| Aromatic hydrocarbons | C | Hydrodynamics, deposition, resuspension, volatilization, phooxidation, decomposition, adsorption, complex formation, (humic acid) | Harris et al., 1984 |
| Oil slicks | C | Transport and spreading, influence of surface tension, gravity and weathering processes | Nihoul, 1984 |
| Persistent organics Chemicals | F | Fate, exposure and human uptake | Paterson and Mackay, 1989 |
| Org. compounds | F | Fate, exposure, ecotoxicity for water and soil | Matthies et al., 1989 |

**Table 3.2 continued** Models of Organic Compounds, other than Pesticides

| Toxic substance | Model class | Model characteristics | Reference |
|---|---|---|---|
| Toxicants, general | D | Effects on populations | Luna and Hallam, 1987 |
| Chemicals hazard | F | Basin-wide ecological fate | Morioka and Chikami , 1986 |
| Mirex and Lindane | F | Fate in Lake Ontario | Halfon, 1986 |
| Photochem. smog | F | Fate and risk | Wratt et al., 1992 |
| Oil spill | F | Fate, spreading | Cronk et al., 1990 |
| Toxicants | D | Effects on populations | Gard, 1990 |
| Toluene, benzene | C | Assess the exposure levels of soil contaminants for human beings | ECETOC, 1990 |
| Ecofate | A | Distribution in aquatic ecosystem, incl food chain | Gobas, 1992 |
| Masas | C | 1D aquatic model incl. adsorption, sedimentation, hydrolysis, photolysis, exchange processes | Ulrich et al., 1994 |
| Cemos | F | Water, soil, air, plume, plant and bioaccumulation | Handbook, 1994 |
| Seabird, oil spill | D | Impact on seabird-populations | Seip et al., 1991 |
| Unifac | - | properties of chemicals | Chen et al., 1992 |
| Hydrophobic chemicals | A | Mass balance equations water, sediment and biota | Endicott et al., 1994 |
| Toxfate | F | Fate in lakes Incl. water, food chains and sediment | Halfon et al., 1990 |
| Simon | C | 3-D model + spreading and | Rajar; see model in section 3.7 |
| Sipor | C | River model with decay, settling, evaporation + adsorption | Rajar: see model in 3.7. |
| Chespill.For | C | spill sinking to the bottom | Christy, 1980 |

The model Simon offers a 3-D simulation of circulation and oil spill in lakes and coastal areas. It is an integrated environmental model for prediction the transport and fate of non-conservative pollutants for instance oil in large aquatic ecosystems. It consists of the model of hydrodynamics, mentioned in Section 2.7, PCFLOW3D, or another similar model, PT3D, to which equations of oil-spill spreading and weathering are added.

The model denoted Sipor is a 1-D transport and dispersion model for prediction of hazardous spill dynamics in rivers. It considers spreading with the river flow, decomposition, adsorption to suspended matter, settling and evaporation.

The model Chespill.For simulates the spill of pure liquids heavier than water. It is described in the model, that it sinks to the bottom when spilled. Spills may be classified into three categories: soluble compounds which mix with the water very rapidly, immiscible (slightly soluble and insoluble) compounds that float on water and immiscible compounds that sink to the bottom. The model focuses on the latter situation.

# 3.7 Models of Other Organic Compounds

**Model identification**
Model name/title: ECOFATE

**Model type**
Media: lake, estuary, river, regional scale
Class: toxicology
Substance(s)/compounds: chemicals, organic compounds
Biological components: ecosystems, aquatic

**Model purpose**
To predict, on an ecosystem-level, the concentration of organic chemicals in water, sediments and aquatic organisms (i.e., phytoplankton, benthos, fish and fish-eating birds) resulting from chemical emissions.

**Short description**
ECOFATE is an integrated environmental fate and food-chain accumulation model for organic chemicals in aquatic ecosystems. The model simulates, in a time-dependent fashion, the transport, transformation and trophic transfer of organic chemicals in aquatic ecosystems (i.e., rivers, lakes and marine systems). The model consists of an environmental fate and a biological food-chain model, which can be used individually or in combination. The model is based on the numerical solution of a set of differential equations, each representing the flux of the chemical in each environmental compartment as a function of the kinetic rate constants for chemical transport and transformation. Important parameters for using the model include (1) data regarding the configuration of the ecosystem (dimensions, flow, temperature, resuspension and sedimentation rates), (2) the weights, lipid contents and feeding preferences of aquatic organisms and (3) the octanol-water partition coefficient, the dissociation constant and the Henry's Law Constant of the chemical substance. The model contains various algorithms to derive feeding rates, gill ventilation rates in fish, uptake and elimination rate constants of organic chemicals in biota, volatilization rate constants, bioavailability and others.

**Model application(s) and State of development**
Application area(s): The model has been successfully tested for PCBs in Lake Ontario, for chlorodibenzo-p-dioxins and chlorodibenzofurans in the Fraser-Thompson River basin, a large river system in British Columbia (Canada) and in Howe Sound, a marine system.
Number of cases studied: 3
State: The model haas been tested and "validated" by comparing chemical concentrations predicted by the model based on observed chemical emissions to observed concentrations. The model needs to be configured for each ecosystem. In our modeling approach we do not calibrate the model.

**Model software and hardware demands**
Platform: PC-based, IBM
OS: DOS
Software demands: Windows 3.0 or 3.1
Hardware demands: 386 or 486

**Model availability**
Purchase: yes, tba. Executable code
Contact person(s) Dr. F. Gobas
               School of Resource and Environmetal Management
               Simon Fraser University
               Burnaby, British Colombia
               Canada, V7K 1S6
Phone: +1 604-291-5928
Fax: +1 604-291-4968
E-mail: gobas@SFU.CA

**Model documentation and major references**
Gobas, F.A.P.C., 1992. *Ecol. Modelling*, 69: 1-17.
Gobas, F.A.P.C., 1992. Modelling the accumulation and toxic impacts of organic chemicals in aquatic food-chains. In: *Chemical Dynamics in Fresh Water Ecosystems* (Gobas, F.A.P.C. and J.A. McCorquodale, Eds.), Lewis Publishers, Boca Ratan, FC, pp. 129-153.

**Other relevant information**

**Model identification**
Model name/title: MASAS - Modelling Anthropogenic Substances in Aquatic Systems.

**Model type**
Media: lake
Class: toxicology
Substance(s)/compounds: chemicals, radionuclides, pesticides, organic compounds
Biological components:

**Model purpose**
MASAS is a user-friendly simulation tool to investigate the dynamic behaviour of organic micropollutants in lakes. The program with a user interface consisting of menus, standard dialog boxes and interactive text and graphic windows, is currently in use both in teaching and research.

## Short description

Underlying the program is the mathematical description of the behaviour of a substance in terms of a one-dimensional vertical model. MASAS computes the concentrations of the substance of interest in the water column (and, optionally, in the sediments) as a dynamic model variable. Transport and transformation processes (loading, adsorption, sedimentation, air-water and sediment-water exchange, chemical reactions) can be defined interactively. The results of simulation runs are output both graphically and on file, and the interactively defined models can be saved for future use.

Library files allow fast access to data on both substances and lakes. Substances are characterized in these files by physico-chemical parameters (vapor pressure, Henry coefficient, etc.) and reactivities (hydrolysis rates, etc.). The lakes are characterized in terms of morphometric (volume, isobath areas, etc.), hydraulic (rate of throughflow, depth of epilimnion, etc.), and physical and chemical (temperature, pH, particle concentration, etc.) parameters.

Models with different degrees of spatial resolution (one-box, two-box, "combi-box", n-box) are available. Processes can be described on various hierarchical levels. Information fields show the user which parameters are missing, and which routines are available to approximate these missing parameters.

The program can be used to solve problems related to model validation, characterization of unknown transport and transformation processes, estimation of input to a lake and of unknown process rates, predictions, etc.

## Model application(s) and State of development

Application area(s): Quantification of transport and transformation processes of organic chemicals in lakes, hazard assessment, predictions.
Number of cases studied: 6
State: conceptualization, verification, calibration, validation - partly, prognosis made?, validation of prognosis - partly.

## Model software and hardware demands

Platform: MAC
OS: MacIntosh - OS
Software demands: Macintosh, English
Hardware demands: 1 Mb RAM, 2 Mb hard disk

## Model availability

Purchase: yes, US $150.00, non-commercial, US $750.00 commercial. Executable code.
Contact person(s) Dr. Markus Ulrich
          Swiss Federal Institute
          for Environmental Science and Technology (EAWAG)
          CH-8600 Duebendorf
          Switzerland
Phone: +41 1 823 54 64
Fax: +41 1 823 5210
E-mail: ulrich@eawag.ch

**Model documentation and major references**

Ulrich, M.M., Imboden, D.M., and Schwarzenbach, R.P., 1994. Modeling of organic pollutants in lakes with the MASAS system - The way from the real system to the user-friendly simulation software. *Environmental Modelling*. In preperation.

Ulrich, M., 1991. Modeling of Chemicals in Lakes - Development and Application of User-friendly Simulation Software (MASAS & CHEMSEE) on Personal Computers. Dissertation ETH No. 9632. Swiss Federal Institute of Technology, Zurich, Switzerland.

Dzambas, Z., and Ulrich, M., 1994. MASAS Library Files for Swiss Lakes. Description of 44 Lake Data Sets for Modeling Organic Substances using the MASAS-Software. Report, EAWAG, Duebendorf, Switzerland.

Ulrich, M., Schwarzenbach, R.P., Imboden, D.M., 1991. MASAS - Modelling of Anthropogenic Substances in Aquatic Systems on Personal Computers - Application to Lakes. *Environmental Software* 6/1, 34-38.

**Other relevant information**

**Model identification**

Model name/title: modeling the physical chemical and toxicological properties of an organic compound to illustrate its multimedia partitioning and quantify its expected approximate environmental and human exposure and severity of impact.

**Model type**

Media: multimedia
Class:
Substance(s)/compounds: organic compounds
Biological components: ecosystem

**Model purpose**

A multimedia equilibrium partitioning model was developed to describe sequentially the environmental distribution of animal and human exposure to, and bioconcentration potential of relatively persistent organic chemicals in southern Ontario.

**Short description**

The model incorporates the physical, chemical and toxicological properties of an organic compound to illustrate its multimedia partitioning and quantify its expected approximate environmental and human exposure and severity of impact. The necessary input data includes molecular weight, aqueous solubility, vapour pressure, octanol/water partition coefficient, reactivity data in the form of half-lives of the chemical in various environmental media, release rates of chemical to the environment, and available toxicological effects levels. The model assumes thermodynamic equilibrium between all environmental media. The media include air, water, soil, sediment, fish, vegetation, meat, dairy products and human adipose tissue. Partition coefficients are calculated for each compartment and combined with reaction and advection data to calculate concentrations in each media.

The computed concentrations can be compared to the effects level concentrations and judgement made on the necessity of reducing emissions. The model illustrates the concept of multiple and critical pathways.

---

## Model application(s) and State of development
Application area(s):
Number of cases studied:
State:

---

## Model software and hardware demands
Platform: PC-based
OS: DOS 3.0
Software demands: BASIC
Hardware demands:

---

## Model availability
Purchase: no, would have to be updated
Contact person(s) Dr. Donald Mackay
               Institute for Environmental Studies
               University of Toronto
               200 College Street
               Toronto, Ontario M5S 1A4
               Canada

Phone:
Fax:
E-mail:

---

## Model documentation and major references
Paterson, S. and Mackay, D., A model illustrating the environmental fate exposure and human uptake of persistent organic chemicals. *Ecol. Modelling*, 47, (1989), 85-114.

---

## Other relevant information

---

## Model identification
Model name/title: CEMOS: Chemical Exposure Model System

---

## Model type
Media: water, river, air, local scale, terrestrial, soil, agricultural, multi-media
Class: biogeochemical
Substance(s)/compounds: chemicals, pesticides, organic compounds
Biological components: ecosystem

**Model purpose**
Simulation of transport and fate of hazardous chemicals in single-medium and multi-media environments; calculations of exposure concentrations in environmental compartments of concern after point and diffuse releases; analysis of the dynamics behaviour (persistance, transfer, mobility, accumulation); exposure predictions as part of the risk assessment of new and existing chemicals. Another feature of CemoS is that it can be used as a shell for models developed by the user. The implementation of the system allows an easy integration of additional model modules.

**Short description**
CEMOS consists of models for various environmental compartments:
> WATER: Box model for water discharges
> PLUME: Plume model for point emissions
> AIR: Box model for diffuse area releases
> SOIL: Vertical transport model for solutes and gases in soil
> PLANT: Compartment model for uptake and translocation in plants
> CHAIN: Food chain transfer model
> LEVEL1: Multi-media model (equilibrium)
> LEVEL2DYN: Multi-media model (non-steady-state, equilibrium)

CEMOS is implemented with a window interface, a substance-database with 12 sample substances, numerical and graphical output, estimation routines for substance data and environmental processes, a two level checking of input data ranges, and an online help. The program system offers facilities to save several scenario-files for each model and to cooperate with other programs (e.g., your standard spread-sheet application). Hard copies of the high resolution graphics can be used for a presentation with transparencies.

**Model application(s) and State of development**
Application area(s): Chemical fate and exposure assessment
Number of cases studied:  10
State: partly validation, prognosis made, partly validation of prognosis

**Model software and hardware demands**
Platform: PC-based IBM
OS: MS-DOS
Software demands: language Pascal
Hardware demands: The minimum hardware requirements, needed to run the program system, are: An IBM or an IBM compatible PC with an XT processor, an EGA graphic adapter (or higher) and a floppy disk drive (for higher performance a harddisk is needed). An available numeric coprocessor will be automatically detected and used.

**Model availability**
Purchase: yes, available as executable code
Contact person(s) Prof. Dr. M. Matthies
      Universität Osnabrück
      Fachbereich Mathematik Informatik
      Albrecht Strasse 28
      D-49069 Osnabrück
      Germany

Phone: +49 541 969 2536
Fax: +49 541 969 2770
E-mail:

**Model documentation and major references**
Handbook CEMOS (English and German). Textbook in preparation.

**Other relevant information**
Outlook on further releases.
(1) A module that allows the usage of numerical solved models is to be implemented.
(2) Estimation procedures for different processes (e.g., bio or photo decomposition) are to be implemented.
(3) A risk assessment module is to be implemented in order to classify the simulation results
(4) The program system is to be adapted to MS-Windows.

**Model identification**
Model name/title: SEABIRD

**Model type**
Media: air, local scale
Class: toxicology
Substance(s)/compounds: organic compounds - oils
Biological components: populations - seabirds

**Model purpose**
To describe seabird populatons, and the impact of oil spills on the distribution, abundance, and recovery of the populations.

**Short description**
The model systématises the most well-known factors which determine seabird's susceptibility to oil spills and the subsequent recovery of bird populations. The model is formulated as a simulation model describing the two-dimensional distribution of seabird populations on the coast and the two-dimensional spreading of an oil slick toward land.

Formulations for population dynamics include seabird age, structure, recruitment, mortality, and migration. Thus, the model requires input parameters for the seabird species studied, an assessment of the effect of oil on population parameters (default values are available for Kittiwake, Guilemot and Eider), and a rudimentary oil spill description, and wind velocity and direction.

## Model application(s) and State of development
Application area(s):
Number of cases studied: 3
State: validation, prognosis made, validation of prognosis

## Model software and hardware demands
Platform: PC-based
OS: DOS
Software demands: PASCAL (source code), Pascal
Hardware demands:

## Model availability
Purchase: yes, free except for handling fee. Executable code, source code (for Norwegian)
Contact person(s) Dr. Knut L. Seip
                 SINTEF SI
                 PB 124
                 N-0134 Blindern, Oslo
                 Norge

Phone: +47 22 06 73 00
Fax: +47 22 06 73 50
E-mail: knut.lehre.seip@si.sintef.no

## Model documentation and major references
Seip et al., 1991. *Ecol. Modelling*, 53, 39-59.

## Other relevant information

## Model identification
Model name/title: Ohio River Oil Spill Model

## Model type
Media: water, river
Class:
Substance(s)/compounds: oil (hydrocarbon)
Biological components: ecosystem - Riverine

## Model purpose
To describe the fate and transport of diesel oil spilled into the Ohio River (USA). To emphasize a user-friendly modeling approach instead of utilizing an already developed model.

## Short description
Estimating fate and transport of a freshwater oil spill loss of oil via sedimentation is an important factor as indicated by the model. Model calibrated by comparing to actual values. High flow conditions in the river influenced re-suspension of oil from sediments. Highest concentrations of oil found downstream where flow rate was slow. Based on oil spill in Ohio River system near Pittsburgh, PA on January 2., 1988.

## Model application(s) and State of development
Application area(s): Freshwater oil spills (fate and transport)
Number of cases studied:
State: calibration

## Model software and hardware demands
Platform: MAC
OS:
Software demands: STELLA, language STELLA
Hardware demands:

## Model availability
Purchase: yes $ 30 to OSU wetland project, available as source code
Contact person(s) Dr. William J. Mitsch
                  The Ohio State University
                  School of Natural Resources
                  2021 Coffey Road
                  Columbus, OH 43210 USA

Phone: +1 614-292-9774
Fax: +1 614-292-9774
E-mail:

## Model documentation and major references
Cronk, J., Mitsch, W. and Sykes, R., Effective modelling of a major inland oil spill on the Ohio River. *Ecol. Modelling*, 51: 161-192.

## Other relevant information

## Model identification
Model name/title: The UNIFAC Model

## Model type
Media: water, air
Class:
Substance(s)/compounds: carbon, organic compounds
Biological components:

## Model purpose
To predict various environmentally-relevant physical-chemical properties of chemicals.

## Short description
The UNIFAC model is based on the group-contribution concept. The basic idea of the group-contribution method is that compounds may be structurally decomposed into functional groups. Therefore, if we assume that a physical property of a fluid is the sum of contributions made by the molecules' functional groups, we obtain a possible technique to predict the properties of chemicals for which no experimental data are available.
In the UNIFAC model the activity coefficient Lambda-i is calculated from two parts:

$$Ln\ LAMBDA\text{-}i = Ln\ LAMBDA\text{-}C\text{-}i + Ln\ LAMBDA\text{-}R\text{-}i$$

where LAMBDA-i is the activity coefficient for the component i, LAMBDA-C-i is the combinatorial part, and LAMBDA-R-i the residual part. The combinatorial term depends on the size and the shape of molecule i. In the combinatorial part, the group surface volume and area for each functional group are used as model parameters. The residual term is mainly governed by the energetic interactions of different groups of molecules. The group-interaction parameter, a-m-n and a-n-m, are introduced in this term. There are a total of 223 pairs of group-interaction parameters, for the 50 main functional groups, available. By virtue of the activity coefficient, the UNIFAC model can be applied to describe many environmentally important properties, among those are 1-octanol/water partition coefficient, water solubility, and the Henry's laws constant.

## Model application(s) and State of development
Application area(s):
Number of cases studied:
State: calibration, validation

## Model software and hardware demands
Platform: PC-based
OS:
Software demands: language FORTRAN 77
Hardware demands: co-processor

## Model availability
Purchase: yes, available as executable code.
Contact person(s) Dr. Fei Chen
>
> Miljøministeriet, Danmarks Miljøundersøgelser
> Frederiksborgvej 399
> DK-4000 Roskilde, Denmark

Phone: +45 46 30 1200
Fax: +45 46 30 11 14
E-mail:

## Model documentation and major references
Chen, F., Holten-Andersen, J., and Tyle, H., 1992. New development of the UNIFAC model for environmental application. *Chemosphere* vol.26, p. 1325-1354.

## Other relevant information

## Model identification
Model name/title: Partitioning, Mass Balance, and Bioaccumulation Model for Hydrophobic Organic Chemicals in Lake Ontario. Integrated lake/sediment/aquatic food chain.

## Model type
Media:
Class: hydrophobic organic compounds.
Substance(s)/compounds:
Biological components: population - phytoplankton, zooplankton, forage fish, predator fish.

## Model purpose
The model is intended to be a simple, screening-level tool to predict the steady-state concentrations of HCOs in water, sediment, and biota, in response to constant atmospheric and tributary loads. It is specifically intended for use in Monte Carlo analysis of parameterization error and predictive uncertainty.

## Short description
The model is an Excel spreadsheet which solves the steady-state mass balance equations for a completely mixed water column and underlying sediment layer, and steady-state bioaccumulation equations for a 4-level food chain. The model requires specification of system-specific parameters including hydrologic and particle transport, chemical-specific physicochemcial and toxicokinetic coefficients, and species-specific bioenergetic parameters.

## Model application(s) and State of development
Application area(s): Great Lakes
Number of cases studied: This model was specifically applied to HCOs in Lake Ontario. However, it is based upon the EPA WASP models, which have seen widespread use in the

Great Lakes throughout the United States.
State: This model has been calibrated with limited validation

---

**Model software and hardware demands**
Platform:
OS:
Software demands: MacIntosh and Microsoft Excel (version 3 or 4)
Hardware demands:

---

**Model availability**
Purchase: available (as a spreadsheet) upon request.
Contact person(s) Dr. Douglas Endicott
          Large Lake Research Station
          9311 Groh Road
          Grosse Ile, MI 48138, USA

Phone: +313-692-7600
Fax: +313-692-7603
E-mail: endicott.douglas@epamail.epa.gov.

---

**Model documentation and major references**
Endicott, D.D., and Cook, P.M., 1994. Modeling the partitioning and bioaccumulation of TCDD and other hydrophobic organic chemicals in Lake Ontario. *Chemosphere*, Vol. 28, pp. 75-87.

---

**Other relevant information**

---

**Model identification**
Model name/title: TOXFATE

---

**Model type**
Media: water, lake, local scale
Class: toxicology
Substance(s)/compounds: chemicals, pesticides, organic compounds
Biological components: ecosystem

---

**Model purpose**
TOXFATE was developed to model and predict the fate of toxic organic contaminants in large lakes. In its present configuration ver. 3.6 which includes a benthic food chain as well as a water food chain, the TOXFATE program can interactively run either steady state or dynamical simulations with an MS-DOS machine.

## Short description

TOXFATE is a contaminant fate model which integrates information on the properties of a chemical with the environment where the chemical is found, such as water circulation, wind speed, the amount of suspended solids, etc. The model can be used to predict temporal and spatial distributions of toxic chemicals in a lake. An interface program TOXSHELL allows execution of the model on a personal computer through a simple to use graphics user interface.

TOXFATE simulates the time-varying concentrations of a toxic contaminant in the water column and in the bottom sediments. The model is formalized as a system of ordinary differential equations; the state variables are organic contaminant concentrations in suspended sediments, water, plankton, fish (alewife and salmonid), bottom sediments and benthos compartments. The physical-chemical properties of the toxic contaminants, such as molecular weight, solubility, octanol-water partition coefficient (Kow), and Henry's Law constant are used as input data to the model. The transport processes of advection in water are handled by a box model. The pollutant is assumed to be immediately and completely mixed within each spatial cell. Model output can be viewed on the screen or exported to spreadsheets for publications.

## Model application(s) and State of development

Application area(s):
Number of cases studied: 1
State: validation, validation of prognosis

## Model software and hardware demands

Platform: PC-based
OS:
Software demands: DOS 5.0, language FORTRAN
Hardware demands: PC 386

## Model availability

Purchase: yes, for free, available both as executable and source code
Contact person(s) Dr. Efraim Halfon
          National Water Research Institute
          Canada Centre for Inland Waters
          P.O. BOX 5050
          Burlington, Ontario
          Canada L7R 4A6

Phone: +1 905-336-4917
Fax: +1 905-336-4972
E-mail: efraim.halfon@cciw.ca

## Model documentation and major references

The model documentation and user guide is included in the floppy disk distributed with the model under the title: "TOXSHELL, a graphic interface to run the fate model TOXFATE on MS-DOS computers".

Halfon, E., 1986. Modelling the pathways in the St. Clair-Detroit River system using the TOXFATE model: The fate of perchloroethylene. *Water Pollution Journal of Canada*, 21: 411-421.

Halfon, E., Brüggemann, R., 1990. Simulation of disulfoton fate in the River Rhine with the TOXFATE model, *Sci. Tot. Environ.*, 97: 385-394.

Halfon, E., and Oliver, B.G., 1990. Simulation and data analysis of four chlorobenzenes in a large lake system (Lake Ontario) with TOXFATE, a contaminant fate model. In: S.E. Jorgensen (Ed.) *Modelling in Ecotoxicology*, Elsevier, pp. 197-214.

Halfon, E., Simons, T.J., and Schertzer, W.M., 1990. Modelling the spatial distribution of seven halocarbons in Lake St. Clair in June 1984 using the TOXFATE model. *J. Great Lakes. Res.*, 16: 90-112.

## Other relevant information
The interface program, TOXSHELL, lets any user run the fate model TOXFATE on MS-DOS microcomputers without having to worry about the specifics of data handling. TOXSHELL is a graphics interface program that uses drop down menus and windows to let any user run TOXFATE easily, modify or add new data files. The format of this interface is by drop down menus and windows. The user is not made aware of the internal mechanism of data handling, this part is automatic. Also, the display of data as XY graphs, for the presentation of computer simulations, and of bar graphs, for display of mass balances, is completely automatic. As an extra feature for advanced users output files subdirectory are created in such a way that they can be entered into any spreadsheet program for easy analysis of numerical results or creation of professional graphs for publication.

## Model identification
Model name/title: SIMON

## Model type
Media: water, lake, ocean, estuary
Class: toxicology (oil), hydrology/hydraulics
Substance(s)/compounds: oil
Biological components: none

## Model purpose
3-D numerical simulation of circulation and oil spills in lakes and coastal seas.

## Short description
SIMON is an integrated environmental model for predicting the transport and fate of non-conservative pollutants (oil) in large aquatic environments (lakes, sea). It consists of the 3-D hydrodynamic model PCFLOW3-D and 3-D Lagrangian (particle tracing) advection-dispersion model PT3D to which equations of oil-spill spreading and weathering are added.

To speed up the calculations and thereby reach the real-time capability, a special system of interpolated flow fields is used. An additional module (interface) has been developed to assure user-friendliness and graphical presentation of the results in AutoCAD (velocity fields and time evolution of the oil slick).

## Model application(s) and State of development
Application area(s): Large surface water bodies (lakes, sea)
Number of cases studied: 1
State: verification, calibration, prognosis made

## Model software and hardware demands
Platform:PC-based - IBM
OS: DOS
Software demands: EDIT, AutoCAD ver. 11, language FORTRAN
Hardware demands: IBM PC 486/8 Mb RAM min.

## Model availability
Purchase: yes, DEM 10,000. Software available as executablde code.
Contact person(s) Prof. Rudi Rajar   and  Asst. prof. Matjaz Cetina
              FAGG-Hydraulic Div.
              Univerza v. Ljubljani
              Hajdrihova 28
              SL-61000 Ljubljana
              Slovenia

Phone: +386 61 12 54 333
Fax: +386 61 219 897
E-mail: rudi.rajar@uni-lj.sl

## Model documentation and major references
For HD part, documentation is the same as for PCFLOW3D. Components of the SIMON are described in two Master's degree works in Slovene. A simple user's manual will soon be available in English.
Rajar, R., Cetina, M., and Zagar, D., 1995. Three-dimensional Model for Oil Spill Simulation, *XXXIVth Congress of CIESM*, La Valette, Malta, 27. -31. 3, 1995.

## Other relevant informations
SIMON is not a general purpose model. Some site-specific problems should be thoroughly discussed with the authors, which leads to the highest efficiency and lowest computer requirements of the model.

## Model identification
Model name/title: SIPOR

## Model type
Media: water, river
Class: toxicology, hydrology/hydraulics
Substance(s)/compounds: chemicals, organic compounds (oil), conservative pollutants (conservative = non-volatile, non-reactive, passive regarding flow velocities)
Biological components: none

## Model purpose
Very fast determination of oil spill (and spills of conservative pollutants) dynamics in rivers.

## Short description
SIPOR is a 1-D transport-dispersion model for prediction of hazardous spill dynamics in rivers. Analytical solution of the 1-D transport-dispersion equation for the case of instantaneous plane source is the basic element of computation. Multiple (succesive) use of this analytical solution (formula) allows for the simulation of the non-uniform continuous plane source. Hydraulic background for the TD equation is the steady, uniform flow in a straight channel of trapezoidal cross-section, computed by Manning equation. Either conservative pollutants or three (non-conservative, volatile) oil derivates can be simulated. The wide-spread formula of Mackay et al. is used for the computation of oil evaporation and formula of Fay is used for the corresponding spill area computation.

Two types of simulations can be performed resulting in either concentration-distance (from source) or concentration-time curve. Graphical and numerical presentation of results is available as well as printer and archivating facilities.

Input: length of computational area, bottom elevations at the beginning and at the end of the computational area, size and shape of the trapezoidal cross-section, discharge, Manning-Strickler friction coefficient, time-discharge function in the case of the non-uniform source, wind velocity and oil type (presently choosing among 3) in the case of oil-spill simulation.

## Model application(s) and State of development
Application area(s): Rivers
Number of cases studied: 0 (SIPOR is a calculator-type simple model)
State: verification, calibration, prognosis made

## Model software and hardware demands
Platform: PC-based - IBM
OS: DOS
Software demands: Windows 3.0, Edit
Hardware demands: IBM PC 486

**Model availability**
Purchase: yes, US $1000. Software available as executable code.
Contact person(s)  Prof. Rudi Rajar    and    Asst. Prof. Matjaz Cetina
                FAGG-Hydraulic Div.
                Univerza v. Ljubljani
                Hajdrihova 28
                Sl-61000 Ljubljana, Slovenia

Phone: +386 61 12 54 333
Fax: +386 61 219 897
E-mail: rudi.rajar@uni-lj.sl

**Model documentation and major references**
Technical Report and Manual are available in Slovene.
Translation of the Manual into English is in process.

**Other relevant information**
Model/program comes on two 3.5" (1.44 mo) diokettco and is easy to install in a Windows environment. The simple structure assures fast work and almost no use of a manual.

**Model identification**
Model name/title: CHESPILL.FOR : Dissolution of Sinkers in Flowing Streams

**Model type**
Media: water, estuary, river, sediment
Class: biogeochemical
Substance(s)/compounds: chemicals, pesticides, organic compounds
Biological components: none

**Model purpose**
Simulates the spills of pure liquids which are heavier than water and sink to the bottom of streams when spilled. Information provided includes length and width of the on-bottom contaminated area, cup-mixing concentration profile at spill site, and dissolution lifetime of the chemcial(s). Inputs include general spill and stream information and pertinent chemical and physical properties of substance(s) spilled.

**Short description**
A model for the short-term fate of sinker liquids as pure substances spilled in flowing water was prepared for the U.S. Department of Transportation. What follows is the executive summary from the final report. Spills into water can be classified into three main categories: soluble compounds which mix with the water very rapidly, immiscible (slightly soluble or insoluble) compounds that float on water, and immiscible (slightly soluble or insoluble) compounds that sink to the bottom of a watercourse.

215

The first two categories of compounds have received much study, and their behaviour is fairly well understood. A significant number of hazardous liquids fall into this last category, which has received minimal study and is the topic of this research.

Decisions associated with water quality, public safety, and the ecosystem must be made immediately after a spill occurs. The expected chemical concentration in the stream is important with respect to the quality of drinking water, industrial water intake and the ecosystem. Estimates of lifetime of the spillage and the duration of water contamination from it are needed. Time is also important with respect to plans for recovery, cleanup, ecosystem (biota) exposure, alternative water sources for industry and the public, and hazard duration. An analytical model of a predictive nature is needed to help provide the necessary information from which to make environmental impact-type decisions.

One objective was to perform laboratory spills in order to develop a database of information that would give insight into the natural processes that occur once the liquid chemical enters the water. The general scope of the work was concerned with organic liquids, more dense than water, entering flowing streams. Specific problems of interest were: a) understanding and quantifying the major processes that occur with respect to the chemical from the time it enters the water until it is dissipated, b) locating the spill on the bottom of the flowing stream, c) developing a mathematical model for estimating the concentration of the chemical dissolved in water, d) developing a mathematical model for estimating the life-time of the chemical on the bottom, and e) constructing a computer version of the models to predict the contaminant consequence and duration in the stream. Much of the laboratory effort was devoted to unraveling the spill mechanism. Numerous spills, employing model chemicals and performed in a quiescent spill tank and in several laboratory scale and larger experimental flowing streams, were made for qualitiative information and quantitative measurements. The spill mechanism has been unraveled and several individual process steps have been identified to occur from the time the liquid enter the water until it is finally dissolved on the wavy sand bottom of the stream. In brief, the major steps are liquid jet break-up, droplet break-up, formation of a stable drop swarm, classification of the drops while in transit to the bottom, accumulation and coalescence on the sand-wave bottom, formation of three geometric shapes (drops, globs, and wave-pools), and finally dissolution. A few of these processes have been identified as being critically important to developing a realistic model and have been chosen for detailed study. Significant experimental effort was aimed at quantifying the size distribution in the descenidng droplet swarm, the dimensions of the zone-of-contamination (ZOC) on the bottom, and the dissolution rate of liquid shapes in the presence of a wavy interface. Approximately 1% of the chemical released at the surface does not arrive at the bottom.

This fraction consisting of tiny drops less than 1 mm in diameter remains suspended by the water turbulence and dissolves quickly. By experiment it was found that drop size was normally distributed with respect to mass fraction and highly dependent upon liquid properties, particularly density and interfacial tension, stream velocity, and injection velocity. A model incorporating the pertinent independent variables was developed to predict size distribution.

The dimensions of the zone-of-contamination on the bottom were obtained from an empirical model based on numerous experiments in a quiescent tank and modified to include the effect of stream velocity and droplet settling velocity. This sub- model includes the parameters associated with the spill source and its dimensions, chemical properties, and stream dimensions. The primary quantitative result is the location and dimension of the ZOC.

Guided by studies of dissolution rates from simple geometric forms an experimental effort was undertaken to measure dissolution rates of pure liquid chemicals in the presence of sand waves of the type that predominate on natural stream bottoms. Rate was found to decrease as wave height increased and to increase with the bottom friction velocity. A rate equation was developed to predict the forced and natural convection dissolution contributions. Dissolution from drops, globs, and wave-pools occurs simultaneously, with drops disappearing first, then globs and finally pools.

A dissolution model based on the three bottom residing geometric form was a key feature that allowed computation of instream "cup-mixing" concentration, mass of chemical remaining and life-times. The dissolution model and associated sub-models were developed into a computer simulation algorithm. Verification of the algorithm was performed with both field and laboratory data that included a massive chloroform spill in the Mississippi River and a small scale spill of furfural in a laboratory scale stream.

A mathematical model and associated computer program for simulating the spill of heavy (rho > 1), immiscible liquid chemicals into rivers and streams have been developed. The information provided by the model includes the length and width of the on-bottom contaminated area, the cup-mixing concentration-time history at the spill site, and the dissolution life-time of the chemical. Inputs into the program include general spill and stream information and pertinent chemical and physical properties of the substance spilled. Verification of the model via comparison with data obtained from both laboratory simulations and documented spills is also presented.

---

## Model application(s) and State of development
Application area(s): spill of heavier than water liquids into flowing streams
Number of cases studied: 6 documented
State: conceptualization, verification, calibration, validation, prognosis made, validation of prognosis - good. Six sets of pilot- scale and field data.

---

## Model software and hardware demands
Platform: PC-based
OS: DOS
Software demands: Fortran Compiler vers. 4.1, language: FORTRAN 77
Hardware demands: 640 Kb minimum RAM - suggested - 1 Mb hard disk: min. size- 4 Mb; suggested size 40 Mb.

---

## Model availability
Purchase: yes, cost of reproduction. CHESPILL is in the public domain. It was developed with federal funds so there is no charge for the right to acquire and use the software. A $20.00 charge to cover diskette plus mailing and copying charges off -sets the expenses incurred by Louisiana State University (LSU). Make the check payable to : LSU- HWRC Foundation Account. Software available both as executable and sourcecode.
Contact person(s) Dr. Louis J. Thibodeaux
        Dept. Chemical Engineering
        Louisiana State University
        Baton Rouge,
        LA 70803-7303, USA
Phone: +1 504-388-6770
Fax: +1 504-388-1476

**Model documentation and major references**
Spill of Soluble, High Density Immiscible Chemicals on Water, Final Report U.S. DOT, 400 7th St., S.W. Washington, D.C., Contract No. DOT-OS-700400, Report No. CG -UOA-80-011, August 1980.
Christy, P.S., and Thibodeaux, L.J., 1982. Spill of soluble high-density immiscible chemicals on water. *Environmental Progress*, Vol. 1, No. 2, May 1982, p. 126-129.
Christy, P.S., 1980. Computer Simulation of Spills of Heavy Immiscible Chemicals in Rivers, M.S. Thesis, University of Arkansas, Fayetteville, AR (1980).

**Other relevant information**

**Model identification**
Model name/title: Quantum chemical estimation of physicochemical compound properties.

**Model type**
Media: water, air, terrestrial
Class:
Substance(s)/compounds: organic compounds
Biological components:

**Model purpose**
Estimation of Henry's constant, water solubility, vapour pressure, and octanol/water partition coefficient of nonelectrolyte compounds from chemical structure calculations.

**Short description**
A quantum chemical approach is used to estimate physicochemical compound properties of nonelectrolytes at 25°C from chemical structure. Only two molecular descriptors are needed that can be calculated in the semiempirical AM1- SM2 continuum solvation model: free energy of aqueous solavtion DELTA Gs, and contact surface area, CSA, that gives the portion of the molecular surface area available for intermolecular interactions.
For Henry's constant H, water solubility Sw, vapour pressure Pv, and octanol/water partition coefficient Kow, simple regression equations of the general form

$$\text{log property - a * DELTA Gs + b * CSA + c}$$

are derived. A test set of 17 benzene derivatives with experimental data for all four properties taken from literature gives explainded variances (corrected for degrees of freedom) between 83 and 95%. Another set of 12 aromatic phosphorothionates yields similar results for log Kow. It is suggested, that the representation of distinct types of intermolecular interaction by molecular descriptors (electrostatic interaction, dispersion interaction and cavity formation energy) will yield new insights into the structural dependence of macroscopic physicochemical compound properties.

## Model application(s) and State of development
Application area(s):
Number of cases studied: 2
State: verification, calibration

## Model Software and hardware demands
Platform: Work-station (e.g., IBM RISC 6000 or Silicon Graphics)
OS: UNIX
Software demands:
Hardware demands:

## Model availability
Purchase: Individual as QCPE programs available, but not as "model".
Contact person(s) Dr. Gerrit Schüürmann
     Centre for Environmental Research
     Department of Chemical Ecotoxicology
     P.O. BOX 2
     D-04301 Leipzig, Germany
Phone: +49 341 235 2309
Fax: +49 341 235 2401
E-mail: gs@theo.ueo.ufz.de

## Model documentation and major references
Cramer, C.J., and D.G. Truhlar, 1992. An SCF solvation model for the hydrophobic effect and absolute free energies of solvation. *Science*, 256: 213-217.
Cramer, C.J., Lynch, G.C., Hawkins, G.D., Truhlar, D.G., and D.A. Liotard, 1993.
AMSOL 4.0: An SCF program for free energies of solvation (QCPE 606). *QCPE Bulletin*, 13: 78.
Pascual-Ahuir, J.L, Silla, E., and Tunon, I., 1994. GEPOL 93: Area and volume of molecules. *QCPE Bulletin*, 14: 14.
Schüürmann, G., 1995. Quantum chemical estimation of octanol/water partition coefficient - First results with aromatic phosphorothionates. *Fresenius Envir. Bull.*, accepted.
Schüürmann, G., submitted 1995. Quantum chemical approach to estimate physicochemical compound properties. *Environ. Toxicol. Chem.*

## Other relevant information

## Model identification
Model name/title: Global Distribution Model for Persistent Organic Chemicals

**Model type**
Media: water, ocean, air, global scale, sediment, soil
Class: toxicology
Substance(s)/compounds: chemicals, pesticides, organic compounds
Biological components:

**Model purpose**
Qualitatively understand and quantitatively describe the zonal distribution and the major trasnport and degradation pathways of persistent organic chemicals such as organochlorinated pesticides in the global environment as influenced by climatic parameters, particularly temperature.

**Short description**
The model is a nonsteady-state, multi-compartmental mass balance model of organic contaminant fate, in which the global environment is represented by nine sequentially arranged climatic zones. Each zone has an air, ocean water, fresh water, fresh water sediment, and two soil compartments, connected by advective and intermedia transport processes. Degradation can take place in every compartment and zone. The time, magnitude, and medium of chemical discharge is specified for each climatic zone. The seasonal variability of some key parameters such as air and ocean water temperature is taken into account. The mass balances for each of the compartments result in a system of 54 differential equations, solved numerically to yield estimates of concentrations, masses, transport fluxes, and reaction rates as a function of time. The model is based on the fugacity approach.

**Model application(s) and State of development**
Application area(s): there is only one, the Earth !
Number of cases studied: so far only illustrative calculations and some preliminary simulations and validation
State:

**Model software and hardware demands**
Platform: PC-based - IBM
OS: Windows
Software demands: Windows
Hardware demands: 486 processor

**Model availability**
Purchase: no
Contact person(s) Dr. Frank Wania
        NILU - Norwegian Institute for Air Research
        P.O. BOX 1245
        N-9001 Tromsø
        Norway

Phone: +47 776 56955
Fax: +47 776 56199
E-mail: frank@nilu.no

## Model documentation and major references

Wania, F., and Mackay, D., 1995. A global distribution model for persistent organic chemicals. *Sci. Total Environ.*, 160/161, pp. 211-232.

## Other relevant information

## 3.8 An Overview: Models of Heavy Metals, Radionuclides and Other Inorganic Compounds in the Environment

Models of heavy metals, radionuclides and other inorganic compounds are often, but not always, simpler than models of organic compounds, because the number of possible processes in the environment is often more limited. The chromium model for Fåborg Fjord developed by Mogensen and Jørgensen in 1979 illustrates the possible simplifications. As chromium in oxidation state 3 precipitates as chromium(III)-hydroxide and therefore settles, only the benthic animals are affected by the chromium contamination of the aquatic ecosystem (a fjord). This implies that the model only needs to consider very few processes.

Mercury, on the other hand is able to react with organic matter and forms methyl- and dimethyl-mercury. A model of mercury contamination is therefore slightly more complex than a chromium model, but will still encompass fewer processes than most models of organic compounds in the environment. It is therefore often possible to develop rather simple models of inorganic compounds in the environment, provided that the modeler has a good knowledge of the environmental chemistry of inorganic compounds, which is the necessary basis for the introduction of simplifications.

Table 3.3. gives an overview of ecotoxicological models of heavy metals, radionuclides and other inorganic compounds. It contains the same information as Tables 3.1 and 3.2: toxic substance to be modeled, model class, characteristics of the model and the main reference.

The Oyster Bioaccumulation Model simulates concentrations of radionuclides, isotopes of silver and cobalt, in oysters maintained in the effluent of nuclear power plants.

The model Polmod.Rad. describes the flow of radionuclides in units of a considered ecosystem: atmosphere, soil, vegetation and surface water. The forcing functions are the meteorological data, the physical and chemical properties of each unit of the focal ecosystem: precipitation, probability of temperature inversion in the atmosphere, wind speed, pH and acidity of soil, soil humus content, cation exchange capacity of soils, calcium cation content, the drainage basin area, calcium ions in water and water temperature.

The model Sensi describes the accumulation and loss of a series of radionuclides as a function of time and the discharge into the sea. It is able to simulate the concentration and turnover of radioactive metal contamination by the brown algae, *Fucus visiculosus*.

Goudey has developed a model, which is able to simulate the toxic effects of metals on phytoplankton growth by steady-state and non-steady-state conditions. Jørgensen has developed a copper model, which is able to describe the concentrations of copper ions, complex-bound copper, copper adsorbed onto suspended matter and copper in the sediment. The forcing functions are discharge of copper and meteorological data. The allocation of copper between copper ions, adsorbed copper, and complex-bound copper is described by the use of equilibrium expressions. In this case is the ionic copper the most crucial state variable, as it is the most toxic copper form.

222

The chromium model developed by Mogensen and Jørgensen is based upon a description of the hydrodynamical distribution of chromium in an aquatic ecosystem and the bioconcentration for the affected benthic organisms. The relationship between the chromium in the sediment and in the benthic organisms is simple first order relation. The concentration in the sediment is based upon the distribution determined by the hydrodynamics.

The mercury model developed by Jørgensen considers the hydrodynamics, the settling of adsorbed mercury, the equilibrium between dissolved mercury and adsorbed mercury, the formation of organic mercury compounds in the sediment and the release of these compounds into the water, the BCF of phytoplankton and fish and the bioaccumulation of mercury in the fish as a result of the contamination of the entire food chain. The mercury concentration of fish is considered the focal state variable, as fish is considered non-edible, if the mercury concentration is more than 1 mg/kg biomass.

Jørgensen has developed a model, which is able to simulate the heavy metal contamination of plants due to the content of heavy metals in natural and artificial fertilizers and soil-conditioners and due to dry and wet deposition from the atmosphere. The state variables are the heavy metal concentration in soil (two fractions are considered: hard bound and less hard bound), the heavy metals in soil-water, in detritus and in the plants. The model has been calibrated and validated for cadmium and lead.

Heavy Rock is a model of the uptake of heavy metals by benthic algae with distinct age class structure. The concentration of heavy metals in the algae is a function of growth rate, mortality and concentration of the heavy metal in the ambient medium.

The model named Scemr simulates one-dimensional migration and redistribution of contaminants in unsaturated soil. The model considers the uptake of the contaminants by roots, adsorption and the leaching processes.

Greenex is a mass flows model developed for an arctic fjord system which receives tailing from a adjacent lead-zinc mine. The model considers lead, cadmium and zinc in dissolved, adsorped and particulate forms.

Young has developed a model for description of the solvent-solid partitioning of ionic solutes in aquatic systems at chemical equilibrium. The model may be applied to effects of acid deposition, including a possible release of heavy metal ions.

**Table 3.3** Models of Heavy Metals, Radionuclides and other Inorganic Compounds in the Environment

| Compounds | Model class | Model characteristics | Reference |
|---|---|---|---|
| Cadmium | A | The entire food chain. | Thomann et al., 1974 |
| Mercury | F | St. var: water. sediment, susp. matter, invertebrates, plant, fish | Miller, 1979 |
| Methyl mercury | D | Trophic level: intake, excretion, metabolism, growth | Fagerstrøm and Aasell, 1973 |
| Heavy metals | C | CF, excretion, bioaccumulation | Aoyama et al., 1978 |
| Copper | E | Complex formation, adsorption, sublethal effects | Orlob et al., 1980 |
| Metals | B | Thermodynamic equilibrium | Jørgensen, 1991 |
| Lead | E | Hydrodynamics, precipitation, toxic effect on algae and fish | Lam and Simons, 1976 |
| Cadmium | B | Settling, sediment, steady state food chain model | Thomann, 1984 |
| Aluminium | D | Survival of fish populations | Jørgensen, 1991 |
| Chromium | B | Hydrodynamics and accumulation in mussels | Jørgensen, 1994 |
| Mercury | C | Hydrodynamics and accumulation in fish | Jørgensen et al., 1991 |
| Radionuclides | C | Hydrodynamics, decay, uptake and release by various aquatic surfaces | Gromiec and Gloyna, 1973 |
| Radionuclides | B | In grass, grains, vegetables, milks eggs, beef and poultry | Kirschner and Whicker, 1984 |
| $SO_2$, $NO_x$ and heavy metals | F | Threshold model for accumulation effect on spruce | Kohlmaier et al., 1984 |

**Table 3.3 (continued)** Models of Heavy Metals, Radionuclides and Other Inorganic Compounds in the Environment

| Compounds | Model class | Model characteristics | Reference |
|---|---|---|---|
| Cadmium | C | Adsorption, complex formation, (humic acid), hydrodynamics | Harris et al., 1984 |
| Nitrate | C | Leaching to groundwater | Wuttke et al., 1991 |
| Zn, Ni, Pb | C | Soil contaminants/human exposure level | ECETOC (1990) |
| Radionuclides | C | Hydrolysis, oxidation biolysis, volatilization and resuspension | Onishi and Wise, 1982 |
| Radionuclides | C | Distribution of radionuclides by a nuclear accident release | A. Simon et al., 1980 |
| Radionuclides | C | Radionuclide conc. in oyster | Rose et al., 1989 |
| Radionuclides | C | Conc. in air, soil, plants and water | Pykh et al., 1991 |
| Radionuclides | C | Conc. in brown algae | Dahlgaard et al., 1992 |
| Heavy metals (Zn) | E | Toxic eff. on phytoplankton | Goudey, 1987 |
| Copper | C | Conc. of copper in form of ions adsorbed, complexes and in sediment | Jørgensen et al., 1982 |
| Heavy metals | C | Conc. in soil, soil-water, detritus and plants | Jørgensen, 1994 |
| Heavy Rock | C | Uptake of heavy metals by algae | Seip; see section 3.9 |
| Scemr1 | C | Migration and redistribution in soil: uptake by plants and leaching | Sheppard et al., 1990 |

**Table 3.3 (continued)** Models of Heavy Metals, Radionculides and Other Inorganic Compounds in the Environment

| Compounds | Model class | Model characteristics | Reference |
|-----------|-------------|-----------------------|-----------|
| Greenex | C | Heavy metals in an arctic fjord dissolved, adsorped, particulate Pb, Zn, Cd. | Nyholm et al., 1983/84 |
| Proton binding | C | Solvent - solid partitioning of ionic solutes at chem. equilibrium | Young: see 3.9 |

# 3.9 Models of Heavy Metals, Radionuclides and Other Inorganic Compounds in the Environment

**Model identification**
Model name/title: Oyster Bioaccumulation Model

**Model type**
Media: estuary
Class: biogeochemical
Substance(s)/compounds: radionuclides
Biological components: organism, oyster

**Model purpose**
Simulate radionuclide (Ag-110m and Co-58) concentrations in oyster maintained in the effluent of a nuclear power plant. Use the model to evaluate alternative release scenarios to minimize radionuclide concentrations in oysters and alternative sampling designs to ensure maximum radionuclide concentrations are detected.

**Short description**
Daily change in radionuclide concentration in an oyster (pico-curies/kg wet weight) is simulated as uptake minus biological excretion minus physical (radioactive) decay minus dilution due to oyster growth. Uptake and biological excretion are allometric functions of oyster weight, modified by a Q10 temperature adjustment. A calibration parameter relates plant releases of radionuclide to uptake. Oyster growth is computed based on season-specific Von Bertalanffy functions. Calibration to Ag-110m and Co-58 was performed using a special depuration study and quarterly measurements of radionuclide concentrations in

oysters maintained for varying numbers of seasons over 8 years in the effluent of a nuclear power plant. Monte Carlo analysis of model inputs for 1-year runs of the calibrated model were performed to evaluate the effects of alternative release scenarios and frequency of sampling.

## Model application(s) and State of development
Application area(s): Contaminant bioaccumulation, risk assessment
Number of cases studied: 1
State: prognosis made?

## Model software and hardware demands
Platform: PC-based IBM
OS:
Software demands: FORTRAN
Hardware demands:

## Model availability
Purchase: no
Contact person(s) Dr. Kenneth Rose
     P.O. Box 2008
     Oak Ridge National Laboratory
     Oak Ridge, TN 37831-6036 USA

Phone: +1 615-574-4815
Fax: +1 615-576-8646
E-mail: ikr@stc1ø.ornl.gov

## Model documentation and major references
Rose et al. (1989), *Ecol. Modelling*, 45: 111-132.

## Other relevant information
Code is no longer available; however the model is ismple and not hard to program.

## Model identification
Model name/title: POLMOD.RAD.: model for simulation of radionuclides' dynamics in the ecosystem

## Model type
Media: local scale
Class: toxicology
Substance(s)/compounds: radionuclides
Biological components: ecosystem - elementary ecosystem

## Model purpose
Model is intended to describe the flow of radionuclides in the units of ecosystem -- atmosphere, soil, vegetation and surface water. Model calculates the level of radioactive pollution accumulated in each unit of the elementary ecosystem.

## Short description
POLMOD.RAD is a software product for simulation the radionuclides in the units of elementary ecosystem. The main term of POLMOD.RAD model is the resistance index. We define the resistance index of ecosystem's compartments as the index of ability to resist the pollutions' flow either due to self-purification ability or due to the decrease of the accumulation rate. POLMOD.RAD consists of four submodels in accordance with the main ecosystems' units - atmosphere, soil, vegetation and surface water submodels. Major state variables are the radionuclide's concentrations in the units of ecosystem enumerated above. Time step equals one year.

Model input includes the date on meteorological, physical and chemical properties of each unit of the ecosystem: precipitations, probability of temperature inversions in the atmosphere, wind speed, pH (soil acidity), soil humus content, cation exchange capacity of soils, calcium cation content, as well as the area of drainage basin, illumination, calcium ions concentration in water and water temperature.

## Model application(s) and State of development
Application area(s): Scientific investigations of pollutants' dynamics in the ecosystems; education and training decisions support system for various kinds of environmental problems.
Number of cases studied: several hundreds
State: validation, validation of predictions

## Model software and hardware demands
Platform: PC-based IBM PC 286 (386)
OS: MS DOS version 3.2 and higher
Software demands: Turbo Pascal compiler, Turbo Pascal version 5.0
Hardware demands: 640k RAM, EGA graphic adapter, math coprocessor

## Model availability
Purchase: yes, available as source code
Contact person(s) Dr. Irina G. Malkina-Pykh
        Center for International Environmental Cooperation (INENCO)
        St. Petersburg, 190000
        Chernomorsky per., 4
        Russia

Phone: 812 311 7622
Fax: 812 311 8523
E-mail: pykh@inenco.spb.su

**Model documentation and major references**
Pykh, Y., and Malkina-Pykh, I.G.. The method of response functions in ecology. *J. Int. Biometerorol.*, 1991, 35: 239-251.
Pykh, Y..A., Malkina-Pykh, I.G., POLMOD (version 1.0) - the model of pollutants' dynamics in the elementary ecosystem. Preprint, Moscow, Institute of Systems Studies, 1992.

Iouri Pykh
International Institute for Applied System Analysis
A-2361 Laxenburg, Austria
Email: pykh@iiasa.ac.at
phone: +43 2236 71521 0
fax: +43 2236 71313

**Other relevant information**

**Model identification**
Model name/title: SENSI

**Model type**
Media:
Class: biogeochemical, toxicology
Substance(s)/compounds: radionuclides, heavy metals - Co, Mn
Biological components: populations, organism - *Fucus vesiculosus*

**Model purpose**
Quantitative estimates of the concentration and turnover of radioactive metal contaminants by the brown algae, *Fucus vesiculosus*, **in situ.**

**Short description**
The model "SENSI" describes the accumulation and loss of a series of radionuclides as a function of month of contaminant discharge into the sea and month of sampling. The concepts behind "SENSI" is separate estimation of the relative accumulation and loss of each radionuclide (or metal) and the inclusion of monthly "growth-dilution factors". The model's capabilities are demonstrated by comparing calculated concentrations of 65-Zn, 60-Co, 58-Co and 54-Mn with measured values in *Fucus vesiculosus*, sampled monthly at a Swedish nuclear power plant. Monthly discharge values were available for the calculations. It is included, that the dilution of contaminant concentrations with new growth is the major cause of decreasing concentrations of previously accumulated zinc, cobalt and manganese in *Fucus*. The actual loss of caesium from *Fucus* is of the same order of magnitude as the growth dilution.

## Model application(s) and State of development
Application area(s): Biomonitoring with Fucus and eco-physiology of radionuclides and metals in *Fucus*. The model can be used in the estimation of uncontrolled discharges and in routine quality control of discharge data
Number of cases studied:
State: calibration, validation, prognosis made, validation of prognosis

## Model software and hardware demands
Platform: PC-based IBM
OS: DOS
Software demands: Q-pro (or LOTUS-123 or similar spreadsheet)
Hardware demands:

## Model availability
Purchase: no
Contact person(s) Dr. Henning Dahlgaard
                       Risø National Laboratory
                       P.O. Box 49
                       DK-4000 Roskilde, Denmark

Phone: +45 4677 4178 (direct) or +45 4677 4170 (department)
Fax: +45 4237 0025
E-mail: Eco-dahl@risoe.dk

## Model documentation and major references
Dahlgaard, H., and Boelskifte, S., 1992. "SENSI": A model describing the accumulation and time-integration of radioactive discharges in the bioindicator *Fucus vesiculosus*. *J. Environ. Radioactivity*, 16, 49-63.

## Other relevant information

## Model identification
Model name/title: Simulation model for the toxic Effects of Metals on phytoplankton Growth

## Model type
Media: water
Class: toxicology
Substance(s)/compounds: heavy metals - Zn, general
Biological components: population

## Model purpose
A mathematical model was developed to simulate the toxiceffects of metals on phytoplankton growth under steady state and non-steady state conditions.

## Short description
A mathematical model was developed to simulate the toxic effects of metals on phytoplankton growth under steady state and non-steady state conditions. Population growth is affected by interactions amongst processes regulating the available metal concentration in solution, uptake, and the dilution of accumulated metal through growth. Model simulations were consistent with published observations of the variable growth responses exhibited by phytoplankton populations exposed to metals including an extended lag phase and a gradual reduction in cell numbers over time. The model also provides new conceptual insights on published results from short-term experiments on metal toxicity and bioaccumulation. Simulations of phytoplankton growth under non-steady state conditions provide additional insight on how interactions amongst physicochemical and biological processes can affect the toxic properties of metals in a variable environment.

## Model application(s) and State of development
Application area(s): Laboratory investigations under constant conditions, non- steady state conditions; mixing zones for wastewater discharges
Number of cases studied: 2
State: prognosis made

## Model software and hardware demands
Platform: PC-based
OS: DOS
Software demands: language BASIC
Hardware demands:

## Model availability
Purchase: no, available as source code
Contact person(s) Dr. J.S. Goudey
    Hydroqual Laboratories LTD
    #3, 6125 12th Street S.E.
    Calgary, Alberta
    Canada T2H 2KI

Phone: +1 403-253-7121
Fax: +1 403-252-9363
E-mail:

## Model documentation and major references
Goudey, J.S., 1987. Modeling the inhibitory effects of metals on phytoplankton growth. aquatic toxicology, 10: 265-278.

## Other relevant information
- The model is based on finite difference equations which can be readily incorporated into other system models as a subroutine.
- The major variables include growth rate, metal speciation, uptake and depuration rates, threshold and toxic concentrations.

- the equations can be adapted to examine effects of metals on different populations and community structure along with addressing bioaccumulation and biomagnification.
- The same approach for simulating metal toxicity under non-steady state conditions can be applied to limiting nutrients.
- The model provides a new conceptual base for establishing testable hypothesis and metal toxicity to phytoplankton.

---

## Model identification
Model name/title: Modelling the control of copper pollution.

---

## Model type
Media: water, lake
Class: biogeochemical
Substance(s)/compounds: heavy metals Cu
Biological components: ecosystem

---

## Model purpose
Control of copper pollution.

---

## Short description
State variables: copper in water as ions, as complexes, and on suspended matter, copper in sediment. Forcing functions: discharge of copper, climate.
Important parameters: equilibrium constant copper/copper complexes, adsorption isotherms $Cu^{2+}$ on suspended matter and settling rate suspended matter.

---

## Model application(s) and State of development
Application area(s):
Number of cases studied: 2
State: 1 calibration, 1 validation

---

## Model software and hardware demands
Platform: PC-based
OS:
Software demands: language SYSL
Hardware demands:

---

## Model availability
Purchase: no
Contact person(s) Dr. S.E. Jørgensen
        Sektion for Miljøkemi
        Danmarks Farmaceutiske Højskole
        Universitetsparken 2
        DK- 2100 København Ø
        Denmark

Phone: +45 35 37 08 50
Fax: +45 35 37 57 44
E-mail: sej@pommes.dfh.dk

## Model documentation and major references
Jørgensen, S.E., Kamp Nielsen, L., Jørgensen, L.A. and Mejer, H., 1982. An environemtal management model of the Upper Nile lake system. *ISEM Journal*, 4: 5-72.

## Other relevant information

## Model identification
Model name/title: SAM: Simulation of Aquatic outdoor Microcosms

## Model type
Media: water pond, microcosm, mesocosm

Class: toxicology, ecology

Substance(s)/compounds: nutrients (simplified nutrient cycle), pesticides (concentration modeled as forcing function (exponential decrease), no detailed fate or exposure model, cadmium (conc. as forcing function, no bioaccumulation, etc.)

Biological components: ecosystem - pond ecosystem, community - plankton, populations- up to 5 phyto- and 5 zooplankton.

## Model purpose
Causal analysis of freshwater field tests (testing of hypotheses).

Interpolation to not tested application rates.

Extrapolation (application rates, time of application, recovery).

Pretesting of designs of microcosm studies.

Risk assessment (probability statements on undesired events).

## Short description
System of differential equations, based on biomasses (dry weight). Forcing functions: test substance (e.g., pesticide, heavy metal), temperature, light, nutrient in- or outflow. State variables: 1 (limiting) nutrient, up to 5 phytoplankton taxa.

Type of parameters: physiological rates for each modeled taxon (max. production, max. consumption, half saturation constants for nutrients or grazing, respiration rates, temperature and light preferences, assimilation rates, prey preferences).

Necessary inputs: starting conditions, taxon specific toxicity values (NOECs, acute and chronic ECs).

Deterministic and stochastic simulations (Monte-Carlo simulations with stratified sampling).

## Model application(s) and State of development
Application area(s): ecology, ecotoxicology, freshwater field tests
Number of cases studied: 4 = 20.
State: verified - sensitivity analyses conducted, calibrated and validated using 4 case studies (1 experiment = 20 with untreated replicates, 2 insecticide studies, 1 cadmium experiment) general risk assessment (e.g., probability of algal blooms after insecticide application), not validated yet to be calibrated and validated using other case studies (herbicide, Dichloroaniline)

## Model software and hardware demands
Platform: PC-based
OS: DOS 3.3 or higher
Software demands: Turbo Pascal (Borland) Version 6.0 or higher
Hardware demands: 286 PC or higher

## Model availability
Purchase: model cannot be purchased at the moment
Contact person(s) Dr. Udo Hommen
                Department of Biology V
                Technical University Aachen
                Worringerweg 1
                D-52056 Aachen
                Germany

Phone: + 49 241 803693
Fax: + 49 241 8888 182
E-mail: Ratte@RWTH. Aachen.de

## Model documentation and major references
Dissertation in preperation. (In German).
Hommen, U., Ratte, H.T., and Poethke, H.J., 1991. Modelling a mesocosm plankton community after insecticide application: a first approach. *System Analysis Modelling Simulation*, 11/12: 821-828.
Hommen, U., and Ratte, H.T., 1994. Application of a plankton simulation model on outdoor-microcosm case studies. In: I.R. Hill, F. Heimbach, P. Leeuwangh, P. Matthiessen (Eds.): *Freshwater Field Tests for Hazard Assessment of Chemicals*. Lewis Publishers, Boca Raton, FL.
Ratte, H.T., and Poethke, H.J., Dumer, U., and Hommen, U., 1994. Modelling aquatic field tests for hazard assessment. In: I.R. Hill, F. Heimbach, P. Leeuwangh, P. Matthiessen (Eds).: *Freshwater Field Tests for Hazard Assessment of Chemicals*. Lewis Publishers, Boca Raton, FL.
Hommen, U., Poethke, H.J., Dumer, U., and Ratte, H.T., 1993. Simulation models to predict ecological risk of toxins in freshwater systems. *ICES Journal of Marine Science*, 50: 337-347.
Hommen, U., Dumer, U., and Ratte, H.T. (in press): Monte-Carlo simulations in ecological risk assessment. To be published in: Proceedings in nonlinear systems. John Wiley & Sons, New York.

## Other relevant information

---

## Model identification
Model name/title: Near sunlight zone model for photodegradation of TCDD in soils containing organic solvents

---

## Model type
Media: soil
Class: toxicology
Substance(s)/compounds: chemicals (dioxins)
Biological components: none

---

## Model purpose
The near sunlight zone model was developed to identify and quantify the controlling factors governing the processes of transport and photolysis of TCDD in soil.

---

## Short description
The model is based on the physical and chemical processes affecting TCDD concentrations near the soil surface. The TCDD concentration is determined by the following four processes: (1) convection of TCDD together with the movement of the organic solution in which TCDD is dissolved; (2) diffusion in the organic solution along a TCDD concentration gradient; (3) photoreaction of TCDD in organic solution within the top layer; (4) photoreaction of TCDD absorbed on soil at the surface. The model needs the following information: porosity of the soil, saturation of solvent in the soil matrix, photoreaction rate constant. Based on these inputs, the model can tell which process is the controlling step, and draw a clear picture of the TCDD concentration variation versus time in the top layer.

---

## Model application(s) and State of development
Application area(s): soil decontamination
Number of cases studied: two
State: verification

---

## Model software and hardware demands
Platform: PC-based or Digit VAX
OS:
Software demands: FORTRAN 77
Hardware demands:

**Model availability**
Purchase: no
Contact person(s) Dr. Yaping Zhong   or M.R. Overcash
　　　　　　　　Department of Chemical Engineering
　　　　　　　　North Carolina State University
　　　　　　　　Raleigh, North Carolina 27695
　　　　　　　　USA
Phone:
Fax:
E-mail:

**Model documentation and major references**
*Chemosphere*, Vol. 26, No. 7, pp 1263-1272, 1993.

**Other relevant information**

**Model identification**
Model name/title: HEAVY ROCK: A Mathematical Model for the uptake of heavy metals in benthic algae

**Model type**
Media: water, estuary
Class: toxicology
Substance(s)/compounds: heavy metals - Cu, Zn
Biological components: ecosystem - rocky shore, community - benthic algae, organism - *Ascophyllium nodosum*

**Model purpose**
Describe the uptake of heavy metals in an organism with distinct age structure (benthic algae).

**Short description**
The model describes uptake of heavy metals (e.g. zink) in the benthic algae *Ascophyllum nodosum*, i.e., an organism with distinct age class- structure. Concentrations in the algae was studied as a function of growth rate, mortality, and concentration of zinc in the ambient medium. The concentration in the algae was found to be an approximately linear function of the mean concentration in the water phase up to about 100 ppb. The result, and the model in itself for other combinations of heavy metals and algal species, can be used to assess biological monitoring programs.

**Model application(s) and State of development**
Application area(s):
Number of cases studied:
State: validation

## Model software and hardware demands
Platform: PC-based
OS:
Software demands: language - FORTRAN
Hardware demands:

## Model availability
Purchase: yes, price - free. Software available as sourcecode.
Contact person(s) Professor, dr. philos Knut L. Seip
          SINTEF
          PB. 124 Blindern
          N-0314 Oslo, Norway
Phone: +47 22 06 7955
Fax: +47 22 06 7350
E-mail: knutlehreseip@si.sintef.no

## Model documentation and major references
Seip, K.L., and Snipen, L.G., 1994. Interaction between phytoplankton and zooplankton in lakes. Verh. *Internat. Verein. Limnol.*, 25: 474-477.
Seip, K.L., Sneek, M., and Snipen, L.-G., 1994. How far do physical factors determine phytoplankton biomass in lakes? *Chemometrics and Intelligent Laboratory Systems.*, 23: 247-258.
Seip, K.L., 1994. Phosphorus and nitrogen limitation of algal biomass across trophic gradients. *Aquatic Sciences*, 56: 16-28.
Seip, K.L., 1994. Restoring water quality in the metal polluted Sørfjorden, Norway. *Ocean and Coastal Management*, 22: 19-43.

## Other relevant information

## Model identification
Model name/title: SCEMR1: Soil Chemical Exchange and Migration of Radionuclides.

## Model type
Media: wetland, local scale, regional scale, terrestrial, soil, forest, agricultural
Class: biogechemical, hydrology
Substance(s)/compounds: radionuclides, heavy metals - Hg, Pb, Cd, Cu, Zn, all elements
Biological components: ecosystem - plants, plant uptake

## Model purpose
The program combines mechanistic models for climatic hydrologic processes with vegetation and soil properties to simulate explicitly: root-zone evaporation, transpiration and drainage, plant and soil water potential, flow of water and radionuclides in soil, and uptake of radionuclides by plant roots.

**Short description**
SCEMR simulates one-dimensional migration and redistribution of contaminants (long-lived radionuclides) in unsaturated soil, root contaminant uptake by mass flow, and contaminant loss by leaching. SCEMR takes into account climate, soil moisture, chemical exchange using a soil solid/liquid partition coefficient (Kd) and soil microhydrologic properties. The model is designed to accomodate contaminants entering the base of the soil profile via the groundwater or at the surface through irrigation water or atmospheric deposition as a pulse or a steady input. Detailed meteorological inputs are used to calculate water contents of soil layers and flows between the layers. SCEMR can also accommodate contaminants placed in one of five layers, including the soil surface layer. Generic application of SCEMR showed that it could be useful in identifying soil types that do not promote upward migration of contaminants, as well as soils that increase leaching losses. SCEMR has been rewritten to decrease its run time requirements, and this new version is called SCEMR1. Reduction comes from the use of stored moisture fluxes after the second year of simulation. Thus, for long-term predictions the moisture flow portion of the model is not called after it attains steady-state. SCEMR1 is appropriate for simulation of radionuclide migration in unsaturated soil contaminated from soil surface, contaminated laterally at any depth below the surface, or contaminated via the groundwater from below, for time spans up to about 10,00 years. A further assessment version of SCEMR1 has been developed that can be used probabilistically where the three key parameters, soil Kd, annual precipitation and soil depth can be sampled from a distribution for four soil types, sand, loam, clay and organic. This version includes radioactive decay, soil degassing for C and I, etc., and loss of contaminants due to agricultural cropping.

**Model application(s) and State of development**
Application area(s): Contaminant migration, high-level nuclear fuel waste management, smelter emmission reduction, soil remediation, soil cleanup criteria
Number of cases studied: more than 5
State: validation

**Model software and hardware demands**
Platform: Mainframe - Digital
OS: 40 Kb
Software demands: language FORTRAN 77
Hardware demands: VAX-8650

**Model availability**
Purchase: no, software available as executable code
Contact person(s) Dr. Marsha I. Sheppard
          Environmental Science Branch
          AECL Research
          Whiteshell Labs
          Pinawa, MB R0E 1L0
          Canada

Phone: +1 204-753-2311 (Ext. 2718)
Fax: +1 204-753-2455
E-mail: sheppardm@wl.aecl.ca

---

**Model documentation and major references**

AECL-9577/COG-91-194 is a major reference.

Sheppard, M.I., 1981. SCEMR: A Model for Soil Chemical Exchange and Migration of Radionuclides in Unsaturated Soil. Atomic Energy of Canada Limited, Technical Record, TR-175.

Huff, D.D., Luxmoore, R.J., Mankin, J.B., and Begovich, C.L., 1977. TEHM: A Terrestrial Ecosystem Hydrology Model. Oak Ridge National Laboratory Reports, EDFB/IBP-76/8 and ORNL/NSF/EATC-27.

Goldstein, R.A., Mankin, J.B., and Luxmoore, R.J., 1974. Documentation of PROSPER: A Model of Atmosphere-Soil-Plant Water Flow. Oak Ridge National Laboratory Report, EDFB/IBP-73-9.

Begovich, C.L., and Jackson, D.R., 1975. Documentation and Application of SCEHM, A Model for Soil Chemical Exchange of Heavy Metals. Oak Ridge National Laboratory Report, ORNL/NSF/EATC-16.

Begovich, C.L., and Luxmoore, R.J., 1979. Some Sensitivity Studies of Chemical Transport Simulated in Models of the Soil-Plant-Litter System. Oak Ridge National Laboratory Report, ORNL/TM-6791.

Sheppard, M.I., and Mitchell, J.H., 1984. SCEMR: Generic Application Using Precambrian Shield Parameter Values and a Proposed SCEMR/SYVAC Interface. Atomic Energy of Canada Limited, Technical Record, TR-240.

Sheppard, M.I., 1911. The Soil Submodel, for the Assessment of Canada's Nuclear Fuel Waste Management Concept, AECL Research Report, AECL-9577.

Bera, K.E., and Sheppard, M.I., 1984. SCEMR1 - Simulation of Soil Chemical Exchange and Migration of Radionuclides on a Geologic Timescale, A User's Manual. Atomic Energy of Canada Limited, Technical Record, TR-249.

Sheppard, M.I., and Bera, K.E., 1984. Sensitivity Analysis of the SCEMR1 Model, Atomic Energy of Canada Limited, Technical Record, TR-318.

Sheppard, M.I., and Sheppard, S.C., 1987. A soil solute transport model evaluated on two experimental systems. *J. Ecol. Modelling*, 37, 191-206.

Sheppard, M.I., and Hawkins, J.L., 1991. A linear sorption/dynamic water flow model applied to the results of a four-year core study. *J. Ecol. Modelling*, in press.

Sheppard, M.I., Stephens, M.E., Davis, P.A., and Wojciechowski, L., 1990. A Model to Predict Element Redistribution in Unsaturated Soil: Its Simplification and Validation. Proceedings of the Symposium and Workshop on the Validity of Environmental Transfer Models, Stockholm, Sweden, October 8-12, 1990.

---

**Other relevant information**

## Model identification
Model name/title: GREENEX

## Model type
Media: water, fjord
Class: biogeochemical, hydrology
Substance(s)/compounds: heavy metals: Pb, Cd, Zn
Biological components:

## Model purpose
To account for the mass flows of heavy metals in an arctic fjord system receiving tailings from a lead-zinc mine.

## Short description
The hydraulic basis of the model has been derived from box model calculations that are input to the metals transport model which includes the state variables dissolved and particulate lead, zinc and cadmium, and particulate matter (described as tabulated measured values). The settling of particulate bound metals and the metals sorption to particulate matter is accounted for: Dissolved metals are released from tailings contaminated sediments. The unknown input of dissolved metals released from discharged tailings is estimated by means of the model as is the transport of metals out of the fjord system.

## Model application(s) and State of development
Application area(s):
Number of cases studied: 1
State: verification, calibration, prognosis made

## Model software and hardware demands
Platform: Mainframe
OS:
Software demands: language: CSMP
Hardware demands:

## Model availability
Purchase: no, software available as sourcecode
Contact person(s) Dr. Niels Nyholm
        Institute of Environmental Science and Engineering
        Bld. 115
        Technical University of Denmark
        DK-2800 Lyngby
        Denmark
Phone: +45 45 93 39 08
Fax: +45 45 93 2150
E-mail:

**Model documentation and major references**
Nyholm, N., Nielsen T.K., and Pedersen, K., 1983/84. Modeling heavy metals transport in an arctic fjord system polluted from mine tailings. *Ecol. Modelling*, (1983/84) Vol. 22, pp. 285-324.

**Other relevant information**
The model is site and application specific, and only the principles as described in the above reference, have general value - not the software.

**Model identification**
Model name/title: Proton Binding Model

**Model type**
Media: sediment
Class: biogeochemical
Substance(s)/compounds: $H^+$, heavy metals: Cu
Biological components:

**Model purpose**
Describe the solvent-solid partitioning of ionic solutes in aquatic systems at equilibrium.

**Short description**
State variables: ionic solute concentration [Me n+], [H+], ....
Forcing functions: solid concentration, volume of solvent, temperature, mixing rate, titraent, normality of solution.

**Model application(s) and State of development**
Application area(s): Acid deposition effects
Number of cases studied: 2
State: calibration

**Model software and hardware demands**
Platform: PC-based
OS: MS/DOS
Software demands: Fortran compiler, language: Wtefor 77
Hardware demands: 512 Kb; floppy

**Model availability**
Purchase: no charge. Software available both as executable and source code
Contact person(s) Dr. T.C. Young
        Box 5715
        Department of Civil and Environmental Engineering
        Clarkson University
        Potsdam, New York 13699, USA

Phone: +1 315-268-4430
Fax: +1 315-268-7636
E-mail: tcyoung@clvm.clarkson.edu

---

**Model documentation and major references**

---

**Other relevant information**

---

**Model identification**
Model name/title: Modelling the control of copper pollution.

---

**Model type**
Media: water, lake
Class: biogeochemical
Substance(s)/compounds: heavy metals Cu
Biological components: ecosystem

---

**Model purpose**
Control of copper pollution.

---

**Short description**
State variables: copper in water as ions, as complexes, and on suspended matter, copper in sediment. Forcing functions: discharge of copper, climate.
Important parameters: equilibrium constant copper/copper complexes, adsorption isotherms $Cu^{2+}$ on suspended matter and settling rate suspended matter

---

**Model application(s) and State of development**
Application area(s):
Number of cases studied: 2
State: 1 calibration, 1 validation

---

**Model software and hardware demands**
Platform: PC-based
OS:
Software demands: language SYSL
Hardware demands:

## Model availability

Purchase: no

Contact person(s) Dr. S.E. Jørgensen
Sektion for Miljøkemi
Danmarks Farmaceutiske Højskole
Universitetsparken 2
DK-2100 København Ø
Denmark

Phone: +45 35 37 08 50
Fax: +45 35 37 57 44
E-mail: sej@pommes.dfh.dk

## Model documentation and major references

Jørgensen, S.E., Kamp Nielsen, L., Jørgensen, L.A. and Mejer, H., 1982. An environemtal management model of the Upper Nile lake system. *ISEM Journal*, 4: 5-72.

## Other relevant information

## Model identification

Model name/title: Chromium-model

## Model type

Media: water, estuary
Class: biogeochemical
Substance(s)/compounds: heavy metals - chromium
Biological components: ecosystem

## Model purpose

Management of chromium pollution.

## Short description

State variables: chromium in water, chromium in sediment and chromium in a selected organisms steady state, but dependent on distance from discharge point. f(x) not f(time).
Forcing functions: discharge of chromium, temperature, geography (morphology).
Important parameter: eddy diffusion coefficient, settling rate suspended chromium, conc. factor sediment/organisms.

## Model application(s) and State of development

Application area(s):
Number of cases studied: 2
State: validation - 1, validation of prognosis - 1.

## Model software and hardware demands
Platform: MAC
OS:
Software demands: STELLA
Hardware demands:

## Model availability
Purchase: no
Contact person(s) Dr. Sv. E. Jørgensen
               Sektion for Miljøkemi
               Danmarks Farmaceutiske Højskole
               Universitetsparken 2
               DK-2100 København Ø
               Denmark

Phone: +45 35 37 08 50
Fax: +45 35 37 57 44
E-mail: sej@pommes.dfh.dk

## Model documentation and major references
Jørgensen, S.E., 1991. A model for the distribution of chromium in Abukir Bay. In: S. E. Jørgensen (Ed.), *Modelling in Environmental Chemistry*, 17, Elsevier, Amsterdam.
Mogensen, B., and Jørgensen, S.E., 1979. Modelling the distribution of chromium in a Danish firth. In: S.E. Jørgensen (Ed.), *Proceedings of 1st International Conference on State of the Art in Ecological Modelling*, Copenhagen, 1978. International Society for Ecological Modelling, Copenhagen, pp. 367-377.

## Other relevant information

## Model identification
Model name/title: Plant contamination by heavy metals.

## Model type
Media: terrestrial, agricultural
Class: biogeochemical
Substance(s)/compounds: heavy metals - Pb, Cd
Biological components: ecosystem

## Model purpose
Management of Pb- and Cd-contamination of plants due to heavy metals in fertilizer sludge and compost.

## Short description
State variables: Pb and Cd in 1) soil, 2) soil water, 3) plants, 4) detritus, 5) soil (hard bound).
Forcing function: input of Pb and Cd from fertilizers sludge and/or compost and from air pollution, climate.
Important parameters: distribution soil/soil water dependent on soil composition and pH; uptake rate of Pb and Cd (dependent on protein content in plants).

## Model application(s) and State of development
Application area(s):
Number of cases studied: 4
State: validation

## Model software and hardware demands
Platform: MAC
OS:
Software demands: STELLA, language STELLA
Hardware demands:

## Model availability
Purchase: no
Contact person(s) Dr. S.E. Jørgensen
                Sektion for Miljøkemi
                Danmarks Farmaceutiske Højskole
                Universitetsparken 2
                DK-2100 København Ø
                Denmark

Phone: +45 35 37 08 50
Fax: +45 35 37 57 44
E-mail: sej@pommes.dfh.dk

## Model documentation and major references
Knudsen, G., and Kristensen, L., 1987. Development of a Model for Cadmium Uptake by Plants. Master's thesis at University of Copenhagen.
Jørgensen, S.E., 1975. Do heavy metals prevent the agricultural use of municipal sludge. *Water Res.*, 9: 163-170.
Jørgensen, S.E., 1976. An ecological model for heavy metal contamination of crops and ground water. *Ecol. Modelling*, 2: 59-67.
Jørgensen, S.E., 1994. *Fundamentals of Ecological Modelling* (2nd edition), Elsevier, Amsterdam.

## Other relevant information

## Model identification
Model name/title: Mercury-pollution

## Model type
Media: water, estuary
Class: biogeochemical
Substance(s)/compounds: heavy metals - Hg
Biological components: ecosystem

## Model purpose
Management of mercury pollution.

## Short description
State variables: mercury in water, phytoplankton, fish, sediment, susp. matter, top carnivorous fish. Steady state (not f(time) but f(x), x- distance from discharge point. Forcing functions: discharge of mercury, climate.
Important parameters: eddy diffusion coefficient, conc. factor water/phytoplankton and water/fish, growth rate fish, settling rate susp. matter.

## Model application(s) and State of development
Application area(s):
Number of cases studied: 1
State: validation

## Model software and hardware demands
Platform: PC-based, MAC
OS:
Software demands: language - SYSL + STELLA
Hardware demands:

## Model availability
Purchase: no
Contact person(s) Dr. S.E. Jørgensen
                Sektion for Miljøkemi
                Danmarks Farmaceutiske Højskole
                Universitetsparken 2
                DK-2100 København Ø
                Denmark

Phone: +45 35 37 08 50
Fax: +45 35 37 57 44
E-mail: sej@pommes.dfh.dk

## Model documentation and major references
Jørgensen, S.E., 1994. *Fundamentals of Ecological Modelling*, (2nd edition), Elsevier, Amsterdam.

## Other relevant information

## Model identification
Model name/title: VAMP-model

## Model type
Media: water, lake
Class: biogeochemical
Substance(s)/compounds: radiocesium, radionuclides
Biological components: ecosystem

## Model purpose
Predict concentrations of Cs-137 in water and predatory fish.

## Short description
Catchment area: inflow areas, outflow areas (= wetland)
Lake: water, suspended matter, active sediment, passive sediment.
Biota: phytoplankton, benthos, prey (zooplankton and small fish), predatory fish.

## Model application(s) and State of development
Application area(s): Radiocesium in lake ecosystems
Number of cases studied: 6
State: validation

## Model software and hardware demands
Platform: MAC
OS:
Software demands:
Hardware demands:

## Model availability
Purchase: no
Contact person(s) Dr. Lars Håkanson
               Institute of Earth Sciences
               Uppsala
               Sweden

Phone: +46 18 183897
Fax: +46 18 555920
E-mail: lars.hakanson@natgeo.uu.se

**Model documentation and major references**
TecDec, IAEA, Vienna, 1995. (in press).

**Other relevant information**
A comparatively large model; mass-balance; target on cesium in water and cesium in predatory fish; validated; gives a high predictive power; based on many new modelling techniques, like seasonal moderators and dimensionless moderators.

# 4. MODELS OF TERRESTRIAL ECOSYSTEMS

## 4.1 General Considerations on Modeling Terrestrial Ecosystems

Terrestrial ecosystems have not been modeled to the same extent as aquatic ecosystems, probably due to the difficulties related to the heterogeneity of soil. While water and air change their properties gradually from one point to another, soil might change its properties radically within a few centimeters, for instance from a clay particle to a small stone to a lump of organic matter.

This change of properties due to the heterogeneity of soil, also very well illustrates the strength and weakness of modeling. When we model the soil properties of a region, for instance, the 150 x 150 km$^2$ used as grid in the acid rain models to account for changes in pH in the soil-water, (see Chapter 5), we cannot consider the small scale heterogeneity, but can only consider average soil properties. It works acceptably for the large scale problems such as the influence of acid rain on the soil properties, but if we want to model a field of, for instance, 5 ha, it becomes very difficult to describe the hydrology properly. Local pockets of sand can, for instance, change the hydraulic conductivity (m/s) significantly, and can thus make essential modifications to the water flows, which are almost impossible to capture in a model. The major effort of modeling terrestrial ecosystems by using biogeochemical models has been invested in models of agricultural land, soil contamination threatening the ground water and forestry. This chapter will illustrate the modeling efforts of terrestrial ecosystems: agricultural land, forestry and other models of terrestrial ecosystems, including soil contamination.

The problem of selecting the right scale is a core issue in ecological and environmental modeling and it has been touched on in Chapter 1. It is particularly essential in modeling a terrestrial ecosystem. The problem may best be realized by introduction of hierarchy theory, which is widely used in ecological system theory. The hierarchical theory has been developed in the context of general systems theory; see for instance Simon (1962, 1969 and 1973), Pattee (1969 and 1972), Mesarovic et al. (1970), Allen and Starr (1982) and O'Neill et al. (1986). The ecological applications are presented by Patten (1978, 1982a and 1985) and Overton (1972 and 1974).

Ecosystems operate under a very wide range of rates, causing great difficulties in modeling these systems. However, usually the rates can be grouped into classes and if the classes are sufficiently distinct, then the system can be considered as a hierarchical system. The structure of ecosystems imposed by differences in rates is sufficient to break down a very complex system into organizational levels and discrete components within each level (Overton, 1974). The interactions between biological components occur at many scales, a dilemma that the hierarchical theory can resolve. The scale of observation determines the organizational level (O'Neill et al., 1986). Higher-level behaviours occur slowly and appear in the descriptions as constants, while lower-level behaviours occur rapidly and appear as averages or steady-state properties in the description - compare to Figure 4.1.

For instance, analyses of annual tree growth need not consider instantaneous changes in stomata's openings, nor long-term changes in regional climate.

From a hierarchical perspective, the definition of the system depends on the window (O' Neill et al., 1986) through which the world is viewed. If one is looking at the effects of nutrients in a five minute pulse of rain, the relevant components are leaves, litter surface, fungi and fine roots, but if the study is concerned with long-term climatic changes, the relevant components may be large pools of organic matter, such as the rain forests or the development in agriculture.

It is important to realize that the entities of ecosystems form networks of selective interactions. Possibilities at each level of identification are limited by the variants produced in lower-level processes and constrained by the selective environment of higher levels; see Figure 4.1.

An ecological system may also be broken down on the basis of spatial discontinuities. The hierarchies of space and time share many properties. For example, the spatial hierarchy, as the temporal one, is nested in the higher level, because the higher level is composed of the lower level; compare again with Figure 4.1.

The real advantage of the hierarchy theory is that it offers an approach to the medium number systems, for instance, ecosystems (O'Neill et al. 1986). Ecosystems are very complex systems and we have to break them down to analyze their behaviour, understand their underlying principles, model them for predictive purposes and so on. The hierarchy theory approaches this problem by searching for a structure that is already there; either it is a rate or a spatial structure.

With these considerations in mind, it should be easier to select the right scale in space and time for a given problem. It is often fruitful to map the hierarchical organization of a considered problem-system complex and to decide of which level we must operate to manage the focal problem. The lower level is then considered by using mean values and the upper level by using constants, as mentioned above. This approach can well be illustrated by comparison of the models presented in this chapter.

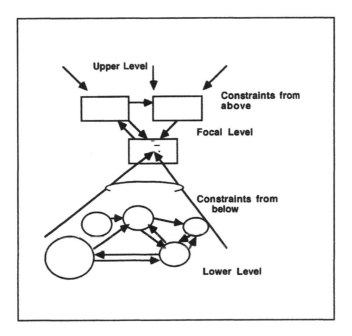

**Figure 4.1:** *A hierarchical network conception of ecological systems. The focal level is assumed to be modeled. The constraints from the upper level occur slowly and are taken into account as constants, while the constraints from the lower levels occur rapidly and are accounted for as averages.*

## 4.2. An Illustration: A Complex Hierarchical Model

The Daisy Model illustrates the typical hierarchical modeling problems associated with a terrestrial ecosystem. It was developed by Hansen et al. (1990) at the Royal Veterinary and Agricultural University, Copenhagen, Denmark. Figure 4.2 shows a schematic representation of the soil-plant system included in the model. The hierarchical construction is indicated by the dividing the two boxes representing plant and soil into submodels.

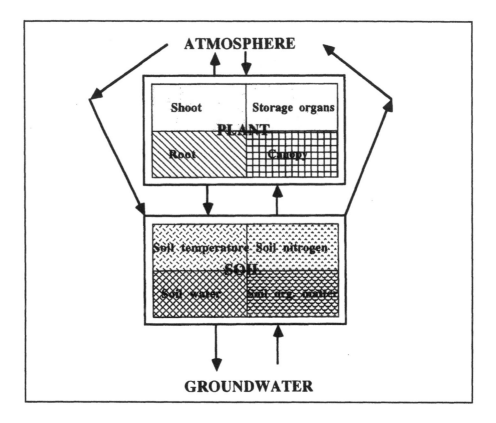

**Figure 4.2:** *A schematic representation of the soil-plant model, named DAISY (Hansen et al., 1990). The models of soil and plant processes can be divided into submodels as shown on the figure.*

The soil water submodel is based upon Darcy's equation:

$$Q = K \, \delta h / \delta x \tag{4.1}$$

where Q is the water flow and dh/dx is the hydraulic potential or gradient. Usually Q is expressed as $m^3/s$ or as $m^3/s*m^2 = m/s$. The hydraulic potential is assumed to consist of a gravitational and pressure potential. The vertical flow can be described by using the following equation:

$$Q = - K ( \partial y / \partial x - 1) \tag{4.2}$$

where the applied unit for potential is m water per meter, "minus" shows that the flow is in the direction of decreasing potential and y is pressure potential.

For non-deformable soil and due to the incompressibility of water the mass conservation principles lead to the following expression:

$$\partial q / \partial t = \partial Q / \partial x - S \tag{4.3}$$

where q is the volumetric water content as $m^3/m^3$, t is time and S is a sink term. A combination of (4.2) and (4.3) yields the Richard equation:

$$Cw \, \partial y / \partial t = \partial / \partial x [ K \, \partial y / \partial x ] - \partial K / \partial x - S \tag{4.4}$$

where Cw is the specific water capacity, defined as dq/dy.

Two functional relationships should be known to solve Richard's equation. These are the relationship between the pressure potential and the volumetric water content and between the hydraulic conductivity and the volumetric water content. These curves are published for a wide range of soil profiles. Richard's equation may be solved by a finite difference scheme.

The model also must include evaporation by calculation of a potential evaporation, dependent on the total crop area index, that is calculated in the plant submodels.

The soil water model operates with three different boundary conditions, dependent on whether water is podding on the surface or not and whether Richard's equation is solved for the whole profile or not.

The soil temperature submodel describes the heat flow with the governing equations parallel to the water flow:

$$q_h = - K_h \, \partial T / \partial x + c* T*Q \tag{4.5}$$

where $q_h$ is the heat flux density in $W/m^2$, $K_h$ is the thermal conductivity in $W/m^* \degree C$, T is the temperature ($\degree C$), x is the depth, c the specific heat capacity of water ($J/kg \degree C$) and Q is the flux of water, here in $kg/m^2$ s.

Conservation of heat yields:

$$\partial H/\partial t = - \partial q_h/\partial x + Sh \qquad (4.6)$$

where H is the heat content of soil ($J/m^2$ ), t is time and Sh is heat sources. The average heat capacity of the soil is found as the weighted average of the heat capacities of its constituents. The complications of frost and thaw and whether the water is liquid water or ice are included in the model. The numerical solution is performed by the same method as for the water flux equations.

**The organic matter submodel** considers three different pools of organic matter: added organic matter (manure for instance), biomass and soil organic matter. Each of the three classes are subdivided into subpools according to their biodegradability. The decay of organic matter is described by first order reaction kinetics. The rate constant is made dependent on: the clay content of the soil, the temperature of the soil and the pressure potential of the water.

**The soil nitrogen submodel** is shown in details in the conceptual diagram in Figure 4.3. One of the major goals of the model is to optimize the use of nitrogen in agriculture and minimize the amount released to the environment. Therefore, this submodel has been given particular attention and is very detailed. Some of the most interesting details of this submodel will be mentioned here.

The nitrification is described by a Michaelis-Menten expression with a rate constant dependent on the temperature and the pressure potential of soil water. Both these relationships are based upon empirical results.

The denitrification rate is related to the amount of easily degradable organic matter, the volume of anaerobic micro-sites within an otherwise aerobic media, the soil temperature and the concentration of nitrate in the soil water.

To cover the first and third factors, a potential denitrification rate is found as a function of temperature and of the carbon dioxide evolution rate:

$$x' = \text{ß} * r \qquad (4.7)$$

where x' is the potential denitrification rate of soil, expressed in kg $N/s^*m^3$, ß is a constant and r is the rate of carbon dioxide evolution, expressed as kg $C/s^*m^3$.

The transport of nitrate to the anaerobic micro-sites is assumed to be by diffusion. The effects of the diffusion coefficient, the surface area of micro-site and diffusion distance to micro-sites are pooled in an empirical constant, Kd. This implies that the transport to the micro-sites becomes a first order reaction expression of the nitrate concentration with Kd as the rate coefficient.

If the transport to the micro-sites exceeds the potential denitrification rate, the actual denitrification rate, x, can be found from:

$$x = x' * f(q) \tag{4.8}$$

where $f(q)$ is a function of the soil water content. Above a water saturation of 80% the function increases from 0, to 1 at 100%, almost by a linear relationship.

If the transport to the micro-sites is insufficient to use the potential denitrification rate, the actual denitrification rate becomes equal to the transport to the micro-sites.

The nitrogen uptake by plants in kg $N/m^2$ is calculated from

$$Nc = \sum c_j * W_j \tag{4.9}$$

where $W_j$ is the amount of dry matter in plant part j per area unit (kg $DM/m^2$) and $c_j$ is the upper limit of nitrogen concentration in plant part j. The movement of nitrate in the root zone and the vertical movement of inorganic nitrogen are both considered in the model. The latter flow follows of course the water flux model.

The crop/plant growth model covers, as shown in Figure 4.3, four submodels. The most characteristic features of these submodels will be presented in this section. Figure 4.4 shows the conceptual diagram of the entire plant model. The plant growth (see Figure 4.4) is very dependent on three factors, that are determined in three of the soil submodels: the temperature, the soil water content and the available nitrogen in soil.

**The crop canopy** is described in terms of total crop area index, $C_{ai}$ ($m^2/m^2$) and green crop area index, $L_{ai}$ ($m^2/m^2$). After emergence the crop canopy expressed as green crop area is assumed to be described by equation (4.10) for the period from emergence and until the top dry matter has reached a given value $Wt^o$. From this point and until a certain temperature sum ($\Omega$) - calculated from the time of emergence - is reached, equation (4.11) is valid. For the remaining part of the growth period the green crop area index and the total crop area index are calculated from equations (4.12) and (4.13), respectively:

$$L_{ai} = 0.5 \, (\exp(2.4 \sum T / ñ) - 1) \tag{4.10}$$

$$L_{ai} = Si * Wt \tag{4.11}$$

## 4.2 A Complex Hierarchical Model

$$L_{ai} = (1 - a^*s/s + L) \, Si \, ^*Wt \tag{4.12}$$

$$L_{ai} = (1 - a'^*s/s + L') \, Si \, ^*Wt \tag{4.13}$$

where: s is an abbreviation for $\Sigma T - \Omega$,
Wt is accumulated top dry matter, $(kg/m^2)$
Si is specific green crop area, $(m^2/kg)$
T is the daily mean temperature, (°C)
ñ is a development parameter, (°C)
$\Omega$ is another canopy development parameter, explained above, (°C)
L is green crop area index damping parameter, (°C)
L' is total crop area index damping parameter, (°C)
a is green crop index damping parameter, (°C)
a' is total crop index damping parameter, (°C).

**Root penetration** is described as a function of the soil temperature:

$$v = \left\{ \begin{array}{l} 0 \\ a(Ts - Tc) \end{array} \right. \quad \text{for} \quad \begin{array}{l} Ts \leq Tc \\ Ts > Tc \end{array} \tag{4.14}$$

where v is the root penetration rate (m/day), a is the root penetration rate parameter m/day *(°C), Tc is the minimum temperature for root growth (°C) and Ts is the soil temperature at the root depth (°C).

**The root density profile** is modeled according to Gerwitzand Page (1974):

$$L_z = L_o \, \exp(-b^*z) \tag{4.15}$$

where: $L_z$ is the root density at soil depth z $(m/m^3)$
$L_o$ is the soil density at the soil surface $(m/m^3)$
b is a root density distribution parameter (1/m) and
z is the soil depth (m)

The total root length is described as:

$$L_t = Sr * Wr \tag{4.16}$$

where Sr is the specific root length (m/kg) andWr is the root dry matter $(kg/m^2)$.

The total root length also may be found from integration of equation (4.15):

$$L_t = L_o (1 - \exp(-b*d))/b \qquad (4.17)$$

where d is the root depth in m.

**Canopy gross photosynthesis** is determined by the photosynthetically active radiation (W/m²), that is considered a fraction of the global radiation.

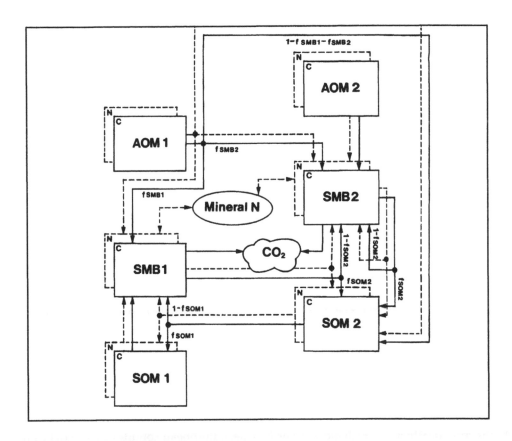

**Figure 4.3:** *Conceptual diagram of the nitrogen and carbon cycles of the model. Rectangles represent state variables, while valves are processes. Solid lines represent flows of matter, while broken lines represent flows of information. Reproduced with permission from Hansen et al. (1990).*

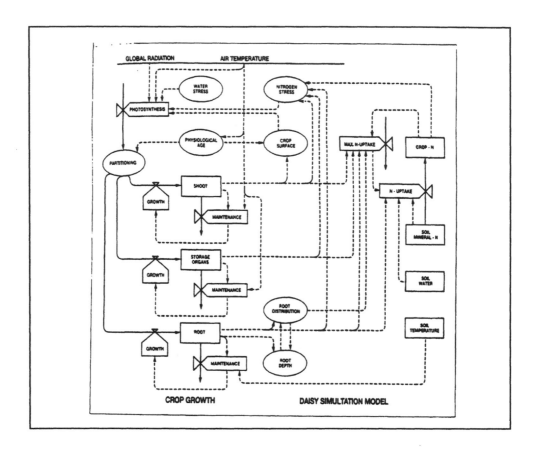

**Figure 4.4:** *Conceptual diagram of the crop part of the presented complex soil-plant model. Rectangles represent state variables, valves designate processes, while ovals indicate auxiliary variables. Solid lines represent flows of matter, while broken lines represent flows of information. Reproduced with permission from Hansen et al. (1990).*

Hansen et al. (1981) has found that 48% of the global radiation can be considered photosynthetically active radiation under Northern European conditions. The distribution of photosynthetically active radiation within a crop canopy is described according to Beer's law deducting the reflected radiation. This means that the reflection coefficient and the extinction coefficient of the canopy must be known.

The relation between gross photosynthesis rate in a single leaf and the photosynthetically active radiation is described by Hansen and Aslyng (1984):

$$F = Fm (1 - Sa / f) \tag{4.18}$$

where F is the gross photosynthesis rate expressed in kg carbon dioxide/m$^2$ s; Fm is a constant depending on the leaf properties, for C$_3$ plants an average value of $8.3* 10^{-7}$ kg carbon dioxide/m$^2$ s can be applied; $f$ is a constant (60 W/m$^2$).

In order to calculate gross photosynthesis of the entire crop, the crop canopy is divided into distinct layers each having a green crop area index.

The efficiency, ef, by which adsorbed radiation energy is converted into chemical energy in plant can now be calculated from:

$$ef = EC\ 44\ F_{ad} / 30\ S_{ad} \tag{4.19}$$

where EC is the energy in carbohydrates = 15.7 MJ/kg CH$_2$O; 30 and 44 are the molecular weight of CH$_2$O and CO$_2$, respectively; $F_{ad}$ is the gross photosynthesis of the crop calculated from equation (4.18) by summarizing over the layers and the time, and $S_{ad}$ is the absorbed photosynthetically active radiation summarizing over the layers and the time, using Beer's law and the reflection coefficient to account for the radiation within the crop canopy.

It may be shown, that ef is dependent on $S_{ad}$ and on $L_{ai}$. Figure 4.5 shows the radiation conversion efficiency in relation to incident photosynthetically active radiation for two values of the green crop area index ($L_{ai}$). These effects are incorporated in the calculation of the gross photosynthesis by using the following equation:

$$ef = 0.15 - (\beta' - \beta * L_{ai} / (3 + L_{ai}))\ S_{ad} / (S_{ad} + 7.8\ 10^6) \tag{4.20}$$

where $\beta'$ and $\beta$ are parameters dependent on the extinction coefficient of the crop canopy. For extinction coefficients between 0.4 and 0.8 they vary from 0.133 to 0.160 and from 0.078 to 0.094, respectively.

When ef has been assessed, the daily gross photosynthesis, Fg, for the crop can be calculated based upon the daily global radiation, Sg, using the following two equations:

$$S_{ad} = (1- A)(1 - \exp(-e\ L_{ai}))\ \pi\ Sg \tag{4.21}$$

$$Fg = ef* S_{ad} / EC \tag{4.22}$$

where A is the reflection coefficient of crop canopy, e is the extinction coefficient, $\pi$ is the above-mentioned ratio between the global radiation and the photosynthetically active radiation (can be set to 0.48; see above).

## 4.2 A Complex Hierarchical Model

**Assimilate partitioning** is the process by which assimilates produced by gross photosynthesis, Fg, are allocated to the various parts of the plant: leaves, stems, roots, storage organs, etc. The temperature sum has been chosen as the variable governing the assimilate partitioning. The model contains specific rules for the allocation according to the temperature sum.

**The net production** is finally found from the following equation:

$$P_n = Y ( Fg - r W) \tag{4.23}$$

where $P_n$ is the net production (kg DM/m$^2$ day), W is the dry matter present (kg DM/m$^2$), r is the respiration coefficient (kg CH$_2$O/kg DM* day ) and Y is the assimilation conversion coefficient (kg DM/kg CH$_2$O). r is strongly dependent on the temperature with a doubling of the rate for each 10°C.

The growth found by the series of equations presented above assumes ample nitrogen. If it is not the case, Fg may be adjusted according to this expression:

$$Fg' = (N_c - N_{co}) / (N_{ca} - N_{co} ) \tag{4.24}$$

where Fg' is the adjusted gross photosynthesis, (Fg is gross photosynthesis by ample nitrogen), $N_c$ is the nitrogen content of crop at actual nitrogen supply (kg N/m$^2$), $N_{co}$ is the nitrogen content of crop at extremely low nitrogen supply (kg N/m$^2$), and $N_{ca}$ is the nitrogen content of crop at ample nitrogen supply (kg N / m$^2$).

The relation between the soil water content and the growth is accounted for in the model by indication of when irrigation is needed.

The model is characterized by its many parameters, that are often based upon empirical relationships. This makes this type of model less causal than for instance eutrophication model.

On the other hand, these parameters are known from many sources; see for instance Hansen et al., 1990, which implies that reasonably good values can be used in the model for most parameters. The model therefore gives good accordance between simulations and measurements, which support the use of this type of model in practice.

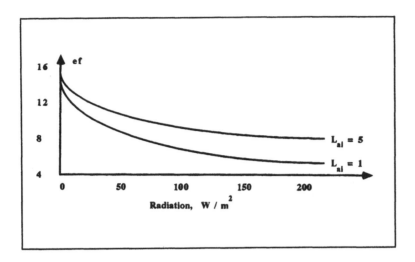

**Figure 4.5:** *Radiation conversion efficiency, ef, in relation to incident photosynthetically active radiation for two values of the green crop area index, $L_{ai}$. Increased green crop area index gives increased efficiency.*

## 4.3 Growth Equations

This section is devoted to a presentation of different equations applied for description of growth, as they are widely used in agricultural and forestry models.

$$dW/dt = k*W \tag{4.25}$$

represents the exponential growth, which is unrealistic for a longer time range for populations as well as for individual organisms, although the growth equation may be used as approximation for a shorter period of time. A regulation will before or later reduce the growth, as it is shown in the equations used to express nutrient and light limitations in models of aquatic ecosystems; see Chapter 2. This regulation is indirect time dependent, as the limitation will be stronger and stronger with the use of the limiting resources for growth. A description will therefore require two or more differential equations, corresponding to the growing population or organisms and the diminishing resources.

An upper limit for W regulates similarly the growth, as shown in the integrated form:

$$\ln((W_{max} - W)/(W_{max} - W_o)) = k*t \tag{4.26}$$

where $W_{max}$ and $W_o$ are the weight (or the number if we consider the number of a population) at maximum respectively at time = 0. If $W_o$ is small and can be approximated by 0, the equation takes the simpler form:

$$W = W_{max} (1 - \exp(-k*t)) \tag{4.27}$$

A graphical illustration of the exponential growth and the growth according to equation (4.27) is given in Figure 4.6.

The most widely applied regulation of the growth is probably made by the introduction of the concept "carrying capacity". The idea, that a system has a carrying capacity or an organism a maximum weight, $W_{max}$, is expressed in the so-called logistic growth equation:

$$dW/dt = \mu*W((W_{max} - W)/W_{max}) \tag{4.28}$$

Integration of equation (4.28) yields:

$$W = W_{max} * \exp(\mu*t))/(W_{max}/W_o - 1 + \exp(\mu*t)) \tag{4.29}$$

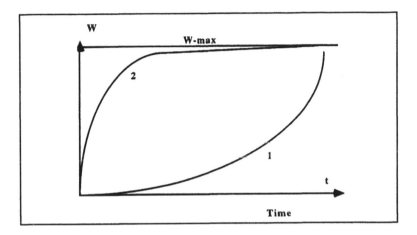

**Figure 4.6:** *The exponential growth, 1, and the growth according to equation (3.53) are shown.*

Gompertz growth equation uses the first order kinetics:

$$dW/dt = \mu*W \qquad (4.30)$$

The specific growth rate $\mu$ is, however, no longer constant but governed by:

$$d\mu/dt = - D*\mu \qquad (4.31)$$

265

where D is an additional parameter, describing the first order decay in the specific growth rate. Integration of these equations leads to:

$$W = W_o \exp (\mu_o (1 - e^{-D^*t}) \tag{4.32}$$

Chanter has devised a function that is a hybrid of the logistic and Gompertz equations. He proposes the following growth equation:

$$dW / dt = \mu^* W ( 1 - W / B ) e^{-D^*t} \tag{4.33}$$

where $\mu$, B and D are constants. Integration of this equation yields:

$$W = W_o * B / ( W_o + (B - W_o ) \exp \{ - [\mu ( 1 - e^{-D^*t} ) / D] \}) \tag{4.34}$$

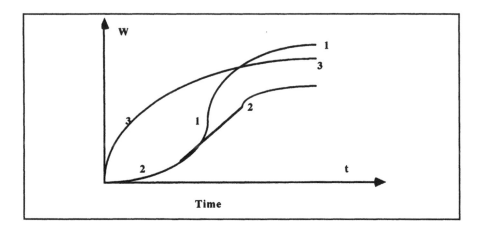

**Figure 4.7:** *The figure shows the logistic growth (1), the Gompertz' growth equation (2) and the Chater's growth equation (3). The shape of the plots is dependent on the values of the constants.*

Growth of organisms is often described by the following equation (Von Bertalanffy's equation):

$$dW / dt = H*W^d - k*W \qquad (4.35)$$

where H, d and k are constants. $H*W^d$ expresses the anabolism, which is considered proportional to the surface, which means that d is equal to 2/3. $k*W$ is the catabolism, which is proportional to the weight. The power 0.75 or 0.8 are used in some editions of this equation, which is better according to the actual observations of the catabolic process. Integration of equation (4.35) may lead to an equation for the growth of the length similar to (4.27), provided that the growth is isometric. Under all circumstances will a higher exponent for the catabolic expression, b, than for the anabolic expression, d, give a maximum value of the weight and the length, corresponding to:

$$(W_{max})^{b-d} = H / k \qquad (dW / dt = 0) \qquad (4.36)$$

## 4.4. Overview of Agricultural Models

Environmental models of agricultural systems simulate the agricultural yield or the pollution caused by agricultural production due to the use of fertilizers and pesticides or both. Only very few models have, however, been developed for an environmental-economic optimization of the agricultural production.

Pesticide models have mainly been covered in Sections 3.4.-3.5., and are only included in this chapter if they are concerned with both pollution and agricultural production. Population dynamic models are included as far as they are concerned with the domestic animals or the interaction between pests and agricultural products, i.e., crops or domestic animals.

Tables 4.1-4.2 give an overview of the agricultural models presented in this chapter and a few other agricultural models published for 7-15 years ago, and which are still of value either directly or as inspiration by development of a new model. Table 4.1 is concerned with plant production and Table 4.2 with livestock.

The model named DASYNEURA describes the population development of the pod midges, *Dasyneura brassicae*, in winter and spring rape. The model simulates the different life stages of the podge midges, and the damage it is able to cause for yield of rape.

The so-called PLATYGASTER simulator is a population dynamic model encompassing three populations: rape, the podge midges, *Dasyneura brassicae*, and its parasites, *Polygaster oebalus*. Both the species have three generations per season. It is the aim of the model to investigate the possibilities to apply biological pest control of the podge midges.

The model ARIDTREE is a olive tree model able to simulate the growth of this tree. The forcing functions are photo-period, fraction overcast, temperature, precipitation, application of irrigation and fertilizers. The state variables are photosynthesis active radiation, gross and net production, organ and reserve biomass. The model consists of a climatic submodel and a biological components submodels, where the carbon assimilation is calculated on basis of the leaf characteristics.

Several rice models have been developed. The model Riceys is a demographic model for rice growth and development. It has been developed to study crop-pest interactions and provide a useful tool for the development of ecologically sound integrated pest management. It encompasses several species of leaf feeding larvae and weeds. The growth of rice is determined by nitrogen, as the only limiting resource, the solar radiation, the temperature and the interactions from larvae and weed.

The model Oryza/Intercoh considers the growth of rice, resulting from the carbon and nitrogen cycle. It is possible in the model to include weed and the corresponding competition for the resources.

The rice model Ricemod is developed as a research tool to understand the soil and plant chemistry and the biophysical processes of the aerial environment. Photosynthesis is a function of solar radiation, canopy structure and the ratio of leaf nitrogen to the leaf area.

Canopy shape is derived from the leaf area index, and the leaf area is assumedproportional to the leaf weight. Photosynthesis is not effected by the temperature but only by the water stress, while respiration is dependent on the temperature. The soil is assumed to consist of four layers, three containing the root-system and the fourth to have a constant water content.

The model Ceres simulates the rice yield determined by the soil water, soil nitrogen solar radiation, temperature and the radiation efficiency use as a function of LAI, which is a function of the leaf appearance and expansion and is reduced by nitrogen deficiency. The user determines the number of soil layers applied to run the model.

The model Macros is intended for educational purposes to illustrate the concept of growth limiting factors in agriculture, including the effect of carbon dioxide on the photosynthesis.

The model Agrotool is intended to estimate water content in a controlled soil layer under crops of several cultures. It aims toward a prediction of plant development and an expected yield. On irrigated fields, the model can be applied to for the control of optimization of soil water regime. It accounts for the weather conditions and the fertilization and irrigation plans. Plant evapotranspiration and soil evaporation are simulated in the model. Weather forecast is considered as an input signal.

Li has developed a model for the simulation of parasitism of *Aphelinus asychis*. It consists of two parts: one for the calculation of an optimal host acceptance decision matrix for the wasp and one for the Monte Carlo simulation of parasitism.

Tamo has developed a model for the growth and development of a photosensitive cow pea variety. The model simulates the age-structures population of leaves, shoots, roots, peduncles and other reproductive organs. It includes a delay subroutine able to emulate stochasticity in growth, aging and losses of population attributes. Growth and yield formation of the plant are driven by the computed ratio between carbohydrates supply and demand. Demand parameters are estimated from detailed dry matter data and developing plant organs.

The model Vimo is a grapevine growth model. The growth occurs per degree-day above the threshold of grapevine. It encompasses the following plant sub-units: the annual population of fruit, leaves, shoots and root development. They are age-structured and have distributed developmental times. The seasonal nitrogen dynamics included in the model are the net results of the processes of new tissue formation by high nitrogen concentrations and the degree-day driven export of nitrogen from aging parts to reserves. The state variables are: fruits, leaves, shoots, roots, perennial frame and reserves. The forcing functions are photosynthetic efficiency of leaves, area/dry matter ratio of leaves, dry and nitrogen demand of grapes, solar radiation, minimum and maximum temperature.

The model Shootgro simulates the development and growth of small grains (winter and spring wheat and winter and spring barley) as response to weather data, nitrogen, light and water availability, based upon accumulation of thermal time.

The model named Eldana investigates the relationship between crop damage by *Eldana saccharina* and the influence on the parasite *Goniozus natalensis* on *Eldana* to investigate the possibilities to use biological control. The state variables are the number of eggs, small larvae, large larvae, pupae and adults for the two populations.

The model Forage is deterministic and is developed as an interface between the grass and beef cattle. It predicts the effect on changing characteristic on feed intake and diet selection of grazing animals. Rate of intake is simulated as the product of bite size and rate of biting, which are sensitive parameters to under estimations. Rate of intake is expressed as a function of quantity of forage offered and the size of the animals. Four inputs are required: forage quality, forage quantity, the amount of forage demanded by the animals and metabolic weight of the animals.

Marsula has developed a model of the screw work fly (*Cochliomyia hominivorax*) which has been an enormous economic importance for the livestock industry. It has been possible to eliminate this insect in USA and Mexico by the sterile insect technique (SIT). The model demonstrates how a fully established pest population spreads into an area in which sterile insects are released. Random spreading of the population is described by diffusion. The model determines the conditions for stopping the invasion by SIT.

Gillman has developed a model to predict the population dynamics of the biennial *Cirsium vulgare* under different grazing management regimes. The model is used for management of pasture weed. The model is stage-structured with a density dependence incorporated at two points in the life cycle. The density dependence is estimated from the field data.

A model developed by Lakshminarayan and Bouzaher is particularly well fitted to deal with the problem of non-point pollution. It is a statistical model based upon input/output analysis, i.e., it is a black box model. The spatial heterogeneity is fully captured in the model which accounts for soil properties, hydrology, weather, crop production, fertilization and other management practices.

The model named Agnps is a grid-based model. All inputs and calculations are made on the cell level. The model accounts for hydrology, erosion, sediment transport, transport of COD, and transport of adsorbed and dissolved nitrogen and phosphorus. The model can be used to estimate agricultural non-point source pollution.

Robertson et al. have developed a model which is able to predict the length of sorghum crop roots as function of soil depth and time. The model accounts for the influence on the root development by soil water and by the general development of the crop.

Texcim is cotton insect pest management model. It integrates four pests, pest control, plant growth, economic and weather. It is a large, complex model with hundreds of variables, functions and parameters.

The model called ISSM (Indiana Soybean System Model) is primarily developed for research. The model simulates the development, growth and yield of soybean crop at a particular location, specified by soil profile and weather data. The model accounts for soil-water balance, photosynthesis, crop stage and characteristics, dry matter production and stage-specific partitioning. The model has been integrated with an age-stage distributed delay model of Mexican bean beetle, a defoliating insect. This version is named ISSM-Crop-Insect Model.

Lang et al. have a model under development which is able to classify agroecosystem-type and to set up scenarios for development of agroecosystem-types for a given area.

Law et al. have developed a mechanistic model to describe the methane production in a rice field. The model accounts for two carbon sources: soil and the rice plant roots. The model assumes that methane is emitted to the atmosphere only by transportation through the plant which is consistent with field observations. The methane transport to the plant is expressed by a simple mass transfer equation. A model denoted DanStress integrates effects of soil, crop and climatic conditions on water relation of field grown crops. The model gives a very detailed picture of the water/crop relationships. It contains, for instance, several layers of the root zone, detailed equations for the root water uptake and the related changes in the crop water storage in stems and leaves. Forcing functions are net radiation, temperature, precipitation, water vapour pressure and wind speed.

**Table 4.1** Overview of Models Related to Plant Production

| Focal component/ name of the model | Characteristic of the model/st. variables | Major references |
|---|---|---|
| Alfalfa | Photosynthesis, growth, resp. and LAI | Holt et al., 1975 |
| Clover | Photosynthesis, growth, resp. and LAI | Fukai et al., 1978 |
| Tobacco | Photosynthesis, growth and senescene | Wann et al., 1978 |
| Sugar bean | Photosynthesis, growth, transp., development, resp. and LAI | Fick et al., 1975 |
| Maize | Photosynthesis, growth, transp., development, nutrient uptake, partitioning resp. and LAI. | de Wit et al., 1978 |
| Grass | Photosynthesis, growth, resp., and enescene | Johnson et al., 1983 |
| Barley | Photosynthesis, growth and resp. | Legg et al., 1979 |
| Soybean | Photosynthesis, growth, transp., development, nutrient uptake, partitioning, resp. and LAI | Mayer et al., 1979 |
| Rape/pod midge | Yield of rape influenced by p.m. | Axelsen, 1993 |
| Platygaster simulator | Biol. control of podge midge | Axelsen, 1995 |
| Aridtree/olive tree | Photosynthesis, production, organ and reserve biomass | Rasik, 1989 |
| Riceys/rice | Integrated pest management, growth limited by nitrogen, photosynthesis | Graf et al., 1990 |
| Oryza/rice | Growth det. by C and N, weed | See model |

**Table 4.1 (Continued)** Overview of Models Related to Plant Production

| Focal component/ name of the model | Characteristic of the model/st. variables | Major references |
|---|---|---|
| Ricemod/rice | Photosynthesis, canopy, resp., LAI, 4 soil layers, water, leaf N/area | McMennamy 1980 |
| Ceres: several cereals (rice) | Yield, water, N, rad., temp., LAI leaf N-deficiency, soil layers | Ceres, 1986 |
| Crayfish and rice | 7 age classes of crayfish, rice growth (density and weight), rice yield | Anastacio et al., 1994 |
| Macros | Growth limiting factors | Contact de Vries |
| Growth/water | Yield/water content in soil | Poluektov, 1979 |
| Wasp | Parasitism by wasp | Li, 1992 |
| Cowpea | Cowpea demographic canopy model | Contact Tamo |
| Grapes | Photosynthesis, N-uptake, growth as f (degree-days) age - structure | Vermelinger et al., 1992 |
| Grain dev. | growth as f (light, N, water, daylength) developmental processes | Wilhelm et al., 1992 |
| Eldana | Biol. contr. prey-predator, 10 st. var. | Hearne, 1995 |
| Calluna | Eutrophication, N deposition, beetle, development of heather | Aerts and Heil, 1993 |
| Nmanage | crop N uptake and mineralization as f(degree days) | Honeycutt et al., 1994 |
| Seed, rodents | 6 st. var. of rodents, avail. + non a. seed | Maurer, 1990 |
| Biol. control | Leslie matrices predator-prey | Selhorst, 1989 |
| Farmspray | Impact pesticides on predator-prey | Sheratt and Jepson, 1993 |
| Cotton rats | 7 st. var. of cotton rats, damage to sugar cane | Montague et al., 1990 |
| Sawfly | Impact of pesticides on sawflies | Aebischer, 1990 |
| MiteSim | Mite predator/prey system | Barry et al. 1991 |
| Daisy | Details see Section 4.2 | Hansen et al., 1990 |

**Table 4.2** Overview of Models Related to Livestock Production

| | | |
|---|---|---|
| Lactation | 4 st. var. secretory cells, hormones, milk prod., time aver., milk prod. | Neal and Thornley, 1983 |
| Egg production | Eggs as f(amino acid intake) | Fischer et al., 1983 |
| Ducks | Reproduction as f(env. and biol. factors | Johnson et al., 1987 |
| Forage / beef cattle | cattle/grass interactions | Baker, 1988 |
| Screw worm fly | control by the use of SIT | Marsula, 1992 |
| Pasture weed | weed/grazing management | Gillman et al., 1993 |

# 4.5 Agriculture Models

**Model identification**
Model name/title: DASYNEURA-population

**Model type**
Media: terrestrial, agricultural
Class:
Substance(s)/compounds:
Biological components: population(s), Pod midge (*Dasyneura brassicae*)

**Model purpose**
The model was built in order to investigate life table more thoroughly than is possible by common life table analysis. In addition to identifying key factors, modeling can be used to quantify changes in these factors.

**Short description**
The model describes the population development of the pod midge (*Dasyneura brassicae Winn.*) in winter rape and spring rape (*Brassica napus L.*) in Denmark. The model works on a degree-day time scale and takes calculations of daily accumulated degree-days above 6-7 °C (egg and larval development) and above 8.1°C (development in cocoons in the soil) as the driving input parameter. Other important parameters are mortalities at the different life-stages, which comes from a life table study. An important procedure in the model is book keeping device, and is used to perform ageing and to some extent also mortality. The model was inspired strongly by the modeling of Gutierrez et al. (1984, 1988).

---

**Model application(s) and State of development**
Application area(s):
Number of cases studied: 1
State: has been calibrated to field data, but not validated on independent data.

---

**Model software and hardware demands**
Platform: PC-based, IBM
OS:
Software demands:
Hardware demands:

---

**Model availability**
Purchase: software available as executable code
Contact person(s) Dr. Jørgen Aagaard Axelsen
                National Environmental Resarch Institute
                Dept. of Terrestrial Ecology
                Vejlsøvej 25
                DK-8600 Silkeborg
                Denmark

Phone:
Fax: +45 89 20 14 14
E-mail:

---

**Model documentation and major references**
1. Axelsen, J.A. 1993. Analysis of the population dynamics of the pod gall midge (Dasyneura brassicae Winn.) in winter rape and spring rape by computer simulation. *Ecol. Modelling*, 69, 443-55.
2. Gutierrez, A.P., Baumgartner, J.U. and Summes C.G. 1984. Multitrophic models of predator-prey energetics. *Can. Entomol.* , 116, 923-963.
3. Gutierrez, A.P., Neuenschwander, P., Schulthess, F., Herren, H.R., Baumgartner, J.U., Wermelinger, B., Löhr, B. and Ellis, C.K. 1988. Analysis of biologycal control of cassava pests in Africa II. Cassava mealybug *Phenacoccus manihoti. J. Appl. Ecol.*, 25, 901-920.
4. Manetch, T.J. 1976. Time-varying distributed delays and their use in aggregative models of large systems. *IEEE Trans. Syst. Man. Cypern.*, 66, 547-553.

---

**Other relevant information**

---

**Model identification**
Model name/title: PLATYGASTER simulator.

---

**Model type**
Media: terrestrial, agricultural
Class:
Substance(s)/compounds:

Biological components: population(s), pod midge (*Dasineura brassicae*) and *Platygaster oebalus*, organism, pod midge (*Dasineura brassicae*) and *Platygaster oebalus*.

## Model purpose
The model was used to simulate the interaction between *Dasineura bnrassicae* and its most important parasitoid *Platygaster oebalus*. The purpose was to investigate the importance of certain behavioral traits, that have been stated in the literature to be important for parasitoids in biological pest control.

## Short description
The model simulates the interactions between the pod midge and the parasitoid in an agricultural ecosystem with both winter rape and spring rape. This makes it possible for both species to have 3 generations per growth season. The population development in the model is driven by temperature and a physiological time scale is used. The population delay development is described by aid of a distributed delay procedure (Manetch, 1976; Vansickle, 1977), that serves as a book keeping device. Important parameters are duration of life-stages in degree-days, mortalities at different life stages, parasitoid handling time, the pod midge batch size and searchefficiency.

## Model application(s) and State of development
Application area(s):
Number of cases studied:
State:

## Model software and hardware demands
Platform: PC-based
OS:
Software demands: Turbo Pascal
Hardware demands:

## Model availability
Purchase: no
Contact person(s) Dr. J.A. Axelsen
　　　　　　　National Environmental Research Institute
　　　　　　　Department of Terrestrial Ecology
　　　　　　　Vejlsøvej 25
　　　　　　　DK-8600 Silkeborg
　　　　　　　Denmark

Phone: +45 89 20 14 14
Fax:
E-mail:

**Model documentation and major references**
1. J.A. Axelsen (in press). Host-parasitoid interactions in an agricultural ecosystem: a computer simulation. *Ecol. Modelling.*
2. T.J. Manetch, 1976. Time-varying distributed delays and their use in aggrgative models of larger systems. *IEEE Trans. Systems, Man, and Cybernetics*, 66, 547-553.
3. J. Vansickle, 1977. Attrition in distributed delaly models. *IEEE Trans. Systems, Man, and Cybernetics*, 7, 635-638.

**Other relevant information**

**Model identification**
Model name/title: ARIDTREE: A model of the productivity of olive trees under optional water and nutrient supply in desert conditions.

**Model type**
Media: terrestrial, agriculture
Class: biogeochemical
Substance(s)/compounds: nutrients, carbon
Biological components: desert agroecosystem, olive trees

**Model purpose**
Computing carbon assimilate production as a function of the daily amount of available radiation, and simulating the allocation dynamics of photosynthate partitioning to respiration, organs tissues, and reserves under conventional cultural practice, with the option to simulate the effects of supplementary irrigation and fertilizer application on the performance of the tree.

**Short description**
The model has two components: (a) climatic submodels, which account for the day length, solar height, transmission characteristics of the sky, and compute the daily amount of available radiation; and (b) biological submodels, where the expected carbon assimilate production is calculated using the leaf characteristics, amount of intercepted light, and extinction coefficient of the crop. The maintenance respiration is calculated as a function of protein and total mineral contents. The resource allocation is simulated as a function of protein of expected maximum relative growth rate of different organs. The biomass growth is computed using the conversion efficiency of net assimilation to structural plant material, than account for loss by death and due to harvesting operations. State variables are photosynthesis active radiation, gross and net photosynthate, organs and reserve biomass. Forcing functions are photoperiod, fraction overcast, temperature regime, maintenance and growth respiration, effect of irrigation and fertilizer application. Important parameters are light use efficiency, photosynthetic rate at light saturation, maximum relative growth and death rates of the crop and of the individual organs, maximum transpiration rate, and the

shape parameters of the different functions. <u>Necessary inputs</u> are latitude, plant density, canopy cover percentage, specific leaf area, nitrogen and total mineral contents in different organs, dates of major phenological events, and organs biomass (biological inputs refer to the first day of simulation). Some other data are contents that can be manipulated for different situations.

## Model application(s) and State of development

Application area(s): Productivity of fruit trees - management of agrosystems
Number of cases studied: 2 (local and regional scales)
State: validation

## Model software and hardware demands

Platform: PC-based
OS: MS - DOS 5
Software demands: Fortran or MS - QBASIC
Hardware demands: PC - IBM compatible

## Model availability

Purchase:  US $1000
Contact person(s) Dr. Mohammed S. Abdel Razik
    Dep. Botany, Fac. Science,
    University of Alexandria
    Egypt

Phone:    + 3 59 72 352, + 3 59 72 628
Fax:   + 3 49 11 794
E-mail:

## Model documentation and major references

Report: Fruit tree cultivation in the northwestern coastal zone of Egypt. Center for Agrobiological Research, Wageningen, The Netherlands, 1987.
Publication: *Ecol. Modelling*, Vol. 45, (1989) pp. 179-204

## Other relevant information

A new version of the model has been developed to predict the herbage production of semiarid rangeland and the consumption by grazing animals as a function of environmental and biological variables. It allows for testing and evaluating the sensitivity of the system to variation in climatic and biological driving forces at the landscape scale, and to implement the output of a geographic information systems (GIS) at the regional scale.

**Model identification**
Model name/title: RICESYS

---

**Model type**
Media: wetland, agricultural
Class: rice yield
Substance(s)/compounds: carbon, nitrogen
Biological components: ecosystem, rice, weeds, insects

---

**Model purpose**
Simulate rice yield.

---

**Short description**
Ricesys (Graf et al., 1990a and b) is a demographic model for rice growth and development as affected by temperature and solar radiation. It was written to study crop-pest interactions and provides a useful tool for the development of sound integrated pest management strategies. It was designed to describe the dynamics of crop-weed competition and to predict dry matter allocation with respect to different weeding strategies (Graf et al., 1990b). The model does not include an effect of ambient $CO_2$ concentrations on plant physiology. A distributed delay model is used to describe the dynamics of tiller production and culm, leaf, root and grain mass growth (Graf et al., 1990a). Ricesys also includes several species of leaffolders feeding on leaves and weeds competing with rice for a limiting resource, solar irradiance or nitrogen.

All nutrients, except nitrogen, are assumed to be non-limiting and water is assumed to be abundant in the irrigated paddies. The model includes a simple soil nitrogen balance. Ricesys does not consider typhoons and assumes no other pests than herbivorous leafhoppers.

The Frazer-Gilbert functional response model is used to predict photosynthesis. This approach allows the simulation of energy acquisition at different trophic levels.

Photosynthesis is a function of solar radiation, temperature, leaf area index and of the plant's demand for carbohydrate. Nitrogen stress (ratio between actual and potential nitrogen content) also modifies photosynthesis. Maintenance respiration is a direct function of temperature. Photosynthate accumulates in a metabolic pool and is used to satisfy respiration first, reproduction, growth and reserves with decreasing priorities. Carbohydrate and nitrogen are partly recycled from dying plant material. The demographic component of the model is a book keeping device for births, deaths, growth and aging of mass and numbers of plant subunits. Ricesys was originally calibrated to represent the variety Makalioka 34 from Madagascar in irrigated paddies that were fertilized before transplanting. It was secondarily validated for high yielding varieties (IR54 and 68) in the Philippines (Graf et al., 1991).

---

**Model application(s) and State of development**
Application area(s): Rice production, Africa and Philippines
Number of cases studied:
State: validation

**Model software and hardware demands**
Platform: PC-based
OS: DOS
Software demands: FTN/Modula 2
Hardware demands:

**Model availability**
Purchase: Benno Graf
Contact person(s)

Phone:
Fax:
E-mail:

**Model documentation and major references**

Graf et al. 1991

**Other relevant information**

**Model identification**
Model name/title: Oryza/Intercoh (Oryza + competition routine)

**Model type**
Media: wetland, agricultural
Class:
Substance(s)/compounds: carbon, nitrogen
Biological components: ecosystem, rice (+ weeds possible)

**Model purpose**
Simulate potential rice yield.

**Short description**
No information.

**Model application(s) and State of development**
Application area(s): rice
Number of cases studied:
State: validation

## Model software and hardware demands

Platform: PC-based
OS: DOS
Software demands: Fortran
Hardware demands:

## Model availability
Purchase: free
Contact person(s) Martin Kropff
        IRRI

## Model documentation and major references

## Other relevant information

## Model identification
Model name/title: Ricemod

## Model type
Media: wetland, agricultural
Class:
Substance(s)/compounds: carbon, nitrogen
Biological components: organism, rice

## Model purpose
Simulate rice yield.

## Short description
Ricemod (McMennamy and O'Toole, 1983; McMennamy, 1980) was developed to assess the state-of-the-art in rice science and guide research by stressing key process that were not well understood. It includes a number of physical parameters, but is also designed to accommodate addition of subroutines dealing with soil and plant chemistry as well as biophysical processes of the aerial environment. Daily weather inputs include precipitation, pan evaporation, maximum and minimum temperatures, daylength and solar radiation. It does not include the influence of $CO_2$. It assumes optimum levels of nutrients and does not take into account typhoons or pest. In this model, photosynthesis is a function of solar radiation, canopy structure and the ratio of leaf nitrogen weight to area. Canopy shape is derived from leaf area index (LAI) (Curry et al., 1975). Leaf area is assumed to be proportional to leaf weight. Leaf nitrogen content is assumed to be proportional to plant age. Photosynthesis is not affected by temperature, but is affected by water stress. Photosynthetic rate is modified linearly by the ratio of actual transpiration over potential transpiration. Maintenance respiration is a direct function of temperature.

Allocation of fixed carbohydrate between plant parts in Ricemod is derived from tables relating allocation to plant age in days. Ricemod also includes a soil-water balance routine. The soil is assumed to consist of four layers, three of which contain roots, and the fourth is assumed to have a constant water content. A two-stage evaporation is calculated based on the method of Ritchie (1972) and uses soil water diffusivity. It relies almost entirely on soil, plant and atmospheric data derived from experiments at IRRI where fields were fertilized.

## Model application(s) and State of development
Application area(s): Rice agriculture
Number of cases studied:
State: validation

## Model software and hardware demands
Platform: PC-based
OS: DOS
Software demands: FTN
Hardware demands:

## Model availability
Purchase: free
Contact person(s)
Phone:
Fax:
E-mail:

## Model documentation and major references
McMennamy & O'Toole, 1983
McMennamy, 1980

## Other relevant information

## Model identification
Model name/title: CERES-Rice

## Model type
Media: wetland, agricultural
Class:
Substance(s)/compounds: carbon, nitrogen
Biological components: organism, rice

## Model purpose
Simulate rice yield and growth.

## Short description
CERES (Crop Environment Resource Synthesis) - Rice emphasizes the effects of management and the influence of soil properties on crop performance. The model was designed to assess yield as constrained by varietal characteristics, soil water and nitrogen, for alternative technology and for new growing sites. It was designed to reduce time and cost of agrotechnology transfer of new varieties and management practices. Weather inputs include daily solar radiation, maximum and minimum temperatures and precipitation. It ignores the potential effect of typhoons and assumes no pests. However, some work has been done to simulate the impact of diseases on yield (J. Ritchie, Michigan State University, East Lansing, MI, USA, personal communication, 1991) and subroutines simulating the impact of blast on rice yield have been developed (P. Teng, International Rice Research Institute, Los Baños, Philippines, personal communication, 1992), but the code was unavailable at the time this paper was written. The influence of $CO_2$ on photosynthesis (enhancement) and transpiration (decrease) has been recently added (scalar modifier) to assess the impact of increased $CO_2$ on rice yield in a U.S. EPA sponsored international project (C. Rosenzweig, NASA Goddard, New York, USA, personal communication, 1991).

In CERES-Rice, photosynthesis is expressed as daily biomass production and calculated as a function of LAI, solar radiation and a radiation use efficiency constant. Biomass production can be reduced by high temperatures and nitrogen deficiency. LAI is a function of the time of leaf appearance and expansion which can also be reduced by nitrogen deficiency. Allocation depends on the phenological stage of the plant which is primarily a function of thermal time. Partitioning between shoots and roots can be modified by water and nutrient deficiencies. Phenological development and crop duration are genotype specific and related to temperature and photoperiod. Phenological stages include sowing, germination, emergence, juvenile period, panicle initiation, heading, beginning of grain fill, end of grain fill and physiological maturity. Photoperiod sensitivity influences thermal time requirement during floral induction. Specific genetic information needed to run the model includes the thermal time (degree-days) between emergence and the end of juvenile period, the photoinduction rate, the optimum photoperiod, the thermal time required for grain fill, the efficiency of sunlight conversion to assimilates, the tillering rate and the grain size.

A modified Priestly and Taylor potential evapotranspiration method is used to drive the water balance separating soil evaporation and transpiration. The soil is characterized by its initial nitrogen content, water holding properties, texture and field topographic position. The user determines the number of soil layers used to run the model. No calibration is required when cultivar characteristics are known.When they are unknown, genetic parameters can be approximated from field results through a calibration exercise.

## Model application(s) and State of development
Application area(s): rice production
Number of cases studied: many
State: validation

## Model software and hardware demands
Platform: PC-based
OS: DOS, FIN
Software demands:
Hardware demands:

## Model availability
Purchase:
Contact person(s) Joe Ritchie
Phone:
Fax:
E-mail:

## Model documentation and major references

## Other relevant information

## Model identification
Model name/title: Updatings of CERES Maize with respect to organic manure, snow, soil temperature etc.

## Model type
Media: local scale, regional scale, terrestrial, soil, agricultural
Class:
Substance(s)/compounds: carbon, nitrogen
Biological components: ecosystem - agroecosystem

## Model purpose
Simulation of impact of agricultural measures and weather variables on growth, development, nitrogen uptake, leaching, etc.

## Short description
The model compares growth, development and nitrogen uptake of maize, temperature, soil water budget, soil nitrogen transformations, soil temperatures, drainings and leaching.

## Model application(s) and State of development
Application area(s): Simulation studies on the behaviour of a maize agroecosystem
Number of cases studied:
State: conceptualization, verification, calibration, prognosis made

---

**Model software and hardware demands**
Platform: PC-based IBM
OS: DOS
Software demands: FORTRAN
Hardware demands: PCAT

---

**Model availability**
Purchase: executable code, source code
Contact person(s) Dr. Friedrich Hoffmann
    Projektzentrum Ökosystemforschung
    Schauenburger Str. 112
    D-24118 Kiel
    Germany

Phone: +49 431 8804084
Fax:
E-mail:

---

**Model documentation and major references**
*J. Agronomy & Crop Science*, 1993, 170: 56-65.
*J. Agronomy & Crop Science*, 1993, 170: 330-340.

---

**Other relevant information**
Former description of the model in Jones, C.H., Kimiry, J.R.: *CERES Maize - A Simulation Model of Maize Growth and Development*. Texas A&M University Press, 1986.

---

**Model identification**
Model name/title: Crayfish and rice dynamics

---

**Model type**
Media: water, wetland, terrestrial, agricultural, rice field
Class:
Substance(s)/compounds: oxygen, pesticides
Biological components: organism - Procambarus clarkil

---

**Model purpose**
To describe a population of crayfish living in a rice field area and to determine the viability of double cropping rice and crayfish.

---

**Short description**
The model allows a better understanding of crayfish population dynamics. The final purpose is the management of a rice field, allowing farmers to obtain more profit. An optimum crop of crayfish and rice should be found, in order to compensate for the damages caused in the rice fields.
State variables: seven size classes for crayfish, rice densities, rice weight

## Model application(s) and State of development
Application area(s):
Number of cases studied: one
State: calibration

## Model software and hardware demands
Platform: PC-based MAC
OS: DOS
Software demands: STELLA 2
Hardware demands:

## Model availability
Purchase: no, available as executable code
Contact person(s) Dr. Pedro M. Anastácio
              Department of Zoology
              University of Coimbra
              3049 Coimbra Codex, Portugal

Phone: + 351 39 23603
Fax: + 351 39 23603
E-mail: anast@gemini.ci.uc.pt

## Model documentation and major references
Anastácio, P.M., Nielsen, S.N., Marquez, J.C. and Jørgensen S.E., 1993. Rice and crayfish production in the lower Mondego river valley (Portugal): A management model for farmers. *Proceedings of the "International Congres on Modelling and Simulation".* (4) 1687- 1692.

## Other relevant information

## Model identification
Model name/title: MACROS

## Model type
Media: wetland, agricultural
Class:
Substance(s)/compounds: carbon, nitrogen
Biological components: organism, rice

## Model purpose
Simulate rice growth.

**Short description**

MACROS is a technology based on the concept of growth-limiting factors (de Witt and Penning de Vries, 1982; Penning de Vries and van Laar, 1982) and is intended for educational purposes, especially in the developing countries. Environmental inputs include solar radiation, maximum and minimum temperatures, precipitation, daylength, vapour pressure and wind speed. It includes the effect of $CO_2$ on photosynthesis. Macros is set up to run in one of three distinct modes in which 1) water and nutrients are in optimal supply, and weeds, pests and diseases are absent; 2) water stress due to limiting water availability.

**Model application(s) and State of development**

Application area(s): Rice production
Number of cases studied: many
State: validation

**Model software and hardware demands**

Platform: PC-based
OS: DOS
Software demands: CSMP/FTN
Hardware demands:

**Model availability**

Purchase:
Contact person(s) Fritz Penning de Vries

Phone:
Fax:
E-mail:

**Model documentation and major references**

**Other relevant information**

**Model identification**

Model name/title: AGROTOOL: model for decision support in agriculture

**Model type**

Media: agricultural
Class: biogeochemical
Substance(s)/compounds: nutrients, carbon, phosphorus, nitrogen, chemicals, organic compounds
Biological components: ecosystem, agro-ecosystem

## Model purpose
Model is intended to estimate water content in a controlled soil layer under crops of several cultures, to forecast a rate of plant development and an expected yield. On irrigated fields the model can be used for control and optimization of soil-water regime.

## Short description
AGROTOOL is a software product for simulation of soil moisture dynamics under crop covers of several cultures as influenced by variable weather conditions and possible cultural measures (fertilization, irrigation). A new method is realized for simulation of plant transpiration and soil evaporation under both soil and water deficit and excess.

It comprises three basic parts:
- dynamic model of yield formation by agricultural crops under natural field conditions;
- model of weather forecast considering as an input signal for simulation of agro-ecosystem dynamics;
- algorithms of the forecast of plant development and yield formation.

Model identification and the estimation of the parameters of weather generator is oriented on the use of a wide-spread information and does not require the carrying out of special field experiments.

Major state variables are soil water content in several (up to ten) layers, biomass of plant organs, leaf area index, root depth, carbon, nitrogen and phosphorus content in soil and plants. Time step equals one day.

Forcing functions are fertilization and irrigation. Model input includes the following daily weather conditions: minimum and maximum air temperature, air humidity, precipitation, sunshine duration and wind speed. Important parameters are hydraulic soil characteristics and coefficients of plant reactions on environment conditions.

## Model application(s) and State of development
Application area(s): Scientific investigation of agro-ecosystem dynamics, education and training, decision support in agriculture
Number of cases studied: several thousand
State: validation, validation of prognosis

## Model software and hardware demands
Platform: PC-based, IBM PC/AT-286 (386)
OS: MS DOS, version 3.2 and higher
Software demands: Turbo-Pascal compiler, Turbo Pascal, version 5.0
Hardware demands: 640 K RAM, EGA graphics adapter, math coprocessor

## Model availability
Purchase: yes, source code
Contact person(s) Dr. Ratmir A. Poluektov
14 Grazhdansky Prospect,
Agrophysical Research Institute
Laboratory of Agro-ecosystem Simulation
195220, St. Petersburg, Russia

Phone: 812-534-46-40
Fax: 812-534-16-44
E-mail: ivl@agrophys.spb.su

---

**Model documentation and major references**

1. Poluektov, R.A., Simulation models of agro-ecosystem productivity. Theoretical base and quantative methods of yield programming. Leningrad: Agrophysical Research Institute, 1979. pp. 14-23. (in Russian).

2. Bondarenko, N.F., E.E. Zhukovsky, I.G. Mushkin, S.V. Nerpin, R.A. Poluektov, and I.B. Uskov. *Simulation of Agro-Ecosystem Productivity* (book, in Russian). Leningrad: Gidrometeoizdat. 1982. 262 pp.

3. Poluektov, R.A. et al. Basic complex model of wheat productivity. Wissenschaftliche Zeitschrift der Humdoldt - Universitaet zu Berlin, Math.-Nat. R. 1984. XXXIII, # 4. pp. 339-342.

4. Poluektov, R.A. et al. Modelling of processes in the soil-plant system in mead - early tomato growing. *Intern. Agrophysics*, 1 (1), 1985. pp. 91-105.

5. Poluektov, R.A. Dynamic models of plant production as a base for elaboration of resource -saving agricultural technologies. Tag.-Ber., Akad. Landwirtsch.-Wiss.DDR, Berlin, 275, 1988, pp. 25-29.

6. Alijev, P.G. and R.A. Poluektov. Simulation of root system growth and development in the framework of basic model of plant production process. *Vestnik Sekjskhozjajstvennoj Nauki*. 1988, #11 (387). pp. 101-105.

7. Poluektov, R.A. et al. Experimental investigation of a growth and development of oats and barley plants with the aim of identification of their dynamic models. *Physiologija i Biokhimija Kuljturnykh Rastenij*, v. 21# 5, 1989. pp. 507-514.

8. Poluektov, R.A. *Dynamic Models of Agro-Ecosystem* (book, in Russian). Leningrad: Gidrometeoizdat, 1991. 312 pp.

9. Poluektov, R.A., G.V. Vasilenko, I.V. Oparina, and A.G. Topazh. Computer system of decision support in agricultural management. *Informatica i Vychisliteljnaja Tekhnica*. 1993, N 1-2. pp. 99-103. (The full text of the report on the 2nd International Conference "Regional Informatics RI -93., St. Petersburg, May 11-14, 1993).

10. Poluektov, R.A. and G.V. Vasilenko. A universal water stress index in cultivation of agricultural crops. *Intern. Agrophysics* (in press).

---

**Other relevant information**

## Model identification
Model name/title: Patch residence time and parasitism of *Aphelinus asychis*

## Model type
Media: local scale, agricultural
Class:
Substance(s)/compounds: nutrients: pea aphid as nutrient on broad bean plant
Biological components: population: parasitoid wasp - *Aphelinus asychis,* and pea aphid

## Model purpose
(1) To summarize the parasitoid wasp's foraging behaviour at within-patches level, including using honeydew as a clue of host's presence, optimal foraging on "decisions" of patch leaving and host acceptance, foraging behaviour adjustment from past experience, and stochastic influences; and (2) to serve as a model component of between-patches parasitism.

## Short description
The model simulates the patch residence time and parasitism rate of the parasitoid wasp, *Aphelinus asychis*, to its pea aphid hosts. The wasp's decision to leave a patch or to accept or reject a host is based on information about previously encountered hosts that are stored in a sliding "memory window" with five units.

The model consists of two parts: one is for the calculation of an optimal host acceptance decision matrix for the wasp, and another is for the Monte Carlo simulation of parasitism. The state variable for calculating the decision matrix is egg reserve of the wasp. The formula for the calculation is the equation (3) in the paper. The main parameters used in the stochastic simulation model are: time for laying an egg; time for feeding a host; time for rejecting a host; probability of not encountering any host; probability of successful attack. A flowchart of the program and detailed description of the simulation is presented in Appendix 1 of the paper.

## Model application(s) and State of development
Application area(s): Behavioral Ecology and Population Ecology (foraging behaviour and its consequences on host population), Biological Control (evaluation and selection of natural enemy), System Ecology (organizing separated pieces of information in a biological framework)
Number of cases studied: 1
State: completed

## Model software and hardware demands
Platform:
OS:
Software demands: Microsoft QuickBASIC for MAC, QuickBASIC
Hardware demands: MAC

---

**Model availability**
Purchase: Since the programs were not prepared for public use at the time they were written, they do not include any explanation and documentation. So, currently they are not available. However, I can work on them and make them available, if there is enough interest and demand for the programs.
Contact person(s) Dr. Chao Li
                 Northern Forestry Centre
                 Canadian Forest Service
                 5320- 122nd Street
                 Edmonton, Alberta
                 Canada T6H 3S5

Phone: +1 403-435-7241
Fax: +1 403-435-7359
E-mail: cli@nofc.forestry.ca

---

**Model documentation and major references**
Model documentation is currently not available. Major references are:
Li, C. 1992. The foraging behavior of a parasitoid wasp, Aphelinus asychis: a modelling approach. Ph.D. dissertation. Simon Fraser University. Burnaby, B.C., Canada.
Roitberg, B.D., Reid, M. and Li, C. 1992. Choosing hosts and mates: the value of learning. In: D.R. Papaj and A.C. Lewis (Eds.), *Insect Learning: Ecological and Evolutionary Perspectives.* Chapman & Hall, London, pp. 174-194.

---

**Other relevant information**
The model is a summary of the parasitoid wasp's foraging behaviour at the within-patches level. This within-patches parasitism is in turn a component of between-patches parasitism. My model for between-patches parasitism is a 3-D Monte Carlo simulation for wasp on the host population dynamics. The model has not been published, see reference (1) above.

---

**Model identification**
Model name/title: (Analysis of the cowpea agro-ecosystem in West Africa. I.) A demographic model for carbon acquisition and allocation in cowpea, *Vigna unguiculata* Walp.

---

**Model type**
Media: agricultural
Class: biogeochemical
Substance(s)/compounds: nutrients, carbon, phosphorus
Biological components: ecosystem - cowpea agro-ecosystem population - cowpea plant organs organism - individual plant organ (leaf, shoot, root, fruit, peduncle)

## Model purpose

This model has been designed for research purposes, to be used as a tool for the exploration of the cowpea agro-ecosystem. It demonstrates the suitability of the demographic approach for the analysis of cowpea growth and development. The model also permits the assessment of yield limiting factors, which is of great value as herbivory will be added to the model in the future.

## Short description

Growth and development of a photosensitive cowpea variety are studied with the help of a demographic canopy model. The major driving variables are abiotic factors such as temperature, solar radiation, soil phosphate and water. The dynamics of age-structured populations of leaves, shoots, roots, peduncles, and other reproductive organs having numbers and dry matter attributes is simulated by a time-varying distributed delay model with attrition. The delay subroutine is used to emulate stochasticity in growth, aging, and loss of population attributes. A modified functional response model from predation theory is used to estimate the daily photosynthates acquisition. Dry matter allocation is simulated by a metabolic pool model. Growth and yield formation of the plant are driven by the computed ratio between carbohydrates supply and demand. Demand parameters were estimated from detailed dry matter data of developing plant organs. Simulation results are compared with four sets of field data. Further, the model has been used to evaluate the effect of drought stress and different levels of available phosphate in the soil.

## Model application(s) and State of development

Application area(s): agriculture
Number of cases studied: 4 validation sets, 2 evaluation sets
State: validation with field data

## Model software and hardware demands

Platform: current - Apple Macintosh Quadra 700, 8 MB RAM min - any 68030 Mac with 68882 FPU, ▢MB RAM
OS: current - system 7.1 min - system 6.0.5
Software demands: Think Pacal 4.0.1
Hardware demands:

## Model availability

Purchase: the source code is available for free if you send me an empty 3.5" disk
Contact person(s) Dr. Manuele Tamo
        IITA-BCCA
        B.P. 08-0932
        Cotonou
        Republic of Benin

Phone:
Fax: +229-30-14-66
E-mail: M.TAMO@CGNET.COM

---

**Model documentation and major references**

---

**Other relevant information**

---

**Model identification**
Model name/title: Grapewine growth model, VIMO

---

**Model type** .
Media: local scale, agricultural
Class: biogeochemical
Substance(s)/compounds: carbon (dry mass), nitrogen
Biological components: ecosystem - agroecosystem, population - grapewine

---

**Model purpose**
The model forms a basis for analysis in the vineyard ecoystem. It is designed as a research tool for explorative studies in multitrophic systems and can be connected to other models such as insect herbivore models.

---

**Short description**
VIMO is a dynamic crop model for dry matter (DM), and nitrogen (N) assimilation and allocation basing on the metabolic-pool model.

Photosynthesis and N uptake from the soil are functional response models and are sink-driven. Growth occurs per degree-day above developmental threshold of grapewine. The plant subunits, i.e., the annual populations of fruits, leaves, shoots and roots developing on a perennial frame are age-structured and have distributed developmental times. Their dynamics are simulated as a time-invariant distributed delay process with attrition. The seasonal N dynamics is the net result of the processes of new tissue formation with high N concentrations and the degree-day-driven export of N from ageing parts to reserves. State variables: fruits, leaves, shoots, roots, perennial frame, reserves. Forcing functions (independent variable: age of individual): photosynthetic efficiency of leaves, area/DM ratio of leaves, DM and N demand of grape berries. Daily driving variables: solar radiation, minimum and maximum temperature. Important parameters: longevity of annual plant subunits, potential leaf growth rate, N translocation rates. Necessary input: initial condition of grapevine in terms of subunit DM and N content, variety and training specific plant characteristics, soil N.

---

**Model application(s) and State of development**
Application area(s): research tool for explorative studies
Number of cases studied: 2
State: validation

## Model software and hardware demands
Platform: PC-based, (IBM compatibles)
OS: DOS
Software demands: Turbo Pascal, English
Hardware demands: 386 + math Co

## Model availability
Purchase: no, source code
Contact person(s) Dr. A. P. Gutierrez
University of California, Berkeley
Division of Biological Control
Albany, CA 94706
USA

Phone: +1 415-642-9186
Fax:
E-mail:

## Model documentation and major references
Wermelinger, B., Baumgartner, J., and Gutierrez, A.P., 1991. A demographic model of assimilation and allocation af carbon and nitrogen in grapevines. *Ecol. Modelling* 53: 1-26.
Wermelinger, B., and Baumgartner, B., 1990. Application of a demographic crop growth model: an explorative study on the influence of nitrogen on grapevine performance. *Acta Hort.*, 276: 113-121.
Wermelinger, B., 1991. Nitrogen dynamics in grapevine: physiology and modeling. In: *Proceedings of the International Symposium on Nitrogen in Grapes and Wine,* Seattle WA, USA (J.M. Rantz, Ed.). American Society for Enology and Viticulture, Davis CA, USA, pp. 23-31.
Wermelinger, B., Candolfi, M.P., and Baumgartner, J., 1992. A model of the European red mite (Acari, Tetranychidae) population dynamics and its linkage to grapevine growth and development. *J. Appl. Entomol.*, 114: 155-166.

## Other relevant information

## Model identification
Model name/title: SHOOTGRO

## Model type
Simulation of small grain development and growth from planting through physiological maturity.
Media:
Class:
Substance(s)/compunds:
Biological components:

## Model purpose
To simulate the development and growth of small grains (winter and spring wheat and winter and spring barley) from easily obtained inputs and weather data in response to several environmental factors (nitrogen, light, and water availability) based on accumulation of thermal time.

## Short description
SHOOTGRO simulates and reports date of appearance, size, date of senescence, and frequency in the canopy of all above-ground parts of the small grain modeled. Events are timed in terms of the phyllochron (the interval between appearance of successiveleaves on a culm) in the model.

The algorithm reported by Baker et al. (1980) is used to predict the phyllochron based on the rate of change in daylength at the time of seedling emergence. Developmental and growth processes are limited by the availability of light, nitrogen, and water. The individual plants in the community are divided into as many as five cohorts (based on time of emergence). Use of cohorts (each have a different time of emergence and therefore somewhat different conditions during each developmental phase) allows the simulation to reflect the variation between plants seen in the field. In addition to growth parameter, the model continuously reports crop development in three commonly used stages -- Feekes, Zadoks-Chang-Konzak, and Haun.

## Model application(s) and State of development
The model has been constructed to run in all small grains production areas of the world (both Northern and Southern hemisphere). The model has been conceptualized and the code verified. No calibration is need to use the model. Validation of the model has been limited to conditions for the Central Grain Plants of the USA and South Africa.
Application area(s):
Number of cases studied:
State:

## Model software and hardware demands
The model was written in standard FORTRAN 77 to run on IBM-compatible PCs with an 80386 processor, math coprocessor, and at least 450 Kbytes of memory. The model will run on less powerful machines with at least 450 Kbytes of memory, but execution time will increase. A hard drive is strongly recommended. The model will run on SUN, Digital, Hewlett-Packard, and IBM work-stations and mainframe as well.
Platform:
OS:
Software demands:
Hardware demands:

## Model availability

Purchase: executable and source code, documentation, users manual, and variable dictionary are available at no cost from:

Contact person(s) Dr. Wallace Wilhelm    or    Dr. Gregory S. McMaster
USDA-ARS                       USDA-ARS
117 Keim Hall              Great Plains System Research Unit
University of Nebraska     P.O. BOX E
Lincoln, NE 68583-0934    Fort Collins, CO 80522
USA                            USA

Phone: +1 402-472-1512          +1 303-490-8340
Fax: +1 402-472-7904             +1 303-490-8310
E-mail: wwilhelm@unl.edu       greg@gpsrv1.gpsr.colostate.edu

## Model documentation and major references

Harrell, D.M., Wilhelm, W.W., and McMaster, G., Sales, S., A computer program to convert among three developmental stage scales for wheat. *Agron. J.* 85: 758-763. 1993.

McMaster, M., Klepper, S., Rickman, B., Wilhelm, W.W. and Willis, W.O.. Simulation of above-ground vegetative development and growt of unstressed winter wheat. *Ecol. Modelling* 53: 189-204, 1991.

McMaster, G.S., Morgan, J.A., and Wilhelm, W.W., Simulating winter wheat spike development and growth. *Agric. For. Meteorol.* 60: 193-220. 1992.

McMaster, G.S., Wilhelm, W.W-., Morgan, J.A., Simulating winter wheat shoot apex phenology. *J. Agric. Sci., Camb.* 119: 1-12. 1992.

Wilhelm, W.W., McMaster, G.S., Richman, R.W., and Klepper, B., Above-ground vegetative development and growth of winter wheat as influenced by nitrogen and water availability. *Ecol. Modelling* 68: 183-203. 1993.

Wilhelm, W.W., McMaster, G.S., Rickman, R.W., and Klepper, B., SHOOTGRO 2.0. Nebraska Agric. Exp. Stn. Software Publication No. CP12., 1992.

## Other relevant information

## Model identification

Model name/title: ELDANA Model

## Model type

Media: terrestrial, agricultural
Class: biogeochemical
Substance(s)/compounds: nutrients, sugar cane
Biological components: population - *Eldana saccharina, Goniozus natalensis*

## Model purpose

The African stalk borer, *Eldana saccharina*, is a serious pest to the sugar industry in southern Africa. The parasitoid *Goniozus natalensis* is a candidate for the biological control of *Eldana*. The model is used to investigate the relationship between crop damage by *Eldana* and the magnitude and frequency of releases of *Goniozus*.

## Short description

The model comprises a system of nonlinear ordinary differential equations. The pest (*Eldana*) is reprensented by state variables of the number of individuals in the lifestages: egg, small larvae, large larvae, pupae, adults. Similarly, state variables for eggs, maggots, pupae, and adult wasps represent the parasitoid (*Goniozus*) population.

The main driving variable is the function of time that gives the rate release of adult parasitoids. Some other inputs and parameters are the monthly average temperatures, the price of laboratory-produced parasitoids and the price fetched for cane at the mill.

## Model application(s) and State of development

Application area(s): Pest management
Number of cases studied: 1
State: conceptualization, calibration, validation

## Model software and hardware demands

Platform: PC-based
OS:
Software demands: Turbo Pascal
Hardware demands: 386/486

## Model availability

Purchase: yes, US $100, executable code, source code
Contact person(s) Professor John Hearne
            Department of Mathematics and Applied Mathematics
            University of Natal
            P.O. Box 375
            Pietermaritzburg
            3200 South Africa

Phone: +27 331 260 5626
Fax: +27 331 260 559
E-mail: hearne@math.unp.ac.za

## Model documentation and major references

## Other relevant information

## Model identification

Model name/title: CALLUNA

## Model type
Media: regional scale, terrestrial
Class: biogeochemical
Substance(s)/compounds: nitrogen
Biological components: ecosystem - dry heathlands society - *Calluna-Genistion pilosae* community - Genisto-Callunetum populations - *Calluna vulgaris, Deschampsia flexuosa, Molinia caerulea* organism - plant

## Model purpose
To evaluate impacts of air pollution (nitrogen deposition) and eutrophication with respect to management and heather beetle plagues in relation to the development/dominance of heather (*Calluna vulgaris*) and grasses (*Deschampsia flexuosa* and *Molinia caerulae*).

## Short description
Air can cause a decline in species through eutrophication of the habitat. In the Netherlands eutrophication largely results from atmospheric nitrogen deposition. The simulation model "Calluna" has been developed for evaluation of the impacts of atmospheric nitrogen deposition on dry heathlands. The model has been applied to three deposition scenarios. Equations are based on Lotka-Volterra kind of relationships. State variables are growth rate, replacement rate, decomposition rate and mineralization rates. Necessary inputs are percentage cover of the plant species at the beginning of a simulation, and amount of atmospheric nitrogen deposition. An important parameter is the occurrence of heather beetle outbreaks.

## Model application(s) and State of development
Application area(s): Distribution area of Genisto-Callunetum
Number of cases studied: 17 acidification areas in the Netherlands
State: validation of prognosis

## Model software and hardware demands
Platform: PC- based
OS: DOS
Software demands: none, BalerXE
Hardware demands: 8086 or higher with 512 K RAM

## Model availability
Purchase: yes, executable code
Contact person(s) Dr. G.W. Heil
            Resource Analysis
            Zuiderstraat 110
            2611 SJ Delft
            The Netherlands

Phone: +31-15-122622
Fax: +31-15-124892
E-mail: gerrith@boev.biol.ruu.nl

## Model documentation and major references

## Other relevant information

Aerts, R. and Heil, G.W. (Eds.) 1993. Heathlands: patterns and processes in a changing environment. *Ecol. Modelling,* 68: 161-182.
Heij, G.J. and Schneider, T., (Eds.) 1991. Acidification research in The Netherlands. Final report of the Dutch Priority Programe on Acidification. Studies in Environmental Science 46, Elsevier, Amsterdam.
The model has been applied within the framework of the Dutch Priority Programme in The Netherlands (see references above).

## Model identification
Model name/title: NMANAGE

## Model type
Media: local scale, soil, agricultural
Class: biogeochemical
Substance(s)/compounds: nitrogen
Biological components: ecosystem - agroecosystem community - plant community, soil microbial community

## Model purpose
To improve the efficiency of N use in crop production and thereby reduce the adverse environmental impacts and excess capital expendiures associated with low N use efficiency.

## Short description
The model employs soil thermal units (degree days) to describe N uptake by a crop and N mineralization from soil organic matter and other organic N sources. Mass balance principles are used to compare crop N uptake from these sources with N mineralization over thermal time. Functions describing N use efficiency dynamics are calculated from this relationship and used to predict N uptake from mineralized organic N in future growing seasons. Additional N required by the crop over thermal time is determined by comparing predicted uptake of N derived from mineralized organic sources with N uptake by a crop when grown at its recommended N fertilizer rate. Important input parameters include; (1) N mineralization vs. thermal unit functions for soil organic matter and any other pertinent organic N sources (i.e., crop residue, manure, sludge), (2) N uptake vs. thermal unit functions for unfertilized and fertilized crops, (3) soil bulk density, and (4) quantities of organic N additions pertinent to the cropping system modeled.

## Model application(s) and State of development
Application area(s): Calculating the quantity and thermal time to apply fertilizer N in potato production
Number of cases studied: one/year starting in 1993
State: validation

## Model software and hardware demands
Platform: PC-based IBM
OS: Currently independent of operating system
Software demands: Any capable of performing mathematical computations, no currently coded into a language
Hardware demands: 512 K RAM

## Model availability
Purchase: no, no currently coded into a language
Contact person(s) C. Wayne Honeycutt, Ph.D, Soil Scientist
        USDA-ARS
        New England Plant, Soil and Water Laboratory
        University of Maine
        Orono, ME 04469-5753 USA

Phone: +1 207-581-3363
Fax: +1 207-866-0464
E-mail: RNE902@MAINE.MAINE.EDU

## Model documentation and major references
Honeycutt, C.W., Clapham, W.M., and Leach, S.S., 1994. A functional approach to efficient nitrogen use in crop production. *Ecol. Modelling*, 72 (in press).

## Other relevant information
The model currently exists only as a set of algorithms.

## Model identification
Model name/title: Mass Flow Model of Community Dynamics

## Model type
Media: terrestrial
Class:
Substance(s)/compounds:
Biological components: community - desert rodents and annual plants in southeastern Arizona, USA

## Model purpose
This model was designed to examine the processes of exploitative and interference competition from the viewpoint of mass flow dynamics in populations of desert rodents and the seeds that they feed upon.

## Short description
State variables include biomasses of adult males, adult females, and juveniles of two species of kangaroo rats, seed banks of available and unavailable seeds, and for one species of kangaroo rat, its seed caches. Important parameters include mass specific rates of foraging, mass specific expenditures on reproduction, and effects of interference interactions on mass specific foraging rates. The model was forced by an empirical precipitation table over several years. Foraging rates were modeled using the Michaelis-Menten kinetic equations. The model consisted of 10 differential equations.

## Model application(s) and State of development
Application area(s): community ecology
Number of cases studied:
State: conceptualization

## Model software and hardware demands
The model was written in FORTRAN. Any machine that ahs a FORTRAN compiler should be able to be used to run simulations.
Platform:
OS:
Software demands:
Hardware demands:

## Model availability
Purchase: model exists as FORTRAN source code. Interested persons can purchase the source code for US $50.00 from
Contact person(s) Dr. Brian A. Maurer
                 Department of Zoology
                 Brigham Young University
                 Provo, UT 84602 USA

Phone: +1 801 378 2426
Fax: +1 801 378 7499
E-mail: maurer@yvax.byu.edu

## Model documentation and major references
Maurer, B.A., 1990. Dipodomys populations as energy-processing systems: regulation, competition, and hierarchical organization. *Ecol. Modelling* 50: 157-176.

## Other relevant information

## Model identification
Model name/title: A model describing the predator prey system *S. longicornis-T. cinnabarinus*

## Model type
Media: agricultural
Class: biogeochemical
Substance(s)/compounds:
Biological components: ecosystem - predator-prey system populations - *Tetranychus cinnabarinus* (prey) *Scolothrips longicornis* (predator)

## Model purpose
The model is intended to improve the biological control of *T. cinnabarinus* using its predator *S. longicornis*. Especially two questions can be answered with the use of the simulation model. First, how many individuals of *S. longicornis* are to be applied? Second, what is the appropriate date for the application?

## Short description
The model is based upon two Leslie matrices, one for each population. The primary parameters, the survival and fertility rates, are parametrized as a function of age using the Weibull distribution and the extended Weibull function. With this large number of primary parameters is reduced to a few secondary ones, which enables to formulate the influence of the predator-prey interaction on the matrix elements by modeling the influence upon the parameters of the describing functions.

The link between the predator and prey population is the predator's eating activity. It influences the predator's fertility in a way that can be modeled using a logistic function. The influence on the survival rates of the prey is modeled by calculating the total number of prey consumed a day, which is substracted from the number of individuals before predation.

The eating activity depends on the predator and prey population density and age structure. The influence of the predator's age is modeled using the extended Weibull function. The dependency between the prey supply and the eating activity is described with Holling's disc equation. This effect was adapted to consider an observed time lag by using a weighted moving average of the prey density.

## Model application(s) and State of development
Application area(s): Biological control of *T. cinnabarinus* with *S. longicornis* at constant conditions of temperature and humidity (greenhouses). The model can be extended to field conditions if experimental data concerning the temperature and humidity effect on the survival and fertility rates of *S. longicornis* are available. Number of cases studied: 25 simulation studies were performed to improve date of performing the biological pest control and the number of predators that are to be applied.
State: validation of single population dynamics
prognosis made : yes
validation of prognosis : no

---

**Model software and hardware demands**
Platform: PC- based, work-station SUN Sparc station 10
OS: DOS, SUN OS
Software demands: Graphic tool for visualization of the simulation results (ASCII data), C++ (OOP)
Hardware demands:

---

**Model availability**
Purchase: model is available for free and is available as executable file
Contact person(s) Dr. T. Selhorst
      Math. Sem. Landw. Fak. University of Bonn
      Nussallee 15
      D-53115 Bonn
      Germany

Phone: +49 228-73-27-19 / 73-27-17
Fax: +49 228-73-27-12
E-mail: sel@ms1202.uni-bonn.de

---

**Model documentation and major references**
Selhorst, Th.; 1989. Mathematische Modelle zur Beschreibung der Dynamik alters-strukturierter Populationen und ihre Anwendung auf die Entwicklung optimaler biologischer Schaedlingsbekaempfungsmassnahmen am Beispiel der Bekaempfung von Tetranychus cinnabarinus (Boisd.) durch gezielten Einsatz von Scolothrips longicornis (Priesner). Ph.D. thesis, Bonn.
Selhorst, T., Soendgerath, D. and Weigand, S.; 1991. A model describing the predator-prey interaction between Scolothrips longicornis and Tetranychus cinnabarinus based upon the Leslie theory. *Ecol. Modelling*, 59: 123-128.

---

**Other relevant information**
Gerlach, S., Sengonca, C.; 1985. Comparative studies on the effectiveness of the predatory mite, Phytoseiulus persimilis Athias-henriot, and the predatory thrips, Scolothrips longicornis Priesner. *J. Plant Dis. Prot.*, 92: 138-146.
Hassell, M.P.; 1978. The dynamics of arthropod predatory-prey systems. Monographs in population biology. Princeton University Press, Princeton, New Jersey.
Hazan, A., Gerson, U. and Tahori, A.S.; 1973. Life history and life tables of the carmine spider mite. *Acarologia*, 15: 414-440.
Leslie, P.H.; 1945. The use of matrices in certain population mathematics. *Biometrica*, 35: 183-212.
Richter, O.; 1985. Simulation des Verhaltens oekologischer Systeme. Mathematische Methoden und Modelle. VCH, Weinheim. Deerfield Beach, Florida.
Richter, O., Soendgerath, D.; 1990. Parameter estimation in Ecology. The link between data and models. VCH, Weinheim, Deerfield Beach, Florida.
Soendgerath, D.; 1989. Estimating parameters in discrete population models. In: D.P.F Moeller (Ed.), 3rd Ebernburger Work. Conf. System Analysis of Biochemical Processes. Vieweg, Braunschweig, pp. 154-165.

## Model identification
Model name/title: FARMSPRAY

## Model type
Media: regional scale, agricutural
Class: toxicology
Substance(s)/compounds: pesticides
Biological components: populations, carabid beetles and cabbage root fly (*Delia radicum*)

## Model purpose
To generate and promote a metapopulation perspective when considering the long-term impact of commercially applied pesticides on invertebrate population dynamics.

## Short description
Our first model was designed to characterize the fundamental attributes of a single metapopulation subject to local episodic mass mortality (pesticides). Parameters included speed of invertebrate movement, permeability of field boundary, reproductive rate of invertebrate, distribution and frequency of pesticide application, and toxicity of pesticide. The second model considered a trophic interaction, in which both the predator and prey metapopulations were exposed to pesticide. Parameters were the same as above (for both metapopulations) and also included a linear functional (predatory) response.

## Model application(s) and State of development
Application area(s):
Number of cases studied:
State: conceptualization, verification, calibration, validation, prognosis made, validation of prognosis, qualitative validation.

## Model software and hardware demands
Platform: PC-based
OS: DOS
Software demands: Turbo Pascal
Hardware demands: 486 or better (for speed)

## Model availability
Purchase: no, executable code
Contact person(s) Dr. Tom Sherratt
        Department of Biology
        University of Southampton
        Bassett Crescent East
        Southampton SO9 3TU
        England

Phone: +44 703-594212
Fax: +44 703-594269
E-mail: T.N.Sherratt@uk.ac.southampton

## Model documentation and major references
Sherratt, T.N. and P.C. Jepson. *Journal of Applied Ecology* (1993), 30, 696-705.

## Other relevant information

## Model identification
Model name/title: RATSIM: Cotton Rat Population Dynamics Simulation

## Model type
Media: agricultural
Class: toxicology
Substance(s)/compounds: pesticides
Biological components: population - cotton rats (*Sigmodon hispidus*)

## Model purpose
To determine the most effective timing of rodenticide applications to sugarcane crops, given uncertainty about the degree of density-dependence involved in cotton rat population cycles in South Florida (USA) sugarcane.

## Short description
RATSIM represents annual growth and death cycles of a population of cotton rats (*Sigmodon hispidus*) composed of 7 state variables: male and female embryos, male and female juveniles, adult males, receptive adult females, and pregnant females. Receptive females become pregnant after a variable delay determined either by time of year (for time-dependent hypotheses of cotton rat cycles) or by population size (for density-dependent hypotheses), according to a switch setting. Likewise, time or density-dependent functions determine the size of the litter so conceived. A specified sex ratio determines male and female embryos. Juveniles are born after a specified gestation time, after which pregnant females return to the receptive category. Juveniles mature into adult males or receptive females after a specified maturation times for each sex. A specified average survival determines death rates due to natural causes for each adult rat state. Embryos die upon the death of pregnant females. Juvenile survival is either a constant or negative exponential function of total population size, depending on a switch setting. A pulse function of a specified efficacy represents the sudden death caused by annual sugarcane harvest. Other similar pulse function (p to two per year at specified times) represent death by rodenticide applications.

**Model application(s) and State of development**
Application area(s):sugarcane fields on Southern Florida (USA)
Number of cases studied: 1
State: conceptualization, verification, calibration, prognosis made

**Model software and hardware demands**
Platform: Mainframe IBM
OS:
Software demands: language DYNAMO
Hardware demands:

**Model availability**
Purchase: no, available as source code
Contact person(s) Dr. Clay L. Montague
               Department of Environmental Engineering Sciences
               University of Florida
               P.O. Box 116450
               Gainesville, FL 32611 USA

Phone: +1 904-392-6222
Fax: +1 904-392-3076
E-mail: cmont@engnet.ufl.edu

**Model documentation and major references**
Montague, C.L., Lefevre, L.W., Decker, D.G. and Holler N.R., 1990. Simulation of cotton rat population dynamics and response to rodenticide applications in Florida sugarcane. *Ecol. Modelling*, 50 (1990): 177-203.

**Other relevant information**
A PC-based version of RATSIM can easily be programmed in almost any high-level computer language from the DYNAMO listing provided in the referenced documentation. A version written in BASIC as available (without additional documentation) from the author for the cost of a diskette, postage, and handling (generally less than US $10).

**Model identification**
Model name/title: SAWFLY

**Model type**
Media: local scale, agricultural
Class: toxicology
Substance(s)/compounds: pesticides
Biological components: population - Sawflies (Hymenoptera: Symphyta) in cereals (Sussex, UK)

---

**Model purpose**
Relate changes in the density of sawflies in cereals to changes in agriculture and to weather, taking density dependence into account.

---

**Short description**
The model was parameterised by fitting purely seasonal transfer-function models to parallel time-series describing density on five farms over 19 years. The input variables are the proportion of cereal fields undersown with grass on each farm, mean daily temperature averaged over the period 10 June-10 July, and total rainfall over the same period, all with a one-year lag. The model incorporates a strong autoregressive component. It accounts for 40% of the variation in the observed time-series. The model was used to predict sawfly density in the 1989 for comparison with the density observed after application of a broad-spectrum insecticide on one farm. Observed density was less than one-tenth of the predicted value and was outside the 95% confidence limits of the prediction; the model predicted furthermore that recovery would take several years.

---

**Model application(s) and State of development**
Application area(s): Agriculture/pesticide use/cereal invertebrates
Number of cases studied: one
State: conceptualization, verification, calibration, prognosis made, validation of prognosis.

---

**Model software and hardware demands**
Platform: other - Any system supporting a compiler or interpreter
OS: N/A
Software demands: N/A, any language (model needs programming)
Hardware demands: N/A

---

**Model availability**
Purchase: no. The model is easily programmed using the equations in Aebischer (1990)
Contact person(s) Dr. N.J. Aebischer
    The Game Conservancy Trust, Fordingbridge
    Hampshire SP6 1EF, UK

Phone: +44 425-652381
Fax: +44 425-655848
E-mail: N/A

---

**Model documentation and major references**
Aebischer, N.J., 1990. Assessing pesticide effects on non-target invertebrates using long-term monitoring and time-series modelling. *Functional Ecology*, 4, 369-373.

---

**Other relevant information**
This model is a case study of using long-term monitoring data to investigate the ecotoxicological effects of pesticide use at the farm level on invertebrates in cereals. In theory, the environmental repercussions of using toxic chemicals at the farm level could be evaluated experimentally, but in practice it is not feasible to carry out experiments in plots measuring several km².

This model illustrates an alternative approach, incorporating agricultural and meteorological variables, that quantifies the amount of between-year variability in the sampled abundance of an insect in order to detect abnormal deviations caused by large-scale pesticide use.

## Model identification
Model name/title: MiteSim: A Simulation Model of the Banks Grass Mite (Acari: Tetranychidae) and the Predatory Mite, *Neoseiulus fallacis* Acari: Phytoseiidae) on Maize.

## Model type
Media: agricultural
Class:
Substance(s)/compounds:
Biological components: populations (Spider mites) organism (maize plant)

## Model purpose
MiteSim can simulate the population dynamics of a mite predator/prey system. The system is directly affected by microenvironment (temperature and humidity) within the boundary layer of air on a maize plant leaf.

## Short description
A simulation model (MiteSim) of the mite predator/prey system consisting of Banks grass mite (BGM), *Oligonychus pratensis* (Banks), and the predatory mite (NEO), *Neosiulus fallacis* (Garman), was developed and validated. This model included the effects of temperature, humidity and predation and was coupled to a comprehensive plant-microenvironment model (Cupid). Leaf temperatures, stomatal and boundary layer resistances, and canopy air temperature and humidity are calculated by Cupid.

A submodel (POPVPD) uses this information from Cupid to calculate leaf surface humidity. Mite developmental rates and fecundity are each calculated as a function of both leaf surface temperature and humidity in MiteSim. Neo-consumption rates are calculated as a function of temperature and prey density. The model predicted the predator/prey population dynamics well for a laboratory microcosm. However, MiteSim consistently overestimated the number of spider mites in field situations, especially toward the end of the simulations.

These results suggest that substantial mortality factors occur in the field that are not accounted for from laboratory data.

## Model application(s) and State of development
Application area(s): ecology, agriculture
Number of cases studied: 1
State:

## Model software and hardware demands
Platform: IBM Mainframe FORTRAN, could be compiled and run on other platforms

---

**Model availability**
Purchase: Source code can be obtained for free. Send a formatted MS-DOS diskette
Contact person(s) Jim Berry
               USDA-ARS
               Rangeland Insect Lab
               Bozeman, MT 59717-0366
               USA

Phone: +1 406-994-3051
Fax: +1 406-994-3566
E-mail: A03LCBOZEMAN(tegn)ATTMAIL.COM

---

**Model documentation and major references**
Berry, J.S., T.O. Holtzer and J.M. Norman. 1991. MiteSim: a simulation model of Banks grass mites (Acari Tetranychidae) and the predatory mite, Neoseiulus fallacis (Acari: Phytoseiidae) on maize: model development and validation. *Ecol. Modelling*, 53: 291-317.
Berry J.S., T.O. Holtzer and J.M. Norman. 1991. Simulation experiments using a simulation model of the Banks grass mite (Acari: Tetranychidae) and the predatory mite, Neoseiulus fallacis (Acari: Phytoseiidae) in a corn plant microenvironment. *Environ. Entomol.*, 20: 1074-1078.

---

**Other relevant informations**

---

**Model identification**
Model name/title: Mallard Productivity Model

---

**Model type**
Media: wetland, terrestrial
Class: not applicable
Substance(s)/compounds: not applicable
Biological components: ecosystem - prairie wetlands population - mallard duck (*Anas platyrhynchos*)

---

**Model purpose**
To model and predict the reproductive success of mallard ducks, based on a variety of environmental measurements and biological relations.

---

**Short description**
This model predicts the reproductive success of mallards during a simulated breeding season. Important components are (1) fraction of birds that attempt to breed, (2) initiation of breeding, (3) nest site selection, (4) clutch size, (5) survival of nest, (6) survival of ducklings, and (7) survival of females during breeding season.

     Important input variables include (1) conditions of wetlands, (2) area, (3) attractiveness, and (4) security of habitats used for nest placement.

Initial average body mass can also be varied, as can limits of arrival date on the breeding area. Most useful output variables include nest success rate, average success of breeding females, recruitment rate, fraction of young in population at end of breeding season, and anticipated change in population size.

## Model application(s) and State of development
Application area(s): Widely used throughout prairie region of north-central United States and southern Canada.
Number of cases studied:
State:

## Model software and hardware demands
Platform: IBM compatible PCs
OS: MS-DOS operating system
Software demands:
Hardware demands:

## Model availability
Purchase: yes, no charge
Contact person(s) Terry L. Shaffer
                   Northern Prairie Wildlife Research Center
                   Route 1, Box 96C
                   Jamestown, ND 58401 USA

Phone: +1 701-252-5363
Fax: +1 701-252-4217
E-mail: ShafferT@mail.fws.gov  (via Internet)

## Model documentation and major references
Johnson, D.H., Sparling, D.W. and Cowardin, L.M., 1987. A model of the productivity of the mallard duck. *Ecol. Modelling*, 38: 257-275.
Johnson, D.H., Cowardin, L.M., and Sparling, D.W., 1986. Evaluation of a mallard productivity model. Pages 23-29 In: J. Verner, M.L. Morrison and C.J. Ralph, Eds. *Widllife 2000: Modeling Habitat Relationships of Terrestrial Vertebrates*. University of Wisconsin Press, Madison.
Cowardin, L.M., Johnson, D.H., Shaffer, T.L, and Sparling, D.W., 1988. Application of a simulation model to decisions in mallard management. U.S. Fish Wildl. Serv. Tech. Rep. 17. 28 pp.
Cowardin, L.M., Johnson, D.H., Frank, A.M., and Klett, A.T., 1983. Simulating results of management actions on mallard production. *Trans. N. Am. Wildl. Conf.*, 48: 257-272.

Other relevant information

**Model identification**
Model name/title: FORAGE: a model of forage intake in beef cattle.

**Model type**
Media: agricultural
Class:
Substance(s)/compounds:
Biological components: organism, beef cattle

**Model purpose**
FORAGE is a deterministic model that was developed as an interface between the plant and animal sub-models of the SPUR2 (Simulation Production and Utilization of Rangelands) grassland ecosystem model. The model predicts forage intake and diet selection of grazing beef cattle by simulating the mechanistic components of grazing behavior.

**Short description**
FORAGE predicts the effect of changing sward characteristics on feed intake and diet selection of grazing animals by simulating changes in the mechanistic components of grazing behavior. Grazing behavior is divided into diet selection and intake, where intake is defined as product of grazing time and rate of intake. Rate of intake in the model is simulated as the product of bite size and rate of biting. The four inputs required by the model are forage quality, forage quantity, the amount of forage demanded by the animal and metabolic weight of the animal. Diet selection is based on the relative quality of the forage offered. Rate of intake is a function quantity of forage on offer and the size of the animal. Grazing time is a function of three limits: 1) amount of forage demanded by the animal; 2) time needed to consume the forage; 3) number of bites needed to consume the forage.

Sensitivity analyses reveal that the model is sensitive to underestimation of the parameters used in the equations that describe bite size and rate of biting. Results from simulations indicate that, when forage availability is limited, intake is limited by rate of intake and grazing time.

**Model application(s) and State of development**
Application area(s): Grazing behavior studies
Number of cases studied: 1
State: validation

**Model software and hardware demands**
Platform: PC-based IBM, Mainframe Digital, Work-station UNIX
OS:
Software demands: Fortran
Hardware demands:

## Model availability
Purchase: Free
Contact person(s)  Dr. Barry B. Baker
    P.O. Box 552
    Fort Collins, CO 80522-0552 USA

Phone: +1 303-490-8322
Fax: +1 303-490-8310
E-mail: Barry (tegn) gpsrv1.gpsr.colostate.edu

## Model documentation and major references
1. Baker, B.B., 1988. A Simulation Model for the Grazing Behavior of Beef Cattle, M.S. Thesis, Department of Animal Sciences, Colorado State University, Fort Collins.
2. Baker, B.B., R.M. Bourdon, and J.D. Hanson. 1992. FORAGE : A simulation model of grazing behavior for beef cattle. *Ecol. Modelling*, 60:257-279.
3. Baker, B.B., J.D. Hanson, R.M. Bourdon and J.B. Eckert. 1933. The potential effects of climate change on ecosystem processes and cattle production on U.S. rangelands. *Climatic Change*, 25:97-117.
4. Hanson, J.D., B.B. Baker, and R.M. Bourdon. 1992. SPUR2 Documentation and Users Guide. U.S. Department of Agric. Great Plains Systems Research Tech. Report-1. Fort Collins, Colorado. 24 pp.
5. Hanson, J.D., B.B. Baker, and R.M. Bourdon. 1992. Comparison of the effects of different climate change scenarios on rangeland livestock production. *Agric. Sys.*, 41:487-502.

## Other relevant information
At this time a stand alone version of FORAGE does not exist. The model is part of the SPUR2 suite of models. These models include a revised version of the USDA SPUR model (Hanson, et al., 1992; citation above) and the Colorado Beef Cattle Production Model (Bourdon, unpublished). The Colorado Beef Cattle Production Model (CBCPM) was developed by Dr. Richard M. Bourdon at Colorado State University in 1991. The CBCPM model is currently undergoing several validation studies. For further information regarding SPUR2 or CBCPM contact:  SPUR2
    Dr. Jon D. Hanson
    GPSR USDA-ARS
    P.O. Box E
    Fort Collins, CO
    +1 303-490-8322 (tel)  +1 303-490-8310 (fax)
    jon@gpsrv1.gpsr.colostate.edu (E-mail)

    CBCPM
    Dr. Richard M. Bourdon
    Dept. of Animal Sciences
    Colorado State University
    Fort Collins, CO
    +1 303-491-6150 (tel)  +1 303-491-5362 (fax)
    rbourdon@chuck.agsci.colostate.edu (E-mail)

## Model identification
Model name/title: Model of the screw worm fly (*Cochliomyia hominivorax*)

## Model type
Media: regional scale, terrestrial, agricultural
Class:
Substance(s)/compounds:
Biological components: population - Insects (*Cochliomyia hominivorax*/ Calliphoridae)

## Model purpose
The screw worm fly (*Cochliomyia hominivorax*) has an enormous economic importance to the livestock industry. It was possible to eliminate this insect in the USA and in Mexico by the sterile insect technique (SIT). To stop the seasonal invasion of insects into Mexico a barrier of sterile insects was established. The model investigates in what conditions spatial barriers are able to control an insect pest invasion.

## Short description
The model shows how a fully established pest population spreads into an area in which steriles are released. Random spreading of the population is described by diffusion. The model determines the conditions for stopping an invasion by a barrier. The optimization with respect to the costs of the barrier is possible. There is a conflict between costs and safety of a barrier. The model allows to make proposals for diminishing this conflict.

## Model application(s) and State of development
Application area(s):
Number of cases studied:
State: conceptualization

## Model software and hardware demands
Platform: PC- based IBM, Digital
OS:
Software demands: language: Pascal
Hardware demands:

## Model availability
Purchase: no, available as executable code
Contact person(s) Dr. Ralf Marsula
　　　　　　　　Permoserstr. 15
　　　　　　　　D-04318 Leipzig, Germany

Phone: +49 341-235-2039
Fax: + 49 341-235-3500
E-mail: rama@pinus.oesa.ufz.de

## Model documentation and major references
Marsula, R., 1992. Modelle zur Kontrolle von Schädlingen durch eine räumliche Barriere. Ph.D. Thesis, University of Marburg, Germany.

## Other relevant information

### Model identification
Model name/title: *Cirsium vulgare* population dynamics

### Model type
Media: local scale, terrestrial, agricultural
Class:
Substance(s)/compounds:
Biological components: population - *Cirsium vulgare*

### Model purpose
To predict the population dynamics of the biennial Cirsium vulgare under different grazing management regimes.

### Short description
Variable and parameters are listed on accompanying sheet. The life-cycle of *Cirsium vulgare* is described on enclosed figure. The model is stage-structured with density dependence incorporated at two points in the life-cycle (seed and seedling). The intensity of density dependence is estimated from the field (this is unusual most similar models assume a form of intensity of density dependence). All other parameters have values estimated from the field. The model is tested against independent data sets and shown to have a high explanatory value ($r^2 = 0.81$, $p < 0.001$).

### Model application(s) and State of development
Application area(s): Management of pasture weeds
Number of cases studied: applications many, only one used in manuscript
State: no information

### Model Software and hardware demands
Platform: model written in Pascal.
OS:
Software demands:
Hardware demands:

### Model availability
Purchase: no, available as source code
Contact person(s) Dr. M. Gillman
        Department of Biology
        The Open University
        Walton Hall, Milton Keynes
        MK7 6AA, England

Phone: +44 908 654068
Fax: +44 908 654167
E-mail:

## Model documentation and major references
Gillman, M., Bullock, J.M., Silvertown, J., Clear Hill, B., 1993. A density-dependent model of Cirsium vulgare population dynamics using field-estimated parameter value. *Oecologia*, 96, 282-289.

## Other relevant information

## Model identification
Model name/title: Agricultural Nonpoint Source Metamodels by Lakshminarayan, P.G. and Aziz Bouzaher

## Model type

Media: soil, surface water, groundwater
Class: biogeophysical, statistical
Substance(s)/compounds: nutrients (N and P), herbicides, soil, soil carbon, soil organic matter
Biological components: watershed, other geographical regions

## Model purpose
Agricultaral nonpoint source metamodels are statistical models that can predict the impact of optimal agricultural production practices on soil and water quality. Metamodeling abstracts away from unneeded detail of a complex process model (biogeophysical model) through statistical response functions, which then allows alternative policy evaluations without the additional simulations of the complex process model.

## Short description
A metamodel is a regression model explaining the input-output relationship of a complex process model. If the complex process model is a tool to approximate the underlying real-life system the analytic metamodel attempts to approximate and aid in the interpretation of the complex process model and ultimately the real-life system. Agricultural nonpoint source metamodels are statistical response functions developed from site-specific simualtion outputs of the biogeophysical models, such as EPIC (Erosion Productivity Impact Calculator), PRZM (Pesticide Root Zone Model), etc. The spatial heterogeneity (variability) of agricultural nonpoint pollution indicators (soil loss, nitrate-nitrogen in runoff, nitrate-nitrogen in percolation, etc.) introduced by soil, hydrology, weather, crop production and management pratices are fully captured in the areawide simulations using a sample distribution of these spatial and management parameters. As a result of this statistical sampling and simulation design, the metamodels developed from areawide simulation outputs have nice properties for out-of-sample prediction. Metamodels are simple regression functions, which are linear or nonlinear depending on the distribution of the dependent variable. The independent variables of the model are represented by the key site-specific parameters that served as inputs to the simulationmodel and also some policy variables and management parameters.

A simple example of a soil loss metamodel is

Soil loss = f(soil properites, hydrology, topography, croping practices, management practices, weather related factors, policy factors).

An important feature of these statistical regression functions is that the independent variables are vectors represented by the underlying unique properties and are not simple qualitative characterizations. It is this quantitative characterization that gives the nice extrapolation property for the estimated metamodels.

## Model application(s) and State of development
Application area(s): nonpoint pollution from crop production, nonpoint pollution from livestock production
Number of cases studied: 1. corn and sorghum herbicide runoff/leaching from crop production in the Midwestern United States. 2. Soil loss from wind and water (sheet and rill) erosion in the Western Prairie Provinces of Canada. 3. Nitrate- Nitrogen runoff/leaching from corn, soybean, sorghum, wheat and hay production in Iowa. State: verification, calibration, validation

## Model software and hardware demands
Platform: PC-based - IBM or DEC workstation
OS:
Software demands: Fortran, C, and SAS (or PC- SAS)
Hardware demands:

## Model availability
Purchase: Model can be purchased: Metamodels have to be developed for specific sites and cases. However, the process models (simulation models) such as EPIC and PRZM are public domain models available as both executable as well as source code.

Contact person(s) For EPIC:  
    Dr. Allan Jones  
    Blackland Research Center  
    USDA-SCS  
    808 East Blackland Road  
    Temple, TX 76502, USA

For PRZM  
Dr. Robert F. Carsel  
    U.S. EPA LAB  
Athens, Georgia  
USA

Phone:  
Fax: +1 817-770-6678  
E-mail:

## Model documentation and major references
Agricultural Nonpoint Source Metamodels: Metamodels and Nonpoint Pollution Policy in Agriculture, by Aziz Bouzaher.
Lakshminarayan, P.G. et al., *Journal of Water Resources Research*, Vol. 29, No. 6, pp 1579-1587, 1993.

Jones et al., EPIC: EPIC: An Operational Model for Evaluation of Agricultural Sustainability, *Agicultural Systems*, 37: 127-137.
PRZM: "User's Manual for the Pesticide Root Zone Model(PRZM): Release 1", EPA-600/3-84-109, U.S. Environmental Protection Agency, Athens, Georgia.

**Other relevant information**

---

**Model identification**
Model name/title: AGNPS : Agricultural Nonpoint Source Pollution Model

---

**Model type**
Media: water, local scale, sediment
Class: hydrology
Substance(s)/compounds: nutrients, phosphorus, nitrogen, chemicals, pesticides, chemical oxygen demand (COD)
Biological components:

---

**Model purpose**

AGNPS is a single-event model intended for use in analyzing water quality and quantity from primarily agricultural watersheds. Model objectives are to estimate relative quantity and quality of watershed outflow in order to assess its pollution potential, evaluate effects on watershed outflow of applying alternative management practices, an to identify critical areas of NPS pollutant production.

---

**Short description**
AGNPS is a rectangular grid-based model which works by dividing the watershed into uniform grids, or cells, for data input to represent the spatial distribution of watershed properties. All characteristic inputs and calculations are made at the cell level. Basic components of the model are hydrology, erosion, sediment transport, chemical oxygen demand transport, and soluble and attached nitrogen and phosphorus transport. The hydrology component estimates runoff volume and concentrated peak flow. The erosion component estimates total upland and channel erosion by five particle, size classes ranging from clay to sand. Point source inputs, such as from feedlots, waste water treatment plant discharges, and stream bank, construction site, and gully erosion (user specified), can also be handled.

AGNPS simulates the generation, transport and deposition of water, sediments, and nutrients from headwaters to the outlet in a step-wise manner so that an assessment can be made for any cell within the watershed boundary. Variable cell sizes can also be included to allow for detailed characterization and analysis of critical areas.

Data required for each cell describe the physical character of each cell as well as the land use and treatment in each cell. Data are obtained from readily available records and maps and from field reconnaissance as needed. The cellular structure of the model allows it to be connected to other software such as Geographical Information System (GIS) and Digital Elevation Models (DEM). This can facilitate the development of a number of the model's input parameters.

316

The model also includes enhanced graphical representations of input and output information.

## Model application(s) and State of development
Application area(s): Watershed assessment, water quality assessment
Number of cases studied: numerous
State: conceptualization, verification, calibration, validation, prognosis made, validation of prognosis

## Model software and hardware demands
Platform: PC-based - IBM, IBM compatible. Workstation - UNIX
OS: DOS/UNIX
Software demands: DOS 3.0 or above
Hardware demands: 286 and above - extended memory

## Model availability
Purchase: free. Software available as executable code
Contact person(s) Dr. Robert A. Young
                 USDA, ARS, MWA
                 North Central Soil Conservation Research Lab
                 USA

Phone: +1 612-589-3411
Fax: +1 612-589-3787
E-mail: eklund@soils.mrars.usda.gov

## Model documentation and major references
Young, R.A., Otterby, M.A., and Ross, A., 1982. An evaluation system to rate feedlot pollution potential. ISSN 0193-3787. USDA Agricultural Research, North Central Region, Peoria, IL. 76 pp.
Lee, Ming T., 1987. Verifications and applications of a nonpoint source pollution model. National Engineering Hydrology ASCE, New York.
Setia, P. and Magleby, R., 1987. An economic analysis of agricultural nonpoint pollution control alternatives. *Journal of Soil and Water Conservation*, 42 (6): 427-431.
U.S. Economic Research Service, 1987. Analysis of Agricultural Nonpoint Pollution Control Options in the St. Albans Bay Watershed. USERS Rep. AGES870423. U.S. Gov. Print. Office, Washington D.C.
Young, R.A., Onstad, C.A., Bosch, D.D. and Anderson, W.P., 1987. AGNPS, Agricultural nonpoint source pollution model - a watershed analysis tool. USDA, ARS, Conservation Research Report 35, 77 pp. 1987.
Naslas, N.A, 1988. The benefits of nonpoint source modeling in watershed planning. A report submitted to the U.S. Environmental Protection Agency, Region V, as supported by the National Network for Environmental Policy Studies grant; and to the Department of Civil, Environmental and Architectural Engineering of the University of Colorado, Boulder, in partial fulfillment of the requirements for the degree of Master of Science.
U.S. Economic Research Service, 1988. Illinois Rural Clean Water Project. ERS Rep. AGES880617. U.S. Gov. Print. Office, Washington, D.C.

American Society of Agrc. Eng., 1989. BMP effectiveness evaluation using AGNPS and a GIS. ASAE paper no. 89-2566. Department of Soil and Water Conservation, Richmond, VA 23219.

Canadian Society of Agric. Eng. 1989. Applicability of AGNPS model for water quality planning. ASAE/CSAE Paper no. 89- 2042. Kansas State University, Manhattan, KS 66506.

Young, R.A., Onstad, C.A., Bosch, D.D., and Anderson, W.P., 1989. A non-point source pollution model for evaluating agricultural watersheds. *J. Soil Water Conserv.*, 44(2): 164-172. 1989.

Prato, T. and Hongqi S., 1990. A comparison of erosion and water pollution control strategies for an agricultural watershed. *Water Resources Research*. 26(2): 199-205.

University of Minnesota, 1990. An evaluation of options for micro-targeting acquisition of cropping rights to reduce nonpoint source water pollution. Inst. of Agric., Forestry and Home Economics. P90-62. University of Minnesota, St. Paul, MN 55108.

American Society of Agric. Eng. 1991. Generating AGNPS input using remote sensing and GIS. ASAE paper no. 91-2622. St. Joseph, MI 49085.

Disrud, L.A., and Yoon, J.W., 1991. Application of the AGNPS (Agrucultural Nonpoint Source) Model in North Dakota. American Society of Agricultural Engineers (ASAE), paper No. MNSK91-103, St. Joseph, MI, 15 p.

University of Idaho, 1991. An integrated ecological-economic framework for assessing agricultural nonpoint source pollution in a watershed system. University of Idaho, Moscow, ID 83483.

Jakubauskas, M.E., Whistler, J.L., and Dillworth, M.E., 1992. Classifying remotely sensed data for use in an agricultural nonpoint-source pollution model. *J. Soil and Water Conservation*, 47(2): 179-183.

Kao, J.J., 1992. Determining drainage pattern using DEM data for nonpoint-source water quality modeling, *Wat. Sci. Tech.*, 26(5-6): 1431- 1438.

Kozloff, K., Taff, S.F., and Wang, Y., 1992. Micro-targeting the aquisition of cropping rights to reduce nonpoint source water pollution. *Water Resources Research*, 28(3): 623-628.

Finney, V.L., 1993. Using the single event model AGNPS to estimate average annual sediment yield from Lindero Canyon, California, USA.

Yoon, J., and Padmanabhan, 1994. A decision support system for nonpoint source pollution management using a distributed model-GIS linkage. The First Congress on Computing in Civil Engineering, June 20-22, 1994, Washington, D.C., American Society of Civil Engineers (ASCE), New York, NY, 9 p.

Lee, M.T., Jehng, J.K., and Ke, Y., Integration of GIS, Remote Sensing, and Digital Elevation Data for a Hydrologic Model.

Patwardhan, A.S., John, C., Beach, P.E., Waggoner, M.A: and Wayne P. Anderson, Evaluation of Conservation Systems Using AGNPS.

---

**Other relevant information**

An annualized, or continuous simulation, of the model is currently under development and should be released within a year. This version will operate on a daily time step and will be driven by either a weather generator or actual historical weather records, thus, allowing more accurate average annual estimates runoff, erosion, sediment yield, and nutrient movement for a representative sequence of weather events for any location.

## Model identification
Model name/title: Sorghum root growth

## Model type
Media: agricultural
Class: biogeochemical
Substance(s)/compounds: carbon
Biological components: community - agricultural crop (grain sorghum)

## Model purpose
Predict the length of sorghum crop roots in soil as a function of soil depth and time.

## Short description
Inputs : daily crop growth rate [g $m^{-2}$ $day^{-1}$], daily rainfall total [mm], soil water characteristics - drained upper limit, lower limit, crop sowing, flowering and maturity date. State variables: root length density [cm $cm^{-3}$soil]
Important parameters: root front velocity, conversion ratio of crop growth rate to new root length, root distribution parameter
Forcing functions: crop growth rate determines new root length growth, root penetration is a function of time and crop growth stage.

## Model application(s) and State of development
Application area(s): Impact of drought on crop yield
Number of cases studied: none
State: validation

## Model software and hardware demands
Platform: PC-based - IBM
OS: MS-DOS
Software demands: language - MS-BASIC
Hardware demands: 386 or higher

## Model availability
Purchase: no
Contact person(s) Dr. M. Robertson
            CSIRO, PMB . Aitkenvale
            Qld 4814, Australia

Phone: +61 77-719-587
Fax: +61 77-251-009
E-mail:

**Model documentation and major references**
Robertson, M.J., Fulcai, S., Hammer, G.L., and Ludlar, M.M., 1993. Modelling root growth of grain sorghum using the CERES approach. *Field Crops Research*, 33: 113-130.

**Other relevant information**

**Model identification**
Model name/title: TEXCIM for Windows

**Model type**
Media: agricultural
Class:
Substance(s)/compounds: pesticides
Biological components: ecosystem - agricultural population - 4 insect pest and cotton plant.

**Model purpose**
A cotton insect pest management model that integrates 4 pests, pest control, plant growth, economics and weather. It forecasts number of pest and fruit and converts this information into comprehensive economic threshold for making management decisions.

**Short description**
This is a large, complex model requiring 3 megabytes of hard disk space and 2 megabytes of memory. Consequently, it has hundreds of variables, functions and parameters. Most of these are available in published papers over the last 20 years. The model is initialzed with counts of insects pest, predators, parasites, weather and economics.

**Model application(s) and State of development**
Application area(s): Global cotton insect pest management
Number of cases studied: some validation on all continents where cotton is grown
State: validation

**Model software and hardware demands**
Platform: PC-based
OS: Windows 3.1 and Windows NT
Software demands: language FORTRAN and Visual Basic
Hardware demands:

**Model availability**
Purchase: yes, US $100 in Texas, US $ 125 elsewhere. Available as executable code.
Contact person(s) Dr. Phil Hurley
                 Extension Computer Applications
                 SpSve Bldg
                 Texas A&M University
                 College Station, TX 77843-2468 USA

Phone: +1 409-845-3929

## Model documentation and major references
Sterling, W.L., et al., 1993. TEXCIM for Windows: the texas cotton insect model. Texas Agric. Exp. Stn. Misc. Publ. MP-1646, College Station, TX.

## Other relevant information
Will send copy of 290 page user's guide and model on floppy disks.

## Model identification
Model name/title: ISSM - Crop Model Indiana Soybean System Model (ISSM): I. Crop Model Evaluation

## Model type
Media: agricultural
Class: biogeochemical (actually Biophysiological crop simulation model)
Substance(s)/compounds:
Biological components: ecosystem - agricultural/crop production system

## Model purpose
The primary purpose of the model is crop management research. The simulation model serves as a stand in for actual crop production to facilite management strategy development and evaluation. Model application for actual crop management use (strategic and tactical) is a goal for further work.

## Short description
ISSM is an adaption of a physiologically-based soybean simulation model for Indiana conditions. The model simulates the development growth and yield of soybean crop at a particular location (specified by soil profile information and daily weather data), under a given set of management actions (planting date, irrigation, etc.). Variety specific crop parameters, that determine the growth and development of the crop, are part of the input specification along with soil, weather and management information. The model uses a daily time-step. The major components of the model (and the driving forces/factors) are: crop phenology (temperature, photoperiod, drought stress), soil-water balance (soil characteristics, rainfall/irrigation, crop stage and characteristics), photosynthesis (crop characteristics, solar radiation, temperature), dry matter production and stage-specific partitioning (crop characteristics, phenology, weather).

Adaption to Indiana conditions included the calibration of varietal coefficients and the use of water retention values to approximate the soil water limits. Field experiments involving two varieties at two locations in Indiana provided the data for model calibration and validation. Results from simulation experiments provided guidance in varietal selection and planting date decisions. In addition, a grouping of weather years, based on soybean yield potential was obtained. This grouping will be useful in simulation analysis aimed at crop and pest management strategy evaluation.

321

## Model application(s) and State of development
Application area(s): Crop management research. Strategy evaluation, variety selection, planting.
Number of cases studied: 13
State: validation

## Model software and hardware demands
Platform: PC-based, IBM
OS: DOS
Software demands: (executable files), language FORTRAN
Hardware demands: IBM compatible computer

## Model availability
Purchase: no. Software available as executable code
Contact person(s) Dr. Kumar Nagarajan
              Purdue University,
              Department of Entomology
              West Lafayette, Indiana, IN 47907, USA

Phone: +1 317-496-2295
Fax: +1 317-494-0535
E-mail: kumar_nagarajan@entm.purdue.edu

## Model documentation and major references
Nagarajan, K., O'Neil, R.J., Lowenberg-DeBoer, J., and Edwards, C.R., 1993. Indiana Soybean System Model (ISSM): I. Crop Model Evaluation. *Agricultural Systems, 43,* 357-379.

## Other relevant informations
ISSM is an adaption of the soybean crop model SOYGRO 5.41 (University of Florida, Agricultural Engineering Dept., Gainesville, FL, USA) for Indiana conditions.

## Model identification
Model name/title: ISSM: Crop-Insect Model Indiana Soybean System Model (ISSM): II. Mexican bena beetle model development, integration and evaluation

## Model type
Media: agricultural
Class: biogeochemical (actually biophysiological simulation model)
Substance(s)/compounds:
Biological components: ecosystem - agricultural/crop production system population - Insect (crop pest) population

## Model purpose

The primary purpose of the model is pest management research. The simulation model serves as a stand in for actual crop production to facilitate pest management strategy development and evaluation. Model application for actual management use (strategic and tactical) is a goal for further work.

## Short description

ISSM crop-insect model is an integration of a physiologically-based soybean crop model (Nagarajan et al., 1993, adapted from Jones et al., 1988) and an age-stage distributed delay model of Mexican bean beetle (MBB), a defoliating insect pest. Number of individuals (Ni) in each of the six (egg, small larva, medium larva, large larva, pupa and adult) stages (i) depends on time (t) and age (a) and stage-specific factors such as mortality (mi) and development (Di).

Change in Ni is expressed as $\partial Ni/\partial t + \partial Ni/\partial a = Ei - mi*Ni - Di*Ni$.

Ei is the number entering stage i. For egg stage Ei represents reproduction (defined by a temperature-dependent fecundity function and age-dependent oviposition rate function); for immature stages it is development from previous stage; and for adults it represents development and immigration. Di represents the stage-specific, temperature-dependent development which uses a cumulative distributed-delay function (Stinner et al., 1975) defined in terms of physiological age and the earliest and latest development durations.

The model uses a 1-day time step. Simulated values of insect population parameters (generation time, net reproductive rate, intrinsic rate of increase) and the effects on crop (defoliation and yield loss), compare favorably with observed data (laboratory and field). Simulation experiments results provide guidelines for varietal selection and planting date decisions.

Model inputs: daily weather (max. and min. temperatures, rainfall and solar radiation), soil profile information, variety-specific crop coefficients (default values are included for most soybean maturity groups), and insect factors (development, repoduction and mortality factors for MBB are included).

## Model application(s) and State of development

Application area(s): Pest management research. IPM strategy evaluation, crop variety selection.
Number of cases studied: 5
State: validation

## Model software and hardware demands

Platform: PC-based IBM
OS: DOS
Software demands: (executable files) Language FORTRAN
Hardware demands: IBM compatible computer

## Model availability
Purchase: no, software available as executable code
Contact person(s) Dr. Kumar Nagarajan
    Purdue University,
    Department of Entomology
      West Lafayette, Indiana, IN 47907 USA

Phone: +1 317-496-2295
Fax: +1 317-494-0535
E-mail: kumar_nagarajan@entm.purdue.edu

## Model documentation and major references
Nagarajan, K., O'Neil, R.J., Edwards, C.R. and Lowenber- DeBoer, J., 1994 Indiana Soybean System Model (ISSM): II. Mexican bean beetle model development, integration and evaluation. *Agricultural Systems*, 45, 291-313.
Nagarajan, K., O'Neil, R.J., Lowenberg-DeBoer, J., and Edwards, C.R., 1993. Indiana Soybean System Model (ISSM): I. Crop Model Evaluation. *Agricultural Systems*, 43, 357-379.

## Other relevant informations
The crop model component, in the ISSM crop-insect model, is an adaption of the soybean crop model SOYGRO 5.41 (Jones et al., 1988, University of Florida, Agricultural Engineering Dept., Gainesville Fl., USA) for Indiana conditions. Mexican bean beetle is a sporadic defoliating insect pest of soybeans in the midwestern USA.

## Model identification
Model name/title: Agroecosystem-Type-Model

## Model type
Media: local scale, agricultural
Class:
Substance(s)/compounds: nutrients
Biological components: ecosystem - agro-ecosystem

## Model purpose
Classifying agroecosystem-type scenarios

## Short description
Still under development.

## Model application(s) and State of development
Application area(s): Research area of the project Forschungverbund Agro-ökosysteme München.
Number of cases studied: 1
State: conceptualization

## Model software and hardware demands
Platform: Work-station, Digital, Alpha
OS: OSF1 2.0
Software demands: ARC/INFO, language - C, ARC/INFO Macrolanguage
Hardware demands: UNIX-workstation

## Model availability
Purchase: not yet
Contact person(s) Dr. Armin Müller
          Lehrstuhl für Landschaftsökologie
          D-85350 Freising
          Germany

Phone: +49 08161-714149
Fax: +49 08161-714427
E-mail: armin@fan.loek.agrar.tu-muenchem.de

## Model documentation and major references


## Other relevant information
Lang, R., Lenz, R., Müller, A., 1993. The Agroecological Information System in the FAM International Congress on Modelling and Simulation. December 6-10, 1993 Perth/Australia Proceedings, Vol.4, p. 1407-1412.

## Model identification
Model name/title: Modeling methane emission from rice soils

## Model type
Media: water, soil
Class: biochemical
Substance(s)/compounds: carbon, methane
Biological components: ecosystem - rice agriculture, organism - methanogens

## Model purpose
Predict methane emission from rice agriculture.

---

**Short description**

Rice paddies are a major source of anthropogenic methane emitted to the atmosphere. Methane is an important greenhouse gas since it is 20 times more absorptive than carbon dioxide and methane emissions have been increasing at a rate of about 1% per year. Modeling the methane production and transport processes in rice soils can yield insight into methods for reducing emissions from rice paddies.

A mechanistic model is developed to describe the production of methane in rice soils and its subsequent transport to the atmosphere. The Monod kinetic model is used for methane production. Two independent sources of carbon are incorporated: the carbon initially in the soil and that given off by the rice plant roots. The model assumes that methane is emitted to the atmosphere only by transport through the plant, which is consistent with field observations, and that methane transport to the plant is by simple mass transfer.

Kinetic parameters are determined by independent experiments involving the decay rate of acetic acid (carbon substrate) in a laboratory culture of a sample of the actual rice soil. Mass transfer coefficients and root carbon exudate function parameters are adjusted so that data from methane flux measurements in an experimental rice field agree approximately with the model simulation.

Results from this initial modeling effort are encouraging. The agreement between model output and field data are well within the precision of the measurement.

---

**Model application(s) and State of development**

Application area(s): methane emission from rice agriculture
Number of cases studied:
State: validation

---

**Model software and hardware demands**

Platform: MAC
OS: System 1
Software demands: STELLA II, language STELLA
Hardware demands: Any MAC

---

**Model availability**

Purchase: freeware - cost of handling. Available as sourcecode.
Contact person(s) Dr. V.J. Law
        Chemical Engineering Department
        Tulane University
        New Orleans, Louisiana 70118 USA

Phone: +1 504-865-5773
Fax: +1 504-865-6744
E-mail: law@che.che.tulane.edu

---

**Model documentation and major references**

Law, J. et al., 1993. *Environmental Software*, 8: 198-207.

## Other relevant information

## Model identification
Model name/title: DanStress

## Model type
Media: agricultural
Class: hydrology
Substance(s)/compounds:
Biological components: ecosystem - crop

## Model purpose
The purpose is to simulate crop water relations, crop water use and root water uptake by a mechanistic model.

## Short description
A simulation model "DanStress" was developed for hourly study (or less) of the integrated effects of soil, crop and climatic conditions on water relations of field grown crops. The root zone was separated into layers. The water potential at the root surface was calculated by a single root model, and the uptake of water was calculated by a root contact model. Crop transpiration was calculated by Monteith's combination equation. Crop conductance was scaled up from an empirical stomatal conductance model used on sunlit and shaded crop surfaces. Transpirational water loss originates from root water uptake and changes in crop water storage. Leaf pressure-volume (PV) curves were used for deriving crop water capacitance, crop water potential, crop osmotic potential, and crop turgor pressure.

Driving variables are net radiation, temperature, water vapour pressure and wind speed. Functions for root water uptake, crop water capacitance, leaf conductance, soil water retention and soil hydraulic conductivity are needed. Inputs are initial crop water content, leaf area index, canopy light extinction coefficient, and root density.

## Model application(s) and State of development
Application area(s): Study the integrated effect of soil, crop and climatic conditions on water relations and water use of crop.
Number of cases studied: 1
State: conceptualization, verification

## Model software and hardware demands
Platform: PC-based IBM
OS: DOS
Software demands: language TURBO PASCAL
Hardware demands:

---

**Model availability**
Purchase: no, available as source code
Contact person(s) Dr. Chr. Ric. Jensen and Dr. H. Svendsen
                The Royal Veterinary and Agricultural University
                Department of Agricultural Sciences
                Section of Soil and Water and Plant Nutrition
                Agrovej 10
                DK- 2630 Taastrup

Phone: +45 35-28-33-96
Fax: +45 35-28-21-75
E-mail:

---

**Model documentation and major references**
Jensen, C.R., Svendsen, H., Andersen, M.N., and Lösch, R., 1993. Use of the root contact concept, an empirical leaf conductance model and pressure-volume curves in simulating crop water relations. *Plant and Soil*, Vol. 149 (1993), pp. 1-26.

---

**Other relevant information**
At the moment we are developing the "DanStress" model for rape crops (*Brassica napus L*).

## 4.6. An Overview of Forestry Models

Forestry managers want to forecast the environmental and economic results of different types of forestry management in order to make decision for the most sound use of the forest as a natural resource. Models are the only tools to apply for the best possible forecast on basis of the available knowledge about forestry and the available data on the focal forest.

In principle, there are no differences between agricultural models and forestry models, but the latter model type uses more long-term forecasts and must also account for the recreational value. The global community demands more sustainable use and a better understanding of forestry. Therefore, growth and yield models for mixed species stand, dealing with complex biophysical interactions, have come into focus as they are better able to include environmental determination of biological diversity as well as wood production.

Forestry researchers are characterized as being reductionists. They have been able to provide sophisticated knowledge of components of forest systems, but have failed to integrate and synthesize information in a way that can offer decision-making tools. Hopefully, the models outlined in this volume are a potentially valuable tool to remedy this situation.

Ironically, the most sophisticated models exist only for the most simple forest ecosystem and mainly from the temperate zone, while the models for the complex tropical rainforest tend to be rather primitive or totally lacking often due to lack of data. It is therefore a challenge for forest modelers to extrapolate the existing models to sites and species for which no or only scarce data are currently available.

The overview of forestry models presented in this and the next section will hopefully inspire to development of new and better models to be applied on complex tropical and temperate forestry accounting simultaneously for sustainability, the recreational value and the wood production of this unique ecosystem.

Table 4.3 gives an overview of the forestry models presented in detail in the next section and a few other important forestry models, which are slightly older. They are included in the table to give a wider spectrum of available models.

The model Gromap indicates the suitability of any perennial plant taxon for cultivation of sites across Africa. The monthly values of maximum temperature, minimum temperature, precipitation, evaporation and solar radiation are estimated for each location. Information describing the selected plant's edaphic and climatic requirements are entered to the model. Color maps of the edaphic, climatic and overall limitations for the focal plant are produced. The program contains a climatic and edaphic database for 10,000 locations.

The model denoted PnEt is able to predict the effects of changes in physical and chemical climate on water quality and quantity and forest production. The model is based on a carbon and water balance and two principal relationships: 1) maximum photosynthetic rate as a function of foliar nitrogen concentration and 2) stomatal conductance as a function of realized photosynthetic rate.

These relationships are combined with standard equations describing the light attenuation in canopies and photosynthetic response to radiation intensity, along with the effect of soil water stress and vapor pressure deficit.

Formix gives a description of the growth dynamics of natural tropical forests under various management conditions and silvicultural treatments. The model is based on an energy balance (as carbon). The model accounts for the mutual interactions of the forest growth and light condition, which cause vertical and horizontal differentiation in the natural forest mosaic. A special version of the model, named Formix2, incorporates multi-storey canopy layer structure of the natural forest.

The model named Treedyn3 describes the growth dynamics of forests under the influence of various pollutants. It is based on a carbon and nitrogen balance. The state variables are tree number, base diameter, tree height, leaf mass, root mass, fruit biomass, carbon and nitrogen in litter, nitrogen and carbon in organic matter in soil and inorganic carbon and nitrogen in soil available for plants. The model is particularly fitted to examine the effect of climate, site, silvicultural management and pollution. It can also be recommended for teaching purposes.

The model Fsyl-mod simulates the perennial courses of carbon and water budget of a beech forest. The state variables are total biomass, leaf area, specific leaf weight, stem, roots in a deciduous forest area.

The model denoted Cardyn aggregates behavior of several deciduous species. The model averages species dependent phenological differences to represent the phenology of a generalized deciduous forest and the resulting carbon dynamics. Four compartments are considered in the model: the green compartment, the non-green compartment and litter compartment and the atmosphere, which is considered an infinite source and sink. The relationship between LAI and leaf carbon is using a conversion coefficient, u, averaging the area to mass relation of light and shade adapted leaves.

Wu et al. have developed a model which is able to predict species richness in response to changes in forest islands area and to generate temporal changes in species richness. Two types of habitats are distinguished in a forest island of sufficient size. interior habitats near the center of edge habitats along the island border. Species in a forest island include both interior and edge species. To model how species richness response for area change in time, the model has tree state variables: interior species richness, edge species richness and area. Simulations show that changes in species richness occur only when the area of forest islands is smaller than a threshold live which is different for the edge and interior species groups.

The model Fagus simulates how silvicultural measures and weather influence growth, development and water and nitrogen balance, including leaching. The model describes the growth and development of trees, the water and nitrogen uptake by trees and the total water and nitrogen budget.

The model named Fed provides a flexible framework for modeling a forest ecosystem, where naturally and anthropogenically induced changes in climate will response in the soil, the vegetation and the radiation regime and how these three factors will modify each other. The framework is able to configure modules using a graphical user interface, for associating ecosystem processes in space, for averaging across incompatible time steps and for interactive display of results. The soil genesis version of Fed is able to simulate a soil profile derived from soil characteristics. The following processes are included in the model to account for the environmental conditions to which soil is subject: fluxes of water, heat and trace gases, ion exchange and leaching. formation and dissolution of coatings, change in soil pH, organic decomposition, changes in density and transformation of clay minerals.

Hybrid predicts carbon and water fluxes between forest and atmosphere to predict tree growth and species succession. Photosynthesis, transpiration, respiration, nitrogen uptake and soil water flows are calculated daily. Carbon and nitrogen are partitioned between leaves, wood and roots on an annual basis.

Hobo is an object-oriented model. The forest is represented by a rectangular grid with 32 x 32 cells of varying size. Each cell can be empty or occupied by a single tree of the two competing species with shifted time of bud break. A distribution of foliage-feeding larvae over the discrete space is simulated. The following factors of system dynamics are included: leaf palatability, growth of larvae resulting in increasing food demand, migration of larvae, localized reproduction of trees via seedlings, ten age classes of trees, death of trees being caused exclusively by defoliation and localized reproduction of larvae.

The model Ecolecon a spatially-explicit, individual based and object-oriented program, which is able to simulate animal population dynamics and economic revenues in response to different forest landscape structures and timber management strategies. The model can generate artificial forest landscape or it can be linked with GIS to run simulations on real landscapes. The model predicts population dynamics, spatial distribution, extinction probability, future landscape structure and income from timber harvesting.

The model named Gendec simulates plant litter decay. It recognizes six pools of carbon and nitrogen: labile plant components, holocellulose, resistant plant components (lignins), live microbial biomass, dead microbial biomass and inorganic nitrogen and carbon.

The model denoted Forac simulate acid and base leaching from forest soil. The state variables are base status of soil and exchangeable base content per rooting space. The forcing functions are atmospheric deposition of acid and base cations and estimates for soil weathering processes.

The model Foret considers the processes that affect the birth, growth and death of individual trees. It accounts for prevailing light, moisture and temperature conditions.

Boerma et al. have developed a spatial model for tropical forests. It covers 300 patches. Each patch is large enough to contain one mature tree. The growth of a tree is influenced by the size of the tree relative to its maximum size, the light and nutrient availability. As state variables are used the height and diameter of the tree and the amount of nutrients in the soil.

The model Temfes couples ecosystem of water, carbon and nitrogen with species population dynamics, i.e., birth, growth and mortality of individual trees. The equations applied build on mass- and energy balances. Temfes consists of five major modules: 1) a physiologically based tree growth model, 2) a biophysical canopy model, 3) a soil water and heat model 4) an organic matter decomposition model and 5) a regeneration and mortality model.

Valentine's almost classical model should also be included in this overview. It is based upon a carbon balance of a tree and the state variables are basal area, length of tree, woody volume and total dry matter.

The model developed by Chen and Gomez is a tree model with many more state variables, particularly to account for environmental stresses. The leaf area, leaf mass and age classes of needles, stem heartwood and sapwood, and coarse and fine roots. All masses have stoichiometric contents of nutrients including Ca, Mg, K, N, C and P. The effects of environmental stresses are defined by equations which modify the rates of the physiological processes. For example ozone and water deficit cause the closure of stomata, resulting in a reduction of the carbon dioxide uptake and photosynthesis rate. A cumulative dose of ozone amplifies the mortality of needles and therefore the rate of litterfall.

**Table 4.3.** An Overview of Forestry Models

| Model component(s)/<br>model name | Model characteristics: st. var.<br>and/or main processes | References |
|---|---|---|
| Competition model | Growth, inter-tree competition, site | Reed et al., 1990 |
| 4 hardwood species | Effect and seasonal growth pattern | |
| 3D-crown geometry | geometry, inter-tree competition | Smith, 1987 |
| Aspen | Productivity influenced by roots<br>and soil/site characteristics | Gale and Grigal,<br>1987 |
| Pollutants/Tree<br>processes | Carbon, water and nutrient balances<br>influenced by pollutants | Weinstein and<br>Beloin, 1990 |
| Gromap | Mapping of edaphic, climatic and<br>overall limitations for plants | Booth, 1991 |
| PnEt | Effect of climate on forest production | Aber et al., 1993 |
| Formix | energy balance (as carbon) | Bossel and<br>Krieger, 1991 |
| Treedyn3 | Carbon and nitrogen balances, tree<br>growth (number, diameter, height,<br>leaf, roots, fruit) | Bossel and<br>Schäfer, 1989 |
| Beech/ Fsyl-mod | Carbon and water budget | Stickan, 1994 |
| Cardyn/deciduous | Carbon dynamics at ecosystem level<br>forest | Veroustraete,<br>1994 |
| Species richness | sp. richness response to changes in<br>forest island area | Wu and<br>Vankat, 1991 |
| Fagus | infl. of silviculture and weather on<br>growth, development and budgets | Hoffmann, 1995 |
| Fed | model framework Fed soil genesis<br>soil profile model | Levine *et al.*,<br>1993 |
| Hybrid | carbon and water exch. between fo-<br>rest and atmosphere | Friend *et al.*,<br>1993 |
| Hobo | two tree species and larvae | Lhotka, 1994 |
| Ecolecon | pop. dyn, and revenue in response<br>to diff. landscape structures | Liu, 1993 |
| Gendec | plant litter decay (N and C) | Moorhead and<br>Reynolds, 1993 |
| Forac | acid and base leaching from soil | Arp, 1983 |
| Foret | factors aff. birth, growth and death | Dale et al., 1988 |
| Tropical forests | two types of trees, spatial structure | Boerma *et al.*1991 |
| Temfes | coupling of water, carbon and<br>nitrogen tree species dynamics | Nikolov and<br>Fox, 1994 |

**Table 4.3. (continued)**

| Model component(s)/ model name | Model characteristics: st. var. and / or main processes | References |
|---|---|---|
| Tree growth | carbon balance | Valentine, 1985 |
| Response to stress | sev. tree st. var., effect of env. stresses on physiological processes | Chen and Gomez, 1990 |
| Forest-BGC | water, carbon and nitrogen cycles, complete coverage of processes | Kremer; see 4.7 |
| Spatial pattern | competition among neighbors | Leps and Kindl-mann, 1987 |

A model called Forest-BGC calculates carbon, water and nitrogen cycles through a forest ecosystem. The model includes canopy interception, evaporation, transpiration, photosynthesis, growth and maintenance, respiration, carbon allocation, litterfall, decomposition , nitrogen mineralization and primary production.

Leps has proposed a model for the development of the spatial pattern of an even-aged plant population. The model accounts for the competition among neighbors and  evaluate the possibilities of spatial pattern intensity as a measure of population organization.

## 4.7 Forestry Models

**Model identification**
Model name/title: GROMAP

**Model type**
Media: regional scale, forest
Class:
Substance(s)/compounds: nutrients
Biological components: organism, any selected perennial plant taxon

**Model purpose**
To indicate the broadscale suitability of any perennial plant taxon for cultivation at sites across Africa.

**Short description**
The program uses a climatic and edaphic database for over 10,000 locations. Mean monthly values of maximum temperature, minimum temperature, precipitation, evaporation and solar radiation were estimated for each location. Information describing the chosen plant's edaphic and climatic requirements are entered into the computer. Colour maps are produced on the microcomputer screen indicating edaphic, climatic and overall limitations for the particular plant. A cursor can be moved over any location on the maps and detailed model calculations examined for that location. Together the database, model and user interface provide a powerful tool for testing descriptions of requirements and identifying suitable regions for particular taxa.

**Model application(s) and State of development**
Application area(s): Broadscale land evaluation tree species and provenance selection
Number of cases studied: one published (see below)
State: conceptualization, verification

**Model software and hardware demands**
Platform: PC-based, IBM
OS: DOS
Software demands: no special requirements, written in Turbo Pascal
Hardware demands: VGA color graphics

**Model availability**
Purchase: no, may be made available at no charge for cooperative projects executable code
Contact person(s) Dr. Trevor H. Booth
        CSIRO Division of Forestry
        P.O. Box 4008
        Queen Victoria Terrace
        Canberra ACT 2600 Australia

Phone: + 61 62 81 82 59
Fax: + 61 62 81 83 12
E-mail: trevorb@cbr.for.csiro.au

## Model documentation and major references

Booth, T.H. (1991). A climatic/edaphic database and plant growth index prediction system for Africa. *Ecol. Modelling*, 56, 127-134.
Booth, T.H. (1991). Where in the world? New climatic methods to assist species and provenance selection for trials. *Unasylva*, 165, 51-57.

## Other relevant information

The African version of GROMAP is one of a family of simulation mapping programs. Another version has been developed for Australia and programs for China and Thailand are currently under development. Simpler climatic mapping programs have also been developed for countries including China (100,000 locations), Australia, Zimbabwe, Thailand and Indonesia, whilst regional programs have been developed for Africa, Central/South America and the whole world (see Booth, (1991), Unasylva, 165, 51-57).

## Model identification
Model name/title: PnEt

## Model type
Media: terrestrial, forest
Class: biogeochemical
Substance(s)/compounds: carbon, nitrogen
Biological components: ecosystem, forest

## Model purpose
Predict the effects of changes in physical and chemical climate on water yield, water quality and forest production.

## Short description
PnET is a simple, lumped-parameter, monthly time-step model of carbon and water balance of forest built on two principal relationships: 1) maximum photosynthetic rate is a function of foliar nitrogen concentration, and 2) stomatal conductance is a function of realized photosynthetic rate. Monthly leaf area display and carbon and water balances are predicted by combining these with standard equations describing light attenuation in canopies and photosynthetic response to diminishing radiation intensity, along with effects of soil water stress and vapor pressure deficit (VPD). PnET has been validated against field data from 10 well-studied temperate and boreal forest ecosystems, supporting our central hypothesis that aggregation of climatic data to the monthly scale and biological data such as foliar characteristics to the ecosystem level does not cause a significant loss of information relative to long-term mean ecosystem responses. Sensitivity analyses reveal a diversity of responses among systems to identical alterations in climatic drivers.

This suggests that great care should be used in developing generalizations as to how forests will respond to a changing climate. Also critical is the degree to which the temperature responses of photosynthesis and respiration might acclimate to changes in mean temperatures at decadal time scales. An extreme climate change simulation (+3C maximum temperature, -25% precipatation with change in minimum temperature or radiation, direct effects of increased atmosphere ignored) suggests that major increases in water stress, and reductions in biomass production (net carbon gain) and water yield would follow such a change.

## Model application(s) and State of development
Application are(s): Northeastern U.S.
Number of cases studied: 10
State: conceptualized, verification, calibration, validation, prognosis made?

## Model software and hardware demands
Platform: PC-based, IBM
OS: DOS 6.0
Software demands: Quickbasic
Hardware demands: Any DOS 6.0 compatible machine

## Model availability
Purchase: Free
Contact person(s) Dr. John D. Aber
        Complex Systems-EOS
        University of New Hampshire
        Durham, NH 03824
        USA

Phone: +1 603-862-1792
Fax: +1 603-862-0188
E-mail: Ida Xandra.UNH.EDU

## Model documentation and major references
Aber, J.D. and C.A. Federer. 1993. A generalized, lumped-parameter model of photo-synthesis, evapotranspiration and net primary production in temperate and boreal forest ecosystems. *Oecologia*, 92: 463-474.

## Other relevant information

## Model identification
Model name/title: FORMIX: Tropical Forest Dynamics Simulator

## Model type
Media: forest
Class: ecophysiological
Substance(s)/compounds: carbon
Biological components: ecosystem, forest, multi-species, natural regeneration

## Model purpose
Comprehensive description of the growth dynamics of natural tropical forests under various management conditions and silvicultural treatments, based on full energy balance (such as carbon) for studying the consequences and long-term regeneration dynamics under different selective logging and silvicultural treatment regimes.

## Short description
The mutual interaction of the forest growth and ligth conditions causes vertical and horizontal differentiation in the natural forest mosaic and must be represented in simulation models describing long-term forest dynamics. The FORMIX2 (tropical) forest simulator incorporates the multi-storey canopy layer structure of the natural tropical forest, the basic processes of gap succession dynamics and gap interaction by tree fall and seed dispersal, several functionally different species groups (emergents, main canopy trees, pioneers, woody treelets, forest floor herbs) and the energy (carbon) balance of assimilation, dissimilation, and increment resulting from the different light conditions for the different canopy layers in the different gap-size areas of a forest mosaic. The vertically and horizontally structured FORMIX2 model produces time-dependent results (average per hectare data) for stem numbers, diameter distributions, basal areas, standing crop and yield, as well as dynamic views of the forest profile and of spatial forest development, as a function of logging method and intensity and silvicultural treatment. The model has been applied in exploratory investigation to compute the long-term development of (a) virgin forest and (b) logged-over forest in Sabah, and (c) peat-swamp forest in West Malaysia using prototypical data. The results obtained agree with empirical data.

## Model application(s) and State of development
Application area(s):
Number of cases studied: about 4
State: calibration, validation, prognosis

## Model software and hardware demands
Platform: PC-based, IBM
OS:
Software demands: Turbo Pascal
Hardware demands:

## Model availability

Purchase: contact author, executable code

Contact person(s)  Dr. Hartmut Bossel or Dr. Andreas Huth
Environmental Systems Research Group
University of Kassel
D-34109 Kassel, Germany

Phone: +49 561-8042519
Fax: +49 561 8043176
E-mail: bossel@usys.informatik.uni-kassel.de

## Model documentation and major references

Hartmut Bossel, and Krieger, H., : Simulation model of natural tropical forest dynamics. *Ecol. Modelling*, 59 (1991) 37-71.

## Other relevant information

Several applications in forest management contexts under way.

## Model identification

Model name/title: TREEDYN3: Forest Dynamics Simulator.

## Model type

Media: forest
Class: ecophysiological
Substance(s)/compounds: nutrients, carbon, nitrogen
Biological components: ecosystem, forest: even-aged, pure stand, deciduous and coniferous trees

## Model purpose

Comprehensive description of the growth dynamics of (planted) forests under various environmental and pollution influence, based on full carbon and nitrogen balances. For teaching purposes and exploration of the effects of climate, site, silvicultural management and pollution.

## Short description

The TREEDYN3 forest simulation model is a process model of tree growth, carbon and nitrogen dynamics in a single-species, even-aged forest stand. The tree/soil system is described by a set of nonlinear ordinary differential equations for the state variables: tree number, base diameter, tree height, leaf mass, fine root mass, fruit biomass, assimilate, carbon and nitrogen in litter, carbon and nitrogen in soil organic matter, and plant-available nitrogen. The model includes explicit formulations of all relevant ecophysiological processes such as: computation of radiation and canopy photoproduction as a function of latitude, seasonal time and daytime; light attenuation in the canopy; respiration of all parts; assimilate allocation and increment formation; self-thinning resulting from light competition; nitrogen fixation, mineralization, humification and leaching; forest manage-

338

ment (thinning, felling, litter removal, fertilization, etc.), temperature effects on respiration and decomposition, and environmental effects (pollution damage to photosynthesis, leaves, and fine roots). Only ecophysiological parameters which can be either directly measured or estimated with reasonable certainty are used.

## Model application(s) and State of development
Application area(s):
Number of cases studied: about 6 (for acacia, eucalyptus, spruce (several), pine)
State: calibration, validation, prognosis made?

## Model Software and hardware demands
Platform: PC-based, IBM, Workstation (Sparc)
OS:
Software demands: Turbo Pascal or C++
Hardware demands:

## Model availability
Purchase: contact author, software available as executable code
Contact person(s) Dr. Hartmut Bossel
                Environmental Systems Research Group
                University of Kassel
                D-34109 Kassel, Germany

Phone: +49 561-8042519
Fax: +49 561-8043176
E-mail: bossel@usys.informatik.uni-kassel.de

## Model documentation and major references
H. Bossel, and H. Schäfer: Generic simulation model of a forest growth, carbon and nitrogen dynamics, and application to tropical acacia and European spruce, *Ecol. Modelling*, 48, (1989), 221-265.

## Other relevant information
The model is written as a Turbo Pascal unit made to be used with the SIMPAS simulator (H. Bossel, Modelling and Simulation. A.K. Peters, Wellesæey, Mass. 1994 (484+xvi pp.+diskette), and H. Bossel: Modellbildung und Simulation. Vieweg Verlag Braunschweig/Wiesbaden 1992/1994 (400 pp.+diskette).

## Model identification
Model name/title: FSYL-MOD

## Model type
Media: local, scale, terrestrial, forest
Class: biogeochemical
Substance(s)/compounds: carbon, water

Biological components: ecosystem - deciduous forest organism - beech saplings (*Fagus sylvatica*)

## Model purpose
FSYL-MOD simulates perennial courses of carbon and water budgets of beech saplings from a temperate deciduous forest. The output gives daily totals of C-input, dark respiration of leaves, respiration of stem, main and lateral roots, net-C-gain and transpiration based on $m^2$ leaf area or plant biomass respectively. Daily relative growth rate (%) and transpiration coefficient based on dry weight production is also given (g $H_2O$ $g^{-1}$).

## Short description
FSYL-MOD is based on field measurements of daily courses of assimilation and transpiration as well as of individual biometrical plant data from 6 harvested sample trees (total biomass, leaf area, specific leaf weight, and of leafs, stem, main and lateral root dry mass) in a deciduous forest in the Solling area, FRG. The driving climatic input variables taken from meteorological standard data are - PPFD (daily total of photosynthetic photon flux density above forest canopy; mol $m^{-2}$) - TEMP (daily mean of air temperature above forest canopy; °C) - VPD (daily mean of vapour pressure deficit above forest canopy; Pa) Site specific data at the forest floor are generated by the model using the measured relationship between field data and meteorological standard data (PPFD, VPD, TEMPair on forest floor and TEMPsoil calculated from TEMP air/soil-relationship).

## Model application(s) and State of development
Application area(s): Deciduous forest in Solling area, FRG
Number of cases studied: several, in total 3 years of field measurement
State: validation

## Model software and hardware demands
Platform: PC-based
OS: MS-DOS
Software demands: Q-BASIC, Q-BASIC
Hardware demands:

## Model availability
Purchase: yes, free of charge, executable code
Contact person(s) Dr. Walter Stickan
                Lehrstühl für Geobotanik
                Untere Karspüle 2
                D-37073 Göttingen
                Germany

Phone:
Fax:
E-mail:

## Model documentation and major references

Stickan, W., Gansert, D., Neeman, G. and Rees, U., 1994. Modelling the influence of climatic variability on carbon and water budgets of trees using field data from beech saplings (Fagus sylvatica L.). *Ecol. Modelling*, in press.

Stickan, W., Der Kohlenstoffhaushalt von Buchen auf ökophysiologischer Basis - ein schlüssel zum Verständnis der Waschstumsdynamik von Waldbeständen. *Scripta Geobotanica*, 20: 207-216.

## Other relevant information

## Model identification

Model name/title: CARDYN: Deciduous Forest Model to evaluate Carbon Dynamics at ecosystem level.

## Model type

Media: air, local scale to regional scale, terrestrial, soil, forest
Class: biogeochemical
Substance(s)/compounds: carbon, oxygen
Biological components: ecosystem - deciduous forest

## Model purpose

The model simulates the aggregated behaviour of several deciduous species. The difference with a single species stand model is located at the phenological level (leaf abscission, shooting, etc;). The model is developed to average species dependent phenological differences to represent the phenology of a generalized deciduous forest and the resultant carbon dynamics associated with it. An approach is also described to model the Net Carbon Exchange (NCE) between the vegetated surface-atmosphere interface of a deciduous forest.

## Short description

Four compartments are considered in the model, between which all carbon fluxes are evaluated. The green compartment (G), representative for organs with large mass fluctuations (photosynthetically active organs; assimilate storage organs, feeder roots, and fruits). The relationship between LAI and leaf carbon (G) mass is formalised by using a conversion coefficient, u, averaging the area to mass relation of light and shade adapted leaves. It is assumed that this relationship is valid during plant development.

The non-green compartment (L) is representative for relatively small mass fluctuations (stems, branches, woody roots).

The litter compartment (L) and the atmosphere, the last one considered here as an infinite source and sink.

Forcing functions: The photosynthetic, respiratory and other carbon fluxes defined in the model, are forced by the photosynthetically active radiation flux (PAR), atmospheric temperature, daylength, carbon dioxide and oxygen mixing ratios.

The carbon fluxes are calculated on an hourly basis. Changes in phytomass and NCE are evaluated by integration on a daily basis. Important parameters S is switching function that allocates assimilates into the green or non-green phytomass compartments according to the phenological phase of the forest. A full description of the formalisation of carbon allocation is given in Janeck et al. (1989), *Ecol. Modelling, 49: 101-124.*

Necessary inputs: $CO_2$ evolution during the two last centuries according to data from Houghton and Woodwell (1989). Scientific American 260: 36-44. A climatological file compiled by Poppe (1911) with monthly mean, maximal and minimal temperatures, was used as input for the model.

Poppe, H., 1991. Klimatologische basiswaarden Vlaanderen, University of Leuven in co-operation with the Royal Meteorological Institute of Belgium, public domain software package.

## Model application(s) and State of development

Application area(s): Global Change research, Ecosystem carbon dynamics simulations.
Number of cases studied: 1. Effects of temperature and $CO_2$ mixing ratio's on young and climax state forests. 2. Prediction of the effect of changes in $CO_2$ mixing ratio's after World War II on forest biomass and Net Carbon Exchange.
State: conceptualization, verification and prognosis were made.

## Model software and hardware demands

Platform: The model was implemented in MS-FORTRAN source code, compiled on a PC-platform (Compaq 386/25e) under MS-DOS 5.0 and the graphical interface Windows 3.1. For long term simulations, compilation and running was performed on a DEC alfa workstation with OSF1 and FORTRAN compilation. Model results and graphical illustrations were processed and generated with a spreadsheet (EXCEL) at PC level starting from a results output file generated by CARDYN.
OS:
Software demands:
Hardware demands:

## Model availability

Purchase: A documented source code of the model and its input climatological file can be obtained on request from the author. The software is available as source code. The price is 30 ECU. The climatological file is free of charge.
Contact person(s) Dr. Frank Veroustraete
        Flemish Institute for Technological Research
        Boeretang 200
        B-2400 Mol
        Belgium

Phone: 0032 14-33-27-80
Fax: 0032 14-32-11-85
E-mail: veroustr@bmlsck11.bitnet

## Model documentation and major references
Veroustraete, F., 1994. On the use of a simple deciduous forest model for the interpretation of climate change effects at the level of carbon dynamics. *Ecol. Modelling,*
Janecek, A., Benderoth, G., Ludecke, M.K.B., Kindermann, J. and Kohlmaier, G., 1989. Model of the seasonal and perennial carbon dynamics in deciduous type forests controlled by climatic variables. *Ecol. Modelling,* 49: 101-124.

## Other relevant information

Houghton, R.H., and G.M. Woodwell, 1989. Globak Climatic Change. Scientific American, 260: 36- 44.

## Model identification
Model name/title: An area-based model of species richness dynamics of forest islands

## Model type
Media: regional scale, terrestrial, forest
Class:
Substance(s)/compounds: species diversity
Biological components: community - forest

## Model purpose
To depict the dynamics of species richness in response to changes in forest islands area, and to generate temporal changes in species richness using static species-area data.

## Short description
Two types of habitats are distinguished in a forest island of sufficient size: interior habitats near the center and edge habitats along the island border. Accordingly, species in such a forest island include both interior and edge species. To model how species richness responds to area change in time, this species richness model has three state variables: interior species richness ($Ri(t)$), edge species richness ($Re(t)$), and area ($Ae(t)$). Other important variables include total richness ($Rt(t)$), interior area($Ai(t)$), edge area($Ae(t)$), and the variables determine the rate variables of edge and interior species richness. The forcing function is the input function of area change rate. The mathematical model consists of a group of differential equations, and a STELLA simulation model on woody species of a deciduous forest island is also presented.

Simulations show that changes in species richness occur only when the area of forest islands is smaller than a threshold level which is different for the edge and interior species groups. And that edge and interior species respond differently to the changes in island area. The STELLA model is capable of simulating the dynamics of edge, interior, and total species richness with a variety of additional scenarios involving the change in area of forest islands.

**Model application(s) and State of development**
Application area(s): Conservation Biology, landscape Ecology
Number of cases studied: 1
State: verification, prognosis made

---

**Model software and hardware demands**
Platform: MAC
OS: MAC
Software demands: STELLA, STELLA
Hardware demands: MAC

---

**Model availability**
Purchase: no, executable code
Contact person(s) Prof. Jianguo Wu, Ph.D.
        Biological Sciences Center
        Desert Research Institute
        P.O. Box 60220
        Reno, NV 89506-0220
        USA

Phone: +1 702-673-7419
Fax: +1 702-673-7397
E-mail: jwu@maxey.unr.edu

---

**Model documentation and major references**
Wu, J., and Vankat, J.L., 1991. An area-based model of species richness dynamics of forest islands. *Ecol. Modelling*, 58: 249-271.
Wu, J., 1991. Dynamics of Landscape Islands: A Systems Simulation Modelling Approach. Ph.D. Dissertation. Miami University, Oxford, OH, USA, 216 pp.

---

**Other relevant information**
Wu, J. and Vankat, J.L., 1991. A System Dynamics model of island biogeography. *Bulletin of Mathematical Biology*, 53: 911-940.
Wu, J., Vankat, J.L. and Barlas, Y., 1993. Effects of patch connectivity and arrangement on animal metapopulation dynamics: a simulation study. *Ecol. Modelling*, 58: 249-271.

---

**Model identification**
Model name/title: FAGUS

---

**Model type**
Media: terrestrial, soil, forest
Class:
Substance(s)/compounds: carbon, nitrogen
Biological components: ecosystem - forest ecosystem

## Model purpose
Simulation of silvicultural measures and weather variables on growth, development, water nitrogen balance, leaching, etc.

## Short description
The model describes growth, development, water and nitrogen uptake, soil water and nitrogen budget, soil temperature, and leaching with a time step of one day.

## Model application(s) and State of development
Application area(s): Simulation studies on the behaviour of Fagus ecosystems
Number of cases studied:
State: conceptualization, verification, calibration, prognosis made

## Model software and hardware demands
Platform: PC-based IBM
OS: DOS
Software demands: FORTRAN
Hardware demands: PC AT

## Model availability
Purchase: executable code, source code
Contact person(s) Dr. Friedrich Hoffmann
                   Projektzentrum Ökosystemforschung
                   Schauenburger Str. 112
                   D-24118 Kiel
                   Germany

Phone:  + 49 431 8804084
Fax:
E-mail:

## Model documentation and major references
Hoffmann, F., FAGUS - a model for growth and development of beech.
*Ecol. Modelling* 1995,

## Other relevant information

## Model identification
Model name/title: Forest Ecosystem Dynamics (FED) Modeling Framework

---

**Model type**
Media: local scale, regional scale, soil, forest
Class: biogeochemical
Substance(s)/compounds: none
Biological components: ecosystem - northern forest

---

**Model purpose**
This framework provides a flexible environment for modeling a forest ecosystem where naturally or anthropogenically induced changes in climate will elicit responses within soil, vegetation, and radiation regime that will effect and modify each other. In this way, ecosystem process representations from different disciplines can be compared and scaling effects tested.

---

**Short description**
When complete, the framework will provide tools for establishing communication among models, for scheduling execution, for configuring modules using a graphical user interface, for associating ecosystem processes in space, for averaging across incompatible time steps, and for iterative display of result. The graphical interface is based on X-windows and the Motif standard. Groups of models can be configured on one workstation and execute on others.

---

**Model application(s) and State of development**
Application area(s): Northern Experimental Forest, Howland, Maine (International Paper)
Number of cases studied:
State: (mixed), conceptualization, verification, calibration, validation

---

**Model software and hardware demands**
Platform: Workstation SUN
OS: UNIX
Software demands: FORTRAN 90, C, C++
Hardware demands:

---

**Model availability**
Purchase: The model cannot be purchased. Software available to potential collaborators as executable code for SUN OS 4.1.3 or source code for portions not restricted by copyright (a less restricted version may be developed depending an available resources).
Contact person(s) Dr. Robert Knox
        NASA/Goddard Space Flight Center
        Code 923/Biospheric Sciences Branch
        Greenbelt, MD 20771 USA

Phone: +1 301-286-4837
Fax: +1 301-286-1757
E-mail: knox@spruce.gsfc.nasa.gov

346

## Model documentation and major references

Levine, E., Ranson, K.J., Williams, D., Shugart, H., Knox, R., and Lawrence, W., 1993. Forest Ecosystem Dynamics; linking forest succession, soil process, and radiation models. *Ecol. Modelling*, 65: 199-219.

Forest Ecosystem Dynamics/OTTER Special issue. Remote Sensing of Environment. 1993, In press.

## Other relevant information

To provide support for intermodel communication and ad hoc queries of values of important state variables, each model brought into the framework will have code added to store and report values of state variables from the most recently completed time step. Each model also gets a set of interface panels for interactive model configuration, and is modified to use an externally generated time step. Other model parts of the framework are shared among models or re-used via inheritance. Although the overall design is highly modular and "object-like", models are brought into framework with the fewest changes feasible. The FED effort could serve the ecological community as a framework for implementing similar simulations of other ecosystems, particularly where remotely sensed parameters are used.

## Model identification

Model name/title: Forest Ecosystem Dynamics (FED) Soil Genesis Model

## Model type

Media: local scale, regional scale, soil, forest
Class: biogeochemical
Substance(s)/compounds: nutrients, carbon, sulfur, phosphorus, oxygen, nitrogen, calcium, magnesium, sodium, potassium, aluminum, iron
Biological components: ecosystem - northern forest organism - soil profile

## Model purpose

This model simulates a soil profile derived from soil characterization data where inputs of matter and energy drive soil processes. Model results show new soil profile characteristics that reflect the environmental conditions to which the soil has been subjected over a given amount of time.

## Short description

The FED soil genesis model simulates processes operating during the formation of soil. Included in the model are processes and properties that will change at a short, moderate, or long time scales and thus affect short- or long-term patterns in vegetative growth or land use. These include fluxes of water, heat, and trace gases, ion exchange and leaching, formation and dissolution of sesquioxides and other coatings, change in soil pH, organic decomposition, changes in horizon density, distribution of particles, and transformation of clay mineralogy. Driving mechanisms are thermal energy, precipitation (chemistry and amount), and inputs of roots and litterfall. Soil units are soil profiles divided into compartments representing master horizons (O, A, E, B, C). Input data required for each horizon within the soil submodel are thickness, particle size, bulk density, % coarse fragments, cation exchange capacity, exchangeable cations (calcium, magnesium, sodium, and potassium), % organic carbon and nitrogen, % iron and aluminum oxide, soil pH, clay mineralogy, and % water holding capacity at 0.03 an 1.5 MPa.

## Model application(s) and State of development
Application area(s): Northern Experimental Forest, Howland, Maine (International Paper)
Number of cases studied: 10 soil associations at the site
State: (mixed), verification, calibration, validation, prognosis made, validation of prognosis

## Model software and hardware demands
Platform: Workstation SUN
OS: UNIX
Software demands: FORTRAN 77, FORTRAN 90
Hardware demands:

## Model availability
Purchase: Software cannot be purchased, available as executable or source code
Contact person(s) Dr. Elissa Levine
              NASA/Goddard Space Flight Center
              Code 923/Biospheric Sciences Branch
              Greenbelt, MD 20771 USA

Phone: +1 301-286-5100
Fax: +1 301-286-1757
E-mail: elissa@lichen.gsfc.nasa.gov

## Model documentation and major references
Levine, E., Ranson, K.J, Williams, D., Shugart, H., Knox, R., and Lawrence, W., 1993. Forest Ecosystem Dynamics: linking forest succession, soil process, and radiation Models. *Ecol. Modelling*, 65: 199-219.
Levine, E.R., and Knox, R.G., 1993. A comprehensive framework for modeling soil genesis. Soil Science Society of America Special Publication. In press.
Forest Ecosystem Dynamics/OTTER special issue. Remote Sensing of Environment. 1993, in press.

## Other relevant information

**Model identification**
Model name/title: HYBRID

**Model type**
Media: water, local scale, terrestrial, forest
Class: biogeochemical
Substance(s)/compounds: nutrients, carbon, nitrogen
Biological components: ecosystem - forest, community - forest, population - tree, organism - tree

**Model purpose**
This model predicts carbon and water fluxes between forest and atmosphere to predict tree growth and species succession.

**Short description**
The HYBRID model has a daily and annual component. Individual trees are treated during a simulation. Each has a set of physiological parameters determined by which species it belongs to. Physiological calculations of photosynthesis, transpiration, and respiration are made each day. Hydrological parameters such as soil water amount, flow, and input are also calculated daily. At the annual time step, carbon is partitioned between leaves, wood, and fine roots. Litterfall is calculated which decomposes to release nitrogen to the soil daily. Nitrogen is taken up by roots each day and distributed within the plant annually. HYBRID is driven by daily climate parameters, and has 20 species-specific parameters. Important parameters include photosynthesis/light response curvature, light extinction coefficient, phenology, and allometric parameters relative to diameter to woody biomass (the pipe model is used to calculate partitioning, with the assumption of fixed foliage to fine root biomass).

**Model application(s) and State of development**
Application area(s): Knoxville, Tennessee; Missoula, Montana; Northern Experimental Forest, Howland, Maine (International Paper)
Number of cases studied: 3
State: validation, prognosis made

**Model software and hardware demands**
Platform: PC-based, Workstation HP710
OS: UNIX or DOS
Software demands: FORTRAN 77
Hardware demands: 4MB RAM, 486 Dx

## Model availability
Purchase: yes, no cost, executable or source code
Contact person(s) Dr. A.D. Friend
>Institute of Terrestrial Ecology
>Edinburgh Research Station
>Bush Estate
>Penicuik
>Midlothian EH26 0QB
>Scotland

Phone: +44 31-445-4843
Fax: +44 31-445-3943
E-mail: adf@unixa.nerc.bush.ac.uk
telex 72579

## Model documentation and major references
Friend, A.D., Shugart, H.H., and Running, S.W., 1993. A physiology-based gap model of forest dynamics. *Ecology*, 79: 792-797.
Friend, A.D., 1994. PGEN: An integrated model of leaf photosynthesis, transpiration, and conductance. *Ecol. Modelling*, (in press).

## Other relevant information


## Model identification
Model name/title: Hobo model of sequential predation

## Model type
Media: local scale, forest
Class: non of th three is applicable
Substance(s)/compounds: none
Biological components: community - two competing tree species with shifted time of bud break, population - foliage-feeding insects, mainly in the larval stage

## Model purpose
Analysis of a special case of apparent competition, i.e., one predator feeding on two competing prey species where the availability of the two preys is not simultaneous. Investigation of the effects of spatial self-structuring as a result of localized prey-predator interaction and reproduction.

## Short description
The model is built using the Hobo object-oriented simulation library (Lhotka, 1994). The forest is represented using a rectangular grid of varying sizes, at most 32 x 32 cells. Each cell can be empty or occupied by a single tree of the two competing species, denoted EARLY and LATE. The EARLY species breaks bud 10 days before the LATE one. A distribution of foliage-feeding larvae over this discrete space is simulated. One generation of the insect per season is assumed. After hatching the larvae can feed only on EARLY tree,

then they can switch to the LATE tree after the latter breaks bud, and shortly before pupating the larvae depend exclusively on the LATE tree because the leaves of the EARLY tree become unpalatable. The following factors of system dynamics are included: decreasing leaf palatability within a season; growth of larvae resulting in increasing food demand; migration of larvae, localized reproduction of trees via seedlings; 10 age classes of trees; death of trees being caused exclusively by defoliation; localized reproduction of larvae.

Model simulations exhibit oscillatory patterns in both time and space with an overall competative advantage for LATE tree. The competition mechanism is largely based upon the existance inhomogeneity in the system.

## Model application(s) and State of development
Application area(s): Theoretical biology, forest ecology, entomology
Number of cases studied: 0
State:

## Model software and hardware demands
Platform: PC-based MAC IBM, Workstation HP900 SUN SPARC IBM RS/6000
OS: Objectworks\smalltalk, v.2.1, smalltalk
Software demands: at least 5MB free RAm
Hardware demands:

## Model availability
Purchase: no, available as source code
Contact person(s) Dr. Ladislav Lhotka
    Laboratory of Biomathematics
    Institute of Entomology CAS
    Branisovska 31
    CS-37005 Ceske Budejovice
    Czech Republic

Phone: +42 38-817 ext. 480
  +42 38-45500
Fax: +42 38-45985
E-mail: lada@entu.cas.cz
TELEX : 114 406 csav c

## Model documentation and major references
Lhotka, L., 1994. Implementation of individual-oriented models in aquatic ecology. *Ecol. Modelling*, 72.
Van der Lann, J., Lhotka, L., and Hogeweg, P., 1994. Sequential predation: a multi-model study. Submitted to *Journal Theor. Biol.*
Lhotka, L., 1994. Hobo User Manual. Institute of Entomology CAS, Ceske Budejovice.

**Other relevant information**
This model was built using the Hobo simulation system. It is an object-oriented library (written in the smalltalk language) for event-driven simulations of ecological systems. Special support is for individual-based models with complex spatial configurations and community structures.

---

**Model identification**
Model name/title: ECOLECON

---

**Model type**
Media: forest
Class:
Substance(s)/compounds:
Biological components: ecosystem - forest, community - birds, population - birds, organism - Bachman's sparrows, pine trees

---

**Model purpose**
To simulate animal population dynamics and economic revenues in response to different artificial landscape structures and timber management scenarios.

---

**Short description**
ECOLECON is a spatially-explicit, individual-based, and object-oriented program. It contains 14 classes or subclasses of ecological and economic information which are hierarchically interlinked. ECOLECON can generate artificial forest landscapes or can link with geographic information systems (GIS) to run simulations on real landscapes. The model inputs include animal habitats, dispersal, survivorship, reproduction, timber growth and yield functions, management costs, timber prices, and management schemes. The model predicts population dynamics, spatial distribution, extinction probability, future landscapes structure, and economic income from timber harvesting based on current tax and timber market situation.

---

**Model application(s) and State of development**
Application area(s): species conservation, multiple use of forests, land use, trade-off between environment protection and economic development.
Number of cases studied: one (done, several in progress)
State: validation

---

**Model software and hardware demands**
Platform: PC-based
OS: DOS
Software demands: Borland C++ (2.0)
Hardware demands:

352

## Model availability

Purchase: yes, mailing costs disk. Executable snd source code.

Contact person(s) Dr. Jianguo Liu
                   Harvard Institute for International Development
                   Harvard University
                   One Eliot Street
                   Cambridge, MA 02138 USA

Phone: +1 617 495 3621
Fax: +1 617 495 0527
E-mail: jliu@husc.harvard.edu

## Model documentation and major references

ECOLOCON: an ECOLogical-ECONomic model for species conservation in complex forest landscapes. *Ecol. Modelling*, 70 (1993), 63-87.

## Other relevant informations

## Model identification

Model name/title: GENDEC: General Decomposition Model

## Model type

Media: local scale, terrestrial, soil
Class: biogeochemical
Substance(s)/compounds: carbon, nitrogen
Biological components: ecosystem - general, community - decomposers

## Model purpose

The purpose of this model is to simulate plant litter decay in a manner that maximizes realistic behaviour of the system at the same time minimizing the amount of information required to obtain model parameters and drive simulats.

## Short description

GENDEC recognizes six pools of carbon (C) and nitrogen (N), including: (1) labile plant compounds; (2) holocellulose (cellulose + hemicellulose); (3) resistant plant compounds (e.g., lignins); (4) live microbial biomass; (5) dead microbial biomass; and (6) mineral N and carbon dioxide.

Nitrogen flows are assumed to balance calculated carbon flows. The loss of carbon from each dead organic matter pool ($C_s$, s=1,2,3,5) is a function of moisture availability ($S_m$), temperature ($S_t$), N limitation ($N_s$) and a maximum intrinsic decay rate ($k_s$): $dC_s/dt = C_s * k_s * S_m * S_t * N_s$; where $S_m$, $S_t$ and $N_s$ are scalar multipliers.

The total quantities of C and N available for microbial use consist of the sum of all losses from dead organic matter pools, while available N includes mineral forms. Decay rate coefficients and functions used to express temperature, moisture and nitrogen limitations are reported by Moorhead and Reynolds (1991, 1993a, 1993b). Microbial growth and respiration are driven by total C losses from the various pools, assuming an assimilation efficiency of 60%. Microbial death primarily consists of a minimum daily mortality of 0.1% of the standing biomass and 20% of incremental growth.

## Model application(s) and State of development
Application area(s): Plant litter decay within a soil environment
Number of cases studied: 2-buried litter decay in a warm desert and a tussock tundra ecosystem
State: validation, validation of prognosis

## Model software and hardware demands
Platform: PC-based MAC IBM, Mainframe DIGITAL IBM
OS: DOS, Macintosh
Software demands: Stella v.II, FORTRAN77, language STELLA (Macintosh) FORTRAN (PC or Macintosh)
Hardware demands: PC-based w/+1 Mb RAM

## Model availability
Purchase: yes, cost of media, shipping ca. US $5.00. Available as source code
Contact person(s) Dr. Daryl Moorhead
                Ecology Program
                Department of Biological Sciences
                Texas Tech University
                Lubbock, Texas 79409-3131 USA

Phone: +1 806 742- 1158
Fax: +1 806-742-2963
E-mail: tudlm@ttacs1.ttu.edu

## Model documentation and major references
Moorhead, D.L. and Reynolds, J.F., 1991. A general model of litter decomposition in the nothern Chihuahuan Desert. *Ecol. Modelling*, 59: 197-219.
Moorhead, D.L. and Reynolds, J.F., 1993a. Changing carbon chemistry during decomposition of creosotebush fine litter in the northern Chihuahuan Desert. *Am Mild Nat.*, 130: 83-89.
Moorhead, D.L. and Reynolds, J.F., 1993b. Effects of climate change on decomposition in artic tussock tundra. A modeling synthesis. *Arctic Alpine Res.*, 25: 403-412.

## Other relevant information

## Model identification
Model name/title: ForAc

## Model type
Media: water, lake, terrestrial, soil, forest
Class: biogeochemical
Substance(s)/compounds: base cations, acid cations, chemicals
Biological components: ecosystem - forest soils

## Model purpose
To simulate acid and base leaching losses from forest soils; to simulate the amount of acidity retained by the soil; to estimate the half-time of the soil response to acidification; to calculate base cation and acid cation concentrations in soil leachates based on atmospheric base and acid cation deposition.

## Short description
State variables: base status of soil, starts with current exchangeable base content per rooting space.
Forcing functions (input): atmospheric deposition of acid and base cations; estimates for soil weathering.
Important model parameter: soil response time (half-time) to soil acidification.
Verification information: changes in soil base saturation; cumulative losses of acid and base cations from soils; acid/base cation concentrations in soil solutions.

## Model application(s) and State of development
Application area(s):
Number of cases studied:
State: conceptualization, verification, calibration, validation

## Model software and hardware demands
Platform: MAC, Work-station any MAC
OS:
Software demands: STELLA
Hardware demands:

## Model availability
Purchase: available as executable code
Contact person(s) Dr. Paul A. Arp
        University of New Brunswick
        Faculty of Forestry and Environmental Management
        Fredericton, New Brunswick
        Canada E3B 6C2

Phone: +1 506-453-4501
Fax: +1 506-453-3538
E-mail: Arp2@unb.ca

**Model documentation and major references**
*Ecol. Modelling*, 19 (1983) 105-117
*Ecol. Modelling* ,19 (1983) 119-138

---

**Other relevant information**
Soil response follows a simple exponential function, suitable for general soil reponse evaluation in the context of landscape dynamics and landscape ecology. The model can be used for surface water and lake acidification studies as well. The ForAc model was the first product of the Forest Modelling Study (ForM-S) at the University of New Brunswick, Faculty of Forestry and Environmental Management.

ther ForM-S models address forest biomass (ForBioM); forest hydrology (ForHyM); forest soil temperatues (ForSTeM); forest radiation (ForRad); forest net photosynthesis (ForNeP); and a generalized forest nutrient cycling mocel (ForSVA, short for forest soil-vegetation-atmosphere model). All these models are formulated in STELLA.

The ForAc model is a precursor of ForSVA. This model is used to assess environmental change and harvesting effects on nutrient (N, S, Ca, Mg, K) storage and cycling in upland forests (Arp and Oja 1992, in Nord 1992:41).

---

**Model identification**
Model name/title: FORET

---

**Model type**
Media: forest
Class: biogeochemical
Substance(s)/compounds: biomass
Biological components: ecosystem - Eastern forest of U.S.

---

**Model purpose**
To project changes in forest stand development and biomass for 1/12 ha plots.

---

**Short description**
The major processes affecting birth, growth and death of individual trees are explicitly included in the model. A sapling or sprout enters the simulated plot depending upon prevailing light, moisture, and temperature conditions. Actual tree growth is determined from the optimal growth as reduced by the effects of the most limiting factor when light, temperature, moisture, and competition are considered. Tree mortality occurs as a result of slow growth or as a stochastic function of the maximum age of a species. Many trees representing a variety of species can occur on any one plot. Forest characteristic are obtained by summing the simulations of many trees and plots.

---

**Model application(s) and State of development**
Application area(s): Eastern U.S.
Number of cases studied:
State: verification, validation

## Model software and hardware demands
Platform: Mainframe IBM
OS:
Software demands: language FORTRAN
Hardware demands:

## Model availability
Purchase: no, available as source code
Contact person(s) Dr. H.H. Shugart
　　　　　　University of Virginia

Phone:
Fax:
E-mail:

## Model documentation and major references
Dale, V.H., Jager, H.I., Gardner, R.H., Rosen, A.E., 1988. Using sensitivity and uncertainty analysis to improve predictions of broad-scale forest development.
*Ecol. Modelling* 42: 165-178.

## Other relevant information

## Model identification
Model name/title: A Spatial Model for Tropical Forests.

## Model type
Media: forest
Class: biogeochemical
Substance(s)/compounds: phosphorus, light, space
Biological components: ecosystem - very simple tropical forest with two types of trees

## Model purpose
The purpose of our model was to investigate whether a very simple set of differences between two types of trees was enough to explain the spatial structure that seems to occur in forests, i.e., a distinct separation between the two along a nutrient gradient.

## Short description
The model simulates the tree growth on a 0.75 ha area, subdivided into 300 patches. Each patch is large enough to contain one mature tree. When a patch is empty, a new occupant is chosen dependent on the seed availability for each species, which in its turn is dependent on the seed production of the direct neighbours. Every time step (year) each tree grows, as a function of the size of the tree relative to its maximum size.

This growth is influenced by two factors reducing growth, light and nutrient availability. Natural mortality of a tree is chosen such that under optimal conditions only 2% of the trees live long enough to reach maximum height and diameter. If a tree dies it may fall, killing some of the other trees in its neighbourhood. The number of state variables per patch is limited: height and diameter of a tree, and the amount of nutrients in the soil. Important parameters are maximum age, height and diameter of the trees, the nutrient concentration in trees, and the saturation constants in nutrient uptake. Moreover, the role of disturbance through tree fall is essential.

### Model application(s) and State of development
Application area(s): none by me
Number of cases studied:
State:

### Model software and hardware demands
Platform:
OS:
Software demands: MAC, the model is written in PASCAL

### Model availability
Purchase: The model cannot be purchased, the software could be made available as sourcecode.
Contact person(s) Dr. Maarten Boersma
                        Netherlands Institute of Ecology
                        Rijksstraatweg 6
                        3631 AC Nieuwersluis
                        The Netherlands

Phone: +31 2943 3599
Fax: +31 2943 2226
E-mail: boersma%niecte@nluug.nl

### Model documentation and major references
Boersma, M., van Schaik, C.P., and Hogeweg, P., 1991. Nutrient gradients and spatial structure in tropical forests: a model study. *Ecol. Modelling*, 55: pp. 219-240.

### Other relevant information

### Model identification
Model name/title: TEMFES: A Temperate Forest Ecosystem Simulation Model

### Model type
Media: local scale, forest
Class: biogeochemcial, hydrology
Substance(s)/compounds: carbon, nitrogen
Biological components: ecosystem - forest, population - tree species

## Model purpose
TEMFES is a process-oriented computer model designed to simulate transient response of unmanaged forest ecosystems to long-term changes in climate and atmospheric $CO_2$ concentration.

## Short description
TEMFES is a forest gap-phase model that couples ecosystem cycles of water, carbon and nitrogen with species population dynamics (i.e., birth, growth, and mortality of individual trees). The model is built on the understanding that incorporating balances of mass and energy is essential in linking short-term biophysical responses to long-term successional patterns of forest ecosystems. TEMFES consists of five major modules: a physiologically-based tree growth model, a biophysical canopy model, a soil water and heat flow model, an organic matter decomposition model, and a regeneration and mortality model. These submodels interact on a number of discrete patches of ground to produce the dynamics of forest ecosystem mosaics. On each patch, individual trees compete for light, soil water, and mineral nitrogen. The model is driven by solar radiation, air temperature, precipitation, atmospheric humididty and $CO_2$ concentration. Input data also include a series of parameters characterizing soil physical properties and physiological and morphological attributes of tree species. Output variables include species composition, tree-size distributions, leaf area index, ecosystem evapotranspiration and $CO_2$ fluxes, soil moisture profiles, N mineralization, and amount of soil organic matter.

## Model application(s) and State of development
Application area(s): climate change impact studies; forest-ecosystem dynamics
Number of cases studied:
State: conceptualization

## Model software and hardware demands
Platform: PC-based
OS: DOS, Windows, Windows NT
Software demands: DOS 5.0 (or higher), Windows 3.0 (or higher), Windows NT 3.1 (or higher)
Hardware demands: 12 MB RAM, hard drive

## Model availability
Purchase: yes (when available); price: unknown yet. Software wil be available both as an executable file and a sourcecode.
Contact person(s) Dr. N.T. Nikolov
        USDA FS/Rocky Mountain Forest and Range Experiment Station
        240 West Prospect Road
        Fort Collins, CO 80526
        USA

Phone: +1 303-498-1315
Fax: +1 303-498-1010
E-mail: /s=n.nikolov/ou1=s28a@mhs.attmail.com

## Model documentation and major references
Nikolov, N.T., and Fox, D.G., 1994. A coupled carbon-water-energy-vegetation model to assess responses of temperate forest ecosystems to changes in climate and atmospheric $CO_2$. Part I. Model concept. *Environmental Pollution*, 83: 251-262.

## Other relevant information
The TEMFES model is currently under development. It will be available for general use in 1996.

## Model identification
Model name/title: Model of tree growth

## Model type
Media:
Class: biogeochemical
Substance(s)/compounds: nutrients, carbon
Biological components: organism - tree

## Model purpose
Tree growth.

## Short description
Carbon balances of a tree.
State variables: basal area, length of tree, woody volume, total dry matter.

## Model application(s) and State of development
Application area(s):
Number of cases studied: several
State: conceptualization, verfication, calibration, validation

## Model software and hardware demands
Platform: PC-based
OS:
Software demands:
Hardware demands:

## Model availability
Purchase: no.
Contact person(s) Dr. H.T. Valentine
　　　　　　　USDA- Forest Service
　　　　　　　Center for Biologic Control
　　　　　　　51 Mill Pond Road,
　　　　　　　Hamden, CT 06514, USA
Phone:
Fax:
E-mail:

## Model documentation and major references
Valentine, H.T., 1985. Tree-growth model. *J. Theoretical Biology*, 117: 579-585.

## Other relevant information

## Model identification
Model name/title: Modelling the effect of environmental stresses on tree growth and development.

## Model type
Media: terrestrial, forest
Class: biogeochemical
Substance(s)/compounds:
Biological components:

## Model purpose
To model the effect of environmental stresses on tree growth and development.

## Short description
State variables for the tree are organized at the whole-plant level (i.e., canopy, stem and root). The canopy has a mass, a leaf area, and an age class of needles. The stem has a mass, a height, and a diameter at breast height (dbh). The mass of the stem is divided into heartwood and sapwood. The root has two classes (coarse and fine) and is distributed among soil horizons. All the masses of canopy, stem, and root have their stoichiometric contents of nutrients (Ca, Mg, K, C, N, P, etc.). The soil is divided into horizons. Each horizon has its mineral composition (including organic matter) and soil solution chemistry.

The model simulates a tree according to its genetic blueprint. The genetic blueprint is expressed by a series of equations that define the rates of various physiological processes, i.e., light interception, photosynthesis, respiration, evapotranspiration, nutrient uptake, water uptake, and the growth and mortality of plants parts (Chen and Gomez, 1988; Chen et al., 1987). These equations are assembled in a computer model and solved simultaneously by a suitable numerical method. The results are the time series of tree diameter, tree height, and biomass of needles, roots, and stem. These time series represent the progression of tree growth.

The effects of environmental stresses are defined by equations which modify the rates of the physiological processes given above. For example, ozone and water deficit, at the sublethal level, cause the closure of stomata, resulting in the reduction of $CO_2$ uptake and photosynthesis rate. The cumulative dose of ozone amplifies the mortality rate of needles and, therefore, the rate of litterfall. Likewise, the acidification of the soil raises the inorganic aluminum concentration at the root zone which, at sublethal concentration, reduces the cation uptake rate and, at lethal concentration, increases the root mortality rate. When the equations of growth physiology are solved together with the equations of stress physiology, the model generates the time series of tree responses that deviate from the normal tree growth. The magnitude of deviations represents the combined impacts of environmental stresses.

---

**Model application(s) and State of development**
Application area(s):
Number of cases studied:
State: conceptualization, verification, calibration, validation

---

**Model software and hardware demands**
Platform: PC-based
OS:
Software demands:
Hardware demands:

---

**Model availability**
Purchase: no.
Contact person(s) Dr. Chen and Mr. Gomez
        Systech Engineering, Inc.,
        3744 Mt. Diablo Boulevard
        Suite 101
        Lafayette, CA 94549 USA
Phone:
Fax:
E-mail:

---

**Model documentation and major references**

## Other relevant information

Chen, C.W., 1986. The development of the Plant- Growth- Stress Model 1: conceptual model and developmental plan. Report to the Electric Power Research Institute. EPRI RP- 2365, Systech Engineering, Inc., Lafayette, CA, USA.

Chen, C.W., Gherini, S.A., Hudson, R.J.M., and Dean, J.D., 1983. The Integrated Lake-Watershed Acidification Study, Vol. I. Model Principles and Application Procedures. EA-3221, Electric Power Research Institute, Palo Alto, CA, USA, 214 pp.

Chen, C.W., and Goldstein, R.A., 1985. Techniques for assessing ecosystem inpacts of air pollutants. In: Legge, A.H., and Krupa, S.V., Eds. *Air Pollutants and Their Effects on the Terrestrial Ecosystem*. John Wiley and Sons, Chichester, *England*, U.K., pp. 603-630.

Chen, C.W., and Gomez, L.E., 1987. Formulations of the Plant-Growth-Stress Model. Progress Report to the Southern California Edison Company and the National Council of the Paper Industry for Air and Stream Improvement. Systech Engineering, Inc., Lafayette, CA, USA, pp. 30.

Chen, C.W., and Gomez, L.E., 1988. Application of the ILWAS Model to Eastern Brook Lake Watershed in the Sierra Nevada Mountains. Reports to the Southern California Edison Company, West Associates, and the Electric Power Research Institute, Palo Alto, CA, USA.

Chen, C.W., Gomez, L.E., Fox, C.A., Goldstein, R.A., and Lucier, A.A., 1987. A tree model with multiple stresses. In: Ek, A.R, Shifley, S.R., and Burk, T.E., Eds., *Forest Growth*
, Vol. 1. General Technical Report NC-120, USDA  Forest Service North Central Forest Experiment Station, St. Paul, MN, USA, pp. 270-277.

## Model identification
Model name/title: FOREST- BGC

## Model type
Media: terrestrial
Class: biogeochemical
Substance(s)/compounds: carbon, nitrogen, water, soil, etc.
Biological components: ecosystem - terrestrial/stand-level

## Model purpose
Calculates carbon, water, and nitrogen cycles through a forest ecosystem. The model treats canopy interception, evaporation, transpiration, photosynthesis, growth and maintenance, respiration carbon allocation, litterfall, decomposition, and nitrogen mineralization in simulating hydrologic balance and primary production.

## Short description
No information available.

## Model application(s) and State of development
Application area(s):
Number of cases studied: Many
State: validation, prognosis made, validation of prognosis

## Model software and hardware demands
Platform: PC-based
OS:
Software demands: TURBO Pascal helpful, not required
Hardware demands: standard minimal

## Model availability
Purchase: no, free upon request. Software available both as executable and sourcecode
Contact person(s) Dr. Robert G. Kremer
        University of Montana
        School of Forestry
        Missoula, MT 59812 USA

Phone: +1 406-243-5616
Fax: +1 406-243-4511
E-mail: kremer@ntsg.umt.edu

## Model documentation and major references


## Other relevant information


## Model identification
Model name/title: The distance dependent forest growth model.

## Model type
Media: forest
Class:
Substance(s)/compounds:
Biological components: population

## Model purpose
The objective of the modeling effort were to evaluate the extent to which the spatial pattern formation may be affected by competition among neighbours, and to evaluate the possibility of spatial pattern intensity as a measure of population organization.

## Short description

The distance dependent forest growth model. Simulated plot contains a population. Each individual is considered separately, and is described by separate state variable (height, others are allometric function of height). Each individual roots in fixed point within the plot. Difference equations are used. Growth of each individual is described by 'logistic-like' function with added term describing distance and relative height dependent influence of neighbours. The probability of survival of an individual to the next step depends also on the neighbours competition, the survival or death is then decided by the Monte Carlo method. The model enables various initial height variabilities and various initial spatial pattern type (including a fexifile generator of aggregated pattern). Spatial pattern of population is evaluated in each step. New (unpublished) developments of the model lead to a multispecies version.

## Model application(s) and State of development

Application area(s): population biology, forest canopy
Number of cases studied: 1
State: verification

## Model software and hardware demands

Platform: IBM
OS: DOS
Software demands: language - BASIC
Hardware demands:

## Model availability

Purchase: no
Contact person(s) Dr. J. Leps
        Faculty of Biological Sciences
        Branisovska 31
        CZ-370 05 Ceske Budejovice
        Czech Republic

Phone: +42 38 817 ext. 658
Fax: +42 38 45985
E-mail: suspa@entu.cas.cz

## Model documentation and major references

Leps, J. and Kindlmann, P., 1987. Models of the development of spatial pattern of an even-aged plant population over time. *Ecol. Modelling*, 39: 45-57.

## Other relevant information

The model exists in several BASIC versions, which are in a form difficult to use by anyone other than the author and are permanently changing.

## 4.8. An Overview: Models of Other Terrestrial Ecosystems

The major modeling effort for terrestrial systems have not surprisingly been made in agriculture and forestry. There is, however, also a need for models of other terrestrial ecosystems: heather, grassland, tundra, alpine ecosystems, savanna and all forms of ecotones. The models presented here do not include the population dynamic models, which are covered in Chapter 7, but all other models of terrestrial ecosystems than agricultural and forest systems, that are already covered in Sections 4.4-4.7. As this section covers a wide range of different ecosystems under a wide range of climatic conditions, it is not surprising that the models presented generally show more differences than the models in Sections 4.4.-4.7. It makes it more difficult to find a model ready to use for a focal problem which implies that the presented models more should be considered as sources of inspiration than as an overview of directly available models. The models presented here supplied by a few slightly older models are summarized in Table 4.4.

The model Cellmatic is a biogeochemical models of sediment processes. It includes the following element cycles: carbon, nitrogen, sulfur and oxygen. The model can have any number of cells in one or two dimensions. Fluxes of chemical compounds between cells are governed by diffusion.

Goto et al. have developed a model which is able to predict the effect of global warming on the terrestrial ecosystems, first of all the carbon cycle and the distribution of vegetation. The terrestrial ecosystems are divided into four compartments: leaf, trunk, detritus and humus. Fluxes among the four compartment and the atmosphere are modeled. The terrestrial ecosystems are classified into five types: tropical forests, temperate forests, boreal forests, grassland and no vegetation. The carbon storage is calculated for a mesh of 1 latitude degree and 1 longitude degree for each of the five types of ecosystems.

The model named Melef solves in the vertical dimension the water flow, transport of heat and transport of several chemical compounds through the unsaturated, saturated, unfrozen and frozen zones of a considered soil.

The model denoted Gem is intended to predict the dynamics of primary producers, microbes, soil fauna and nitrogen availability and the effects of climate change on these dynamic processes. The state variables are soil water, live shoots, dead shoots, roots, bacteria, saprophytic fungi, mycorrhizal fungi, labile and resistant dead organic matter, consumers of bacteria, consumers of fungi, above and below ground herbivore, predators, ammonium and nitrate.

Polmod.hum is a model developed for simulation of humus dynamics under the influence of climate change. The model describes the humus dynamics in natural eluvial soil types under natural vegetation. The main factors (temperature, humidity, pH, etc.), determining the humus formation is taken into account.

The model named Smass computes water, solute and oxygen transport in one dimension. The model describes the chemical reactions in an acid sulfate profile divided into compartment: oxidation of pyrites and organic matter, precipitation, dissolution, adsorption, desorption and reduction.

The model Steppe describes successional dynamics of grassland populations and communities in response to environmental factors including climate change. It models the recruitment, growth and mortality of individual plants as a function of time. It keeps track of size and age of each plant. The model has recently been generalized to simulate short-grass, mixed grass and tall grass grasslands in semiarid regions.

Several models have been develop to cover the processes in soil. The model Soilheat5 estimates the influence of climate change and agricultural practice on soil heat regimes. It comprises four basic parts: dynamic model of heat transfer in soil, a model of applied agricultural technology, forecast of weather and forecast of the influence of plant cover on heat and mass transfer between soil and atmosphere.

Zaslavsky has developed a model for optimal control of soil water content. It provides the tool for simulation of soil water dynamics, using the following input information: soil texture, bulk density, organic matter content, water content at saturation, residual water content, field capacity, wilting point, hydraulic conductivity at saturation, wind speed, relative air humidity, net radiation, average daily temperature, rainfall of irrigation and LAI.

SoilFug is used to calculate concentrations in soil and adjacent surface water of neutral chemical on a basin scale. The model predicts concentrations of runoff water.

NCSoil describes carbon and nitrogen flows in soil. It distinguishes three different pools of organic carbon with different half life times. The model also accounts for the microbial succession to a certain extent.

**Table 4.4** Overview: Models of other Terrestrial Systems

| Model component(s)/ model name | Model characteristics: st. var. and/or main processes | References |
|---|---|---|
| Coast erosion | Eff. of tides, streams, waves and wind | Gagliardi, 1978 |
| Waste disposal | accumulation of waste, economic considerations | Friedman, 1978 |
| Cellmatic | Cycles of C, N, S and O in sediment | See model |
| C-cycle and vegetation distribution | leaf, trunk, detritus and humus, storage and C-fluxes | Goto et al., 1994 |
| Melef | Unsaturated zone vertical fluxes of water, heat and chem. compounds | Padilla et al., 1992 |
| Gem | Eff. change on dynamics of grass, microbes and fauna | Hunt, 1995 |
| Pol.mod.hum. | Humus dynamics by various land-use practice | Pykh et al., 1994 |
| Smass | Acidification and deacidification of acid sulfate soil | Van Wijk et al., 1994 |
| Steppe | Successional dynamics of grassland population, response to env. factors | Coffin et al., 1994 |
| Soilheat5 | Inf. of climate and agriculture on soil heat regime | Kurtener and Uskov, 1988 |
| Soil water control | Soil water flows, soil properties | See model |
| SoilFug | Predict conc. in soil and surface water | Di Guardo, 1994 |
| NCSoil | C + N flows in soil, 3 C-pools with diff. half life, microbial succession | Hada and Molina, 1993 |
| SoilN | Nitrogen processes and leaching | Eckersten et al., 1994 |
| Momo | Several organic carbon forms and dynamics | Sallih and Pansu, 1993 |
| Slim | Predicts effect of liming on terrestrial + aquatic systems as f (catchment area) | Dalziel et al., 1992 |

**Table 4.4. (continued)**

| Model component(s)/ model name | Model characteristics: st. var. and/or main processes | References |
|---|---|---|
| Spur2 | 15 plant species on 36 grassland sites primary production | Baker et al., 1993 |
| Tlaloc | Dynamics of vegetation mosaic | Miller, 1981 |
| Energy flows | Energy flows and storages of a prairie | Jeffries, 1989 |

SoilN focuses on nitrogen processes and nitrogen leaching from agricultural land and forest. The model covers a detailed description of the nitrogen processes in soil.

The model Soil simulates water and heat processes in many types of soils: bare soils and soils covered by vegetation. The structure of the model is a depth profile of the soil.

Momo is looks into the various carbon forms in soil. It is able to account for several carbon forms as state variables in the model.

The purpose of the model Slim is to predict the effect of soil liming on terrestrial and aquatic ecosystems as function of catchment characteristics (hydrology, hydrochemistry and morphology), soil characteristics (CEC, density) and forcing functions (climate, application rate of lime, properties of the applied lime)

Spur 2 is a grassland model which is able to simulates the production of up to 15 species on 36 heterogeneous grassland sites. It includes grassland hydrology, nitrogen cycles and soil organic matter on grazed ecosystems. It is aimed for simulation of the primary production under different climatic regimes, environmental conditions and management alternatives.

The model named Tlaloc simulates the dynamics of a vegetation mosaic in arid or semiarid regions. It has two groups of state variables: the soil and the plants. It simulates the water relations in a simple functional unit and analyzes the central role of water in the system dynamics.

The ABC model is a mechanistic model of plant production, organic matter decomposition and nitrogen cycling in a tundra ecosystem.

Jeffries presents in his book, *Mathematical Modeling in Ecology*, 1989, an energy flow model for a short-grass prairie ecosystem. The model accounts for the energy of green shoot, three species of grasshoppers and sparrows at various stages from eggs to adults.

## 4.9 Models of Other Terrestrial Ecosystems

---

**Model identification**
Model name/title: Cellmatic

---

**Model type**
Media: sediment
Class: biogeochemical
Substance(s)/compounds: nutrients, carbon, nitrogen, sulfur, oxygen, organic compounds
Biological components: ecosystem, sediment

---

**Model purpose**
To model interrelationships between carbon, nitrogen, sulfur and oxygen cycles in sediments.

---

**Short description**
The model is based on system dynamics principles. It contains the following agents (stocks): POC, DOC, $CO_2$, PON, DON, $NH_4$, $NO_3$, $N_2$, $SO_4$, HS, porosity, variable cell height. The model can have any number of cells, in one or two dimensions. Movement between cells is by diffusion.

---

**Model application(s) and State of development**
Application area(s): Sediment diagnosis
Number of cases studied: 3
State: conceptualization, verification, calibration, validation, prognosis made, validation of prognosis.

---

**Model software and hardware demands**
Platform: MAC
OS: Hypercard
Software demands:
Hardware demands:

---

**Model availability**
Purchase: yes, no charge
Contact person(s) Dr. T. H. Blackburn
        Dept. Microbial Ecology
        Institut of Biological Science
        Aarhus Universitet
        DK-8000 Aarhus C, Denmark

Phone: +45 89 42 32 46
Fax: + 45 86 12 71 91
E-mail: Henry(@)weinberg.pop.bio.aau.dk

---

**Model documentation and major references**

---

**Other relevant information**

---

**Model identification**

Model name/title: Carbon Cycle Model between Atmosphere and Terrestrial Ecosystems.

---

**Model type**

Media: global scale, terrestrial, soil, forest

Class:

Substance(s)/compounds: carbon

Biological components: ecosystems, terrestrial ecosystems (forest, grassland, etc.)

---

**Model purpose**

Model prupose is to predict the effects of global warming on the terrestrial ecosystems, for example, carbon cycle and vegetation distribution.

---

**Short description**

This model divides the terrestrial ecosystems into four compartments, leaf, trunk, dead biomass and humus. Fluxes among the four compartments and the atmosphere are modeled by mathematical equations. This model can calculate the carbon storages of each compartment and the carbon fluxes among them. The terrestrial ecosystems are classified into 5 types: tropical forest, temperate forest, boreal forest, grassland and no vegetation. The world land is divided into some meshes (latitude 1° x longitude 1°), and the carbon storages in three vegetations, tropical, temperate and boreal forest, are calculated in each mesh. The calculation needs temperature, precipitation, solar radiation, soil distribution and so on in each mesh. In comparison with the carbon storage in the three and in every mesh, we assume that the vegetation having the maximum carbon storage is the vegetation domination the mesh. When the carbon storage in the resultant dominant vegetation is less than 0.5 $(kgC/m^2)$, the dominant vegetation is replaced by no vegetation. Also, when it is still less than 5 $(kgC/m^2)$, it is considered to be grassland. Then we can get the global distributions of not only carbon storage, but also vegetation. Furthermore, by using future climate conditions derived from some GCMs, the changes of the distributions can be predicted.

---

**Model application(s) and State of development**

Application area(s): Global environmental problems (greenhouse effect)

Number of cases studie: 1

State: prognosis made

## Model software and hardware demands
Platform: Mainframe, Work-station, SUN
OS: Solaris
Software demands: SPARC FORTRAN, FORTRAN
Hardware demands: SPARC station

## Model availability
Purchase: no, executable code
Contact person(s) Dr. Naohiro Goto
Inst. of Industrial Science
University of Tokyo
7-22-1 Roppongi, Minato-ku
Tokyo 106
Japan

Phone: 81-3-3402-6231 ext. 2413
Fax: 81-3-3408-0593
E-mail:

## Model documentation and major references
M. Suzuki, N. Goto and A. Sakoda. Simplified dynamic model on carbon exchange between atmosphere and terrestrial ecosystems. *Ecol. Modelling*, 1994, in press.
N. Goto, A. Sakoda and M. Suzuki. Modelling of soil carbon dynamics as a part of carbon cycle in terrestrial ecosystems. *Ecol. Modellling*, 1994, in press.

## Other relevant information
My address will change from April or May.
Research Institute of Innovative Technology for the Earth (RITE)
9-2, Kizugawadai, Kizu-cho, Sorak-gun. Kyoto 619-02 Japan
tel. 81-7747-5-2304
fax 81-7747-5-2317

## Model identification
Model name/title: MELEF

## Model type
Media: water, air, soil, ice
Class: hydrology
Substance(s)/compounds: nitrogen, pesticides, 2-3 compounds, organic compounds, dissolved compounds
Biological components: organism, biodegradation

## Model purpose
MELEF is a scientific and academic conceptual model. This model has been conceived in order to be easily adapted for application to different problems occurring in the unsaturated zone of the soil.

## Short description
MELEF solves in the vertical dimension the water flow as well as the transport of heat and several chemical compounds through the unsaturated, saturated, unfrozen or frozen zones of the soils. The model had been applied to different mathematical and environmental problems.

In the application of the finite element method to diffusion and convection-dispersion equations over a groundwater domain, the Galerkin technique has been used to incorporate Dirichlet, Neuman, Cauchy and Open boundary conditions. Theoretical and numerical proofs are given about the accuracy of the model in numerous transport and diffusion conditions.

The numerical model has been mainly applied to organic compounds (pesticides and nitrogen) and it considers different transformations in the soil: adsorption, degradation, nitrification, denitrification, as well as immobilization and mineralization between species (ammonium and nitrate) and organic-N. Based upon thermodynamical and experimental relationships, the model considers water uptake by plant roots as well as the effect of temperature and water content on the equilibrium constants of reaction and on the transformation rates in the soil.

The behaviour of frost-susceptible soils is analyzed using the Clapeyron equation and the discontinuity concept of the Stefan problem in order to solve for water quality, phase changes, frost heaving and the different conditions of ice in the soil.

## Model application(s) and State of development
Application area(s): Hydrogeology, soil unsaturated zone, Nordic hydrology, geotechnics, contaminant hydrogeology, frost heave of roads, organic compounds fate, agriculture
Number of cases studied: 4-5
State: conceptualization, verification, calibration, validation

## Model software and hardware demands
Platform: PC-based, IBM
OS: MS-DOS
Software demands: Fortran compiler, FORTRAN and C
Hardware demands: PC-386, PC-486, etc.

## Model availability
Purchase: yes, $ 200 (cost of preparation and mail delivery), executable code sourcecode
Contact person(s) Dr. Francisco Padilla
        Universite du Quebec
        INRS-Eau
        2700 rue Einstein
        Quebec, Canada G1V 4C7

Phone: +1 418-654-2598
Fax: +1 418-654-2600
E-mail: paco-padilla@inrs-eau.uquebec.ca

## Model documentation and major references

Padilla, F. and J.P. Villeneuve (1992). Modeling and experimental studies of frost heave including solute effects. *Cold Regions Science and Technology*, 20:183- 194.

Padilla, F., Leclerc, M. and J.P. Villeneuve (1990). A formal finite element approach for open boundaries in transport and diffusion ground water problems. *International Journal Numerical Methods in Fluids*, 11:287-301.

Padilla, F., Lafrance P., Robert, C. and J.P. Villeneuve (1988). Modeling the transport and the fate of pesticides in the unsaturated zone considering temperature effects. *Ecol. Modelling*, 44: 73-88.

Padilla, F., Camara, O. and D. Cluis (1992). Modeling nitrogen species transformations and transport in the unsaturated zone considering temperature and water content effects. HYDROSOFT 92, Universidad Polytécnica, Valencia, Spain 21-23 July 1992: 15-26.

## Other relevant information

The program which runs the MELEF model is at research stage and it is still being polished for practical applications. We are presently working on the adaptation of the two-dimensional model to the above mentioned and other types of environmental problems. At the present time, we do not have another useful documentation or user's manual.

## Model identification

Model name/title: GEM

## Model type

Media: local scale, terrestrial, soil, agricultural
Class: biogeochemical, hydrology
Substance(s)/compounds: carbon, nitrogen
Biological components: ecosystem - grassland, cultivated grain

## Model purpose

The model is intended to represent the physiological ecology of grass plants, microbes, and fauna. The objectives are to predict dynamics of primary producers, microbes, soil fauna and nitrogen availability, and the effects of climate change on these dynamics.

## Short description

The model includes mechanism necessary to represent (1) the effects of temperature, water and $CO_2$ on primary production and decomposition, and (2) nutrient cycling feedbacks involving substarte quality and nitrogen mineralization.

State variables include soil water, live shoots, dead shoots, roots, bacteria, saprophytic fungi, mycorrhizal fungi, ammonium and nitrate. Organic residues are divided into the labile and resistant portions of fresh material, and into two fractions of soil organic matter with turnover times of 50 and 1000 years.

Trophic structure of the fauna recognizes consumers of bacteria, consumers of fungi, above ground herbivores, below ground herbivores, and predators.

The most important driving variables are daily precipitation and max/min air temperatures. Additional driving variables are soil temperatures, wind speed and relative humidity. Adaption of the model to a new grassland ecosystem requires information about soil texture and fertility. The plant model can be adapted to a new graas species using data on the seasonal growth dynamics or data on the responses of primary production to treatments such as precipitation or elevated $CO_2$.

### Model application(s) and State of development
Application area(s): Effects of elevated $CO_2$ and climate change
Number of cases studied: 4
State: conceptualization, verification, calibration, validation

### Model software and hardware demands
Platform: Workstation SUN, Adaptible to PC
OS: UNIX
Software demands: none, FORTRAN
Hardware demands:

### Model availability
Purchase: no, source code
Contact person(s) Dr. H.W. Hunt
              Natural Resources Ecology Laboratory
              Colorado State University
              Fort Collins, CO 80523 USA

Phone: +1 303-491-1985
Fax: +1 303-491-1965
E-mail: billh@poa.NREL.Colostate.EDU

### Model documentation and major references
*Ecol. Modelling*, 53: 205-246. ms. in prep. Full description of model equations available on request.

### Other relevant information

### Model identification
Model name/title: POLMOD.RAD.: model for simulation of radionuclides' dynamics in the ecosystem

## Model type
Media: local scale
Class: toxicology
Substance(s)/compounds: radionuclides
Biological components: ecosystem - elementary ecosystem

## Model purpose
Model is intended to describe the flow of radionuclides in the units of ecosystem -- atmosphere, soil, vegetation and surface water. Model calculates the level of radioactive pollution accumulated in each unit of the elementary ecosystem.

## Short description
POLMOD.RAD is a software product for simulation the radionuclides in the units of elementary ecosystem. The main term of POLMOD.RAD model is the resistance index. We define the resistance index of ecosystem's compartments as the index of ability to resist the pollutions' flow either due to self-purification ability or due to the decrease of the accumulation rate. POLMOD.RAD consists of four submodels in accordance with the main ecosystems' units - atmosphere, soil, vegetation and surface water submodels.

Major state variables are the radionuclide's concentrations in the units of ecosystem enumerated above. Time step equals one year.

Model input includes the date on meteorological, physical and chemical properties of each unit of the ecosystem: precipitations, probability of temperature inversions in the atmosphere, wind speed, pH (soil acidity), soil humus content, cation exchange capacity of soils, calcium cation content, as well as the area of drainage basin, illumination, calcium ions concentration in water and water temperature.

## Model application(s) and State of development
Application area(s): Scientific investigations of pollutants' dynamics in the ecosystems; education and training decisions support system for various kinds of environmental problems.
Number of cases studied: several hundred
State: validation, validation of predictions

## Model software and hardware demands
Platform: PC-based IBM PC 286 (386)
OS: MS DOS version 3.2 and higher
Software demands: Turbo Pascal compiler, Turbo Pascal version 5.0
Hardware demands: 640 k RAM, EGA graphic adapter, math coprocessor

## Model availability
Purchase: yes, available as source code
Contact person(s) Dr. Irina G. Malkina-Pykh
                Center for International Environmental Cooperation (INENCO)
                St. Petersburg, 190000
                Chernomorsky per., 4
                Russia

Phone: 812 311 7622
Fax: 812 311 8523
E-mail: pykh@inenco.spb.su

---

## Model documentation and major references

Pykh, Yu., and Malkina-Pykh, I.G.. The method of response functions in ecology. *J. Int. Biometerorol.*, 1991, 35: 239-251.
Pykh, Yu.A., and Malkina-Pykh, I.G.. POLMOD (version 1.0) - the model of pollutants' dynamics in the elementary ecosystem. Preprint, Moscow, Inst. of Systems Studies, 1992.

Iouri Pykh
International Institute for Applied System Analysis
A- 2361 Laxenburg, Austria
Email: pykh@iiasa.ac.at
phone: +43 2236 71521 0
fax: +43 2236 71313

---

## Other relevant information

---

## Model identification
Model name/title: SMASS: a Simulation Model for Acid Sulphate Soils

---

## Model type
Media: soil
Class: biogeochemical
Substance(s)/compunds: oxygen, sulfur, acidity, all major cations and anions (K, Na, Mg, Ca, Fe, Al, $HCO_3$, Cl, $SO_4$)
Biological components: none

---

## Model purpose
To predict effects of water management (e.g., irrigation, drainage) on acidification and de-acidification of acid sulfate soils.

---

## Short description
The SMASS model computes one-dimensional vertical water, solute and oxygen transport, pyrite oxidation and chemical reactions in an acid sulphate soil profile. To carry out the numerical computations, the vertical soil profile is divided into compartments.

The vertical water and solute transport in the soil profile is computed according to the existing SWAP model, depending on the initial conditions in the various compartments, and the boundary conditions. This yields the water content and solute concentration profile in the soil. The air contents in the soil at various depths are complementary to the water contents. From the air content profile, oxygen diffusion coefficients are calculated. The oxygen consumption is derived from the pyrite and organic matter contents. The oxygen diffusion and consumption together determine the oxygen concentration profile in the soil.

The rate of pyrite oxidation at a given depth is calculated depending on the oxygen concentration at that depth. The oxidized amount of pyrite gives the amount of H+, $Fe^{3+}$ and $SO_4^{2-}$ released into the soil solution. The remaining amount of pyrite in the soil is used for calculations in the next time step.

As a result of pyrite oxidation and solute transport a number of chemical processes will occur in each soil layer: precipitation-dissolution, oxidation-reduction, adsorption-desorption. At the end of one model time step these processes are all computed. This results in computed concentrations of chemical compounds in the soil solution, composition of exchange complex, amount of precipitates, and the concentration of elements in the flux to the groundwater.

Time steps for model simulations are in the order of hours. The output of the SMASS model and its submodels is generally given on a daily basis. Model predictions can be done over periods of tens of years, so that long-term effects of various water management strategies can be evaluated quantitatively.

---

**Model application(s) and State of development**
Application area(s): Development of ecologically sound water management strategies in coastal lowlands in Kalimantan, Indonesia
Number of cases studied: 2
State: validation, prognosis made, no validation of prognosis due to long-term predictions.

---

**Model software and hardware demands**
Platform: Mainframe, (PC under development)
OS: VAX-VMS
Software demands: language FORTRAN 77
Hardware demands:

---

**Model availability**
Purchase: yes, price US $300, available as executable code
Contact person(s) Dr. K. Oostindie
                DLO-Winand Staring Centre
                P.O. Box 125
                NL-6700 AA Wageningen, The netherlands

Phone: 31 8370 74200
Fax: 31 8370 24812
E-mail: K.Oostindie@SC.AGRO.NL

---

**Model documentation and major references**
AARD/LAWOO, 1993. Research on acid sulphate soils in the humid tropics. Volume 4: A simulation model to redict effects of water management strategies. ILRI, Wageningen, The Netherlands, 242 pp.
Bronswijk, J.J.B., Groenenberg, J.E., 1993. SMASS: A Simulation Model for Acid Sulfate Soils, I: Basic principles. In: Dent, D., and van Mensvoort, M.E.F. (Eds): Selected Papers of the Ho Chi Minh City Symposium on Acid Sulphate Soils. ILRI Publication, 53, 341-355, International Institute for Land Reclamation and Improvement, Wageningen, The Netherlands.

Bronswijk, J.J.B., Groenenberg, C.J., Ritsema, van Wijk, A.L.M., Nugroho, K., 1993. Evaluation of water management options for acid sulphate soils using a simulation model: A case study in Indonesia. *Agricultural Wat. Manag.* (in press).

Bronswijk, J.J.B., Nugroho, K., Aribawa, I.B., Groeneneberg, J.E., Ritsema, C.J., 1993. Modeling of oxygen transport and pyrite oxidation in acid sulphate soils. *J. Env. Qual.* 22: 544-554.

Bronswijk, J.J.B., van den Bosch, H., Nugroho, K., Groenenberg, J.E., Ritsema, C.J., 1994. SMASS Users manual. SC report. DLO Winand Staring Centre, Wageningen, The Netherlands.

Van Wijk, A.L.M., Gedjer Widjaja-Adhi, I.P., Ritsema, C.J., Konsten, C.J.M., 1993. SMASS: A simulation model for acid sulphate soils II: Validation and application in evaluation of water management strategies. In: Dent, D., van Mensvoort, M.E.F. (Eds): Selected Papers of the Ho Chi Minh City Symposium on Acid Sulphate Soils. ILRI Publication 53, 357- 367, International Institute for Land Reclamation and Improvement, Wageningen, The Netherlands.

Van Wijk, A.L.M., Bronswijk, J.J.B., Groenenberg, J.E., Ritsema, C.J., 1994. Model of physical and chemical processes in acid sulphate soils: principles validation and application. Proceedings ISSS Congress, Acapulco, Mexico, July 1994.

**Other relevant information**

**Model identification**
Model name/title: STEPPE

**Model type**
Media: grassland
Class: vegetation
Substance(s)/compounds:
Biological components: community - grasslands in central U.S. in particular arid and semi-arid grasslands

**Model purpose**
To evaluate successional dynamics of grasslands populations and communities, the response of grasslands to environmental factors, including climate change and disturbance.

**Short description**
The STEPPE model simulates the recruitment, growth, and mortality of individual plants on a small plot (0.1 m²) through time at an annual time step. Size and age of each plant are kept track of through time. Driving variables are precipitation, temperature, soil texture, and disturbance. The STEPPE model is currently linked to a daily time step, multi-layer soil water model. STEPPE simulated vegetation dynamics and the soil water model simulates soil water dynamics. Information is passed between the models on the appropriate temporal and spatial scales either by aggregation or disaggregation. Required inputs are daily precipitation, daily minimum and maximum temperatures, disturbance sizes and

frequencies of occurrence, silt texture (% sand, silt and clay) by depth in the soil profile, monthly relative humidity, cloud cover and wind speed, and species-specific life history traits associated with the recruitment, growth, and mortality of individual plants. Recruitment and mortality have stochastic features, whereas growth is determined by competition for below ground resources, in particular soil water. The model was originally developed for semiarid grasslands in northcentral Colorado, U.S.A.; it has been applied to dry alpine grasslands of the Rocky Mountains and the desert grasslands in New Mexico. Recently the model was generalized to simulated shortgrass, mixed grass and tall grass grasslands in the central region of the U.S. This model includes competition for light and nitrogen in addition to water.

## Model application(s) and State of development
Application area(s): Three (alpine grassland, desert grassland, shortgrass steppe)
Number of cases studied: three
State: validation

## Model software and hardware demands
Platform: Workstation SUN
OS:
Software demands: language FORTRAN
Hardware demands:

## Model availability
Purchase: no, available as executable code
Contact person(s) Dr. Debra Coffin
                 Rangelnad Ecosystem Science Dept.
                 Colorado State University
                 Fort Collins, CO 80523 USA

Phone: +1 303-491-7662
Fax: +1 303-491-2156
E-mail: deb@aristida.criv.colostate.edu

## Model documentation and major references
Coffin, D.P., and Lauenroth, W.K., 1989. Disturbances and gap dynamics in a semiarid grassland: A landscape-level approach. *Landscape Ecology* 3(1): 19-27.
Coffin, D.P., and Lauenroth, W.K., 1990. A gap dynamics simulation model of succession in the short grass steppe. *Ecol. Modelling* 49: 229-266.
Lauenroth, W.K., Urban, D.L., Coffin, D.P., Parton, W.J., Shugart, H.H., Kirchner, T.B., Smith, T.M., 1993. Modeling vegetation structure-ecosystem process interactions across sites and ecosystems. *Ecol. Modelling* 67: 49-80.
Coffin, D.P., and Lauenroth, W.K., 1994. Successional dynamics of a semiarid grassland: Effects of soil texture and disturbance size. *Vegetation* 110: 67-82.
Coffin, D.P., Lauenroth, W.K., and Burke, I.C., 1994. Spatial dynamics in recovery of shortgrass steppe ecosystems. In: R. Gardner, Ed. Theoretical Approaches to Predicting Spatial Effects in Ecological Systems. *Lectures on Mathematics in the Life Sciences* 23: 75-108.
Humphries, H.C., Coffin, D.P., and Lauenroth, W.K. An individual-based model of alpine plant distributions. *Ecol. Modelling* (in press).

Coffin, D.P., Lauenroth, W.K.. Regional analysis of transient responses of grasslands to climate change. *Climate Change* (in press).

## Other relevant information
A general version of the model is being developed and verified for all grasslands in the central U.S.

Coffin, D.P., and Lauenroth, W.K.. Regional analysis of transient responses of grasslands to climate change. *Climate Change* (in press).

## Model identification
Model name/title: SOILHEAT5 - model for estimate infuence of climate change and agricultural technology on soil heat regime.

## Model type
Media: soil
Class: biogeochemical
Substance(s)/compounds: chemicals
Biological components: ecosystems agro-ecosystem

## Model purpose
Model for estimate influence of climate change and agricultural technology on soil heat regime. Model identification and the estimation of the parameters of weather generator is oriented on the use of widespread information and does not require the carrying out of special field experiments.

## Short description
SOILHEAT5 is a software products for simulation of soil heat regime under crop covers of several cultures as influenced by variable soil and weather conditions and parameters of agricultural technology. SOILHEAT5 comprises four basic parts:
- dynamic model of heat transfer in soil;
- applied model of agricultural technology;
- algorithms of the forecast of weather;
- algorithms of the forecast of influence of plant cover on process of heat and mass transfer between soil and atmosphere.

Model identification and the estimation of the parameters of weather generator is oriented on the use of widespread information and does not require the carrying out of special field experiments. Major state variables are soil water and heat content, soil density and thermal conductivity, soil surface albedo, leaf area index, root depth, air temperature, air humidity, precipitation, global radiation and wind speed.

## Model application(s) and State of development
Application area(s):Scientific investigation of agroecosystem dynamics, education and training, decision support in agriculture.
Number of cases studied: several thousand
State: validation, validation of prognosis

**Model software and hardware demands**
Platform: PC-based, IBM PC/AT-286 (386);
OS: MS DOS, version 3.2 and higher;
Software demands: Turbo - Basical compiler;
Hardware demands: 640K RAM, EGA graphics adapter, math coprocessor; Turbo - Basic, version 3.0

**Model availability**
Purchase: yes, software available as sourcecode.
Contact person(s) Dmitry Kurtener
        14 Grazhdansky Prospect
        Agrophysical Research Institute Laboratory of Agroclimat Simulation
        195220, St. Petersburg, Russia

Phone: (812) 534-46-30
Fax:   (812) 534-16-44
E-mail: ivl agrophys.spb.su

**Model documentation and major references**

**Other relevant information**

**Model identification**
Model name/title: Optimal Control of Soil Water Content

**Model type**
Media: agricultural
Class: hydrology
Substance(s)/compounds: none
Biological components: ecosystem - cultivated plans

**Model purpose**
The dialog system provides the tools for the simulation of soil water dynamics and optimal control of soil water content. The system uses the following input information: soil texture, bulk density, organic matter content, saturation water content, residual water content, field capacity, wilting point, saturation hydraulic conductivity, average daily temperature, wind speed, relative air humidity, net radiation above the canopy, rainfall or irrigation, and leaf area index. A user can calculate optimal strategy and compare alternative strategies of irrigation for the various weather scenarios.

**Short description**
The modeling of soil water flux is supported by the module generating the soil water retention curve and hydraulic conductivity curve. The other module of the system generates the dynamic models of the soil water content. The module has two blocks. The first one provides the numerical solution af the partial differential equation of soil water flow. The second block uses the minimal realization method to generate linear models of soil water flow. The microclimate block transforms the weather conditions scenarios into the transpiration and evaporation from the soil surface. The output of the microclimate block is an input of the linear models simulating soil water content. The special module of the system calculates optimal regimes of irrigation for specified weather conditions.

**Model application(s) and State of development**
Application area(s): agricultural, forecasting and control
Number of cases studie:
State: prognosis made

**Model software and hardware demands**
Platform: PC-based
OS: MS-DOS version 3.0 and higher
Software demands: 400 KB free RAM, Turbo Pascal
Hardware demands: 640 KB RAM, EGA/VGA display

**Model availability**
Purchase: yes for $1000.00, Software is available as executable code
Contact person(s) Dr. B.G. Zaslavsky
                21, Drezdenskaja k 144
                St. Petersburg, 194017
                Russia

Phone: 7 (812) 55277 16
Fax:
E-mail: andrew@boris.spb.su

**Model documentation and major references**
Hanks, R.J., and Ashcroft, G.L., 1980. *Applied Soil Physics. Soil Water and Temperature Application.* Springer-Verlag, Berlin.
Zaslavsky, B.G., Oparina I.V. and Terlejev V.V., 1988. Bank of soil hydrophysical characteristics of soil. *Reports Soviet Agricu. Acad. Sci.*, 11, 40-42.
Thompson, L.M. and Troeh, F.R., *Soil and Soil Fertility.* McGraw-Hill, 4th ed.
Zaslavsky, B.G. and Zaharov, M. Yu., 1989. The Software Package of Applied Models of Soil Water Dynamics. *Bull. Agrophysical Research Institute,* 76, 9-14.

**Other relevant information**

## Model identification
Model name/title: SoilFug: A Fugacity Model for Chemical Runoff in Agricultural Basins Version 1.0 (February 1994) Written by Antonio Di Guardo, Davide Calamari (University of Milan, Italy) and Donald Mackay (University of Toronto, Canada).

## Model type
Media: soil, agricultural, river, local scale, regional scale
Class: biogeochemical
Substance(s)/compounds: neutral pesticides, and neutral organic chemicals
Biological components:

## Model purpose
The model can be used to calculate predicted environmental concentrations (PEC) in soil and in surface water of neutral chemicals on a basin scale. The predicted concentrations of runoff water generally lie within an order of magnitude of observed mean values, being the predictions generally useful for risk assessment purposes.

## Short description
The SoilFug model is an unsteady-state, but equilibrium model, based on the fugacity approach. The equilibrium events are the rain event simulations, which are periods of time initiated by a rainfall and ending when the water level in the adjacent streams returns to the background conditions. During the rain events, partitioning calculations are performed, while the unsteady-state term is referred to the calculations of the dissipation of the pesticide following its application, mainly due to degradation reactions, volatilization and runoff/leaching. In short, this model calculates the partition and the concentrations in different media (soil air, soil water, organic matter, mineral matter and outflowing water) of the chemical applied and the amount of chemical lost by means of volatilization, degradation and runoff phenomena at different times, and the rain events. In particular, it utilizes the available data for rainfall and outflowing water for the calculation of the amount of chemical lost at each rain event. The total amount of water outflowing from the basin is then utilized for the calculation of the concentration of the chemical in water at the outlet. The program can perform the calculations for one application at a time or for multiple applications (in different fields); the outflowing water concentrations are calculated as sum of the respective contributions of the different treated fields. The input of the model are the physico-chemical properties of the chemical (molecular weight, vapour pressure, water solubility, log Kow or Koc, mineral/water partition coefficient, half-life in soil), the soil properties (temperatures, soil depth, volume fraction of air and water in soil at field capacity, organic carbon fraction), the water data (number of rain events, their duration, the dry period between them, the amount of rainfall and outflowing water) and the area and application data (total area of the basin, the area(s) treated, the dosage(s)). It is also possible to input measured water concentrations, to be compared with the calculated results.

## Model application(s) and State of development

Application area(s): The model has been applied to four basins, of different sizes and properties: two areas in Northern Italy (3 km² and 17 km²), one in Southern England (1.5 km²) and one in Chile (106 km²), confirming the accuracy of the predictions usually within an order of magnitude.
Number of cases studied:
State: validation

## Model software and hardware demands

Platform: PC-based, running in Windows
OS: MS-DOS 3.0 and up
Software demands: Windows 3.1x, language not required; it is a functional compiled Windows program
Hardware demands: 386 coprocessor or up, 4 Mb of RAM, 1 MB of hard disk space.

## Model availability

Purchase: The model is available at no charge, except mail expenses (US $20)
Contact person(s) Dr. Antonio Di Guardo, Prof. Davide Calamari
> Institute of Agricultural Entomology
> University of Milan
> Via Celoria 2
> I-20133 Milano MI, Italy

Phone: +39 2 236 2880   or   +39 2 236 3439
Fax: +39 2 266 80320
E-mail: entom@imiucca.csi.unimi.it

## Model documentation and major references

Di Guardo, A., 1994a. User's Manual of SoilFug, version 1.0.
Di Guardo, A., Calamari, D., Zanin, G., Consalter, A., Mackay, D., 1994b. Fugacity Model of pesticide runoff to surface water: development and validation, *Chemosphere*, 28, 511-531.
Di Guardo, A., Williams, R.J., Matthiessen, P., Brooke, D.N., Calamari, D., 1994c. Simulation of pesticide runoff at Rosemaund Farm (UK) Using the SoilFug model, *Environmental Science and Pollution Research*, in press.
Barra, R., Vighi, M., and Di Guardo, A., 1994d. Prediction of runoff of chloridazon and chlopyrifos in an agricultural watershed in Chile. Submitted to *Chemosphere*.

## Other relevant information

## Model identification

Model name/title: NCSOIL - Nitrogen and Carbon Transformation in Soil

## Model type
Media: soil
Class: biogeochemical
Substance(s)/compounds: carbon, nitrogen, organic compounds, $NH_4^+$, $NO_3^-$, $CO_2$, $N_2$ with tracer C and N, Biological components: ecosystem - soil, organism - microbial biomass

## Model purpose
The process model NCSOIL describes C and N flows in soil.

## Short description
Carbon from chemicals added to soil is stabilized into two distinct organic pools: the more reactive Pool II with a half-life of 115 days and the more stable Pool III with half-life of 190 years. The decomposition of organic substrate, Pool II, and Pool III is microbially mediated. The primary decomposers utilize organic chemicals as energy sources. They are cannibalized by other microbes, which are in turn attacked by other microbes, forming a process from which microbial successions emerge. As carbon flows through the microbial successions, a fraction of the organic carbon goes in the bioreactive Pool II. Further stabilization also occurs with the channeling of a fraction of the decomposition products from Pool II to the more stable Pool III.

## Model application(s) and State of development
Application area(s): $CO_2$ exchange on the soil/air interface. Inorganic N availability to plants.
Number of cases studied: 11
State: validation, prognosis made, validation of prognosis

## Model software and hardware demands
Platform: PC-based
OS: DOS
Software demands: Compiler-linker ("PROFORTRAN, IBM"), language - FORTRAN
Hardware demands: Math co-processor NOT needed

## Model availability
Purchase: yes, at no charge
Contact person(s) Dr. J.A.E. Molina
        University of Minnesota
        439 Borlaug Hall
        1991 Upper Buford Circle
        St. Paul, MN 55108-6028
        USA

Phone: +1 612-625-1244
Fax: +1 612-625-2208
E-mail: gvb6082@vx.cis.umn.edu

**Model documentation and major references**
Molina, et al., 1983. *Soil Sci. Soc. Am. J.*, 47: 85-91.
Houet, et al., 1989. *Soil Sci. Soc. Am. J.*, 53: 451-453.
Barak, et al., 1990. *Ecol. Modelling*, 51: 251-263.
Hadas and Molina, 1993. *Physiol. Plantar.*, 87: 528-534.
Nicolardot et al., 1994. *Soil Biol. Biochem.*, 26: 235-243.

**Other relevant information**
NCSOIL is a stand alone program. It is also a subroutine in the model NCSWAP where it is integrated with water dynamics plant growth; managerial and climatic driving variables.

**Model identification**
Model name/title: SOILN

**Model type**
Media: soil, forest, agricultural, local scale, regional scale, water, air
Class: hydrology
Substance(s)/compounds: nitrogen, carbon
Biological components: ecosystem - plant/soil

**Model purpose**
To quantify and increase the understanding concerning nitrogen processes and nitrogen leaching from agricultural land and forest.

**Short description**
This model simulates the carbon (or biomass) and nitrogen flows both in the soil and the crops of an agricultural land. The soil mineral nitrogen is located in the pools of ammonium nitrogen and nitrate nitrogen. The ammonium pool is increased by nitrogen supplied from the mineralization of litter and humus and atmospheric deposition and decreases by immobilization to litter, nitrification to the nitrate pool and plant uptake. The nitrate pool is increased through nitrification of the ammonium pool, fertilization and atmospheric deposition, and is decreased by leaching, denitrification and plant uptake. It is also affected by the vertical redistribution of nitrogen due to water flows between layers. All these processes depend on soil water and heat conditions and give the amount of nitrate and ammonium in different soil layers readily available for plant uptake. The soil is divided into horizontal layers from which nitrogen is taken at various rates, depending on plant demand and root uptake efficiency. The potential root uptake in a given layer is given by the total daily nitrogen demand of the plant and the fraction of the total root- surface area present in the layer. However, if the available amount is lower the the potential uptake, a fraction of the deficit could be added to the potential root uptake of another layer with a surplus of nitrogen. Both the vertical distribution of root-surface area and root depth are calculated by the model. Uptake is the smallest value of the potential uptake and the available amount.The latter is a fraction of the total mineral nitrogen in the layer concerned.

**Model application(s) and State of development**
Application area(s):
Number of cases studied: around 50 case studies
State: validation

**Model software and hardware demands**
Platform: PC-based - IBM
OS: DOS
Software demands: Preferable 486 processor, Language: FORTRAN and C
Hardware demands:

**Model availability**
Purchase: available free of charge as executable code. Distribution from FTP-server FTP.SUNET.SE dirctory PUB\PC\SIMULATE\HYDROLOGY
Contact person(s) Dr. Per-Erik Jansson
          Department of Soil Sciences
          P.O. BOX 7014
          S-75007 Uppsala
          Sweden

Phone: +46 18 67 11 74
Fax: +46 18 67 27 95
E-mail: Per-Erik.Jansson@mv.slu.se

**Model documentation and major references**
Eckersten, H., and Jansson, P.-E., 1991. Modelling water flow, nitrogen uptake and production for wheat. *Fertilizer Research*, 27: 313-329.
Eckersten, H., Jansson, P.-E., and Johsson, H., 1994. SOILN model, user's manual. 2nd edition, Division of Agricultural Hydrotechnics Communications 94:4, Department of Soil Sciences, Swedish University of Agricultural Sciences, Uppsala. 58 pp.
Johnsson, H., Bergstrøm, L., Jansson, P.-E., and Paustain, K., 1987. Simulation of nitrogen dynamics and losses in a layered agricultural soil. *Agriculture, Ecosystems & Environment*, 18: 333-356.

**Other relevant information**

**Model identification**
Model name/title: SOIL

**Model type**
Media: soil, forest, agricultural, local scale, regional scale, water, air
Class: hydrology
Substance(s)/compounds:
Biological components: ecosystem - plant/soil

## Model purpose
To quantify and increase the understanding concerning hydrological and/or thermal processes in the soil.

## Short description
The model simulates soil water and heat processes in many types of soils; bare soils or soils covered by vegetation. The basic structure of the model is a depth profile of the soil. Processes such as snow-melt, interception of precipitation and evapotranspiration are examples of important interfaces between soil and atmosphere. Two coupled differential equations for water and heat flow represent the central part of the model. These equations are solved with an explicit numerical method. The basic assumptions behind these equations are very simple: (i) The law of conservation of mass and energy and (ii) flows occur as a result of gradients in water potential (Darcy's Law) or temperature (Fourier's Law).

The calculations of water and heat flows are based on soil properties such as: the water retention curve, functions for unsaturated and saturated hydraulic conductivity, the heat capacity including the latent heat at thawing/melting and functions for the thermal conductivity.

The most important plant properties are: development of vertical root distributions, the surface resistance for water flow between plant and atmosphere during periods with a non limiting water storage in the soil, how the plants regulate water uptake from the soil and transpiration when stress occurs, how the plant cover influences both aerodynamic conditions in the atmosphere and the radiation balance at the soil surface.

All of the soil-plant-atmosphere system properties are represented as parameter values. Meteorological data are driving variables to the model. The most important of these are precipitation and air temperature, but also air humidity, wind speed and cloudiness are of great interest.

Results of a simulation are such as: temperature, content of ice, content of unfrozen water, water potential, vertical and horizontal flows of heat and water, water uptake by roots, storages of water and heat, snow depth, water equivalent of snow, frost depth, surface runoff, drainage flow and deep percolation to groundwater.

## Model application(s) and State of development
Application area(s):
Number of cases studied: around 50 cases
State: validation

## Model software and hardware demands
Platform: PC-based - IBM
OS: DOS
Software demands: 486 processor preferable. language: FORTRAN and C
Hardware demands:

## Model availability
Purchase: available free of charge. Available as executable code. Distribution from FTP-server FTP.SUNET.SE directory PUB\PC\SIMULATE\HYDROLOGY.
Contact person(s) Dr. Per-Erik Jansson
Department of Soil Sciences
P.O. BOX 7014
S-75007 Uppsala
Sweden

Phone: +46 18 67 11 74
Fax: +46 18 67 27 95
E-mail: Per-Erik.Jansson@mv.slu.se

## Model documentation and major references
Eckersten, H., and Jansson, P.-E., 1991. Modelling water flow, nitrogen uptake and production for wheat. *Fertilizer Research*, 27: 313-329.

Jansson, P.-E., 1991. Soil water and heat model. Technical description. Division of Agricultural Hydrotechnics Report 165, Department of Soil Sciences, Swedish University of Agricultural Sciences, Uppsala. 72 pp.

Jansson, P.-E., 1994. SOIL model, user's manual. 3rd edition Division of Agricultural Hydrotechnics Communications 94:3, Department of Soil Sciences, Swedish University of Agricultural Sciences, Uppsala, 66 pp.

## Other relevant information

## Model identification
Model name/title: MOMO'S : MOdélisation (de la) Matière Organique (des) Sols Modelling (of) Organic matter Of Soils

## Model type
Media: soil
Class: biogeochemical
Substance(s)/compounds: carbon, nitrogen, organic compounds
Biological components: ecosystem - natural, agricultural

## Model purpose
A general equation permits the description of the simultaneous evolution of organic carbon forms (C- compartments) in soils. The equation appears valid whatever the number of C-compartments taken into account: one, two, three or four, five and perhaps with more compartments.

## Short description

Evolution of a given C- compartment m among i compartments is directed by:

$$dCm/dt = -km*Cm + Pm*(\text{summation af}) \, ki*Ci + f(t)*f(m,l)*w$$

Cm, Ci = C- contents of compartments;
km, ki = rate constants of compartments [T i -1];
Pm = input proportion of C in m compartment;
w = external input of C (constant or variable);
f(t) = Boolean function of time regulating external input;
f(m,l) = function of repartition of external input in the given compartment m;
l = labile fraction of the amendment.

The equation is given for the following models:
1) - one compartment model: total C, m=i; P=0 (Hénin et Dupuis, 1945. - *Annales Agron.*, French, 15, 17- 20). 2) - two compartment model: m=i tilhører {L,H}; L = labile organic matter, H = stable organic matter; 3) - three compartment model: m,i tilhører {V,L,H}; V = Plant fragments- C (decomposition law not necessarily a first order function); L,H = labile and stable organic matter except plant fragments; 4) - four compartment model: m,i tilhører {Vl, Vr, L, H} as for 3 except Vl, Vr respectively labile and resistant plant fragments. 5) - five compartment model: m,i tilhører {Vl, Vr, A, B, H}: as for 4 but with L divided into A= labile not living-C and B= microbial biomass-C.
With plant debris amendments:
In model 1 we have f(m,l) = 1- l (isohumic coefficient); in model 2 we have f(m,l) = l for m=L and f(m,l) = 1-l for m=H; in model 3 we have f(m,l) = 1 for m=V and f(m,l) = 0 for m tilhører {L;H}; in models 4 and 5 we have : f(m,l) for m=Vl; f(m,l) = 1-l for m=Vr; f(m,l) = 0 for m tilhører {A,B,H};
An approximation of l can be given by the following formula (Parton et al., 1987 - *Soil Sci. Soc. Am. J.*, 51, 1173-1179): l = 0.85 - 0.018*L/N with L and respectively the lignin and N content of the external input.
With amendments other than plant fragments, it is possible to take : f(m,l) = 0 for m tilhører {Vl, Vr, B} and f(m,l) =l for m=A and f(m,l) = 1-l for m=H.
Taking into account the climatic conditions, ki and km must be replaced by:

$$ki, eff. = ki*teta*cM$$

with teta and cM correction factors of temperature and moisture, respectively (could be analogous to those provided by Van Veen et Paul (*Can. J. Soil Sci.*, 1981, 61: 185-201) or Parton et al., ibid. op. cit.).

The time-step can be a day, week or month, by adapting the values of the rate constants ki and km. The model of Jenkinson and Rayner (1977 - *Soil Sci.*, 123, 298-303) for long-term predictions, can be written with the same equation, but with the time-step as a year. Although this model was conceived for the C-compartments, it also appears to provide a good description of the simultaneous evolution of the N-compartments when N mineral forms added as new state variables (our actual research, not yet published).

## Model application(s) and State of development
Application area(s): fitting and prediction of organic evolution of soils
Number of cases studied: 5 experimental cases (3 for 2 and 3 compartment model; 2 for 5 compartment model).
State: calibration, validation

## Model software and hardware demands
Platform: PC-based IBM
OS: Windows 3.1
Software demands: Versions (Ventana Systems Inc.), language - that of Versions
Hardware demands: processor 486 DX or better

## Model availability
Purchase: yes, purchase of version software gratis for model program except citation if scientific publication. Software available as sourcecode.
Contact person(s) Dr. Marc Pansu
　　　　　　　　ORSTOM
　　　　　　　　BP 5045
　　　　　　　　F-34032 Montpellier Cedex
　　　　　　　　France

Phone: +33 67 61 75 62
Fax: +33 67 54 78 00
E-mail: pansu@orstom.orstom.fr

## Model documentation and major references
Sallih, Z., and Pansu, M., 1993. Modelling of soil carbon forms after organic amendment under controlled conditions. *Soil Biol. Biochem.*, 25: 1755-1762.
Pansu, M., and Sidi, H., 1987. Cinitique d'humufication et de minéralisation des mélanges sols - résidus végétaux. *Science du Sol*, 25: 247-265.

## Other relevant information

## Model identification
Model name/title: SLIM : Soil Liming Model

## Model type
Media: water, local scale, soil
Class: biogeochemical
Substance(s)/compounds: nutrients, Ca, Al
Biological components:

## Model purpose
The purpose of the model is to predict the effects of soil liming on terrestrial and aquatic systems as a function of catchment characteristics (hydrology, hydrochemistry, morphology), soil properties (CEC, density) and design parameters (application rate, material properties).

## Short description
The model is dynamic, determinisrtic and has moderate structural complexity.
Mixing model: Continuous stirred tank reactors representing limed soil horizon and water body. Key reactions are dissolution of calcite in the soil, soultion acid-base reactions. State variables are amount of undissolved calcite, exchangeable bases, solution Ca and ANC. Important parameters are: CEC exchangeable bases, dissolution rate parameters, solution equilibrium constants. Forcing functions include: Element input to soil layer, seasonal patterns of tempepature, water content and water flux.

## Model application(s) and State of development
Application area(s):
Number of cases studied: 3
State: calibration, validation, validation of prognosis

## Model software and hardware demands
Platform: MAC
OS:
Software demands:
Hardware demands:

## Model availability
Purchase: no
Contact person(s) Dr. Per Warfvinge
                Dept. Chemical Engineering II
                Lund Institute of Technology
                P.O. BOX 124
                S-22100 Lund
                Sweden

Phone: +46 46 10 36 26
Fax: +46 46 10 82 74
E-mail: per.warfvinge@chemeng.lth.se

## Model documentation and major references
Dalziel, T.R.K., Dickson, A., Warfvinge, P., and Protector, M.V., 1992. Targets and time-scales of liming treatments, in G. Howells and T. Dalziel (Eds.), *Restoring Acid Waters:Loch Fleet 1984-1990*, New York, pp.365-391. ISBN 1-85166-663-X.
Warfvinge, P., and Sverdrup, H., 1989. Modelling limestone dissolution in soils, *Soil Science Society of America Journal*, 53, 44-51.

---

**Other relevant information**

---

**Model identification**

Model name/title: SPUR2: Simulation of Production and Utilization of Rangelands Version 2.0

---

**Model type**

Media: local scale, regional scale, agricultural

Class: biogeochemical

Substance(s)/compounds: nutrients, carbon, nitrogen

Biological components: ecosystem - temperate grasslands and shrublands

---

**Model purpose**

To determine beef cattle performance and production by simultaneously simulating production of up to 15 plant species on 36 heterogeneous grassland sites. To simulate grassland hydrology, nitrogen cycling, and soil organic matter on grazed ecosystems. To simulate production under different climatic regimes, environmental conditions, and management alternatives.

---

**Short description**

SPUR2 is a general grassland ecosystem model that simulates carbon and nitrogen cycling through aboveground live and dead vegetation, live and dead roots, litter, and soil organic matter. Effect of stocking rate, season of grazing, and type of grazing system have been incorporated into the model. SPUR2 has been modified to simulate the direct effects of $CO_2$ on plant production. A user interface has been developed to assist with model parameterization. The model is driven by daily inputs of precipitation, maximum and minimum temperatures, solar radiation, and daily wind run. These variables are derived either from existing weather records or from use of a stochastic weather generator. The hydrology component calculates upland surface run-off volumes, peak-flow, snow-melt, upland sediment yield, and channel stream flow and sediment yield. Soil-water tensions, used to control various aspects of plant growth, are generated using a soil-water balance equation. Surface run-off is estimated by the SCS curve number procedure and soil loss is computed by the modified universal soil loss equation. The mechanistic process-oriented structure of SPUR2 makes the model well suited for examining the interactions between management decisions and climatic influences on short-term ecological processes and evaluating possible adaptive management strategies.

---

**Model application(s) and State of development**

Application area(s): Grasslands and shrublands of the United States, pastures in Argentina, grasslands and pastures in Canada

Number of cases studied: 8

State: conceptualization, verification, calibration, validation, prognosis made

## Model software and hardware demands
Platform: PC-based IBM, Mainframe digital, Workstation HP 9000/750, SUN Sparc
OS: DOS, MS-Windows, VMS, UNIX
Software demands:
Hardware demands: 4 Mb RAM for DOS or 8 Mb RAM for MS-Windows; 20 Mb disk

## Model availability
Purchase: yes, send diskette to contact person, executable code, source code
Contact person(s) Dr. Jon D. Hanson
        USDA, ARS
        P.O. Box E
        301 S. Howes
        Fort Collins, Colorado 80522, USA

Phone: +1 303-490-8323
Fax: +1 303-490-8310
E-mail: jon@gpsrv1.gpsr.colostate.edu

## Model documentation and major references
Baker, B.B., Hanson, J.D., Bourdon, R.M., and Eckert, J.B., 1993. Analysis of the potential effects of climate change on rangeland ecosystems. *Climate Change*, 25: 97-117.
Hanson, J.D., Baker, B.B., and Bourdon, R.M., 1992. SPUR2 Documentation and User Guide. GPSR Technical Report No. 1, 24 pp., 1 diskette.
Hanson, J.D., Baker, B.B., and Bourdon, R.M., 1993. Comparison of the effects of different climate change scenarios on rangeland livestock production. *Agricultural Systems*, 41: 487-502.
Hanson, J.D., Skiles, J.W., and Parton, W.J., 1988. A multispecies model for rangeland plant communities. *Ecol. Modelling*, 44: 89-123.
MacNeil, M.D., Skiles, J.W. and Hanson, J.D., 1985. Sensitivity analysis of a general rangeland model. *Ecol. Modelling*, 29: 57-76.

## Other relevant information

## Model identification
Model name/title: TLALOC: Simulating the dynamics of a vegetation mosaic: a spatialized functional model

## Model type
Media: local scale, terrestrial, soil
Class: hydrology
Substance(s)/compounds: carbon, water
Biological components: ecosystem - Arid zone vegetation, community - Vegetation patches, populations - Shrub and grass species

## Model purpose
In arid areas, vegetation patches depends on surface water redistribution and are characterised by a slow movement in the opposite direction to the water flow. The model was designed to simulate the water relations in a simple functional unit and to analyse the central role of water in the whole systems dynamics.

## Short description
TLALOC is a spatialised model describing a transect folowing the main slope. All state variables correspond to 1 m² plots. There are 2 groups of state variables:
(1) the soil is described to a depth of 2 m mainly by its volumetric water content. It changes according to infiltration, soil evaporation and plant transpiration.
(2) the plants are characterised by their cover, i.e., their transpiring surface for each species and age class. Transpiration determines photosynthesis and growth (PS-maintenance) taking into account a negative effect of light extinction. Transpiration is the link between the plants and the soil. The seed production, dispersal and germination are also simulated. The necessary soil parameters are minimum soil water content and field capacity (that is water available to plants) and the parameter of the Ritchie evaporation model, depending on texture. Plant parameters include water use efficiency, biomass/cover relationships, temperature dependence of photosynthesis (which make the difference between C3 and C4 species) and maintenance costs. Seed dispersal distances and minimum soil water content for germination are needed.
    Climatic inputs are daily rainfall and potential evapotranspiration and an initial state of the vegetation cover for each species is required (including in particular the ratio of bare areas to vegetated areas).

## Model application(s) and State of development
Application area(s): All arid or semi-arid systems were an horizontal redistribution of rain water induces spatial heterogeneity.
Number of cases studied: 1
State: conceptualization, verification, calibration

## Model software and hardware demands
Platform: PC-based IBM
OS:
Software demands: language FORTRAN
Hardware demands:

## Model availability
Purchase: no, but freely available as source code
Contact person(s) Dr. A. Mauchamp or Dr. S. Rambal
                        Centre d'Ecologie Fonctionelle et Evolutive
                        L. Emberger, C.N.R.S
                        B.P. 5051
                        F-34033 Montpellier, France

Phone: +33 67 61 32 89
Fax: +33 67 41 21 38
E-mail: mauchamkfrmop22.cccnusc.fr

## Model documentation and major references
Monta, A., C., Lopez Portillo, J., and Mauchamp, A., 1990. The response of two woody species to the conditions created by a shifting ecotone in an arid ecosystem. *J. Ecol.*, 78: 789-798.

Mauchamp, A., Monta A.C., Lepart, J., and Rambal, S., 1993. Ecotone dependent recruitment of a desert shrub (Flourensia cernua) in vegetation stripes. *Oikos*, 68: 107-116.

Fisher, R.A., and Turner, N.C., 1978. Plant productivity in the arid and semi arid zones. *Annu. Rev. Plant. Physiol.*, 29: 277-317.

Molz, F.J., 1981. Models of water transport in the soil plant system: a review. *Water Resour. Res.*, 17: 1245-1260.

Miller, P.C., 1981. *Resource Use by Chapparal and Matorral: A Comparison of Vegetation Function in Two Mediterranean Type Ecosystems*. Springer-Verlag, Berlin.

Rambal, S., and Cornet, A., 1982. Simulation de l'utilisation de l'eau et de la production végétale d'une phytocénose sahélienne du Sénégal. Acta Oecologica, *Oecol. Plant.*, 3: 381-397.

Ritchie, J.T., 1972. Model for predicting evaporation from a row crop with incomplete cover. *Water Resour. Res.*, 8: 1204-1212.

## Other relevant information
The different processes considered in TLALOC were simulated with different time steps ranging from the day (transpiration, rainfall and infiltration) to the year (seed production and dispersal) according to the level of precision we needed and to a necessary compromise between precision and realism.

## Model identification
Model name/title: Artic Biophysical Carbon Model (ABC - Model)

**Model type**
Media: wetland, air, regional scale, terrestrial, soil
Class: biogeochemical
Substance(s)/compounds: nutrients, carbon, nitrogen (organic and minerâl), oxygen (organic and mineral), chemicals, organic compounds
Biological components: ecosystem - tundra (wed sedge tundra, tussack tundra)

**Model purpose**
ABC is a mechanistic model of plant production, organic matter decomposition and nitrogen cycling in tundra ecosystems. It is designed as a predative tool of the effect of climate change on the net $CO_2$ flux and C cycle in tundra ecosystems.

**Short description**
ABC simulates the daily NPP, plant growth, litter production and decomposition as a function of a few climatic variables and vegetation parameters. It is therefore easily applicable to any site included the tundra biomass. In order to account for the influence of permafrost on decomposition rates and nutrient availability, the NNP/decomposition model is coupled to a physical model of the soil temperature and moisture in a permafrost terrain.
Inputs: mean daily or monthly air temperature, total precipitation, cloudiness, wind speed and relative humidity, annual maximum snow cover depth.
Important parameters: initial thickness of the organic layer, mineral soil texture, organic and mineral soil bulk density, moisture capacity and thermal coefficients, mean leaf stomatal resistance, maximum photosynthesis rate, respiration rate, death rate and growth rate, plant lignin/N ratio, plant and soil C/N ratios.
State variables: soil and surface temperature, soil moisture content, soil organic C and plant biomass, soil mineral N and plant labile N.
Outputs: all $CO_2$ fluxes between the ecosystem and the atmosphere, C and N stocks in soil and vegetation.

**Model application(s) and State of development**
Application area(s): predictions of the effect of climate change on tundra ecosystems, diagnostic tool to understand the current trends in measurement of $CO_2$ fluxes, biomass, etc., in relation to the climatic conditions.
Number of cases studied:
State: validated, some prognosis made

**Model software and hardware demands**
Platform: IBM, SUN, Work-station SUN
OS: UNIX
Software demands: Language FORTRAN 77
Hardware demands:

## Model availability
Purchase: not yet
Contact person(s) Dr. Claire Waelbroeck or Dr. Patrick Monfray
        LMCE-CEA Saclay
        L'Orme des Merisiers, bât. 709
        F-91191 Gif-sur-Yvette CEDEX
        France

Phone: +33 1 69 08 3112 or +33 1 69 07 7724
Fax: +33 1 69 08 7716
E-mail: clairew@asterix.saclay.cea.fr

## Model documentation and major references
Waelbroeck, C., 1993. The soil physics of ABC, *Ecol. Modelling*, 69: 185-225.

## Other relevant information
The code itself is commented. The NNP/decomposition submodel of ABC is described in a paper in preparation - please contact C. Waelbroeck for copies.

## Model identification
Model name/title: Modelling the energy flow in an ecosystem.

## Model type
Media:
Class: bioenergetic
Substance(s)/compounds:
Biological components: ecosystem - prairie

## Model purpose
Energy flow in an ecosystem.

## Short description
State variables: (all as energy per unit of area) green shoot biomass, biomass of nymphs and adult of 3 grasshopper species, adult male and female sparrows, sparrow eggs, embryos within female sparrows, biomass of sparrow nestings, biomass of sparrow fledglings. Unit $J/m^2$ biomass, $J/m^2h$ flows.

## Model application(s) and State of development
Application area(s):
Number of cases studied: 1
State: conceptualization, verification, calibration, validation

---

**Model software and hardware demands**
Platform: PC-based, Mainframe
OS:
Software demands:
Hardware demands:

---

**Model availability**
Purchase:
Contact person(s) Dr. Clark Jeffries
          Dept. of Math. Sciences
          Clewson University
          Clewson, SC 29634-1907, USA
Phone:
Fax:
E-mail:

---

**Model documentation and major references**
Jeffries, C., 1989. *Mathematical Modeling in Ecology*, Birkhäusen, Boston, 1989.

---

**Other relevant information**

# 5. MODELS OF ATMOSPHERIC POLLUTION AND CLIMATE

# 5.1 General Considerations on Modeling Air Pollution and Climatic Changes

Many air pollution models contain a huge system of equations to cover aerodynamic processes. In principle these equations are not different from hydrodynamics equations, but in practical modeling situations the aerodynamic models are quite different from the hydrodynamics ones, as the considerations on possible simplification will be different.

By the presentation the reader will get an overview of the problems in air pollution modeling and get an idea of the state of the art in the field. Furthermore, the reader will hopefully understand the basis for the more complex models and learn to develop simple or complex air pollution models himself by the use of parts of the presented models.

Section 5.1 presents a model of acid rains and Section 5.3 a models of plume dispersions. They represent some of the most basic models in the area of modeling atmospheric pollution. Models of acid rains show how atmospheric models can be used to solve complex regional pollution problems, while the models of plume dispersion are very useful to solve local pollution problems, originated from one or a few sources.

Dennis (1983) distinguished between three classes of air pollution models: analytical models, numerical models and statistical models.

If it is assumed that a single point source of air pollution exists, that there are no sinks and the source strength is Q, the mass conservation equation has the following solution:

$$C(x,y,z) = \frac{Q}{2\pi\delta_y\delta_z \cdot U} \exp\left[ -\frac{1}{2}\left(\frac{y}{\delta_y}\right)^2 -\frac{1}{2}\left(\frac{z-H}{\delta_z}\right)^2 -\frac{1}{2}\left(\frac{z+H}{\delta_z}\right)^2 \right] \tag{5.1}$$

where z is height above the ground, H is the source of the source $d_y$ and $d_z$ are dispersion variances, U, is average wind speed and C, is the pollutant concentration.

This equation is Gaussian in form, which means that the distribution of pollutant plume about the plume center is Gaussian. Therefore equation (5.1) is named the Gaussian plume model. It is the basis for all plume dispersion models.

The major strength of this type of model is that it is easy to use. It is well working in simple situations, i.e., flat terrain and short travel time from the source ($\leq 10$ km). The weakness of this type of models is that they cannot treat or only very poorly treat dispersion problems and chemical transformations beyond simple first order reactions. Because it is a steady state formulation they cannot account for build up of pollutants with time.

Numerical models attempt to overcome these weaknesses. They account for wind shear and eddy diffusivity shear and non-linear chemistry. Numerical models go back to the turbulent diffusion equation and specify the vertical and horizontal diffusivities. It results in partial differential equations, which must be solved numerically.

There are two major groups of numerical models: Eulerian multiple box models, which use a fixed coordinate system and Lagrangian trajectory models, which use a moving coordinate system. The strength of numerical models is on the one hand that they eliminate the weakness of analytical models. They also can be used rather generally for urban air quality and for long-range transport of pollutants, as they are easy to adapt to different areas. The numerical models require significant computer time - a disadvantage that is steadily reduced as the computer technology develops. At the more theoretical level, the gradient transfer hypothesis is inconsistent with observations (this affects, of course, also the analytical models), Numerical models do not consider for point sources that the diffusivity is a function of travel time. Further development of the numerical models should be expected in the coming years to eliminate these theoretical disadvantages.

Analytical and numerical models are deterministic and they will predict concentration, that are ensemble averages at a given time and place. The statistical models recognize and accommodate the description of the stochastic part of the atmospheric diffusion processes. They treat the data as a time series and set up auto correlations. The observed time series is taken as the realization of some underlying stochastic processes. This realization is used to build a model of the process, that generated the time series. The model is built as a single time series model, a multiple time series model with or without deterministic elements.

Statistical models are cost-effective and can as mentioned account to a certain extent for stochastic processes, but they are non-causal and might therefore have a large uncertainty of predictions associated with them. This type of model will not be treated in this chapter.

Many models of climatic changes have been developed during the last two decades due to our concern for the consequences of the greenhouse effect, and the destruction of the ozone layer. It is difficult in this volume to give a comprehensive survey of the many global models of climatic changes, but a few will be presented in details and several more will be included in the overview given in Table 5.2. Most of the greenhouse and ozone layer models have been developed during the last ten years, but already in 1982 an overview was published of the greenhouse models existing 13 years ago; see C. Clark Carbon Dioxide Review: 1982, Oxford University Press, Oxford, 1982. The most relevant models from this volume are included in Table 5.4.

There four basic types of climate models:
1.    Energy balance models (EBM): one-dimensional models predicting the variation of the surface temperature with latitude.
2.    Radiative-convective models (RCM): one-dimensional model able to compute the vertical temperature profile.
3.    Statistical dynamical models (SDM): two-dimensional models dealing with surface processes and convective adjustment.

4.  General circulation models (GCM): three-dimensional models of the atmosphere and oceans. These models include most important, physical processes. The aim of GCM is the calculation of the full three-dimensional character of climate.

## 5.2 Models of the Distribution and Effect of Acidic Rain

Several models must be coupled to relate emission of sulphur and nitrogen to the effect of acid rain on the ecosystems. Figure 5.1 shows how a chain of models can be built. A given energy and environmental protection policy is related with emission of sulphur and nitrogen compounds, which again determine the air chemical processes and by that the concentration of various sulphur and nitrogen compounds. The next model step in the chain is a long-range transport model, which gives the emission of acidic components. A soil model uses this as input and describes changes in soil water composition including pH changes.

The result of this model could be used to direct the energy and environmental protection policy, as we can set a minimum pH-value of f.ins. 4.2 for soil and soil water. But the results of the soil model also can be used as input to a model concerned with the effect on lakes, streams, forests and agriculture, as the composition of soil water influences these ecosystems significantly. The output from the effect models can then be used as feedbacks, see Figure 5.1, to the political decisions.

The total S-emission per unit of time, $S_r$ can be found from the following simple equation:

$$S_r = E_k \cdot S_k \cdot P_k(l - r_k)$$ 
(5.2)

where $E_k$ is energy consumption, $S_k$ is the sulphur concentration in fossil fuel, $P_k$ is the fraction of sulphur emitted by the applied environmental technology and $r_k$ is the fraction of sulphur removed by ash and slags. $E_k$ and $S_k$ are political decisions. The energy consumption is determined by the GNP per capita *and* the energy policy. $S_k$ is dependent on the combination of energy sources, as shown in Table 5.1. $r_k$ is known for most industrial processes and $P_k$ is dependent on the legislation, i.e., which equipment is the industry and power stations forced to purchase to reduce the sulphur emission.

The long-range transport model considers a grid of the geographical area modeled. In Figure 5.2 it is shown how the so-called EMEP-model (see Eliassen and Saltbones (1983), Fischer (1984) and Lamp (1984)) has divided Europe in parcels of 150 x 150 km². The following equations are used to account for the emission, the chemistry and the deposition:

$$\frac{dC_{SO_2}}{dt} = Q_{SO_2} - D_{SO_2} - Ox$$
(5.3)

$$\frac{dC_{SO_4^{2-}}}{dt} = Q_{SO_4^{2-}} - D_{SO_4^{2-}} - Ox \tag{5.4}$$

$$Ox = k \cdot C_{SO_2} \tag{5.5}$$

$$Q_{SO_2} + Q_{SO_4^{2-}} = S_k \tag{5.6}$$

$$D_{SO_2} = \delta_{SO_2} C_{SO_2} \quad D_{SO_4^{2-}} = S_{SO_4^{2-}} \cdot C_{SO_4^{2-}} \tag{5.7}$$

where Q represents sources, D depositions, k and d rate constants Ox represents the rate of oxidation $SO_2 \rightarrow SO_3$

**Table 5.1.** Approximate Average Emission of Acid Components per TJ for Various Types of Fossil Fuel

| Energy source | keq hydrogen ions per TJ |
|---|---|
| Coal | 8.9 |
| Heavy fuel oil | 8.5 |
| Light fuel oil | 6.1 |
| Gasoline | 5.8 |
| Natural gas | 2.6 |
| Biogas from organic waste | 0.4 |
| Straw | 0.6 |
| Wood | 0.2 |

At this stage of the model chain it is possible to predict sulphur deposition as $g/m^2$ y, see Figure 5.3 for a given set of political decisions. Figure 5.3 is based upon an average meteorological year and a total sulphur emission in Europe equal to the emission in 1970, but with 20% higher energy consumption, which implies that $S_k \times P_k$ must be reduced with a factor 1.5.

The emission used in Figure 5.3 is unacceptable, however, because it is desirable to reach an emission of 0.5 g S/$m^2$ year for the major part of Europe to assure that pH in soil is $\geq 4.2$.

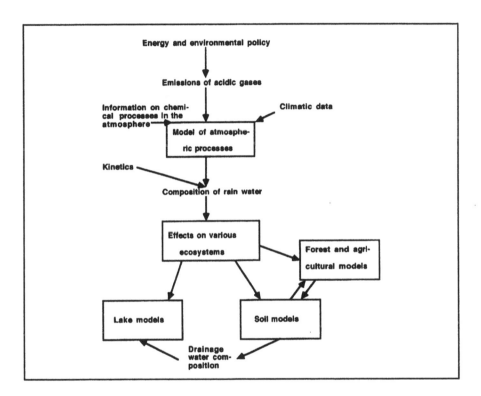

**Figure 5.1:** *A series of models must be used to relate energy - and environmental protection policy to effects on soil, plants, water chemistry and aquatic ecosystems, which serve as feedback to political issues.*

The Rains model (Alcamo et al., 1990) - one the most recent development of acid rain models in Europe - covers the entire chain of submodels, presented in Figure 5.1 and uses the grid (or similar) shown in Figure 5.2. The atmospheric composition is found on basis of computation of the sulphur, the nitrogen oxide and the ammonia emissions separately, as illustrated in Figure 5.4.

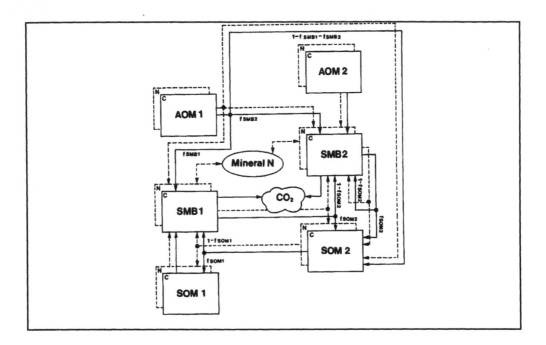

**Figure 5.2:** *Parcels in the EMEP-model.*

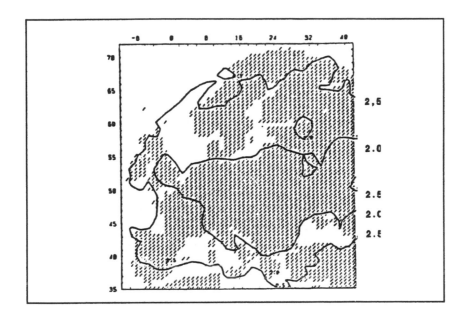

**Figure 5.3:** *Prognosis of sulphur deposition (g/m² year) in assistance with EMEP.*

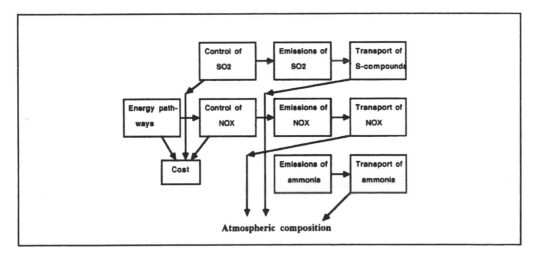

**Figure 5.4:** *The chains of submodels for emissions, transport and atmospheric processes.*

409

Two submodels are concerned with the control strategies of sulphur dioxide and nitrogen oxides. They give the input to the emissions of these two acidic gases. The ammonia emission is computed in parallel and is according to the measured ammonia emission. It is, in other words, not considered to reduce the ammonia emission. It could, however, be considered in the future to include a strategy model for the ammonia emission, which mainly comes from agriculture.

The emissions of sulphur compounds and NOX are based on equation (5.5). The transport models of the RAINS model are based upon the atmospheric transport due to climatic conditions. Mass conservation equations are set up for each box for the three components, taking into consideration the deposition (the atmospheric fall out), the transport to other boxes and the chemical transformations mainly caused by oxidations for instance of sulphur dioxide to sulphur trioxide, which again forms sulphuric acid by a reaction with water.

Six species are considered for the NOX-model: nitric oxide, NO, ozone, uncoupled oxygen, nitrogen dioxide, $NO_2$, nitric acid, $HNO_3$ and nitrate aerosols. Only ammonia and ammonium are considered for the ammonia model, where the ratio of the two species is determined by the pH. The results become the composition of the dry and wet depositions, which need to be translated into environmental impacts.

Figures 5.5-5.6 illustrate the use of the submodels summarized in Figure 5.4. Figure 5.5 gives the present long-term volume-weighted pH in precipitation in Europe (1978-1982) and North Eastern America (1980-1984). Figure 5.6 shows the long-term volume-weighted average of sulphate as mg S/l in precipitation for the same regions and periods. The results of the current European reduction plans are illustrated in Figure 5.7, where the isolines for 1 g S/y*$m^2$ year 1980 and year 2000 are shown. An improvement is clearly seen, but the isolines should be compared with the fully acceptable level of 0.5 g S/y*$m^2$ and the Canadian aim of 0.7 g S/y*$m^2$. The results are, in other words, similar to the one referred above.

It is therefore understandable that further improvements are discussed on a long-term basis. In the year 2012, a further reduction of the sulphur emission of 30% is foreseen according to a not yet decided, but still rather realistic, reduction plan, which in addition considers a 20% increase in the energy consumption from the year 1990 to year 2012. This plan would almost meet the Canadian aim for the whole of Europe. Several models have been developed to translate the emission of sulphur- and nitrogen compounds into changes in soil chemistry in the first hand to changes in soil-water-pH. Kauppi et al. (1984) has used knowledge about the buffer-capacity and velocity to relate the emission with soil-water-pH.

From the results of the deposition it is possible to estimate the acid load in equivalents of hydrogen ions per square meter per year. In the European model RAINS, acid load is computed from the deposition after accounting for forest filtering and atmospheric deposition of cations (for details; see Alcamo et al., 1990).

To organize soil data from soil maps either the grid shown in Figure 5.2 or the grid used by RAINS using 1o ¹ongitude x 0.5 o latitude may be applied.

The Soil Map of the World classifies European soils into 80 soil types. The fraction of each soil type within each grid element is computerized to an accuracy of 5%. The resolution of the RAINS model is such that each grid element includes one to seven soil types, with a mean of 2.2.

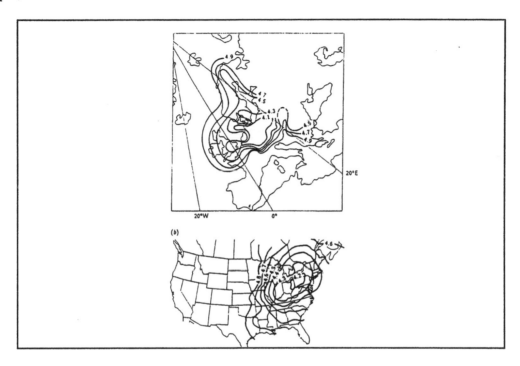

**Figure 5.5:** *Long-term, volume-weighted averages of pH in precipitation in Europe (1978-1982) and Eastern North America (1980-1984). Sources EMEP/CCC, 1984; NAPAP, 1987. Reproduced from Alcamo et al., 1990 with permission.*

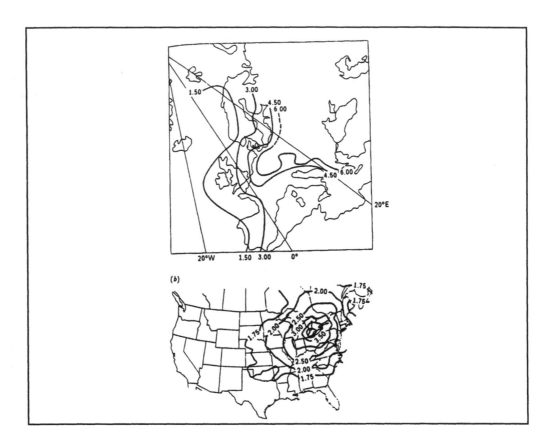

**Figure 5.6:** *Long-term, volume-weighted averages of sulphur (mg/l) in precipitation in Europe (1978-1982) and Eastern North America (1980-1984). Sources EMEP/CCC, 1984; NAPAP, 1987. Reproduced from Alcamo et al., 1990 with permission.*

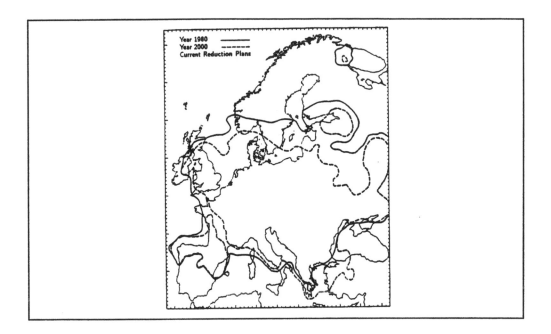

**Figure 5.7:** *Isolines of 1 gS/y\*m2 for year 1980 and year 2000 under current reduction plans. Reproduced with permission from Alcamo et al., 1990 with permission.*

The following parameters are given for each soil type:

1) Buffering capacity of the carbonate range, BCCA. Unit: keq/ha
2) uffer capacity of the cation exchange, BCCE, which is the product of total cation exchange capacity, CEC, and base saturation, ß. It is the amount of exchangeable cation reserves. Unit: keq/ha.
3) Buffer rate or the actual dissolution rate of base cations such as magnesium and calcium ions and acid cations such as aluminum ions. In general, the buffer rate corresponds to either the rate of dissolution of calcium carbonate or the dissolution rate of aluminum hydroxide. Unit keq/ha*y.
4) Silicate weathering rate according to the guideline of Ulrich (1981 and 1983). He reports weathering rates in European soil between 0.2 and 2.0 keq/ha*y.

**The basic computations of the model follow six steps:**

1. The weathering rate of silicate minerals is subtracted from the annual acid load.
2. The result of step one is then subtracted from the soil buffering capacity to account for the depletion of the acid neutralizing capacity of the soil. Base saturation and pH are computed based upon these comparisons.

3.  In calcareous soil, containing free carbonate, base saturation is not depleted and the soil pH is assumed to stay above 6.2. In non-calcareous soil, as long as silicate weathering keeps pace with the acid load, no decline of pH and base saturation is assumed to occur.

4.  If the acid load exceeds the buffer rate of the silicates, the soil shifts into the cation exchange buffer range and base saturation begins to decline accordingly. The capacity of the cation exchange rate is depleted at a rate equal to the difference between the acid load rate and the silicate buffer rate. pH is computed to decrease according to the decline of base saturation.

5.  When base cations are almost depleted, an equilibrium is computed between solid phase aluminum and soil $H^+$.

6.  A recovery of soils is computed in terms of base saturation and pH, when silicate buffer rate exceeds the rate of acid deposition.

The aim of the model is to keep track of the development of soil pH and buffer capacity. Figure 5.8 illustrates some typical results, that can be obtained by use of the model. Two situations are compared here, soil pH year 2000, if there would be no control and if the current reduction plans are applied.

To simulate flows within the catchment area, the soil is vertically segmented into two soil layers: the uppermost 0.5 m and the deeper parts of soils. The run-off is divided into two flows named quick flow and base flow. The quick flow is mainly in contact with the upper mineral and humus layers, while the base flow is assumed to come from the saturated soil zone. To compute ion concentrations of the catchment flows, the same approach as used in the soil acidification model; see above the six steps of computations. Complete mixing is assumed and chemical equilibrium is assumed to be reached according to computed saturation. The acid load in the soil participates in two processes, cation exchange and release of inorganic aluminum species, the net effect being to provide a buffer for hydrogen ions in the soil solution.

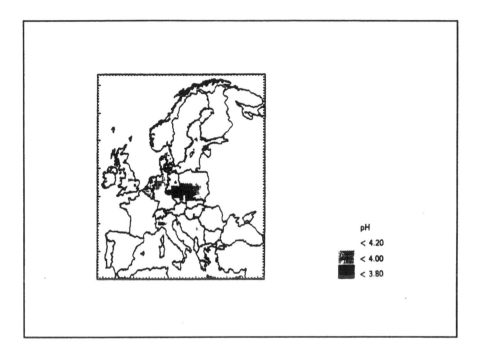

**Figure 5.8:** *Map of Europe showing pH level of forest soils for no control scenario, year 2000. Reproduced from Alcamo et al. (1990) with permission.*

Figure 5.9 gives the summary of processes used in the RAINS catchment model, including its links to the model of lake water chemistry.

Arp (1983) has developed an even more complex model, which is working with n soil layers (n is in practice between 10 and 50) and which takes into considerations the carbon, nitrogen and sulphur cycles. This model considers many chemical reactions in the soil as result of the acid rain. Further development of soil models is, still, needed as the soil model must be considered the weakest model in the chain. It can be concluded that the models of soil processes and of the chemical composition, including the concentrations of various ions and pH, of the drainage water are complex. It has not been the intention here to present the models in detail - they are far too complex to do so - but rather to present the difficulties and the basic ideas behind the models and give some understanding of the considerations behind very complex regional models. Further details can be found in the references given in this section.

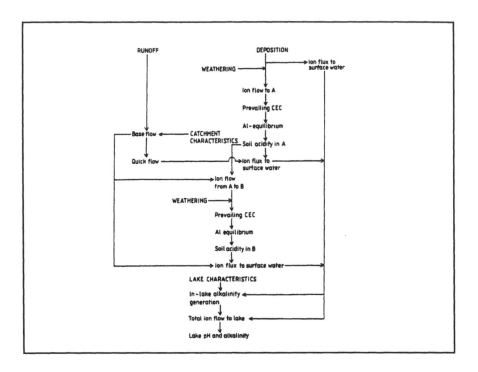

**Figure 5.9:** *Summary of processes used in the RAINS catchment model. Reproduced from Alcamo et al. (1990) with permission.*

Almer et al., (1978) have made an empirical model for the relationship between lake water pH and sulphur load; see Figure 5.10. They plot related values of pH and sulphur loads for a considered catchment area with a given sensitivity for acidification. The curves look like a titration curve. In Figure 5.10, two plots representing a sensitive (low calcium concentration) and a medium-sensitive area (medium to high concentration of calcium) are shown. It can be seen from the plots, that for a sensitive area a sulphur load of less than 0.7 or maybe 0.5 gS/m$^2$*y is needed to assure lake pH at 5.3 or above.

Henriksen (1980) has developed a model based upon the assumption that a titration of hydrogen carbonate buffered lakes takes place. Hydrogen carbonate ions are replaced by sulphate ions, which will mobilize the aluminum ions with a corresponding biological effect. The buffer capacity is estimated on basis of the concentrations of non-marine calcium and magnesium ions, while the acid load is estimated from concentration of the non-marine sulphate. Henriksen distinguishes three classes of lakes: hydrogen carbonate lakes, transition lakes and acid lakes with pH and low buffering capacity and consequently low concentrations of non-marine calcium and magnesium ions; see Figure 5.11.

416

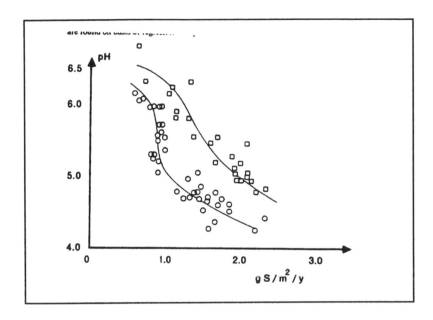

**Figure 5.10:** *The relationship between pH and sulphur load in gS/m²/y. Squares represent a catchment area with medium sensitivity to acidification, while circles are from a catchment area with high sensitivity to acidification.*

The diagram, Figure 5.11, can be used to translate concentration of sulphate in the lake water and the concentrations of non-marine calcium and magnesium ions into pH of rain water and to the type of lake (the three classes mentioned above) or to translate pH of rain water and concentrations of calcium and magnesium ions into sulphate concentration and type of lake.

The simple empirical approaches presented in Figures 5.10 and 5.11 are used without consideration of the drainage water. Weathering processes can, however, be incorporated in the empirical approaches, although the results have little generality, as they are found on basis of regression analysis of local measurements.

RAINS is using a more complex modeling approach to consider the inputs of ions from the catchment area. A nonlinear relationship between the base saturation, ß, and the soil pH is used regarding soil chemistry (Reuss, 1983):

$$- \log [H^+] \; = \; 4.0 \; + \; 1.6* \; ß^{3/4} \tag{5.8}$$

where the base saturation is given by ß = buffer capacity (BC)/total cation exchange capacity (CEC).

If the cation exchange system does not play any role in buffering the inputs to the soil solution, i.e., BC is zero, it is assumed that the equilibrium with gibbsite controls soil buffering. Aluminum is dissolved or precipitated until gibbsite equilibrium is reached:

$$[Al^{3+}] = Kg^* [H^+]^3 \tag{5.9}$$

where the gibbsite equilibrium constant = $10^{2.5.}$

Over the long term, the flux of acids and bases into and out of the soil changes the chemical state of the soil. In non-calcareous soil the weathering rate of base cations, wr, largely determines the long-term response of the catchment. As long as the input of base cations from weathering, wr, and from base deposition, db, is larger than the acid load, there will be no change in pH in the soil.

If, however, the acid load exceeds the base cation input, the capacity of the cation exchange buffer system is depleted by the rate:

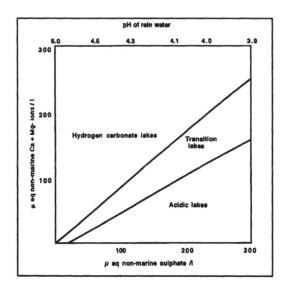

**Figure 5.11:** *Diagram relating pH of rain water, concentrations of non-marine calcium and magnesium ions and concentrations of non-marine sulphate ions in lake water.*

$$dBC/dt = db + wr - df \tag{5.10}$$

where df is the acid load equal to the total deposition of acid and dt, corrected for the through-fall deposition on forested land and for the fraction of forested land in the relevant grid.

The lake water pH and alkalinity are found from the convective flows Qa and Qb from the two soil reservoirs, and the direct input from the atmosphere to the lake:

$$FH = Qa^* [H^+] + Qb^* [H^+] + df^* AL \tag{5.11}$$

where FH is the flux of hydrogen ions to the lake and AL is the area of the lake.

The fluxes of hydrogen carbonates originating from the terrestrial catchment, FHC, will contribute to the alkalinity of the lake:

$$FHC = db + wr - df \tag{5.12}$$

The contribution from the lakes internal alkalinity generation, FL, must be added to FHC to find the total flux of hydrogen carbonates. FL is found with the following equation:

$$FL = dt * ks/(Qt /AL + ks) \tag{5.13}$$

where ks is the sulphate retention rate coefficient and Qt is the total flow of water to the lake.

The fluxes of ions mix with the lake water cause a change in the hydrogen carbonate and hydrogen ion concentrations until an equilibrium is reached:

$$[HCO_3^-] = K1 * H * pCO_2/[H^+] \tag{5.14}$$

where K1 is the first dissociation constant for carbonic acid, H is Henry's law constant for carbon dioxide and $pCO_2$ is the partial pressure of carbon dioxide in the lake water.

The presented modeling approach has been used on Scandinavian lakes. Figure 5.12 gives the results of the application on the region between Göteborg and Stockholm, named Götaland. It can be seen that the effect of applying control of the emissions on the lake water chemistry is unambiguous. Several approaches have been developed to relate the acid load or pH of the water to the ecological impact, particularly on the fish populations.

Muniz and Seip (1982) have developed an empirical model, in which they distinguish between lakes of different conductivity.

Kohlmaier et al. (1984) have developed a model, which relates the atmospheric composition and the soil water composition, including pH and the concentration of aluminium ions, with the effect on trees. The model is a black box approach as it is based upon a statistical analysis of these relations.

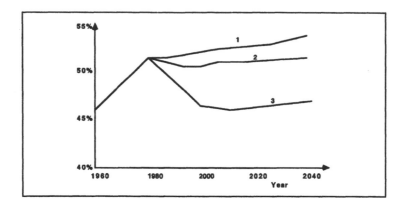

**Fig. 5.12ure** *Time trend of percentage of lakes in the Swedish region, Götaland, between Göteborg and Stockholm with pH <5.3. 1 corresponds to a scenario based on no control, 2 to the scenario of current reduction plans and 3 to the scenario of best available technology. (Alcamo et al., 1990).*

The chain of models and the entire modeling approach presented here illustrates clearly the difficulties involved in development of reliable regional models of the effect by air pollution, but the solutions on the other hand illustrate also how it is possible to overcome these practical and theoretical modeling problems.

## 5.3. Plume Dispersion

Determination of the atmospheric concentration of pollutants emitted from point sources is an important example of the use of the mass conservation principle and of a modeling approach to a local air pollution problem. The model presented here is classical. It is the basis for development of other models of air pollution originated from point sources and used for local air pollution management.

Figure 5.13 illustrates a plume from a source at $x = 0$, $y = 0$. As the plume moves downwind it grows through the action of turbulent eddies. The instantaneous plume has a high concentration over a narrow width. Over ten minutes for instance, the plume will touch a much broader area, but the concentration will, of course, be correspondingly lower. Likewise, in two hours the plume will stretch out in the cross wind direction and its concentration will be further reduced. The concentration distribution perpendicular to the axis of the wind is Gaussian or, rather, the Gaussian distribution seems a useful model for the calculation of plume concentration. The shortcomings of this model will be discussed later.

The aim of the plume dispersion model is to determine C, the concentration of the pollutant, as a function of its position downwind from the source. C is inversely proportional to the wind speed, U (m s$^{-1}$):

$$C = 1/U \qquad\qquad (5.15)$$

where U is the wind speed.

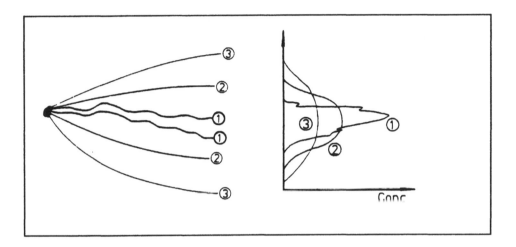

**Figure 5.13**: *Plume from a source (1) corresponds to the instantaneous plume. (2) is 10 minutes average plume and (3) is the 2 hour average plume. The diagram represents the cross-plume distribution patterns.*

Thus we might model the plume as a Gaussian function in vertical, z, and horizontal, y, co-ordinates. The behaviour in the wind direction, the x co-ordinate, should be directly proportional to the source strength and to 1/U.

The Gaussian function in the y direction is expressed as follows:

$$C = A * \exp\left[ -\frac{1}{2} \left( \frac{y}{\partial_y} \right)^2 \right] \qquad\qquad (5.16)$$

where A is a constant. If the Gaussian function is normalized, so that the area under the curve has a unit value, by setting A to $1/\sqrt{2\pi} * \partial_y$, the integral of concentration must be equal to the total amount of pollutant emitted. Thus:

$$C = \frac{1}{\sqrt{2\pi} * \partial y} [-1/2 - (-)^2] \quad (5.17)$$

(5.17) gives, however, the distribution in one direction, and we are concerned with the distribution in three dimensions, as shown in Figure 5.14.

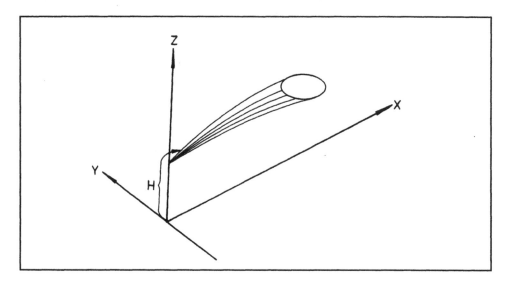

**Fig. 5.14:** *Co-ordinate system applied in the plume model.*

The Gaussian function in 3 dimensions, considering that the integral of concentration over x, y and z coordinates must be equal to the total emission, takes the form of equation (5.1):

$$C(x,y,z) = \frac{Q}{2\pi\partial_y\partial_z{}^*U} \exp\left[-1/2\left(\frac{y}{\partial_y}\right)^2\right]\left[\exp\left(-1/2\left(\frac{z-H}{\partial_z}\right)^2\right)+\exp\left(-1/2\left(\frac{z+H}{\partial_z}\right)^2\right)\right] \quad (5.18)$$

where:

C = the concentration $(g\ m^{-3})$

Q = source strength $(g\ sec^{-1})$

U = average wind speed $(m\ sec^{-1})$

H = the effective stack height (m) (see below and Figure 5.15 for explanation)

Notice that z - H accounts for the real above-ground source and the term z + H accounts for the reflection of the plume, corresponding to an imaginary source below the ground.

The maximum concentration, Cmax, at ground level can be shown to be approximately proportional to the emission and to follow approximately this expression:

$$Cmax\ =\ k * Q / H^2 \tag{5.19}$$

where Q is the emission (expressed as g particulate matter per unit of time), H is the effective stack height and k is a constant.

The definition of the effective stack height is illustrated in Fig. 5.15. It is the physical stack height, h, + the plume rise, Δh, and it can be calculated from the following equation:

$$H\ =\ h\ +\ 0.28 * V_s * D_s\ [\ 1.5\ +\ 2.7\ \frac{(T_s\ -\ 273)}{T_s}\ *\ D_s\ ] \tag{5.20}$$

where:

$V_s$   = stack exit velocity in m per second

$D_s$   = stack exit inside diameter in m

$T_s$   = stack exit temperature in degrees Kelvin

h   = physical stack height above ground level in m

H   = effective stack height in m

These equations explain why a lower ground-level concentration is obtained when many small stacks are replaced by one very high stack. In addition to this effect, it is always easier to reduce and control one large emission than many small emissions, and it is more feasible to install and apply the necessary environmental technology in one big installation.

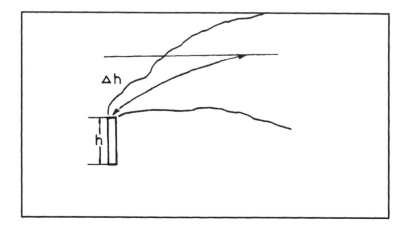

**Figure 5.15:** *Effective stack height H = h + Δh.*

For $z = 0$, at ground level, we find by using equation (5.18):

$$C(x,y,0) = \frac{Q}{\pi \partial_y \partial_z {}^* U} \exp\left[ -\left( \frac{y^2}{2\partial_y{}^2} + \frac{H^2}{2\partial_z{}^2} \right) \right] \qquad (5.21)$$

For the concentration along the centerline (y=0) and at ground level, we have:

$$C(x,0,0) = \frac{Q}{\pi \partial_y \partial_z {}^* U} \exp\left[ -\frac{H^2}{2\partial_z{}^2} \right] \qquad (5.22)$$

Compare (5.22) with (5.17).

U is a function of z but usually the average wind speed at the effective stack height is used. If the wind speed $U_1$ at level $z_1$ is known, U can be estimated by the following equation:

$$U = U_1^* \{ H / z_1 \}^n \tag{5.23}$$

Smith (1968) recommends n = 0.25 for unstable and n = 0.50 for stable conditions. For further discussion on this subject see Turner (1970).

$\partial_y$ and $\partial_z$ can be considered as diffusion coefficients and can be found as functions of atmospheric conditions (Smith, 1968).

These relationships are demonstrated in Table 5.1 and Figures 5.16 and 5.17. Uncertainties in the estimation of $\partial_y$ are generally fewer than those of $\partial_z$. However, wide errors in the estimate of $\partial_z$ occur over longer distances. In some cases $\partial_z$ may be expected to be correct within a factor of 2. These cases are 1) stability for distance of travel to a few hundred metres, 2) neutral to moderately unstable conditions for distances to a few kilometers, 3) unstable conditions in the lower 1000 metres of the atmosphere with a marked inversion above for a distance of 10 km or more.

It can be shown that the maximum ground-level concentration occurs where $\partial_z = 0.707$ H, provided that the ratio $\partial_4/\partial_z$ is constant with downwind distance X. Turner (1970) discusses a procedure for handling diffusion, when the plume expansion is limited by an upper level inversion. The principle of the method is to calculate the concentrations as if they were distributed uniformly throughout the layer of height $H_i$ - the distance from ground level to the inversion.

In this case the expression for C at ground level becomes:

$$C = \frac{Q}{\sqrt{2\pi}\, UH_i \partial_y} \exp\left[-1/2\left(\frac{y}{\partial_y}\right)^2\right] \tag{5.24}$$

This equation can be applied downwind, of when the plume first reaches the $H_i$ elevation, which can be estimated when $2.15\, \partial_z = H_i - H$. Turner (1970) recommends using equation (5.24) for $x > 2x_i$, where $x_i$ is the point at which the plume reaches the inversion. Before $x_i$ the regular diffusion result applies and between $x_i$ and $2x_i$ an interpolation between the results at $x_i$ and $2x_i$ is recommended.

The same equation can be applied under fumigation conditions, in which case $H_i$ is the height to which the unstable air risen. (For an explanation of inversion and fumigation, see Figure 5.18). However, Q must then be corrected to the fraction of the plume that can be carried down to the ground. Note that the stable classifications used to calculate $\partial_y$ and $\partial_z$ are also used to predict fumigation conditions. Turner, however, uses a correction factor to account for the plume spreading in the y direction:

$$\partial y, \text{fum} = \partial y, \text{stable} + H/8 \tag{5.25}$$

**Table 5.1** Key to Air Stability Categories

| Surface wind speed (at 10 m) m sec⁻¹ | Day Incoming solar radiation | | | Night Thinly overcast or | |
|---|---|---|---|---|---|
| | Strong | Moderate | Slight | >4/8 low cloud | <3/8 cloud |
| < 2 | A | A-B | B | | |
| 2-3 | A-B | B | C | E | F |
| 3-5 | B | B-C | C | D | E |
| 5-6 | C | C-D | D | D | D |
| > 6 | C | D | D | D | D |

The neutral class, D, should be assumed for overcast conditions during day or night. Source: Turner, (1970). It is easy to use the plume model in practice, but, of course, such a somewhat simple approach has some limitations:

1. The diffusion coefficients applied are not very accurate,
2. The turning of the wind with height owing to friction effects is neglected,
3. Adsorption or deposition of pollutants is neglected, but could easily be included, f required,
4. Chemical reactions along the plume path have been omitted, but might be included if the necessary knowledge were available,
5. Shifts in wind direction are not considered.

In view of these limitations, the plume dispertion model should only be used as a first estimate of pollutant concentration (see Scorer, 1968).

Atmospheric dispersion depends primarily on horizontal and vertical transport. The horizontal transport depends on the turbulent structure of the wind field. As the wind velocity increases so does the degree of dispersion and there is a corresponding decrease in the ground level concentration of the contaminant at the receptor site.

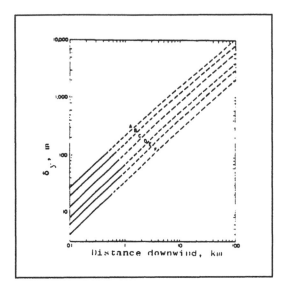

**Figure 5.16.** *Horizontal dispersion coefficient as a function of downwind distance from the source. (Turner, 1970).*

The emissions are mixed into larger volumes, of air and the diluted emission is carried out into essentially unoccupied terrain away from any receptors. Depending on the wind direction, the diluted effluent may be funnelled down a river valley or between mountain ranges. Horizontal transport is sometimes prevented by surrounding hills forming a natural pocket for locally generated pollutants. This topographical situation occurs in the Los Angeles area, which suffers heavily from air pollution.

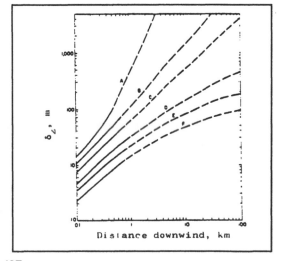

***Figure 5.17:*** *Vertical dispersion coefficient as a function of downwind distance from the source. (Turner, 1970).*

427

The vertical transport depends on the rate of changes of ambient temperature with altitude. The dry adiabatic lapse rate is defined as a decrease in air temperature of 1°C per 100 m. This is the rate at which, under natural conditions, a rising parcel of unpolluted air will decrease in temperature with elevation into the troposphere up to approximately 10,000 m. Under so-called isothermal conditions the temperature does not change with elevation. Vertical transport can be hindered under stable atmospheric conditions, which occur when the actual environmental lapse rate is less than the dry adiabatic lapse rate. A negative lapse rate is an increase in air temperature with latitude. This effectively prevents vertical mixing and is known as inversion.

These different atmospheric conditions are illustrated in Figure 5.18, where stack gas behaviour under the various conditions is shown. Further explanations are given in Table 5.2.

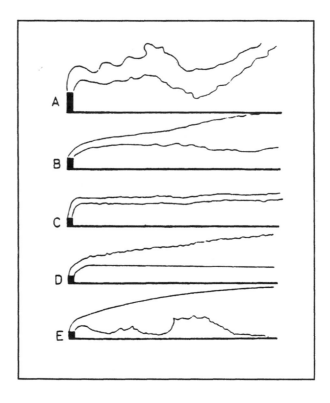

**Figure 5.18:** *Stack gas behaviour under various conditions. A) Strong lapse (looping), B) Weak lapse (coning), C) Inversion (fanning), D) Inversion below, lapse aloft (lofting), E) Lapse below, inversion aloft (fumigation).*

**Table 5.2** Various atmospheric conditions

| | |
|---|---|
| Strong lapse (looping) | Environmental lapse rate > adiabatic lapse rate |
| Weak lapse (coning) | Environmental lapse rate < adiabatic lapse rate |
| Inversion (fanning) | Increasing temperature with height |
| Inversion below, | Increasing temperature below, |
| lapse aloft (lofting) | App. adiabatic lapse rate aloft |
| Lapse below, inversion | App. adiabatic lapse rate below, |
| aloft (fumigation) | Increasing temperature aloft |

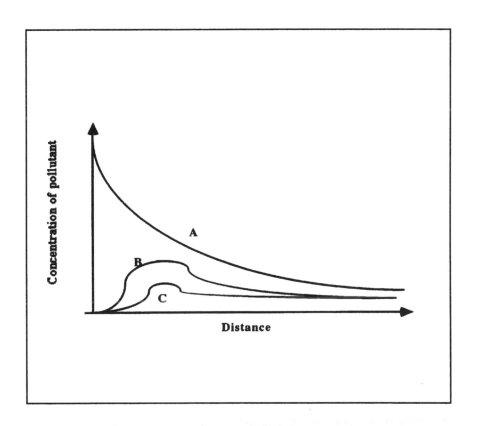

**Figure 5.19:** *Distribution of emission from stacks of different heights under coning conditions.*
*A) height 0 m,  B) height 50 m,  C) height 75 m.*

Figure 5.19 illustrates the distribution of material emitted from a stack at three different heights under coning conditions. The figure demonstrates that the distribution of particulate material is more effective, the higher the stack.

## 5.4 An Overview: Models of Air Pollution

Models of air pollution may be classified in accordance with the scale of the problem, that the model attempts to solve: local air pollution models, regional air pollution models and global air pollution models.

The first mentioned class is represented by the plume model, and a handful of model included in this survey may be considered modifications and improvements of this classical model.

The regional air pollution models are illustrated by the chain of models presented in Section 5.2. The model named Rains represents to a certain extent the state of the art of this type of models. Most regional models are in principle similar to the models presented in Section 5.2. and only a few regional models are actually significant different from these models on the most important components. Therefore, only a few regional models, selected to represent the differences from the models in Section 5.2, are included in the overview given in this and the next section. The global models are not illustrated by an important example as the two other classes of models, because global models are too different to easily be presented by one illustrative example. There are many different modeling ideas published in this field during recent years. It is attempted to represent the most important of these ideas in this and the next section.

Table 5.3 gives an overview of the models included in Section 5.5 and some other useful, illustrative air pollution models, selected to present the entire spectrum of models in the area of air pollution management.

The Yearley has developed global carbon cycle model. It contains the following state variables: carbon in biota, soil, atmosphere, cold ocean surface, warm ocean surface, two intermediate ocean layers, ten deep ocean layers and ocean sediment. The forcing function is the anthropogenic carbon.

The ABC-model considers the influence of climate change on the net carbon dioxide flux and carbon cycle in tundra ecosystems. The model includes plant production, organic matter decomposition and nitrogen cycling. The model attempts to answer the question: if the climate is changed due to the greenhouse effect, which changes will be observed in tundra ecosystems? and which influence will these changes have on the carbon cycle? and therefore further on the climate? Several models have been developed to consider similar feedback chains: the carbon dioxide concentration in the atmosphere increases, it implies change in the climate, which influences the rates of many biogeochemical processes, which may change the carbon cycle, and thereby the climate and so on.

The Osnabruck Biosphere Model 3m simulates the cycling of carbon through the terrestrial vegetation on continental to global scale. The model accounts for the response to changes of climate and in the emission of carbon dioxide. Six state variables based on carbon are included in the model: agricultural biomass, natural herbaceous and woody living biomass, herbaceous and woody litter and soil organic carbon.

The OBC model calculate oceanic carbon uptake. The oceans are divided into 10 latitude bands and 11 vertical layers. The model encompasses the following processes: transfer of carbon, nutrients and alkalinity as function of transport processes by advection and diffusion, biological production and air-sea exchange processes (dissolution).

The so-called FBM-model describe the seasonal and long-term carbon dynamics of the major types of ecosystem in order to calculate the global carbon dioxide source/sink pattern in a 1 x 1 degree spatial resolution. 33 different terrestrial ecosystems and combined with seven soil types.

Chemifog illustrates the type of models that considers many chemical reactions, as it is also known from the models of the decomposition of the ozone layer. The model contains an aerodynamic two dimensional description of the transport processes and an environmental chemical description of 110 atmospheric reactions and 32 aquatic reactions.

A Lagrangian long range model air pollution model has been developed for Eastern North America. It considers the atmospheric chemistry of nitrogen and is able to predict spatial concentrations and depositions of nitrogen oxides, PAN, nitric acid and nitrate in the atmosphere.

Bunce et al. have developed a model which can estimate the tropospheric concentrations of the hydroxide radicals at any location and time and to use this information to estimate the tropospheric half-lives of organic pollutants whose principal sink is by chemical reaction with hydroxide radicals. The model has been applied on chlorinated aromatic compounds, polycyclic aromatic compounds and chlorinated aliphatic compounds.

Table 5.3 Models of Air Pollution

| Model component/ model name | Characteristics of the model | References |
| --- | --- | --- |
| Wind shear effect | Appl. of Monte-Carlo scheme for Lagr. Part. diffusion with wind shear eff. | Zanetti, 1981 |
| 3-D Lagrangian Model | Transport model in 3-D | Blondin, 1980 |
| Plume transport | Trajectory model, | Crabtree and Marsh , 1981 |
| Heavy gas model | Simulation of atmospheric releases of of heavy gases over variable terrain | Chan et al., 1982 |

431

**Table 5.3 (Contiuned)** Models of Air Pollution

| Model component/ model name | Characteristics of the model | References |
|---|---|---|
| PAQSM | Urban photo-chemistry, ozone and nitrogen oxides | Schreffler 1982 |
| Photo-chemical smog | 3-D grid model,ozone, nitrogen oxides | Stern and Scherer, 1980 |
| Oxidant formation | 3 layers, several photo-chemical react. formation of ozone, nitrogen oxides and PAN | Gram et al., 1980 |
| Vertical diffusion | Sources near ground, surface layer diffusion model | Holtslag et al., 1981 |
| Global Carbon Cycle | Carbon in biota, soil, atmosphere, sev. oceanic compartment, ocean sediment | Yearsley 1989 |
| ABC | C-cycle and C-fluxes in tundra eco-systems | Waelbroeck, 1993 |
| Biosphere C-model | 6 C-pools in each grid cell | Esser et al., 1982 |
| OBC | Carbon uptake by ocean (modeled as 10 degrees bands), 11 vertical layers | Klepper 1995 |
| FBM | Carbon dynamics of ecosystems: 33 vegetation types and 7 soil types | Luedeke et al., 1994 |
| Chemifog | 1-D to 2-D model with 110 possible atmosph.and 32 aquatic reactions. | Bott, 1991 |
| Nitrogen compounds | Distribution of air pollutants for a region, nitrogen oxides, PAN, nitric acid, nitrate | Bhumralkar et al., 1982 |
| Organic compounds | Trophic conc. of hydroxide radicals are used to estimate the half lives of org. pollutants | Bunce et al., 1991 |
| Samus | Massbalances of nutrients and pollutants in a forest exposed to air | Brodersen, 1995 |
| Pollution | Profile effect of acid rain on soil chemistry as function soil chemistry, climate and acid deposition | Warfvinge and Sverdrup, 1992 |
| Safe | Dynamical counterpart to Profile | Warfvinge et.al., 1993 |

The model Samus is based upon mass balances for major nutrients and some pollutants in trees and upper soil layer in a forest exposed to air pollution. The model consists of the following state variables: crown, stem, roots litter layer, and four soil layers. Various pools of organic matter may be considered. The forcing functions are rain fall. deposition rates, evapotranspiration and climatic conditions determining the growth of plants.

Two models are concerned with the effects of acid rain. The model Profile predicts the effect of atmospheric deposition on the soil chemistry as a function of soil properties, nutrient uptake, water balance and temperature. The results of the model is a steady state profile of the  soil chemistry which is changed due to acid rain. Critical load of acid deposition can be calculated regionally. The model Safe is the dynamical counterpart of Profile.

## 5.5 Models of Air Pollution

**Model identification**
Model name/title: Global Carbon Cycle

**Model type**
Media: water, ocean, air, global scale, terrestrial, soil, forest
Class: biogeochemical
Substance(s)/compounds: carbon
Biological components: ecosystem - phytoplankton, terrestrial biota.

**Model purpose**
The model was developed as a research tool to assess the tradeoff between increasing model complexity and collecting data of higher quality or quantity. The global carbon model is used in conjunction with the Kalman filter to test hypotheses regarding modelling performance.

**Short description**
State variables: Carbon in - Biota, Soil, Atmosphere, Cold Ocean Surface, Warm Ocean Surface, Two Intermediate Ocean Layers, Ten Deep Ocean Layers, Ocean Sediments.
Forcing functions: Anthropogenic carbon.

**Model application(s) and State of development**
Application area(s): Evaluation of value of complexity in global carbon models compared to value of data collection programs.
Number of cases studied: one case study with 40 scenarios
State: conceptualization, simple parameter estimation

**Model software and hardware demands**
Platform: DEC Microvax, DG Avion Workstation
OS: DEC VMS, Unix
Software demands: FORTRAN 77 compiler, FORTRAN 77
Hardware demands: 4 Mb RAM, 200 Mb storage

**Model availability**
Purchase: none, software available as FORTRAN 77 source code
Contact person(s) Dr. John R. Yearsley
    EPA Region 10 ES097
    1200 Sixth Avenue
    Seattle, WA 98101 USA

Phone: +1 206-553-1532
Fax: +1 206-553-0119
E-mail: jyearsle@r0dg09.r10.epa.gov
   yjd@sequoia.nesc.epa.gov

**Model documentation and major references**
Yearsley, J.R., 1989. State estimation and hypothesis testing: A framework for the assessment of model complexity and data worth in environmental systems. Technical Report No. 116, University of Washington, Seattle, WA. 232 pp.
Yearsley, J.R. and Lettenmaier, D.P., 1987. Model complexity and data worth: An assessment of changes in the global carbon budget. *Ecol. Modelling*, 39: 201-206.

**Other relevant information**

**Model identification**
Model name/title: Artic Biophysical Carbon Model (ABC - Model)

**Model type**
Media: wetland, air, regional scale, terrestrial, soil
Class: biogeochemical
Substance(s)/compounds: nutrients, carbon, nitrogen (organic and mineral), oxygen (organic and mineral), chemicals, organic compounds
Biological components: ecosystem - tundra (wed sedge tundra, tussack tundra)

**Model purpose**
ABC is a mechanistic model of plant production, organic matter decomposition and nitrogen cycling in tundra ecosystems. It is designed as a predictive tool of the effect of climate change on the net $CO_2$ flux and C cycle in tundra ecosystems.

**Short description**
ABC simulates the daily NPP, plant growth, litter production and decomposition as a function of a few climatic variables and vegetation parameters. It is therefore easily applicable to any site included the tundra biomass. In order to account for the influence of permafrost on decomposition rates and nutrient availability, the NNP/decomposition model is coupled to a physical model of the soil temperature and moisture in a permafrost terrain.
Inputs: mean daily or monthly air temperature, total precipitation., cloudiness, wind speed and relative humidity, annual maximum snow cover depth.
Important parameters: initial thickness of the organic layer, mineral soil texture, organic and mineral soil bulk density, moisture capacity and thermal coefficients, mean leaf stomatal resistance, maximum photosynthesis rate, respiration rate, death rate and growth rate, plant lignin/N ratio, plant and soil C/N ratios.
State variables: soil and surface temperature, soil moisture content, soil organic C and plant biomass, soil mineral N and plant labile N.
Outputs: all $CO_2$ fluxes between the ecosystem and the atmosphere, C and N stocks in soil and vegetation.

**Model application(s) and State of development**
Application area(s): predictions of the effect of climate change on tundra ecosystems, diagnostic tool to understand the current trends in measurement of $CO_2$ fluxes, biomass, etc., in relation to the climatic conditions.
Number of cases studied:
State: validated, some prognosis made

**Model software and hardware demands**
Platform: IBM, SUN, Work-station SUN
OS: UNIX
Software demands: Language FORTRAN 77
Hardware demands:

**Model availability**
Purchase: not yet
Contact person(s) Dr. Claire Waelbroeck or Dr. Patrick Monfray
                LMCE-CEA Saclay
                L'Orme des Merisiers, bât. 709
                F-91191 Gif-sur-Yvette CEDEX
                France

Phone: +33 1 69 08 3112 or +33 1 69 07 7724
Fax: +33 1 69 08 7716
E-mail: clairew@asterix.saclay.cea.fr

**Model documentation and major references**
Waelbroeck, C., 1993. The soil physics of ABC : *Ecol. Modelling*, 69: 185-225.

**Other relevant information**
The code itself is commented.
The NNP/decomposition submodel of ABC is described in a paper in preparation - please contact C. Waelbroeck for copies.

**Model identification**
Model name/title: Osnabrück Biosphere Model 3m

**Model type**
Media: global scale, terrestrial
Class: biogeochemical
Substance(s)/compounds: carbon
Biological components: biome

## Model purpose

The model simulates the cycling of carbon through terrestrial vegetation on continental to global scale in response to a climate and $CO_2$ forcing. Implicitly the model assumes a contemporary steady state of climate and vegetation.

## Short description

Annual time step. 2.5 by 2.5 spatial resolution horizontally. Land areas only, Antarctica excluded. Model dynamics depend on local vegetation and soil types for each grid cell. The formulation of net primary productivity (NPP) is regression based and includes a $CO_2$ fertilization effect. Agricultural and potential natural vegetation are considered separately in every grid cell. Changes in agricultural area and relative agricultural productivity (as compared to potential natural NPP) are prescribed from statistics. The model can simulate oceanic $CO_2$ uptake employing a response function formulation with the response of the Hamburg Model of the Oceanic Carbon Cycle (HAMOCC).

State Variables: six carbon pools in each grid cell - agricultural, natural herbaceous and woody living biomass, herbaceous and woody litter and soil organic carbon (SOC).

Forcing: Gridded mean annual temperature, annual precipitation. Mean annual atmospheric $CO_2$ concentration. Fraction of agricultural used area per grid element. Per country relative agricultural productivity. Important parameters: Soil factor - being a grid mean soil factor derived from soil types and NPP measurements. Stand age - grid mean vegetation stand age (biome dependent). Fractions for partitioning natural NPP into the pools of living natural herbaceous and woody biomass (biome dependent). Fraction for partitioning produced herbaceous and woody dead biomass into the pools herbaceous, woody litter and SOC (global constant). One gram of dry living biomass is assumed to contain 0.45 g of carbon.

## Model application(s) and State of development

Application area(s): Global simulations of biospheric carbon fluxes 1860- 2000
Number of cases studied: 6
State: prognosis made, validation of prognosis

## Model software and hardware demands

Platform: no specific machine necessary
OS: no specific operative system necessary
Software demands: FORTRAN (F77) compiler, FORTRAN
Hardware demands: 4 Mb RAM, 10-30 Mb Disk space for input data, ca. 30 Mb disk space for detailed output of 100 years model simulation.

## Model availability

Purchase: no, it is not possible to purchase, but it is possible to obtain permission for use of Dr. G. Esser only (Address: Dr. G. Esser, Institute für PflanzenÖkologie, Justus Liebig Universität Giessen, Heinrich-Buff-Ring 38, D-35392 Giessen) as source code.
Contact person(s) Mr. J. Kaduk
    Max-Planck-Institute for Meteorology
    Bundesstr. 55
    D-20146 Hamburg, Germany

Phone: +49 41173 282
Fax: +49 41173 298
E-mail: kaduk@dkrz.d400.de

## Model documentation and major references
Esser, G., Aselman, I., and Lieth, H., 1982. Modelling the Carbon reservoir in the System Compartment "Litter", in Transport of carbon and minerals in major world rivers, Part. 1. Eds. Degens and Kempe, Mitt. Geolog.--Paläontolog. Insti. Univ. Hamburg, SCOPE/UNEP Sonderband, 52, 39-58.
Esser, G., 1987. Sensitivity of global carbon pools and fluxes to human and potential climatic impacts, *Tellus*, 39B, pp. 245-260, 1987.
Esser, G., 1991. The Osnabrück Biosphere Model: construction, structure, results, in Esser, G. and D. Overdieck (Eds.), in: *Modern Ecology: Basic and Applied Aspects*, Elsevier, Amsterdam, 844 pp.
Kaduk, J. and Heimann, M., 1994. The climate sensitivity of the Osnabrück Biosphere Model on the ENSO time scale, *Ecol. Modelling*, in press.

## Other relevant informations
OBM3m is a slightly modified version of OBM3. The basic differences are : (1) No river transport of carbon, (2) Optionally oceanic $CO_2$ uptake simulated by a response function, (3) Removal of a minor inconsistency in the formulation of litter production. Generally equilibrium model. Data of agricultural land fraction and agricultural productivity 1860-1980 are available with the model. Scenarios of these can be chosen for the period after 1981. However these scenarios might be outdated. All necessary input data (climate, $CO_2$ and parameters) are available as input files.

## Model identification
Model name/title: High Resolution Biosphere Model (HRBM)

## Model type
Media: global scale, regional scale, terrestrial
Class: biogeochemical
Substance(s)/compounds: carbon
Biological components: biome

## Model purpose
The model simulates the cycling of carbon through terrestrial vegetation on regional to global scale in response to a climate and $CO_2$ forcing. The models assumes a fixed biome distribution.

## Short description
6 day time step. 0.5 by 0.5 spatial resolution horizontally. Land areas only, Antarctica excluded. The model considers potential natural vegetation only. The biome distribution is prescribed from a BIOME I model run (Prentice et al., 1992).

Model dynamics depend on local vegetation and soil water capacity for each grid cell. Parts of litter production as well as soil organic carbon (SOC) decay is modeled as in the HRBM (Esser et al., 1994). The photosynthesis scheme is physiologically based employing the photosynthesis model of Farquhar et al. (1980). Climate driven phenology scheme.

The model includes a simple bucket model for simulation of soil water dynamics and the ratio of actual to potential evapotranspiration.

State variables: six carbon pools in each grid cell - an assimilate pool within plants, natural herbaceous and woody living biomass, herbaceous and woody litter and soil organic carbon (SOC). Soil water.

Forcing: Gridded mean monthly temperature, monthly precipitation, mean monthly fraction of possible sunshine. Mean annual atmospheric $CO_2$ concentration.

Important parameters: Minimal, optimal and maximal assimilation temperatures, maximal assimilation rates. Stand age-grid vegetation stand age (biome dependent). ratio of herbaceous to woody living biomass. Fraction for partioning natural NPP into the pools of living natural herbaceous and woody living biomass (climate driven).

Fractions for partioning produced herbaceous and woody dead biomass into the pools herbaceous, woody litter and SOC (globally constant). Soil water capacity derived from soil types and soil texture. One gram of dry living biomass is assumed to contain 0.45 g of Carbon. Necessary inputs: biome distribution, gridded soil water capacity, forcing as above.

## Model application(s) and State of development
Application area(s): Global simulations of natural biospheric carbon fluxes 1860- 2100
Number of cases studied: 1
State: verification, calibration

## Model software and hardware demands
Platform: Machine, one year model simulation needs 4 min CPU-time on a CRAY 2S
OS: No specific operative system necessary
Software demands: FORTRAN (F77) compiler, FORTRAN
Hardware demands: 16 Mb RAM, 10-30 Mb disk space for input data, ca. 100 Mb disk space for detailed output of 1 year model simulation

## Model availability
Purchase: no
Contact person(s) Mr. J.Kaduk
        Max-Planck-Institute for Meteorology
        Bundesstr. 55
        D-20146 Hamburg
        Germany

Phone: +49 41173 298
Fax: +49 41173 298
E-mail: kaduk@dkrz.d400.de

**Model documentation and major references**

Esser, G., Hoffstadt, J., Mack, F., and Wittenberg, U., 1994. High Resolution Biosphere Model - Dokumentation. In: Esser, G.: *Mitteilungen aus dem Institut für Pflanzenökologie der Justus Liebig Universität Giessen*, Heft 2, 71 pp.

Farquhar, G.D., Von Caemmerer, S., and Berry, J.A., 1980. A biochemical model of photosynthetic $CO_2$ assimilation in leaves of C3 Species, *Planta*, 149, 78-90.

Prentice, I.C., Sykes, M.T., and Cramer, W., 1993. A simulation model for the transient effects of climate change on forest landscapes, *Ecol. Modelling*, 65: 51-70.

Prentice, I.C., Cramer, W., Harrison, S.P., Monserud, R.A., and Solomon, A.M., 1992. A global biome model based on plant physiology and dominance, soil properties and climate, *Journal of Biogeography*, 19, 117-134.

Kaduk, J., and Heimann, M., Development of a global prognostic model of carbon cycling in the terrestrial biosphere, in preparation.

**Other relevant information**

**Model identification**

Model name/title: OBC: Ocean Biosphere and Chemistry

**Model type**

Media: ocean, global scale

Class: biogeochemical

Substance(s)/compounds: carbon, nitrogen, oxygen, radionuclides, organic compounds, Ca

Biological components:

**Model purpose**

Calculate oceanic carbon uptake during next century.

**Short description**

Ocean is divided into 10° latitude bands (Atlantic and Indo-Pacific separately) and 11 vertical layers. Transfer of carbon, nutrients and alkalinity as a function of transport (advection/dispersion), biological production and dissolution (air-sea exchange) is calculated.

Forcing functions are temperature and atmospheric $CO_2$ concentrations.

Important parameters are biological productivity coefficients (light, nutrient affinity) and transport rates (upwelling , exchange). Good (steady-stable) starting values are essential.

**Model application(s) and State of development**

Application area(s): World ocean

Number of cases studied: 1

State: verification, calibration, prognosis made

**Model software and hardware demands**
Platform: PC-based, Workstation SUN
OS:
Software demands: FORTRAN
Hardware demands:

**Model availability**
Purchase: no, source code. "I hesitate to sell it commercially, anybody is welcome to use the source code of course."
Contact person(s) Dr. O. Klepper
National Institute for Public Health and Environmental Protection
Center for Mathematical Methods
NL-3720 BA Bilthoven
The Netherlands
Phone:
Fax:
E-mail:

**Model documentation and major references**
Klepper, O. et al., Biological feedbacks in the oceanic carbon cycle. *Ecol. Modelling*. in press.
Klepper, O., Modelling the oceanic food wels using a quasi steady state approach. *Ecol. Modelling*, in press.

**Other relevant information**

**Model identification**
Model name/title: Frankfurt Biosphere Model (FBM)

**Model type**
Media: global scale, regional scale, terrestrial, soil
Class: biogeochemical
Substance(s)/compounds: carbon, water
Biological components: ecosystems - decidous/evergreen forests, shrublands, grasslands

**Model purpose**
Description and prediction of: the seasonal and long term carbon dynamics of the main ecosystem types in order to calculate the global biotic $CO_2$-source/sink pattern in a 1 x 1° spatial resolution; the seasonal and long term LAI development (boundary condition for GCMs).

**Short description**
The Frankfurt Biosphere Model (FBM) is a global model for calculating the seasonal pattern of uptake and release of $CO_2$ by the vegetation and soil in a steady state climate simulation as well as the long term development in a changing environment. Within the terrestrial ecosystems 33 vegetation types are distinguished and combined with 7 distinct soil types with respect to their holding capacities. Within each vegetation type the living biomass is divided into two compartments, one with a short (seasonal) turnover containing the photosynthesizing tissue, feeder roots, and assimilate store, and another one with a long turnover mainly consisting of structural plant material. The mathematical description is based on three hypotheses: (1) vegetation tends to maximize photosynthesizing tissue, (2) a minimum amount of structural tissue is needed to support and maintain the productive parts, described by an allometric relation, (3) the phenology is controlled by the (potential) carbon flux balance of the living parts of the vegetation.

The fluxes are modeled using standard equations for gross photosynthesis of the canopy, autotrophic respiration, and decomposition of dead organic matter depending on surface temperature and soil moisture including irradiation for the photosynthetic process. Within the system of differential equations the free parameters for each vegetation type are calibrated on the basis of a characteristic seasonal climate which was derived from the monthly values of mean surface temperature, cloudiness and precipitation using the method of overlapping the driving variables' maxima of all grid elements belonging to this vegetation type. It could be shown that the model yields satisfactory results with respect to phenology, and gradients in net primary production, and standing biomass and thus holds the promise to also yield good global results.

**Model application(s) and State of development**
Application area(s): global carbon cycle modelling, greenhouse climate prognosis
Number of cases studied: all annual fluxes and states: 1
                 NPP, biomass: about 20
                 Phenology: about 30
                 Long-term dynamics: 3
State: validation, 3 x $CO_2$ response prognosis for temperate forests and tundra ecosystems

**Model software and hardware demands**
Platform: PC (regional applications up to one ecosystem type)
        high performance systems (global calculations)
OS: OS2, UNIX
Software demands: FORTRAN
Hardware demands:

**Model availability**
Purchase:  available as source code
Contact person(s) Dr. Matthias K.B. Luedeke  or   Dr. Ralf. D. Otto
                 IPTC der J.-W. -Goethe Universistät SAME ADDRESS
                 Marie Curie Str. 11
                 D-60439 Frankfurt/Main
                 Germany

Phone: + 49 69 5800 9416
Fax: +49 69 5800 9445
E-mail:Luedeke@chemie.uni-frankfurt.d400.de or Otto@chemie.uni-frankfurt.d400.de

## Model documentation and major references

Janecek, A., Benderoth, G., Luedeke, M.K.B., Kindermann, J., and Kohlmaier, G.H., 1989. Model of seasonal and perennial carbon dynamics in decidous-type forests controlled by climatic variables. *Ecol. Modelling*, 49: 101-124.

Kindermann, J., Luedeke, M.K.B., Badeck, F.-W., Otto, R.D., Kluadius, A., Haeger, Ch., Wuerth, G., Lang, T., Doenges, S., Habermehl, S., and Kohlmaier, G.H., 1993. Structure of a global and seasonal carbon exchange model for the terrestrial biosphere - the Frankfurt Biosphere Model (FBM), WASP 70. 675-684.

Luedeke, M.K.B.N., Badeck, F.-W., Otto, R.D., Haeger, Ch., Doenges, S., Kindermann, J., Wuerth, G., Lang, T., Jaekel, U., Kluadius, A., Ramge, P., Habermehl, S., and Kohlmaier, G.H., 1994. The Frankfurt Biosphere Model. A process oriented model for the seasonal and long term $CO_2$ exchange between terrestrial ecosystems and the atmosphere. Submitted to *Climate Research*.

Luedeke, M.K.B.N., Doenges, S., Otto, R.D., Kindermann, J., Badeck, F.-W., Ramge, P., Jaekel, U., Kohlmaier, G.H., 1994. Physiological, ecological and net carbon response of temperate and boreal forests as well as tundra ecosystems to a $CO_2$ induced climate change as evaluated by the Frankfurt Biosphere Model (FBM). Submitted to *Tellus B*.

## Other relevant information

## Model identification
Model name/title: CHEMIFOG

## Model type
Media: air
Class: hydrology
Substance(s)/compounds: chemicals
Biological components:

## Model purpose
Description of: life cycle of radiation fogs, microphysics radiation, gas phase and aqueous phase chemistry, sulphur oxidation, acid fog.

## Short description
1-D model spectral microphysics with 2-dimensional particle spectrum, shortwave and longwave radiation calculation with S-two stream approximation.
Higher order closure turbulence. Gas phase chemistry with 110 reactions, aqueous phase chemistry with 32 reactions, equidistant grid spacing DELTA Z = 10 m, timestep DELTA t = 10s (microphysics), DELTA t = 1 s (chemistry) DELTA t = 60 s (radiation).

## Model application(s) and State of development
Application area(s):
Number of cases studied:
State: verification

## Model software and hardware demands
Platform: Workstation - DEC, ALFA-processor
OS:
Software demands: language: FORTRAN
Hardware demands:

## Model availability
Purchase: no, available as source code
Contact person(s) Dr. A. Bott
          Inst. Atm. Phys.
          Universität Mainz
          D-55099 Mainz, Germany

Phone: + 49 6131 392862
Fax: +49 6131 393532
E-mail: bott@mzdmza.zdv.uni-mainz.de

## Model documentation and major references
Bott et al., 1990. J AS, 47, 2153-2166.
Bott, 1991. *Bound Layer Meteor.*, 56, 1-31.
Bott and Carmichael, 1993. *Atmos. Environ.*, 27A, 503-522.

## Other relevant information

## Model identification
Model name/title: Lagrangian Long Range Air Pollution Model for Eastern North America

## Model type
Media: air, regional scale
Class: biogeochemical
Substance(s)/compounds: nitrogen
Biological components:

## Model purpose
Distribution of air pollutants, particularly nitrogen compounds.

## Short description
State variables: $NO_2$, $NO$, $P$, $N$, $HNO_3$ and $NO_3^-$ .
Results: deposition and concentration of these 5 compounds.

## Model application(s) and State of development
Application area(s):
Number of cases studied:
State: conceptualization, verification, calibration, validation

## Model software and hardware demands
Platform: Mainframe
OS:
Software demands:
Hardware demands:

## Model availability
Purchase: no.
Contact person(s) Dr. C.M. Bhumralkar
        SRI International
        Menlo Park
        California, USA
Phone:
Fax:
E-mail:

## Model documentation and major references
Bhumralkar, C.M., Mancuso, R.L., Wolf, D.E., Nitz, K.C., and Johnson, W.B., 1980. Adaptation and application of the ENAMAP-1 model to eastern North America --Phase II, Final report, Contract 68-02-2959, SRI International, Menlo Park, CA.

Bhumralkar, C.M., Mancuso, R.L., Wolf, D.E., and Johnson, W.B., 1981. Regional air pollution model for calculating short-term (daily) patterns and transfrontier exchanges of airborne sulphur in Europe, *Tellus*, 33: 142.

Bhumralkar, C.M., Endlich, R.M., Brodzinsky, R., Nitz, K.C., and Johnson, W.B., 1982. Further studies to develop and apply long- and short-term models to calculate regional patterns and transfrontier exchanges of airborne pollution in Europe. Final report to Umweltsbundesamt, Federal Republic of Germany, Project 8365, SRI International, Menlo Park, CA.

## Other relevant information

## Model identification
Model name/title: Model to estimate tropospheric concentrations of the hydroxyl radical at any location, date, and time of day

## Model type
Media: air, regional
Class: geochemical
Substance(s)/compounds: organic compounds
Biological components:

## Model purpose
To estimate tropospheric concentrations of the hydroxyl radical at any location, date, and time of day, and to use this information to estimate the tropospheric half- lives of organic pollutants whose principal sink is by chemical reaction with hydroxyl radicals.

## Short description
The concentrations of hydroxyl is calculated through the use of (1) astronomical parameters to determine the zenith angle of the Sun; (2) interpolation of tabulated data on the solar photon flux as a function of wavelength at defined zenith angles; (3) the relationship inferred from work of Platt et al. that the concentration of hydroxyl parallels the rate of photolysis of ozone. Calculation of reaction rates requires in addition the rate constant for the reaction of the substrate with OH. The limitations include neglect of cloud cover, so that the hydroxyl concentrations are maximum estimates. In cases where diurnal data on the concentrations ozone and nitrogen dioxide are available, the model can be used to simulate the urban troposphere. The rate of direct photolysis of the substrate can also be modelled, but is limited by paucity of data in most cases on the wavelength dependence of the quantum yield of direct photolysis.

## Model application(s) and State of development
Application area(s): Chlorinated aromatic compounds
　　　　　　　　Polycyclic aromatic compounds
　　　　　　　　Chlorinated aliphatic compounds
Number of cases studied:
State:

## Model software and hardware demands
Platform:
OS:
Software demands: XT-compatible or better DOS-based PC, runs on Lotus 1- 2- 3 or compatible spreadsheet
Hardware demands:

## Model availability
Purchase:
Contact person(s) Professor of Chemistry Nigel J. Bunce
　　　　　　　　Department of Chemistry and Biochemistry
　　　　　　　　University of Guelph
　　　　　　　　Guelph, Ontario, Canada N1G 2W1

Phone: (519) 824 4120 Ext. 3962
Fax: (519) 766 1499
E-mail: bunce@chembio.uoguelph.ca

## Model documentation and major references
Bunce, N.J., Nakai, J.S., and Yawching, M., 1991. *Chemosphere*, 22, pp. 305-315.
Bunce, N.J., Nakai, J.S., and Yawching, M., 1991. *J. Photochem. Photobiol.* (A), 57, pp. 429-439.
Bunce, N.J., and Dryfhout, H.G., 1992. *Can. J. Chem.*, 70, pp. 1966-1970.
Bunce, N.J., and Schneider, U.A., 1994. *J. Photochem. Photobiol.* (A), in press.

## Other relevant information

## Model identification
Model name/title: SAMUS: Single Area Model of Unsaturated Soil

## Model type
Media: terrestrial, local scale, soil, forest
Class: biogeochemical
Substance(s)/compounds: nutrients, carbon, nitrogen
Biological components: ecosystem - soil microorganisms, society - Forest trees

## Model purpose
Material balances for the turn over of major nutrients and some pollutants in trees and upper soil layers in a forest exposed to air pollution.

## Short description
SAMUS is a one-dimensional model describing vertical transport (deposition, throughfall, percolation and evapotranspiration) for a system consisting of a plant (a conifer subdivided into crown, stem and root), a litter layer and four soil layers belonging to the unsaturated zone: rainfall, deposition velocities, evapotranspiration and growth rates for the plant must be specified together with transfer factors for the root uptake/soil solution concentration. Microbial decay rates for degradation of various organic pools must be provided. Conversion of nitrogen containing species is handled somewhat similarly. Simple models for gas diffusion ($O_2$, $N_2O$, $NH_4$) are provided. Two types of simple hydrology models can be used with specification of parameters. Soil chemistry is depended on ion exchange properties subdivided into permanent and pH depended capacities. Values for the capacities must be provided together with equilibrium constants. Weathering is included in a tentative manner. The state of K, Na, Ca, Mg, Al, Cl, S, N, organics and pH is modelled. A few heavy metals may be included. Output is loss with percolate and the concentrations in the various pools of the system.

## Model application(s) and State of development
Application area(s): Effects of pollution or management procedures
Number of cases studied: So far only some test examples
State: conceptualization, verification

**Model software and hardware demands**
Platform: PC-based, IBM
OS: MS-DOS
Software demands: language: TURBO5 PASCAL
Hardware demands: 80486 coprocessor, VCA/CGA graphics, 3.5" 1.44 MB disks

**Model availability**
Purchase: no, software available as executable code.
Contact person(s) Dr. Knud Brodersen
                  Risø National Laboratory
                  DK-4000 Roskilde
                  Denmark

Phone: +45 4677 4940
Fax:   +45 4675 3533
E-mail:

**Model documentation and major references**
Description in preparation, to be issued as a Risø report in 1995.

**Other relevant information**
The ECCES model referred to in *Developments in Environmental Modelling* 15, Elsevier, is not being operated any more, although in principle it is still available at the System Analysis Department at Risø. The soil chemistry part of ECCES have been incorporated and extended in SAMUS which, however, is restricted to the modeling of a single area where ECCES was set up as a regional model. SAMUS is developed as a part of the NECO projects under the Danish Environmental Strategic Research program.

**Model identification**
Model name/title: PROFILE

**Model type**
Media: local scale, regional scale, terrestrial, soil
Class: biogeochemical
Substance(s)/compounds: nutrients, nitrogen, sulphur, Ca, Mg, K, Al , H+, OH-, $HCO_3^-$, $CO_3^{2-}$, DOC
Biological components:

**Model purpose**
The purpose of the model is to predict the effects of atmospheric deposition and nutrient uptake on the chemistry of soil systems as a function of soil properties (mineralogy, morphology) and boundary conditions (nutrient uptake, water balance, temperature). Regional application for calculation of critical loads of acid deposition.

**Short description**

The model is steady-state, deterministic and has moderate structural complexity. Large emphasis on weathering reactions.

Mixing model: A series of continuous stirred tank reactors each representing a soil horizon. Key reactions are chemical weathering, solution acid-base reactions.

State variables are soil solution ANC (pH, Al, organic acids), base cation concentration.

Important parameters are mineralogy, dissolution rate constants, mineral surface area, water content, Al-solubility, DOC, soil solution equilibrium constants.

# 6. LANDSCAPE MODELS AND MODELING TOOLS

6.1 Introduction

6.2 An Overview: Models of Regions and Landscapes

6.3 Models of Regions and Landscapes

6.4 An Overview: Auxiliary Models as Modeling Tools

6.5 Auxiliary Models as Modeling Tools

## 6.1 Introduction

The development on the software front has been overwhelming during the last decade and has, for instance, made it possible on personal computers to develop regional models and models of entire landscape. Introduction of GIS, geographical information system, and various sophisticated graphical software packages has facilitated these progresses in modeling. It is therefore not surprising that it today is possible to produce excellent overviews of the environmental quality for entire regions and landscapes, based on the application of modeling environmental processes with a spatial in- and output. The first section in this chapter covers these new modeling approaches, which in principle are not different from the other models presented in the previous chapters, except that they have introduced a spatial dimension into modeling. Another development during the last decade is included in this chapter in Section 6.2, namely the introduction of various modeling tools. The development of artificial intelligence and very advanced statistical packages such as for instance SAS, has made it possible to offer to the modelers several very useful tools which facilitate the estimation of parameters and process equations and the coupling of hydrology and aerodynamics with model descriptions of chemical and ecological processes.

## 6.2 An Overview: Models of Regions and Landscape

Several models of air quality and climate are regional models, for instance the model Rains. They have (to a certain extent) been included in Chapter 5, but could also have been included in this Chapter 6, as they are coping with regional pollution questions. A few air quality and climate models are included in this chapter, as they are characterized by a typical geographical presentation of the results.

Table 6.1 presents the model reviewed in Section 6.3 plus a few slightly older models of general interest. The table gives information about the model components, mainly which region or landscape is in focus, the main characteristics of the model and the major reference. The model Dispatch is a typical landscape model. It focuses on the temporal changes of the spatial structure of landscapes subject to natural disturbances. The state variables are map of landscapes and about 60 measures of its spatial structure, which are altered by human impacts and climatic changes. Grant has developed a hierarchical simulation model that is able to account for the response of an alpine tundra to changing climate. The model consists of three hierarchical nested submodels. The model represents the possibilities to model landscape response due to changes in climate (or pollution) by the use of a hierarchical approach which in several situations can be strongly recommended to apply, as it represent the organization of nature. The model denoted Climex predict the potential geographical distribution of species based on it climatic preferences. It is a dynamic simulation model which enables the estimation of an animal or plant geographical distribution and relative abundance as determined by the climate. A population growth index describes the potential growth of a population during a favourable season and four stress indices describe the probability of the population surviving through the unfavourable season. The growth index and the stress indices are combined into an ecoclimatic index to give an overall measure of how favourable the location is for permanent occupation by the target species. The results are presented as tables, graphs and maps.

Wilkie has developed a spatial model of land use and forest regeneration in the Ituri forest of Northeastern Zaire. It simulates how cultivation decisions within a tropical moist forest impact the spatial distribution of cleared land, forest regeneration and post fallow reclearing within the context of human settlement patterns, determined by population growth and other factors. The model uses a two dimensional representation of the focal area.

The model named Celss is used for spatial ecosystem modeling. Five or six different models of different study areas have been developed using the same architecture. The models have focused on the spatial distribution of a wide range of environmental factors, including nutrients, population and hydrology.

The model Firesum simulates the effect of different fire regimes on tree composition, stand structure and fuel loading in forest of the inland portion of Northwestern United States. The model is deterministic containing stochastic properties (tree establishment and tree mortality).

The model Finich simulates the response of Californian Shrub-lands to the interaction of climatic changes and fire intensity and frequency. Fire intensity is calculated on basis of fuel accumulated since last fire. Regeneration depends on fire intensity. Growth and survival are functions of annual precipitation and mean January and July temperatures; these functions are derived through geographical distribution analysis. Climatic forcing functions alter species responses, changes fuel conditions and thus regeneration.

The model Mosel is developed to include dispersal and a spatially explicit landscape for existing models of forest dynamics. Dispersal among cells is simulated by using Monte Carlo methods.

SHE, the European Hydrologic System, is composed of several hydrological modules: evapotranspiration, overland and change flow, unsaturated zone, saturated zone, snow melt, frames for organizing other components. It is able to account for a complex hydrological scheme with a superstructure of environmental variables. It has been applied on a wide range of ecosystems and landscapes. The application of this approach has been expanded via the model MIKE SHE; see Section 6.3 for details.

Ecosys is another huge landscape/region model which is able to include simulation of the spatial distribution. It is prepared for and has been tested on a very wide range of environmental variables.

Hough has developed a model which provides estimations of the growing and grazing season for managed grassland in a temperate climate. It is based on estimation of the dry matter growth of grass. The forcing functions are meteorological and climatic information, while the major state variables are dry biomass and leaf area.

Exe is a modeling tool to study the responses of ecosystems and hydrology to climate perturbations. It couples water budget model with an explicit treatment of ecological dynamics. The forcing functions are the climatic conditions and the modeling results are a complete picture of micro-climate and the ecology of a model forest.

Lang has developed an abiotic landscape model, based upon nutrients and agroecosystems.

Targets is a tool to assess regional and global environmental health targets for sustainability. It allows the user to perform analysis and assessment on a global scale of the linkages among social and economic processes, biophysical processes and effects on ecosystems and humans from an integrated system dynamics perspective. It consists of a human system including population, health, available resources and economics, and environmental system which contains an element cycle module, a land module and a water module.

**Table 6.1** Models of Landscapes and Regions

| Model name<br>Model component | Characteristics | References |
|---|---|---|
| Hydrology | Flow pattern of a sea | Rizolli and Berga-masco, 1983 |
| Air quality | Distribution of acid rain | Seman, 1989 |
| Echam/atmosphere | Atmospheric climate model | IPCC, 1990 |
| Lsg/oceanocean | Circulation model, temp. salinity | Maier-Reimer et al., 1987 |
| Opyc/ocean | Dynamical thermodynamic model | Oberhuber, 1990 Dispatch |
| 7 landscape | Spatial structure of landscape | Baker et al., 1991 |
| Alpine tundra | Hierarchical model of landscape | Grant; see model |
| Climes/species | Geograph. distribution of species due to climatic pre-ferences | Maywald and Sutherst, 1991 |
| Land use | Spatial model of land use, forest regeneration in Zaire | Wilkie; see model |
| Celss/sev. st. var. | Spatial model of sev. variables | Maxwell; see model |
| Firesum/fire | Fire/tree composition | Keane et al., 1989 |
| Finich/shrub-land | Population response to fire and climate | Malanson et al., 1992 |
| Mosel/forest | Spatial landscape of forest dynamics | Hansson et al., 1990 |

**Table 6.1 (Continued)** Models of Landscapes and Regions

| Model name Model component | Characteristics | References |
| --- | --- | --- |
| SHE/hydrology | Coupling hydrology and env. factors | DHI; see model |
| MIKE SHE | Coupling hydrology and env. factors | DHI; see model |
| Ecosys/sev. env. | Var. spatial distribution of sev. env. var. | Grant; see model |
| Grassland | Growing, grazing, grassland | Hough, 1993 |

Many examples of large scale models are related to the green house effect. Table 6.1 includes three models, Echam, Lsg and Opyc, which are used as a huge coupled model complex to make predictions of the global warming resulting from the increased concentrations of green house gases in the atmosphere. These types of models usually requires the application of super-computers. They do not usually contain new approaches and new equations which can inspire other modelers, but their enormous complexity is due to a high number of repetitions of the same equations to account for a spatial distribution of a huge area.

# 6.3 Models of Regions and Landscape

**Model identification**
Model name/title: DISPATCH/Disturbance Patch

**Model type**
Media: landscape
Class:
Substance(s)/compounds:
Biological components:

**Model purpose**
Model temporal change in the spatial structure of landscapes subject to natural disturbances.

**Short description**
Forcing functions and parameters: synoptic (climatic) type distribution, disturbance size distribution and its parameters, disturbance interval distribution and its parameters, landscape disturbability equation.
State variables: map of landscapes and about 60 measures (e.g. patch size) of its spatial structures.
Necessary inputs: initial map of landscape, parameter values.

**Model application(s) and State of development**
Application area(s): Changes in disturbance regimes by human alteration and climatic change and the effect on landscapes.
Number of cases studied: 2
State: conceptualization, verification, calibration, validation

**Model software and hardware demands**
Platform: Workstation SUN
OS: Sunos (Unix)
Software demands: >16 Mb RAM  C compiler
Hardware demands: 7 Mb

**Model availability**
Purchase: no
Contact person(s) William L. Baker
           Dept. of Geography
           University of Wyoming
           Laramie, WY 82071
           USA

Phone: +1 307-766-3311
Fax: +1 307-766-2697
E-mail: bakerwl@uwyo.edu

## Model documentation and major references
Baker et al., 1991. *Ecol. Modelling*, 56: pp. 109-125.
Baker, W.L. (in press) Restoration of landscape structure altered by fire suppression. *Conservation Biology*, in press.

## Other relevant information

## Model identification
Model name/title: Response of Alpine tundra to a changing climate: A hierarchical simulation model

## Model type
Media: local scale, regional scale, global scale, terrestrial
Class:
Substance(s)/compounds: biomass
Biological components: ecosystem, community, Alpine tundra

## Model purpose
The present paper describes an alpine tundra simulation model that has been developed within a hierarchical framework based primarily on data from Niwot Ridge and adjacent areas of the Colorado Front Range. Our objective is to simulate the response of alpine tundra to climatic change. The model consists of three hierarchically nested submodels. The Level I submodel represents the alpine zone of the Front Range Colorado (16.6 km²) and operates on a time scale of centuries. The Level II submodel represents Niwot Ridge alpine (6.8 km²) and operates on a time scale of decades. The Level III submodel, which is itself divided into six submodels, represents each of the six major soil-plant communities that occur on Niwot Ridge (several hectares each) and operates on a time scale of years.
Simulations were run representing tundra system dynamics under (a) constant environmental temperatures and (b) a cooling trend of 0.07°C per year. Simulation results show shifts in the elevation of treeline (Level I), shift in the proportion of tundra occupied by each of six different soil-plant communities (Level II), and changes in rates primary production and decomposition in each community type (Level III).

## Short description
The model consists on three hierarchically nested submodels, each representing different spatial and temporal scales.
    The Level I submodel represents the Front Range of Colorado (16.6 km² of alpine area) and operates on a time scale of centuries. The main process represented is the altitudinal movement of treeline in response to changing environmental temperatures.

The Level II submodel represents Niwot Ridge (6.8 km$^2$) and operates on a time scale of decades. The main process represented is the shift in proportional areal coverage of the six different soil-plant communities that occur on Niwot Ridge in response to the average number of snow-free days annually, which is a function of environmental temperature.

The Level III submodel, which is itself divided into six submodels, represents each of the six major soil-plant communities that occur on Niwot Ridge (several hectares each), and operates on a time scale of years. The main process represented is biomass transfer through the primary producer and litter compartments of each community, again in response to snow-free days and temperature. The model was implemented with the aid of STELLA software.

## Model application(s) and State of development
Application area(s):
Number of cases studied:
State: prognosis made

## Model software and hardware demands
Platform: MAC
OS:
Software demands: STELLA
Hardware demands:

## Model availability
Purchase: no
Contact person(s) Dr. W.E. Grant
                 Dept. of Wildlife and Fisheries Sciences
                 Texas A&M University
                 College Station, TX 77845 USA

Phone: +1 409-845-5702
Fax: +1 409-845-3786
E-mail: wegrant@orca.tamu.edu

## Model documentation and major references

## Other relevant information

## Model identification
Model name/title: CLIMEX

## Model type
Media: regional scale, global scale, terrestrial, agricultural
Class:

Substance(s)/compounds:
Biological components: community - generic, population - generic, organism - generic

## Model purpose
To predict the potential geographic distribution of a species based on its climatic preferences.

## Short description
CLIMEX is a dynamic simulation model which enables the estimation of an animal or plant's geographic distribution and relative abundance as determined by climate. CLIMEX is applied to different species by selecting the values of a series of parameters which describe the species' response to temperature and moisture. A population "Growth Index" (GI) describes the potential for growth of a population during a favourable season and four stress indices

(Cold, Hot, Wet, Dry) describe the probability of the population surviving through the unfavourable season. The GI and Stress Indices are combined into an "Ecoclimatic Index", EI, to give an overall measure of favourableness of the location or year for permanent occupation by the target species. The results are presented as tables, graphs or maps.

A species' climatic requirements are inferred from its known geographical distribution (its native range or in another location where it has been introduced), relative abundance and seasonal phenology. Some life cycle data, such as developmental threshold temperatures, can be used to fine tune or interpret the CLIMEX parameter values. Initial estimates of parameter values are refined by comparing the indices with known presence or absence, or preferably relative abundance, of the species in each location. Once parameter values have been estimated, CLIMEX can be used to make predictions for other, independent locations.

## Model application(s) and State of development
Application area(s):
Number of cases studied: +/- 100
State: validation

## Model software and hardware demands
Platform: PC-based, Workstation - SUN
OS: Windows
Software demands: C++
Hardware demands: 486, 8 Mb RAM

## Model availability
Purchase: yes, For MS DOS version and Windows version Australian $250,00.
Contact person(s) Software Applications Officer
        Centre for Tropical Pest Management
        The University of Queensland
        Brisbane, Australia 4072

Phone: +61-7-3651574
Fax: + 61-7-3651855
E-mail: bryce@ctpm.uq.oz.au

## Model documentation and major references

Maywald, G.F. and Sutherst, R.W., (1991). User's guide to CLIMEX a computer program for comparing climates in ecology. 2nd edition. CSIRO Australia, Division of Entomology Report No. 48, 51 pp.
Sutherst, R.W. and Maywald, G.F., (1985). A computerised system for matching climates in ecology. *Agriculture Ecosystems & Environment* 13, pp. 281-299.

## Other relevant information

## Model identification

Model name/title: A Spatial Model of Land use and Forest regeneration in the Ituri Forest of Northeastern Zaïre.

## Model type

Media: terrestrial, forest, agricultural, local scale
Class: not applicable
Substance(s)/compounds: not applicable
Biological components: tropical forest

## Model purpose

To simulate how cultivation decisions by small scale horticulturalists within a tropical moist forest impact the spatial distribution of cleared land, forest regeneration and post fallow reclearing within the context of human settlement patterns, usufructuary land tenure, and population growth.

## Short description

This is a tessellation space model, based on a stylized 2-dimensional cellular representation of a 10 km² area surrounding a section of road in Northeastern Zaïre. The road and villages are spatially accurate, rivers or topographic faetures were not included. the model uses eight land-use categories and each was assigned a travel cost in minutes. Forest succession is determined deterministically by time since cultivation. Transition from old seral to mature forest depends on age if the cell was cleared once (root resprouting possible), otherwise the cell must remain adjacent to a mature forest cell for 50 years in addition to the 50 years post cultivation. To simulate natural forest gaps, climax forest cells return to open 'farm-bush' with a probability of 0.0025 per year. Field selection decisions are based on the age of the cell travel time to the cell from the village and from the previous year's field, and whether the cell is in the village's usufruct. Productivity of the field cell is a function of the number of years allowed. Initial land-use composition. Population density and growth rate, fallow period, and presence or absence of land tenure are user defined.

---

**Model application(s) and State of development**
Application area(s): Impacts of human land-use on landscape composition, patch size and distribution
Number of cases studied: 1, the Ituri Forest of Northeastern Zaïre
State: preliminary validation using Landsat imagery

---

**Model software and hardware demands**
Platform: PC-based, IBM, Mainframe-based
OS:
Software demands: FORTRAN, C
Hardware demands: Size of cell grid is limited by RAM

---

**Model availability**
Purchase: no - available free on request, Source code in FORTRAN or C
Contact person(s) Dr. David S. Wilkie
                645 Centre Street,
                Newton, MA 02158-2340 USA

Phone: +1 617-527-0926
Fax:
E-mail: dwilkie@pearl.tufts.edu

---

**Model documentation and major references**

---

**Other relevant information**

---

**Model identification**
Model name/title: CELSS/Barataria

---

**Model type**
Media: wetland
Class: hydrology
Substance(s)/compounds: nutrients, phosphorus, nitrogen
Biological components: ecosystem - population.

---

**Model purpose**
The spatial modeling workstation ia an environment for spatial modeling. It enables scientists to utilize advance computer architectures without requiring computer programming.

---

**Short description**
5 -6 different models of different study areas have been built using the architecture.

## Model application(s) and State of development
Application area(s): Spatial ecosystem modeling
Number of cases studied: 5-6 example applications
State: validation

## Model software and hardware demands
Platform: Workstation MAC
OS: UNIX or MAC-OS
Software demands: C plus extensions
Hardware demands: workstation on MAC or connection machine or transputers

## Model availability
Purchase: yes, free via ftp, available as source code
Contact person(s) Dr. Tom Maxwell
               Box 548
               Salomons, MD 20688 USA
Phone: +1 410-326-7388
Fax: +1 410-326-7394
E-mail: maxwell@kabir.umd.edu

## Model documentation and major references

URL: http://kabir.umd.edu/smp/mvd/cø.htm

## Other relevant information

## Model identification
Model name/title: FIRESUM: a FIRE Succession Model

## Model type
Media: forest
Class: biogeochemical
Substance(s)/compounds:
Biological components: ecosystem - Forest, primarily ponderosa pine, whitebark pine and Douglas fir

## Model purpose
We developed a computer model, called FIRESUM (FIRE Succession Model), to simulate the effect of different fire regimes on tree composition, stand structure and fuel loading in forests of the inland portion of the northwestern United States. Comparison on long-term fire effects prediction under different fire regimes could prove useful for developing fire management prescriptions to meet resource management objectives.

## Short description

FIRESUM is a deterministic model containing stochastic properties. Tree growth, woody fuel accumulation, and litterfall are simulated deterministically, whereas tree establishment and mortality algorithms are stochastic. The model simulates all processes on an individual tree level in a 400 m² area called simulation plot. Since the particular combination of stochastic events occurring within a fiven FIRESUM simulation represent only one case among the set of many possible simulation outcomes, the model repeats simulations many times to obtain an average of simulated results.

FIRESUM is a gap-replacement model following the approach used for JABOWA in which individual trees are grown deterministically using an annual time step, difference equation. Tree growth is affected by several site factors including available light, water stress, and growing season warmth. Tree establishment and mortality are modeled stochastically using Monte-Carlo techniques. Fuel loadings are calculated yearly. Fires are introduced at various intervals and effects of each fire are simulated by reduction of litter, duff, and down woody fuels; and by tree mortality and by post-fire tree regeneration and growth.

## Model application(s) and State of development

Application area(s):
Number of cases studied:
State: calibration

## Model software and hardware demands

Platform: PC-based IBM, Workstation SUN
OS: DOS
Software demands: FORTRAN
Hardware demands: 386, math coprocessor

## Model availability

Purchase: yes, free. Both executable and source code
Contact person(s) Dr. Bob Keane
                     Intermountain Fire Science Laboratory
                     Forest Service
                     U.S. Department of Agriculture
                     P.O. Box 8089
                     Missoula, MT 59807 USA

Phone: +1 406-329-4846
Fax: +1 406-329-4877
E-mail:

## Model documentation and major references

Keane, R.E., Arno, S.F., and Brown, J.K., 1989. FIRESUM - An Ecological Process Model for Fire Succession in Western Conifer Forests. Res. Pap. INT-266. Ogden, UT: U.S. Department of Agriculture, Forest Service, Intermountain Research Station.

**Other relevant information**
FIRESUM was created by extensively modifying the process model SILVA (Kercher and Axelrod 1981), which simulates forest succession involving fire in mixed conifer forests of the California Sierra Nevada. Parameters and algoritms within SILVA were revised, deleted and added to reflect current knowledge of ecological processes inherent in various types of forests. Currently, FIRESUM, can be applied to ponderosa pine/ Douglas fir and whitebark pine/subalpine fire forest of the inland Northwest and the northern Rocky Mountains.

**Model identification**
Model name/title: FINICH

**Model type**
Media: local scale, terrestrial, shrubland
Class:
Substance(s)/compounds:
Biological components: community - chaparral, coastal sage scrub population - shrub species

**Model purpose**
To project community and population response of California Shrublands to the interaction of climatic change and fire intensity and frequency.

**Short description**
Fire intensity is calculated based on fuel accumulated since last fire. Regeneration depends on fire intensity. Growth and survival are functions of annual precipitation and mean January and July temperatures; these functions are presently derived through Geographic Distribution Analysis. Growth and death after the fuel conditions for the next fire. Climate forcing alters species responses, changes fuel conditions, and thus regeneration.

**Model application(s) and State of development**
Application area(s): Southern California
Number of cases studied:
State: conceptualization

**Model software and hardware demands**
Platform: Workstation HP 720
OS: HP-UX (UNIX)
Software demands: C
Hardware demands:

## Model availability
Purchase: no, available as source code
Contact person(s) Dr. George Malanson
                                   Dept. of Geography
                                   University of Iowa
                                   Iowa City, IA 52242 USA

Phone: +1 319-335-0151
Fax: +1 319-335-2725
E-mail: george-malanson@uiowa.edu

## Model documentation and major references
Malanson, G.P., Westman, W.E., and Yan, Y.-L., 1992. Realized versus fundamental niche functions in a model of chaparral response to climatic change. *Ecol. Modelling*, 64: 261-277.

## Other relevant information
Work is in progress to change climate response parameters based on physiologically mechanistic models and to link to a landscape scale fire behaviour model (REFIRES of F. Davis, University of California, Santa Barbara, CA).

## Model identification
Model name/title: MOSEL

## Model type
Media: local scale, regional scale, terrestrial, forest
Class:
Substance(s)/compounds:
Biological components: community - forests of eastern North America and northern Rocky Mountains

## Model purpose
To include dispersal and a spatially explicit landscape for existing forest dynamics models.

## Short description
A JABOWA-FORET type model is run on multi-cell grid. Dispersal among cells is simulated using Monte Carlo method where the probabilities differ among propagule morphologies.

    The grid may be defined to include barriers and environmental gradients; it is designed to use GIS input for initial environmental conditions.

## Model application(s) and State of development
Application area(s): Upper Midwest, USA. Montana, USA
Number of cases studied:
State: conceptualization

**Model software and hardware demands**
Platform: Workstation HP 720
OS: HP-UX UNIX
Software demands: C
Hardware demands: lots

**Model availability**
Purchase: no, available as source code
Contact person(s) Dr. George Malanson
                Dept. of Geography
                University of Iowa
                Iowa City, IA 52242 USA

Phone: +1 319-335-0151
Fax: +1 319-335-2725
E-mail: george-malanson@uiowa.edu

**Model documentation and major references**
Malanson, G.P., Armstrong, M.P. and Bennett, D.A., 1993. Fragmented forest response to climatic warming and disturbance, in proceedings of the 2nd International Conference/Workshop on Integrating GIS and Environmental Modelling. NCGIA, Santa Barbara, CA.

**Other relevant information**
Also see: Hanson, V.S., Malanson, G.P., and Armstrong, M.P., 1990. Landscape fragmentation and dispersal in a model of riparion Forest dynamics. *Ecol. Modelling*, 49: pp. 277-296 - an earlier version of MOSEL. Ongoing work is examining scaling questions.

**Model identification**
Model name/title: SHE (European Hydrologic System/Systéme Hydrogique Europeèn)

**Model type**
Media: Water, lake, river, wetland, local scale, regional scale, terrestrial, sediment, soil, forest, agricultural
Class: Hydrology
Substance(s)/compounds: nutrients, phosphorus, nitrogen, chemicals
Biological components:

**Model purpose**
A wide range of environmental applications.

**Short description**
The SHE is composed of a number of modules, at primarily the following: Interception/ Evapotranspiration, Overland and change flow, Unsaturated Zone, Saturated zone, Snow melt, Frame, for organising other components.
It is available as a fourth-generation modeling system, as 'MIKE SHE'.

**Model application(s) and State of development**
Application area(s): Hydrology, with environmental superstructures
Number of cases studied: 20-30 major studies
State: validation, validation of prognosis, fully production status

**Model software and hardware demands**
Platform: Workstation, MOST
OS:
Software demands:
Hardware demands:

**Model availability**
Purchase: DKR 500.000,00 (+/-)
Contact person(s) Dr. J.C. Refsgaard
              D.H.I.
              Agern Allé 5
              DK-2920 Hørsholm, Denmark

Phone: +45 42 86 80 33
Fax:
E-mail:

**Model documentation and major references**
Available from DHI

**Other relevant information**

**Model identification**
Model name/title: European Hydrologic System/Système Hydrogique Européen

**Model type**
Media: water, lake, river, wetland, local scale, regional scale, terrestrial, sediment, soil, forest, agricultural
Class: biogeochemical, toxicology, hydrology
Substance(s)/compounds: nutrients, carbon, phosphorus, nitrogen, sulphur, oxygen chemicals, radionuclides, pesticides, organic compound, heavy metals
Biological components: organism - zooplankton, algae, grass

## Model purpose
Full range of applications to environment, agronomic and other problems, especially theses involving conjunctive uses and notes.

## Short description
The SHE is a complete hydrologic modeling system that has now been raised largely to a present generation status (as MIKE-SHE). Initiated in 1977, it has been developed by three European organizations working is cooperation. Over the last five years, each of these organizations has followed an own development trajectory. A recent brochure is enclosed.

## Model application(s) and State of development
Application area(s): see the enclosed brochure for the DHI development, other developments on request.
Number of cases studied: about 100
State: verification, calibration, validation

## Model software and hardware demands
Platform: all principal platforms accommodated
OS: all common systems
Software demands:
Hardware demands:

## Model availability
Purchase: yes, sold on modular basis. Total about D.Kr. 500.000 available as executable code
Contact person(s) Dr. M. B. Abbott, project manager, ASHE
    v.d. Houvensstraat 63
    NL-2596 P. M., Den Haag
    The Netherlands

Phone: + 31 70 3243511 or + 31 15 151812
Fax: + 31 70 3282069
E-mail:

## Model documentation and major references
Available from DHI and possibly from other ASHE partners.

## Other relevant information

## Model identification
Model name/title: 'ECOSYS'

## Model type
Media: wetland, local scale, terrestrial, soil, forest, agricultural
Class: biogeochemical, hydrology
Substance(s)/compounds: nutrients, carbon, phosphorus, nitrogen, oxygen, saltions: $CO_2$, $Mg^{2+}$, $K^+$, $Na^+$, $SO_2^{2-}$, $Cl^-$, $H^+$, $OH^-$, $HCO_3^-$, $CO_3^{2-}$
Biological components: ecosystem - terrestrial

## Model purpose

## Short description
All enclose documentation refers to the single model 'ecosys', in which published literature is cited.

## Model application(s) and State of development
Application area(s): Resource management, climate change
Number of cases studied:
State: conceptualization, verification, validation, prognosis made, validation of prognosis

## Model software and hardware demands
Platform: Mainframe - CRAY-2
OS: UNIX
Software demands: none, language FORTRAN 77
Hardware demands: at least 20 MB RAM

## Model availability
Purchase: no, available as source code
Contact person(s) Dr. Robert Grant
             Dept. of Soil Science
             University of Alberta
             Edmonton, Alberta T6G 2E3
             Canada

Phone: +1 403 492 6609
Fax: +1 403 492 1767
E-mail: sccc@mts.ucs.valberta.ca

## Model documentation and major references

## Other relevant information

## Model identification
Model name/title: The Growing and Grazing Season in the United Kingdom.

## Model type
Media: agricultural
Class:
Substance(s)/compounds:
Biological components: ecosystem - managed grassland

## Model purpose
To provide estimates of the growing and grazing season for managed grassland in a temperate climate based upon estimates of the dry matter growth of grass.

## Short description
Dry matter is estimated from absorbed solar radiation by means of a temperature-dependent conversion efficiency. Leaf area is related to leaf extension. Leaf extensions is a function of temperature.
Input variables: daily mean air temperature, daily solar radiation total (or sun hours).

## Model application(s) and State of development
Application area(s): The model is a research tool and needs testing against observations.
Number of cases studied:
State:

## Model software and hardware demands
Platform: Mainframe
OS:
Software demands: language FORTRAN
Hardware demands:

## Model availability
Purchase: no
Contact person(s) Dr. M.N. Hough
    Meteorological Office
    Johnson House, London Road
    Bracknell
    Berks, U.K. RG12 2SY

Phone: +44 344 856205
Fax:+44 344 854588
E-mail:

---

**Model documentation and major references**
*Grass and Forage Science*, 1993, 48: 26-37.

---

**Other relevant information**

---

**Model identification**
Model name/title: EXE (Energy, Water, and Momentum Exchange and Ecological Dynamics)

---

**Model type**
Media: local scale, terrestrial, soil, forest
Class: biogeochemical, hydrology
Substance(s)/compounds: nutrients, carbon
Biological components: ecosystem - forest, organism - trees

---

**Model purpose**
To provide a tool to study ecosystem and hydrological responses to climate perturbations.

---

**Short description**
The Energy, Water, and Momentum Exchange and Ecological Dynamics Model (EXE) is a climatically-sensitive gap-phase forest dynamics model. EXE couples a physically- and physiologically-based water budget with an explicit treatment of ecological dynamics. As input, EXE requires climatic (radiation, air, temperature, precipitation, air humidity, and wind) data - which constitute key control variables - and sylivicultural information on the plants species to be simulated - which determines the choice crucial parameters. In principle, EXE could be forced by atmospheric general circulation model output. As output, EXE provides a complete picture of the microclimate and the ecology of a model forest (the main state variable is diameter at breast height for each individual of each species). EXE is made of two modules; an ecological one and a physical one. The biological routines in LINKAGES, a model developed by Pastor and Post, provided the framework for the ecological module. Significant changes were made to couple physiology to physics in an effort to make the model realistically sensitive to climate. The microclimate physics module is based on physical principles. This module was built from scratch. Evaporation is an explicit function of the amount of water available in the soil which can be taken up at different depths by plant roots and of the atmospheric demand for water vapour. Forest hydrology and forest microclimate are computed once during daytime and once during nighttime for each day of the year. The information regarding the hydrology and the microclimate is transferred to the ecological module. On the basis of this environmental information, the ecological dynamics of the forest are then simulated. Over time, ecological dynamics results in new physical and physiological forest characteristics. When appropriate these characteristics are changed. This "physics-ecology-physics" cycle is repeated each simulation year.

## Model application(s) and State of development
Application area(s): forest and hydrological dynamics, global change, biogenic trace gas emissions.
Number of cases studied:
State: validation of prognosis

## Model software and hardware demands
Platform: Workstation - IBM RISC/6000 580
OS: UNIX
Software demands: F77 or XLF, language FORTRAN
Hardware demands:

## Model availability
Purchase: no, software available both as executable and sourcecode (on a one- to- one basis).
Contact person(s) Dr. Philippe Martin
European Commission, TP440
I-21020 ISPRA (Varese), Italy

Phone: +39 332 785868
Fax: +39 332 789073
E-mail: philippe.martin@jrc.it

## Model documentation and major references
Martin, Ph., 1992. EXE: A climatically sensitive model to study climate change and $CO_2$ enrichment effects on forests. *Australian Journal of Botany*, 40, 717-735.
Martin, Ph., 1990. Forest Succession and Climate Change: Coupling Land-Surface Processes and Ecological Dynamics. Cooperative Thesis No. 131 between the University of California at Berkeley and the National Center for Atmospheric Research (NCAR), Boulder, CO. University Microfilms International (UMI) Catalog Number 9126692, 342 pp.

## Other relevant information

## Model identification
Model name/title: Abiotic Landscape Model

## Model type
Media: regional scale, agricultural
Class:
Substance(s)/compounds: nutrients, phosphorus, nitrogen
Biological components: ecosystem - agro-ecosystem

**Model purpose**
Classification of landscape-types, scenarios.

**Short description**
Still under construction.

**Model application(s) and State of development**
Application area(s): Tertiarres Húgelland, Bovania
Number of cases studied: 1
State: conceptualization

**Model software and hardware demands**
Platform: Digital
OS: OSF1 2.0
Software demands: ARC/INFO, language ARC/INFO Macrolanguage
Hardware demands: UNIX Workstation

**Model availability**
Purchase: not yet
Contact person(s) Dr. Ruth Lang
Lehrstuhl für Landschaftsökologie
D-85350 Freising
Germany

Phone: +49 8161 714149
Fax: +49 8161 714427
E-mail: ruth@fam.loek.agrar.tu-muenchem.de

**Model documentation and major references**

**Other relevant information**
Lang, R., Lenz, R., and Müller, A., 1993. The Agroecological Information System in the FAM. International Congress on Modelling and Simulation. Dec. 6-10, 1993, Perth/ Australia. Proceedings, Vol. 4, pp. 1407-1412.

**Model identification**
Model name/title: TARGETS : Tool to Assess Regional and Global Environmental and health Targets for Sustainability

**Model type**
Media: water, lake, ocean, river, wetland, air, regional scale, global scale, terrestrial, sediment, soil, forest, agricultural
Class: biogeochemical, toxicology, hydrology

Substance(s)/compounds: nutrients, carbon, sulphur, phosphorus, nitrogen, others (greenhouse gases: $CO_2$, $CH_4$, $N_2O$, CO, OH, $O_3$, CFC11, CFC12, CFC113, $CCl_4$, $CH_3CCl_3$, HCFC22, HCFCs, HFCs) chemicals, pesticides, organic compounds, heavy metals (Hg, Pb, Cd, Cu, Zn)

Biological components: 14 land types : rain-fed arable land (tropical and temperate), forest (tropical and temperate), grass (tropical and temperate), irrigated land (tropical and temperate), human area (tropical and temperate), other (e.g., desert, wetlands) (tropical and temperate), forest cold, grass cold.

## Model purpose
TARGETS might be used to perform an analysis and assessment on a global scale of the quantitive and qualitative linkages among social and economic processes, biophysical processes and effects on ecosystems and humans from an integrated system dynamics perspective.

## Short description
The whole TARGETS model consists of a human system (including population and health dynamics and energy/resources and economics) and an environmental system which contains an element cycle module, a land module and a water module.

The element cycle module simulates the biogeochemical and physical processes that result in the cycling of carbon, nitrogen, phosphorus and sulphur between the atmosphere, the terrestrial biosphere, the lithosphere (soils) and the hydrosphere (rivers and oceans). Furthermore it describes the cycles of toxic substances such as heavy metals, pesticides and organic micro-pollutants.

The land module simulates land cover changes, land productivity changes and food demand and supply. The land module consists of an agricultural module (focusing on irrigation, land clearing, intensification, food production and degradation) and a land use change module (which deals with deforestation and urbanisation).

The water model describes the casual relations between pressures on the water system (water demand and emission) the dynamics of the water system (hydrology and water quality) and the performance of societal and ecological functions of the water system.

## Model application(s) and State of development
Application area(s): policy analysis on a global scale. Will be used for the World environmental outlook of the UNEP

Number of cases studied: by now only a river basin study of Ganges-Brahmaputra basin and Yangtze basin

State: conceptualization, calibration, validation, prognosis made

## Model software and hardware demands
Platform: PC-version under development, runs under UNIX systems (SUN, SILICON Graphics)

OS:

Software demands: language - C

Hardware demands: UNIX workstation

**Model availability**
Purchase: no (but yes at the end of 1995).
Contact person(s) Dr. M.G.J. den Elzen (land module and element cycles module)
     or Dr. A.Y. Hoekstra (water module)
     Global Dynamics and Sustainable Development
     P.O. Box 1
     NL-3720 BA Bilthoven
     The Netherlands

Phone: +31 30 743320
Fax: +31 30 252973
E-mail: GLOBO@rivm.n1

**Model documentation and major references**
Elzen, M.G.J. den, 1993. *Global Environmental Change: An Integrated Modelling Approach*, International Books, Utrecht, The Netherlands.
Elzen, M.G.J. den, and Rotmans, J., 1994. Modelling global biogeochemical cycles: an integrated modelling approach, RIVM Report No. 461502007, Bilthoven, The Netherlands,
Elzen, M.G.J. den, Koster, H.W., Martens, W.J.M., Hilderink, H., 1994. The global land model of TARGETS, RIVM Report No. 461502005, Bilthoven, The Netherlands, in preparation.
Hoekstra, A.Y., 1994. AQUA: an integrated tool for global water policy making, RIVM-Report No. 461502006, in preparation, Bilthoven, The Netherlands.
Rotmans, J., and Elzen, M.G.J. den, 1993. Modelling feedback processes in the carbon cycle: balancing the carbon budget, *Tellus* 45B.
Rotmans, J. et al., 1994. Global change and Sustainable Developments: a modelling perspective for the next decade, RIVM Report No. 461502004, Bilthoven, The Netherlands.

**Other relevant information**

## 6.4 An Overview: Auxiliary Models as Modeling Tools

This section covers a spectrum of various modeling tools or auxiliary models which are able to assist the modeler in such preparatory tasks as, for instance, estimations of parameters and equations. These auxiliary models are based on the use of statistics, artificial intelligence or similar tools able to facilitate the interpretation of data or they are models of a very general character, which may be used as a supporting model component. This section can obviously not cover the wide range of statistical software or software packages for the application of artificial intelligence methods, but includes only "supporting models" which are of some use for environmental modelers.

GES is model or software able to estimate generally equation for ecosystem processes, based on observations (data). It is developed as a statistical regression analysis, but includes several features, which makes it better fitted for environmental observations; see the description in detail in Section 6.5.

Menard has developed a stochastic model for ordering categorical series of data with covariance. Parameters of the model are estimated by maximum likelihood. The order of the process and the influence of exogenous covariates are assessed by the likelihood ratio statistics.

Dejak et al. have developed a model for estimation of the average space-time annual pattern for the water temperature in a water basin, where thermocline stratification occurs. The stratification is simulated by modeling the process of vertical diffusion of heat. It has been applied on the Lagoon of Venice, where stratification occurs in a few deep channels between the shallow waters.

Minimum negentropy criterion can used to determine the optimal order of a harmonic model, aimed to reproduce seasonal behaviour of environmental variables, also with stochastic fluctuations. A model has been developed to carry out this task. A Fourier series expansion is used to determine the seasonal component and a minimum negentropy is used as a goal function to determine the optimum model. An overview of various structural models using different goal functions are reviewed in Chapter 8. Negentropy is one of the possible goal functions, but no modelling tool has been developed so far for the direct coupling of optimization by the use of a goal functions with other possible modeling components.

Simpson has developed a thermocline model for a fjord system. The model is based upon a description of the actual hydrodynamics and not upon an energy balance as most thermocline models for lakes are.

Schernewski et al. present a thermocline model based upon the heat balance at the lake surface. It accounts for the net global radiation, long-wave and atmospheric back radiation, lake evaporation and convection.

Vassiliou et al., 1988 have developed a model describing seasonal changes of biotic and abiotic variables. It consists of transition matrices denoting the periodical variation of the system structure. The model is applied to describe changes in species richness and population density of zoobenthos in a heavily polluted marine ecosystem.

Tie et al. have developed a 3-D model for the global transport of chemical trace gases. It is based upon a grid system and includes horizontal and vertical advection and diffusion.

Mmgep is a global model for ecological processes. It focuses on the hierarchy of biogeochemical, biocenotic and hydrological processes in the biosphere. It contains a prognosis of the prolonged consequences of the anthropogenic impacts on these processes.

The model denoted Mrf is able generally to simulate the transport of solute in heterogeneous environments. The task is accomplished by a class of stochastic models, termed mass response functions, which incorporate simplified concepts of chemical/physical non-equilibrium kinetics in the formulation of transport by travel time distribution.

Kirsta has developed an information model of natural hierarchical systems which is able to simulate natural system organization, exchange of information between hierarchical levels and formation of information products by these processes. The model is a useful tool to simulate hierarchical systems.

Wu has developed a model which is able to simulate the effects of patch connectivity, metapopulation dynamics and persistence in a variety of scenarios. The state variables in the model are the sizes of local populations in habitat patches.

Loehle and Johnson have proposed a microbial subsurface transport model. It links percolation models with microbial dynamics in a multi-chemostat formalism.

**Table 6.2.** Models as Modeling Tools

| Model name<br>Model components | Model characteristics | References |
|---|---|---|
| GES | Regression/equations of ecological processes | Liang et al., 1992 |
| Time series | Stochastic model ordered time series with covariates | Menard et al., 1993 |
| Thermocline | Vertical temperature profile | Dejak et al., 1992 |
| Negentropy | Negentropy applied as goal function | See Section 6.6. |
| Stratification | Stratification determined by hydrology | Simpson; see model |
| Thermocline | Heat balance at lake surface | Schernewski et al., 1994 |
| Mmgep | Global ecological processes | Krapivin, 1993 |
| Mrf | Transport of solute species in heterogeneous environment | Rinaldi and Marani, 1989 |
| Seasonal changes | Biotic and abiotic variables. Transition matrices denote the variation of system | Vassiliou et al., 1988 |
| Trace gases | 3-D distribution of trace gases | Tie et al., 1992 |
| hierarchical systems | information and exchange of information between hierarchical levels | Kirsta, 1994 |
| Patch connectivity | Effects of subpopulation size, inter-patch immigration and pop. dynamics | Wu and Vankat, 1991 |
| Subsurface transport | Links percolation model with microbial dynamics | Loehle and Johnson 1995 |

## 6.5 Auxiliary Models as Modeling Tools

**Model identification**
Model name/title: Generalized Estimating Equations

**Model type**
Media: water, lake, ocean, estuary, river, wetland, terrestrial, sediment, soil, forest, agricultural
Class:
Substance(s)/compounds:
Biological components: ecosystem - all, community - multispecies, populations - all, organism - longitudinal data

**Model purpose**
To fit regression models to correlated data.

**Short description**
Generalized estimating equations (GEE) are a methodology suited for the statistical regression analysis of correlated data. For example, observations on twins or pairs of eyes, ecological multispecies data, longitudinal data, etc. It can be fitted when the correlations are either of primary interest (e.g., in synecological studies) or are a nuisance. With correlated data, GEE is superior to the usual generalized linear models (GLMs) because the standard errors do not necessarily depend on an independence assumption, therefore inference are more reliable. Unlike GLMs, the response may vector-valued. That is, GEE operates on data $(y_i, X_i)$, $i=1,....,n$, where $y_i$ is the response vector of the i'th cluster and $x_i$ the explanatory variables. The response $y_i$ may be, e.g., continuous, binary, count, etc. - just like for GLMs.

**Model application(s) and State of development**
Application area(s): Medicine, biology
Number of cases studied: many
State: conceptualization, verification, calibration, validation

**Model software and hardware demands**
Platform: PC-based IBM, Mainframe IBM, Workstation all
OS:
Software demands:
Hardware demands:

**Model availability**
Purchase:
Contact person(s)
Phone:
Fax:
E-mail:

## Model documentation and major references
Liang, Zeger and Qagish, 1992. *J. Roy. Statist. Soc. B.*, 54: 3-40.

## Other relevant information
We, (Yee and Mitchell) are currently preparing a paper entitled "Generalized estimating equations in plant ecology", that describes some of the ideas and their potential use in plant ecology. We intend to submit it to *J. of Vegetation Science* in about 3 months time (ca., October 1994).

Also, Yee and Wild are preparing a paper describing nonparametric GEE. This will be submitted to *J. Roy. Statist. Soc. B.* soon. Nonparametric GEE will be briefly described in Yee and Mitchell.

## Model identification
Model name/title: Stochastic model for ordered categorical time series with covariates

## Model type
Media:
Class:
Substance(s)/compounds:
Biological components:

## Model purpose
Analysis of semi-quantitative time series. The model allows to assess the influence of possibly continuous and time-dependent stochastic covariates on the basic process.

## Short description
An observed time series is considered as a particular realization of a stochastic process in discrete time $(Y t) t < $ or $ = 1$, where t is time of observation. $Y_t$ takes its values in a set of ordered categories. Concomitantly, a stochastic process is observed, which corresponds to the covariates of interest. The model is written in terms of the cumulative probability function of the basic process $Y_t$, conditionally on the observed past. The cumulative probabilities are a function of linear combination of the previous state of the basic process (the number of previous states determines the order of the process), and the exogenous covariates taken at previous times. The logistic form of McCullagh's generalized linear model for ordered categorical outcomes has been chosen. To account for any kind of periodic modulation in the data, one assumes that the transition probabilities are perodic functions of time. Parameters of the model and their asymptotic standard deviations are estimated by maximum likelihood. The order of the process, and the influence of exogenous covariates, are assessed by the likelihood ratio statistics.

## Model application(s) and State of development
Application area(s): MARINE ECOLOGY: temporal fluctuations of zooplankton populations
Number of cases studied: 2 Mediterranean salp populations. Analysis of the influence of environmental variables.
State: conceptualization

---

**Model software and hardware demands**
Platform: DIGITAL
OS:
Software demands: language FORTRAN
Hardware demands:

---

**Model availability**
Purchase: no, available as source code
Contact person(s) Dr. Frédéric Ménard
　　　　　　　　Département de biostatistique et informatique médicale
　　　　　　　　1 av. Claude Vellfaux
　　　　　　　　F-75475 Paris Cedex 10
　　　　　　　　France
Phone: +33 1 42 49 97 98
Fax: +33 1 42 49 97 45
E-mail:

---

**Model documentation and major references**
Ménard, F., Dallot, S., and Thomas, G., 1993. A stochastic model for ordered categorical series. Application to planktonic abundance data. *Ecol. Modelling*, 66: 101-112.
Ménard, F., Dallot, S., Thomas, G., and Braconnot, J.C. (in press). Temporal fluctuations of two mediterranean salp populations from 1967 to 1990. Analysis of the influence of environmental variables using a Markov chain model. *Marine Ecology Progress Series*.

---

**Other relevant information**
Although the model was developed to analyze the influences of environmental variables on salp abundance, it is in effect on all-purpose regression model for the analysis of ordered categorical time series with stochastic covariates.

---

**Model identification**
Model name/title: 1 D vertical model

---

**Model type**
Media: water, lake, estuary
Class: heat balance
Substance(s)/compounds: temperature
Biological components:

---

**Model purpose**
The seasonal dynamic of spatially differentiated aquatic ecosystem is driven by two important forcing functions: sun radiation and water temperature. This model provides a way for estimating the average space-time annual pattern for the water temperature in a average shallow water basin, where thermocline stratification occurs in a few deep channels. In this case, even a broad data base of experimental measurement might not provide a reliable input data set for a fine grid mathematical model of the system.

**Short description**

The problem can be approached by using time series of metereological data commonly collected by weather forecast sampling station, for evaluating the average heat balance for a shallow water basin. Stratification is simulated by modeling the process of vertical diffusion of heat. Once the balance has been written in differential form, water temperature is found as the asymtotic periodic solution of a ordinary differential equation, for a completely mixed basin, or of a partial differential equation, if transport phenomena within the basin need to be taken into account. For the case of Venice Lagoon, a set of vertical diffusivities has been estimated, based on the statistical analysis of about a thousand of thermal profile. Data needed for the calculation are incident solar radiation, wet and dry bulb air temperature, albedo, wind speed and cloudiness. Once applied to a single water column, the 1-D version of the model can be used for improving the estimation of the open boundary conditions for 2-D and #D models of a polluted area, as a subprogram which follows the behaviour of the same state variables under unperturbed conditions can be easily implemented.

**Model application(s) and State of development**

Application area(s): Venice Lagoon
Number of cases studied:
State: calibration, validation

**Model software and hardware demands**

Platform: PC-based, IBM
OS: UNIX, DOS
Software demands: FORTRAN
Hardware demands:

**Model availability**

Purchase: no, executable code and sourcecode
Contact person(s) Dejak Camillo, Pastres Roberto, Pecenik Giovanni
        Dorsoduro 2137
        Universit' di Venezia
        Dip. Chimica Fisica
        Calla Larga S.Marta 27
        I-30123 Venezia, Italy
Phone:
Fax:
E-mail: pastres@unive.it

**Model documentation and major references**

Dejak, C., Franco D., Pastres R., Pecenik G., and Solidoro C., 1992. Thermal exchanges at air-water interfacies and reproduction of temperature vertical profiles in water columns. *J. Marine Systems*, Elsevier, 3:465-476.

**Other relevant information**

**Model identification**
Model name/title: Environmental periodic time series

**Model type**
Media: water, lake, river, estuary
Class: hydrology
Substance(s)/compounds: nutrients, phosphorus, nitrogen, Chl a.
Biological components:

**Model purpose**
A minimum negentropy (cross-entropy) criterion is utilized to determine the optimal order of a harmonic model aimed to reproduce the seasonal behaviour of environmental variables, also with short time series and high stochastic fluctuations.

**Short description**
The procedure, which involves the computation of the prediction error, represents a unifying approach of coventional and inferential statistics. At the same time it accounts for all informational content embodied in the data and it avoids both underfitting and overfitting of experimental observations, thus guaranteeing the principle of model's parameter parsimony and balanced accuracy.

The method is implemented in a package for the analysis of environmental data, and for the determination of their seasonal component. Through the transformation of the raw data, their interpolation, and the detrending with moving averages, and on the basis of the Wold's theorem, the data are divided into a periodical component and a stochastic one.

A Fourier series expansion is used to determine the seasonal component, and a minimum negetropy method based on a Kullback-Leibler formula is employed to find output.

The model output, i.e., the optimal harmonic interpolation, can be used for a recalibration or anic model.

**Model application(s) and State of development**
Application area(s):
Number of cases studied:
State:

**Model software and hardware demands**
Platform:
OS:
Software demands:
Hardware demands:

**Model availability**
Purchase: no
Contact person(s) Dr. Camillo Dejak
                   Dorsoduro 2137
                   Universit di Venezia
                   Dip Chemica Fisica
                   Calla Larga S.Marta 2137
                   I-30123 Venezia, Italy
Phone:
Fax:
E-mail:pastres@unive.it

**Model documentation and major references**

**Other relevant information**

**Model identification**
Model name/title: The Clyde Sea: a Model of the Seasonal Cycle of Stratification and Mixing.

**Model type**
Media: ocean, estuary
Class: physics
Substance(s)/compounds: nutrients, oxygen, temperature, salinity
Biological components: ecosystem, primary production

**Model purpose**
1) To demonstrate understanding of key processes controlling the seasonal cycle of stratification in the Clyde Sea fjord system. 2) To predict the complete mixing and bottom water removal. 3) To provide a basic for nutrient/primary production models.

**Short description**
The processes controlling stratification are represented in a filling box model of the Clyde system in which the exchange flow with the North Channel is related to the surface density difference across the sill. The stratifying effect of buoyancy inputs as heat and freshwater is opposed by mechanical stirring due to (i) wind-stress and (ii) the barotrophic tide. Additional stirring contributions from (iii) convective effects associated with deep-water replacement and (iv) the internal wave mechanism proposed by Stigebrandt (1976) may also contribute.

      Forcing by (i) freshwater input from rivers, (ii) surface heating/cooling, (iii) wind-stress, (iv) tide, (v) boundary conditions in the North Channel.

485

---

**Model application(s) and State of development**
Application area(s):
Number of cases studied:
State:

---

**Model software and hardware demands**
Platform:
OS:
Software demands:
Hardware demands:

---

**Model availability**
Purchase:
Contact person(s) Dr. J.H. Simpson
           School of Ocean Science
           University of Wales
           Bangor Menai Bridge
           Gwynedd, England
Phone:
Fax:
E-mail:

---

**Model documentation and major references**

---

**Other relevant information**

---

**Model identification**
Model name/title: Thermocline Model

---

**Model type**
Media: lake
Class: biogeochemical (heat)
Substance(s)/compounds: heat energy
Biological components:

---

**Model purpose**
Modelling the depth of the thermocline and the vertical temperature distribution as well as the surface heat energy balance in stratified lakes in the course of a year. This can serve as a basis for studying and modeling the eutrophication process in stratified lakes.

**Short description**
In a first step the heat energy balance at the lake surface is set up, including net global radiation, longwave atmospheric and back radiation, lake evaporation and convection. This acts as an input to the heating algorithm of the model which utilizes a forward difference scheme in time for solving the standard heat equation (parabolic differential equation) in a simple form numerically. Wind mixing is introduced into the model using a comparison between the potential energy of the stratification and the kinetic energy entering the lake by wind. These steps of successive heating and wind mixing are iterated each time step of the model (1 day) using the surface heat energy flux as a control.
State variables: water temperature in epilimnion, depth of the thermocline, atmospheric radiation, back radiation, convection.
Forcing functions: wind velocity (in 10 m height), global shortwave radiation, lake evaporation, air density.
Important parameters: wind shear coefficient, proportion of global shortwave and longwave radiation, respectively, reflected at the lake surface, coefficient of cloud type, psychometric constant, coefficient of heat diffusion of water, heat capacity of water, maximal density of water.
Necessary inputs: cross sectional area of the lake (depending on the depth).

**Model application(s) and State of development**
Application area(s): lakes
Number of cases studied: 1
State: verification, calibration

**Model software and hardware demands**
Platform: PC-based - IBM
OS: DOS 5
Software demands: language - TURBO PASCAL
Hardware demands:

**Model availability**
Purchase: yes, no price. Software available as source code.
Contact person(s) Dr. Lars Theesen
        Ecosystem Research Center
        University of Kiel
        Schauenberger Str. 112
        D-24118 Kiel
        Germany

Phone: +49 4321 880 4034
Fax: +49 4321 880 4083
E-mail: lars-oekosys.uni-kiel.d400.de

**Model documentation and major references**

Schernewski, G., Theesen, L., and Kerger, K.E., 1994. Modelling thermal stratification and calcite precipitation of Lake Belau (Northern Germany). In: Proc. 8th ISEM Conference, Kiel 1992. *Ecol. Modelling.*

Hurley Octavio, K.A., Jirka, G.H., and Harleman, D.R.F., 1977. Vertical Heat Transport Mechanisms in Lakes and Reservoirs. Massachusetts Institute of Technology, Rep. No. 227.

**Other relevant information**

**Model identification**

Model name/title: MMGEP: Mathematical Model of Global Ecological Processes

**Model type**

Media: global scale
Class: biogeochemical
Substance(s)/compounds: nutrients, carbon, phosphorus, nitrogen, sulphur, oxygen
Biological components: ecosystem - oceanic, terrestrial

**Model purpose**

A purpose of MMGEP is the parametrization of the hierarchy of the biogeochemical, biocenotic and hydrological processes in the biosphere. It gives a prognosis of the consequences of the anthropogenic impacts on these processes.

**Short description**

MMGEP describes the interaction of the atmosphere with the land and ocean ecosystems. It comprises blocks describing biogeochemical cycles of carbon, nitrogen, sulphur, phosphorus and oxygen; global hydrologic balance in liquid, gaseous and solid phases; productivity of soil-plant formation with 30 types defined; photosynthesis in ocean ecosystems taking into account its depth and surface inhomogeneity; demographic processes and anthropogenic changes. The model is designed to be connected to a global climate model.

MMGEP includes the blocks describing the function of the world ocean pelagic, arctic and shelf ecosystems. The world ocean is divided in four parts: Arctic, Pacific, Atlantic and Indian Oceans.

The spatial structure of the MMGEP is determined by the databases available. Spatial inhomogeneity is provided for by the various forms of space discretization. The basic type of spatial discretization of the earth surface is a uniform geographic grid with arbitrary latitude and longitude steps.

MMGEP has about 50 forcing functions, up to 70 state variables and more 400 parameters. The inputs of MMGEP include the spatial initial data on the parametrized processes. Also MMGEP demands the forming a scenario for the human activity.

**Model application(s) and State of development**

Application area(s): global ecology, global changes
Number of cases studie: more than 30
State: conceptualization, verification, calibration, validation, prognosis made.

**Model software and hardware demands**
Platform: PC-based
OS: DOS
Software demands: standard library, C++
Hardware demands: IBM PC-486 with standard configuration

**Model availability**
Purchase: no
Contact person(s) Dr. Vladimir F. Krapivin
              Institute of Radioengineering and Electronics
              Russian Academy of Science
              Vvedensky sq. 1,
              Moscow reg., 141120 Russia

Phone: 095 526-9276
Fax: 095 203-8414
E-mail: cherepen%ire.phys.msu.su@neotext.ca

**Model documentation and major references**
Krapivin, V.F., Mathematical model for global ecological investigations. *Ecol. Modelling*, 67, N 2-4, 1993 : 103-127.

**Other relevant information**
MMGEP may help in formulation of requirements to be placed on the contents and structures of databases. Since the MMGEP includes regional scale models we may study the interactions between regional systems, a region being a land or ocean part within a certain set of geographic grid cells. The adaption of the MMGEP to the concrete problem demands the additional modification of different blocks of the MMGEP.

**Model identification**
Model name/title: MRF Mass Response Function

**Model type**
Media: river, regional scale, agricultural
Class: hydrology
Substance(s)/compounds: phosphorus, nitrogen, radionuclides, pesticides
Biological components: ecosystem - hydrologic cycle (surface water, groundwater)

**Model purpose**
The model is aimed to the simulation of the transport of passive and reactive solutes in surface and subsurface waters.

**Short description**
This model is concerned with the transport of solute species in heterogenous environments. The task is accomplished through a class of stochastic models, termed mass response functions, (MRFs), which incorporate simplified concepts of chemical/physical nonequilibrium kinetics in the formulation of transport by travel time distributions. MRF's are probability density functions (pdfs) associated with solute particles travel time within transport volumes. The model links the approaches of surface hydrologists with recent subsurface transport models. The water discharge and the solute flux is given by the convolution between the rainfall intensity and suitable transfer functions.

The transfer function of the solute flux is assumed proportional to the travel time pdf through a function which is in turn related to the solute concentration at equilibrium.

The input parameters are: i) parameters that identifies the travel time pdf: mean seepage velocity and dispersion coefficient, for groundwater applications. Mean convection velocity, hydrodynamics dispersion and mean residence time outside the channel network, for surface water. These parameters need to be computed for each channel and they are in turn related, through algebraic relations, to some basin properties as mean channel slope and contributing area.

Two parameters are also used for the division of the rainfall into net and lost parts. ii) Finally, the computation of the equilibrium concentration requires another parameter, termed distribution coefficients, that relates the equilibrium concentration to the solute mass instantaneously present into the control volume.
The applications of the model have shown a good agreement with experimental observations demonstrating the MRFs are rational models of the complex chain of events occurring in large-scale solute sorption and transport, and that they may be validly employed for quantitative studies of environmental impacts to the release of chemical species in surface or subsurface waters.

**Model application(s) and State of development**
Application area(s): Surface and subsurface water pollution
Number of cases studied: 2
State: validation

**Model software and hardware demands**
Platform: PC-based, Workstation
OS: DOS-UNIX
Software demands: no other software is necessary, language ANSI FORTRAN 77
Hardware demands: PC 486; any model of workstation

**Model availability**
Purchase: no
Contact person(s) Dr. Andrea Rinaldo (*)
                Istituto di idraulica G. Poleni, Universita' di Padova
                Via Loredan 20
                I-35131 Padova, Italy

or             Dr. Alberto Bellin
Dipartimento di Ingengeria Civile ed Ambientale
Via Mesiano 77
I-38050 Trento, Italy

Phone: +39 49 831448 (*)  or +39 461 882620
Fax: +39 49 831463 (*)  ·  or +39 461 882672
E-mail: rinaldo@idra.unipd.it (*)  or alberto@itncal.ing.unitn.it

---

## Model documentation and major references
Rinaldo, A., and Marani, A., 1987. Basin scale model of solute transport, *Water Resour. Res.* 23, 2107-2118,
Rinaldo, A., Marani, A., and Bellin, A., 1989. On mass-response functions, *Water Resour. Res.*, 25 (7), pp. 1603-1617, .
Rinaldo, A., Bellin, A., and Marani, A., 1989. A study on solute $NO_3^-$ N transport in the hydrologic response by an MRF model, *Ecol. Modelling*, 48, pp. 159-191.

---

## Other relevant information

---

## Model identification
Model name/title: Modelling the seasonal changes of biotic and abiotic variables

---

## Model type
Media: water
Class: biogeochemical, hydrology
Substance(s)/compounds:
Biological components: ecosystem - marine, community - zoobenthos

---

## Model purpose
To describe seasonal changes of biotic and abiotic variables.

---

## Short description
This is a non-homogeneous Markovian model aiming to describe seasonal changes of biotic and abiotic variables. This model focuses on the periodical changes of the system structure rather than the forcing factors. It consists of transition matrices denoting the periodical variation of the system's structure. In the general case input and output vectors could be used. The model was applied to describe changes in species richness and population density of zoobenthos in a heavily polluted marine ecosystem.

---

## Model application(s) and State of development
Application area(s):
Number of cases studied:
State: conceptualization, verification, prognosis made

---

**Model software and hardware demands**
Platform: PC-based
OS:
Software demands:
Hardware demands:

---

**Model availability**
Purchase: yes, available as sourcecode.
Contact person(s) Dr. D.P. Patouches       and   Dr. G. Stamou
                  Div. of Ecology
                  Dept. of Biology
                  UPB 119
                  University of Thessaloniki
                  Thessaloniki 54006, Greece

Phone: +031 998316
Fax: +031 998379
E-mail:

---

**Model documentation and major references**
Vassiliou, P.- C.G., 1984. *J. Appl. Prob.*, 21, pp. 315-325.
Vassiliou, P. - C.G., and Tsaklidis, G., 1988. *J. Appl. Prob.*, 27, pp. 766-783.

---

**Other relevant information**

---

**Model identification**
Model name/title: Three-dimensional Global Transport Chemical Tracer Model

---

**Model type**
Media: air, global scale
Class:
Substance(s)/compounds: chemicals
Biological components:

---

**Model purpose**
To simulate global distribution of trace gases and their potential changes due to natural and anthropogenic sources.

---

**Short description**
The Los Alamos CTM is a grid point model that uses dynamical and thermodynamical input from GCM to simulate the global atmospheric transport of chemical species. There are 40 latitudinal grid points with 4.5°-interval and 48 longitudinal grid points with a 7.5°-interval. In the vertical dimension, the model has 20 levels in the sigma coordinate system. The model top is about 9 mb.

Model transport includes horizontal and vertical advection; horizontal and vertical diffusion. Five precalculated dynamical and thermodynamical variables are required: the horizontal winds (u and v), the vertical wind (w), temperature (T), and the vertical diffusion coefficient (K). These variables are calculated once each model-day by the GCM and are interpolated in 30-minutes time step for the simulations of chemical tracers.

## Model application(s) and State of development
Application area(s): Global distribution of FCF-11, $CH_3CCl_3$ and $CH_4$
Number of cases studied: 3
State: verification

## Model software and hardware demands
Platform: Mainframe, CRAFT-YMP
OS:
Software demands: FORTRAN
Hardware demands:

## Model availability
Purchase: no
Contact person(s)

Phone:
Fax:
E-mail:

## Model documentation and major references
Tie, X., and Mroz, E.J., The potential changes of methane due to an assumed increased use of natural gas : A global three-dimensional model study, *Chemosphere*, 26, pp. 769-776, 1993.
Kao, C.-Y., Tie, X., Mroz, E., Cunnold, D., and Alyea, F.. Simulation of the global CFC-11. Using the Loa Alomos chemical tracer model, *J. Geophy. Res.*, 97, pp. 15827-15838, 1992.
Tie, X., Kao, C.-Y., Mroz, E., Cicerone, R., Alyea, F.N., and Cunnold, D.M.. Three-dimensional simulations of atmospheric methyl chloroform: effect of an ocean sink, *J. Geophys. Res.*,, 97, pp. 20751-20769, 1992.

## Other relevant information

## Model identification
Model name/title: Information models of natural hierarchical systems

## Model type
Media: lake, ocean, wetland, local scale, regional scale, global scale, terrestrial, forest, agricultural
Class:
Substance(s)/compounds: nutrients

Biological components: ecosystem - any type, community - any type, organism - any type

## Model purpose
The simulation of the information "core" of natural system organization, including the splitting of the last into consecutive hierarchical levels, exchange of information between levels, cyclic dynamics of level basic processes, formation of information products by these processes, and prognosis of system dynamics.

## Short description
The models are based on the established information principle of matter evolution: matter arranges itself in molecular-genetic, ecological and biospheric systems with hierarchical organization. The time cycles of each level includes a number of lower level time cycles according to certain time-quantization schemes. Models are developed for each hierarchical level and characterize: its basic process(es) as forcing function(s); four types of its information (entropy, ordering and two their derivatives) as state parameters; energy-matter and time products (transferring its information) as state variables; time as state variable and input. Models show substantial and time types of errors defined by simulated level information.

## Model application(s) and State of development
Application area(s): dynamics of any evolutionary developed hierarchical systems
Number of cases studied: more than ten
State: validation of prognosis

## Model software and hardware demands
Platform:
OS:
Software demands:
Hardware demands:

## Model availability
Purchase: no
Contact person(s) Dr. Yuri B. Kirsta
                Inst. for Water and Environ. Problems
                Russian Academy of Science
                P.O. Box 656
                656038 Barnaul, Russia

Phone: 3852 252140
Fax: 3852 240396
E-mail: iwep@iwep.altai.su

## Model documentation and major references
Kirsta, Y.B., 1992. Time-dynamic quantization of molecular-genetic, photosynthesis and ecosystem hierarchical levels of the biosphere. *Ecol. Modelling*, 62, pp. 259-274.
Kirsta, Y.B., 1994. Exchange of information in natural hierarchical systems. *Ecol. Modelling*.

**Other relevant information**
All suggestions for collaboration will be carefully considered.

**Model identification**
Model name/title: Effects of patch connectivity and arrangement on animal metapopulation dynamics: a simulation study

**Model type**
Media: regional scale, terrestrial
Class: population dynamics
Substance(s)/compounds:
Biological components: population - animal

**Model purpose**
To simulate the effects of subpopulation size, interpatch immigration, and patch configurations on metapopulation dynamics and persistence in a variety of scenarios.

**Short description**
The metapopulation models are composed of two and three single-patch population models. The state variables in these models are the sizes of local populations in habitat patches. It is assumed that local populations are subject to density-dependent regulation and that there exists a threshold population size below which local extinction would occur immediately. Each habitat patch has a certain carrying capacity which might be in turn determined by its area and habitat diversity. The degree of local crowding is modeled as the ratio of population size to the carrying capacity, which in turn determines per capita net growth rate. The general shape of the curve is agreeable with those reported for some field end experimental populations including the ring-necked pheasant, the fruit-fly, and Antarctic fin whales. Time delays in populational regulation, reflecting time-lagged populational responses in per capita net growth rate to crowding effect, are also incorporated in the model. The metapopulation dynamics are the totality of changes in its component subpopulations, while the size of each subpopulation is determined by its net growth rate and net interpatch immigration rate.

**Model application(s) and State of development**
Application area(s): Population dynamics, Landscape Ecology, Conservation Biology
Number of cases studied: 2
State: calibration

**Model software and hardware demands**
Platform: MAC
OS: MAC
Software demands: STELLA,
Hardware demands: Macintosh compiler

---

**Model availability**
Purchase: no, executable code.
Contact person(s) Prof. Jianguo Wu, Ph.D
                Biological Sciences Center
                Desert Research Institute
                P.O. Box 60220
                Reno, NV 89506-0220
                USA
Phone: +1 702-673-7323
Fax: +1 702-673-7397
E-mail: wu@eno.princeton.edu

---

**Model documentation and major references**
Wu, J., Vankat, J.L., and Barlas, Y., 1993. Effects of patch connectivity and arrangement on animal metapopulation dynamics: a simulation study. *Ecol. Modelling*, 221-254.

---

**Other relevant information**
Wu, J., and Vankat, J.L., 1991. A system dynamics model of island biogeography. *Bulletin of Mathematical Biology*, 53, pp. 911-940.
Wu, J., Barlas, Y., and Vankat, J.L., 1990. Modelling Patchy Ecological Systems Using the System Dynamics Approach. In: D.F. Andersen, G.P. Richardson and J.D. Sterman (Eds). *System Dynamics*, Vol. III. The System Dynamics Society, MIT, Cambridge, MA, pp. 1355-1369.

---

**Model identification**
Model name/title: Microbial Subsurface Transport Model

---

**Model type**
Media: local scale, soil
Class: toxicology, hydrology
Substance(s)/compounds: microbes
Biological components: organism - microbes

---

**Model purpose**
Establishes a spatial modelling framework for subsurface transport of microbes.

---

**Short description**
Links percolation models with microbial dynamics in a multi-chemostat formalism. Model is conceptual, no code exists.

---

**Model application(s) and State of development**
Application area(s): Microbial transport
Number of cases studied: 0
State: conceptualization

**Model software and hardware demands**
Platform: NONE
OS:
Software demands:
Hardware demands:

**Model availability**
Purchase: no, software not available.
Contact person(s) Dr. C. Loehle
          Env. Research Div
          Argonne National Lab.
          Chicago, IL 60439 USA

Phone:
Fax:
E-mail:

**Model documentation and major references**
Loehle, C., and Johnson, P., A framework for modelling microbial transport and dynamics in the subsurface. *Ecol. Modelling.* 1995.

**Other relevant information**

# 7. MODELING POPULATION DYNAMICS

## 7.1 Introduction

This chapter covers population models, where state variables are numbers of individuals or species. The famous Lotka-Volterra model as well as several more realistic predator-prey and parasitism models are presented in Section 7.2. Age distribution is introduced in Section 7.3.

The last part of the chapter is devoted to an overview of the available models of plant growth, i.e., models of primary production, and to an overview of modeling population dynamics.

The chapter presents several models in this area to facilitate the development of population dynamic models for the readers. This chapter deals with biodemographic models, characterized by numbers of individuals or species as units for state variables.

Already in the 1920s Lotka and Volterra developed the first population model, which is still widely used today (see Lotka, 1956 and Volterra, 1926). Most population models have been developed, tested and analyzed since, and it will not be possible in this context to give a comprehensive review of all these models.

The chapter will mainly focus on models of primary production, age distributions, growth, and species interactions, developed during the last decade. Mainly deterministic models will be mentioned.

Those who are interested in stochastic models can refer to Pielou (1977), which gives a very comprehensive treatment of this type of model.

A population is defined as a collective group of organisms of the same species. Each population has several characteristic properties, such as population density (population size relative to available space), natality (birth rate), mortality (death rate), age distribution, dispersion growth forms and others.

We distinguish (see above) between models of plants and other models of population dynamics, because the growth of plants is dependent on other factors (solar radiation, available water and nutrients) than the growth of animals. They are widely used for management of natural resources, for instance in fishery and wild life sciences, as these fields are characterized by a high complexity. See the textbook by W. Grant, 1986, targeted at graduate and upperclass undergraduate students.

A population is a changing entity, and we are therefore interested in its size and growth. If N represents the number of organisms and t the time, then $dN/dt$ = the rate of change in the number of organisms per unit time at a particular instant (t) and $dN/(Ndt)$ = the rate of change in the number of organisms per unit time per individual at a particular instant (t). If the population is plotted against time a straight line tangential to the curve at any point represents the growth rate.

**Natality** is the number of new individuals appearing per unit of time and per unit of population.

We have to distinguish between absolute natality and relative natality, denoted $B_a$ and $B_s$ respectively:

$$B_a = \frac{\Delta N_n}{\Delta t}$$

(7.1)

$$B_s = \frac{\Delta N_n}{N \Delta t}$$

(7.2)

where $DN_n$= production of new individuals in the population.

**Mortality** refers to the death of individuals in the population. The absolute mortality rate, $M_a$, is defined as:

$$M_a = \frac{\Delta N_m}{\Delta t}$$

(7.3)

where $DN_m$ = number of organisms in the population, that died during the time interval $Dt$, and the relative mortality, $M_s$, is defined as:

$$M_s = \frac{\Delta N_m}{\Delta t \cdot N}$$

(7.4)

Growth models consider only one population, for instance when the primary production is considered. Interactions with other populations are taking into consideration by the specific growth rate and the mortality, which might be dependent on the magnitude of the considered population but independent of other populations. In other words, we consider only one population as state variable.

The simplest growth model assumes *unlimited resources and exponential population growth*. See also Section 4.3., where this simple case was treated as a submodel. Figure 7.1 gives a logarithmic representation of the exponential growth equation.

The net reproductive rate, $R_0$ is defined as the average number of age class zero offspring produced by an average newborn organism during its entire lifetime. Survivorship $l_x$ is the fraction surviving at age x. It is the probability that an average newborn will survive to that age, designed x. The number of offspring produced by an average organism of age x during the age period is designated $m_x$. This is termed **fecundity**, while the product of $l_x$ and $m_x$ is called the realized fecundity. According to its definition $R_0$, can be found as:

$$R_0 = \int_0^\infty l_x m_x \, dx \tag{7.6}$$

If we in equation (6.7) set $R_0$ to 1 and call the generation time T, then $R_0$ can be found as:

$$\ln R_0 = \ln 1 + r \times T = r \times T \tag{7.7}$$

A curve that indicates $l_x$ as function of age is called a survivor ship curve. Such curves differ significantly for various species as illustrated in Figure 7.2.

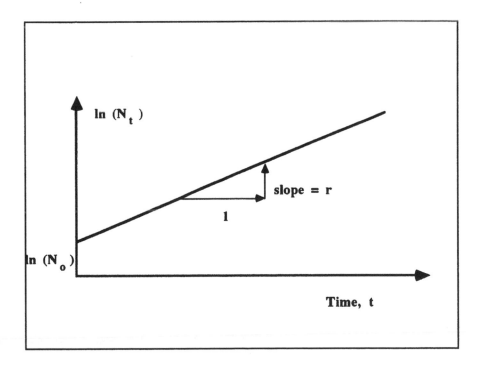

**Figure 7.1:** *ln N$_t$ is plotted versus time, t.*

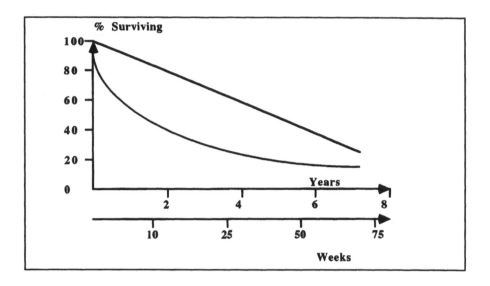

**Figure7.2:** *Survivorships of (1) the lizard Uta (the lower x-axis) and (2) the lizard Xantusia (the upper x-axis). After Deevey (1947), Tinkle (1967), Zweife and Lowe (1966).*

The so-called **intrinsic rate of natural increase, r,** is like $l_x$ and $m_x$ dependent on the age distribution, and only time independent when the age distribution is stable. When $R_o$ is as high as possible - it means under optimal conditions and with a stable age distribution, the maximal rate of natural increase is realized and designated $r_{max}$. Among various animals it ranges over several orders of magnitude, see Table 7.1.

The value of r is cumbersome to calculate, but can be determined by use of the equation:

$$\sum_x e^{-rx} \cdot l_x \cdot m_x = 1 \tag{7.8}$$

Derivation of this equation can be found in Mertz (1970) or Emlen (1973).

If $R_o$ is close to one, r can be estimated using the following approximate formula:

$$r \cong \frac{\ln R_0}{T} \tag{7.9}$$

504

**The Reproductive value,** $V_x$, is defined as the age-specific expectation of future offspring. In a stable population at equilibrium, it is implied that:

$$V_x = \int_x^\infty \frac{l_t}{l_x} \cdot m_t \cdot dt \qquad (7.10)$$

The exponential growth is a simplification, which is only valid in a certain time interval. Sooner or later every population must encounter the limitation of food, water, air or space, as the world is finite. To account for this, we introduce the concept of **carrying capacity, K,** defined as the density of organisms at which r is zero. At zero density $R_0$ is maximal and r becomes $r_{max}$. This growth equation is already treated in Section 4.3. The application of the logistic growth equation requires three assumptions:
1) That all individuals are equivalent.
2) That K and r are immutable constants independent of time, age distribution etc.
3) That there is no time lag in the response of the actual rate of increase per individual to changes in N.

All three assumptions are unrealistic and can be strongly critized. Nevertheless a number of population phenomena can be nicely illustrated by use of the logistic growth equation.

This simple situation in which there is a linear increase in the environmental resistance with density, i.e., logistic growth is valid, seems to hold good only for organisms that have a very simple life history.

**In populations of higher plants and animals, that have more complicated life histories, there is likely to be a delayed response.** Wangersky and Cunningham (1956 and 1957) have suggested a modification of the logistic equation to include two kinds of time lags: 1) The time needed for an organism to start increasing, when conditions are favorable, and 2) The time required for organisms to react to unfavourable crowding by altering birth and death rates.

**Table 7.1** Estimated Maximal Instantaneous Rate of Increase ($r_{max}$, per Capita per Day) and Mean Generation Times (in Days) for a Variety of Organisms

| Taxon | Species | $r_{max}$ | Generation time |
|---|---|---|---|
| Bacterium | *Escherichia coli* | ca. 60.0 | 0.014 |
| Algae | *Scenedesmus* | 1.5 | 0.3 |
| Protozoa | *Paramecium aurelia* | 1.24 | 0.33 - 0.50 |
| Protozoa | *Paramecium caudatum* | 0.94 | 0.10 - 0.50 |
| Zooplankton | *Daphnia pulex* | 0.25 | 0.8 - 2.5 |
| Insect | *Tribolium confusum* | 0.120 | ca. 80 |

**Table 7.1 (Continued)** Estimated Maximal Instantaneous Rate of Increase ($r_{max}$, per Capita per Day) and Mean Generation Times (in Days) for a Variety of Organisms

| Taxon | Species | $r_{max}$ | Generation time |
|---|---|---|---|
| Insect | *Calandra oryzae* | 0.110(.09-.11) | 58 |
| Insect | *Rhizopertha dominica* | 0.085(.07-.10) | ca. 100 |
| Insect | *Ptinus tectus* | 0.057 | 102 |
| Insect | *Gibbium psylloides* | 0.034 | 129 |
| Insect | *Trigonogenius globules* | 0.032 | 119 |
| Insect | *Stethomezium squamosum* | 0.025 | 147 |
| Insect | *Mezium affine* | 0.022 | 183 |
| Insect | *Ptinus fur* | 0.014 | 179 |
| Insect | *Eurostus hilleri* | 0.010 | 110 |
| Insect | *Ptinus sexpunctatus* | 0.006 | 215 |
| Insect | *Niptus hololeucus* | 0.006 | 154 |
| Octopus | – | 0.01 | 150 |
| Mammal | *Rattus norwegicus* | 0.015 | 150 |
| Mammal | *Microtus aggrestis* | 0.013 | 171 |
| Mammal | *Canis domesticus* | 0.009 | ca. 1000 |
| Insect | *Magicicada septendecim* | 0.001 | 6050 |
| Mammal | *Homo sapiens* | 0.0003 | ca. 7000 |

If these time lags are $t - t_1$ and $t - t_2$ respectively, we get:

$$\frac{dn(t)}{dt} = r \, N_{t-t_1} \cdot \frac{K - N_{t-t_2}}{K} \tag{7.11}$$

Population density tends to fluctuate as a result of seasonal changes in environmental factors or due to factors within the populations themselves (so-called intrinsic factors). We shall not go into detail here, but just mention that the growth coefficient is often temperature dependent and since temperature shows seasonal fluctuations, it is possible to explain some of the seasonal population fluctuation in density in that way. Another modified logistic equation was introduced by Smith (1963):

$$\frac{dN}{dt} = r \cdot N \cdot \frac{K - N}{K + \frac{r}{C} \cdot N} \tag{7.12}$$

where $C$ = the rate of replacement of biomass in the population at saturation density.

## 7.2 Interaction between Populations

The growth models presented in Section 4.3 might have a constant influence from other populations reflected in the selection of parameters. It is, however, unrealistic to assume that interactions between populations are constant. A more realistic model must therefore contain the interacting populations (species) as state variables:

$$\frac{dN_1}{dt} = r_1 \cdot N_1 \left( \frac{K_1 - N_1 - \alpha_{12}N_2}{K_1} \right) \tag{7.13}$$

$$\frac{dN_2}{dt} = r_2 \cdot N_2 \left( \frac{K_2 - N_2 - \alpha_{21}N_1}{K_2} \right) \tag{7.14}$$

where $a_{12}$ and $a_{21}$ are competition coefficients. $K_1$ and $K_2$ are carrying capacities for species 1 resp. 2. $N_1$ and $N_2$ are numbers of species 1 and 2, while $r_1$ and $r_2$ are the corresponding maximum intrinsic rate of natural increase.

The steady-state situation is found by setting (7.13) and (7.14) equal to zero. We get:

$$N_1 = K_1 - a_{12} \cdot N_2$$

$$N_2 = K_2 - a_{21} \cdot N_1, \tag{7.15}$$

The four cases are also summarized in Table 7.2. The equation can also be written in a more general form for a community composed of n different species:

$$\frac{dN_i}{dt} = r_i N_i \left( \frac{k_i - N_i - \left( \sum_{j \neq i}^{n} \alpha_{ij}N_j \right)}{k_i} \right) \tag{7.16}$$

where is and is are subscript species and range from 1 to n. At steady state $dN_i/dt$ is equal to zero for all i and

$$N_i \equiv N_{ie} = k_i - \sum_{j \neq i}^{n} \alpha_{ij}N_j \tag{7.17}$$

Lotka-Volterra wrote a simple pair of **predation equations:**

$$\frac{dN_1}{dt} = r_1 \cdot N_1 - p_1 N_1 \cdot N_2 \tag{7.18}$$

$$\frac{dN_2}{dt} = p_2 \cdot N_1 \cdot N_2 - d_2 \cdot N_2 \tag{7.19}$$

where $N_1$ is prey population density, $N_2$ predator population density, $r_1$ is the instantaneous rate of increase of the prey population (per head), $d_2$ is the mortality of the predator (per head) and $p_1$ and $P_2$ are predation coefficients.

Each population is limited by the other and in absence of the predator the prey population increases exponentially. By setting the two differential equations to zero, we find:

$$N_2 = \frac{r_1}{p_1} \tag{7.20}$$

$$N_1 = \frac{d_2}{p_2} \tag{7.21}$$

**Table 7.2** Summary of the Four Possible Cases of Lotka-Volterra Competition Equations

| | Species 1 Can Contain Species 2 $(K_2/a_{21}<K_1)$ | Species 2 Cannot Contain Species 2 $(K_2/a_{21}>K_1)$ |
|---|---|---|
| Species 2 can contain Species 1 $(K_1/a_{12}<K_2)$ | Either species can will (Case 3) | Species 2 always wins (Case 2) |
| Species 2 cannot contain Species 1 $(K_1/a_{12}>K_2)$ | Species 1 always wins (Case 1) | Neither species can contain the other stable coexistence (Case 4) |

Thus, each species isocline corresponds to a particular density of the other species. Below some threshold prey density, predators always decrease, whereas above that threshold they increase. Similarly, prey increase below a particular predator density, but decrease above it, see Figure 7.3. A joint equilibrium exists where the two isoclines cross, but prey and predator densities do not converge on this point.

Any given initial pair of densities results in oscillations of a certain magnitude. The amplitude of fluctuations depends on the initial conditions.

These equations are unrealistic since most populations encounter either *self-regulations density-dependent feedbacks, or both*. Addition of a simple self-damping term to the prey equation results either in a rapid approach to equilibrium or in damped oscillations. Perhaps a more realistic pair of simple equations for modeling the **prey-predator relationship** is

$$\frac{dN_1}{dt} = r_1 \cdot N_1 - z_1 \cdot N_1{}^2 - \beta_{12} \cdot N_1 \cdot N_2$$

(7.22)

$$\frac{dN_2}{dt} = \gamma_{21} \cdot N_1 \cdot N_2 - \beta_2 \cdot \frac{N_2^{\,2}}{N_1}$$

(7.23)

As seen, the prey equation is a simple logistic expression combined with the effect of the predator. The predator expression considers a carrying capacity, which is dependent on the prey concentration.

However, these equations also can easily be criticized. The growth term for the predator is as seen, just a linear function of the prey concentration of density. Other possible relations are shown in Figure 7.4. The first relation (a) corresponds to a *Michaelis-Menten expression*, while the second relation (b) only approximates a Michaelis-Menten expression by use of *a first order expression in one interval and a zero order expression in another*. The third relation (c) shown in Figure 7.4 corresponds to a *logistic expression*: with increasing prey density the predator density first grows exponentially and afterward a damping takes place. This relation is observed in nature and might be explained as follows: the energy and time used by the predator to capture a prey is decreasing with increasing density of the prey. This implies that the predator not only can capture more prey due to increasing density, but also less of the energy consumed is used to capture the next prey. Thus, the density of the predator increases more than proportionally to the prey density in this phase. However, there is a limit to the food (energy) that the predator can consume and at a certain density of the prey, further decrease in the energy used to capture the prey cannot be obtained. Consequently, the increase in predator density slows down as it reaches a saturation point at a certain prey density.

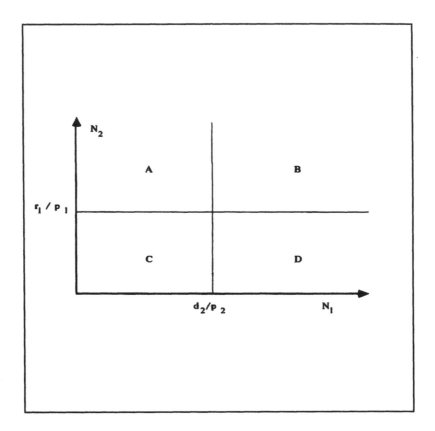

**Figure 7.3:** *Prey-predator isoclines for Lotka-Volterra prey-predator equation. A: both species decrease B: predator increase, prey decrease, C: prey increase, predator decrease, D: both species increase.*

The fourth relationship (d) corresponds to the often found relation between growth and pH or temperature. It is here characteristic that the *predator density decreases above a certain prey density*. This response might be explained by effect on the predator of the waste produced by the prey. At a certain prey density the concentration of waste is sufficiently high to have a pronounced negative effect on the predator growth.

510

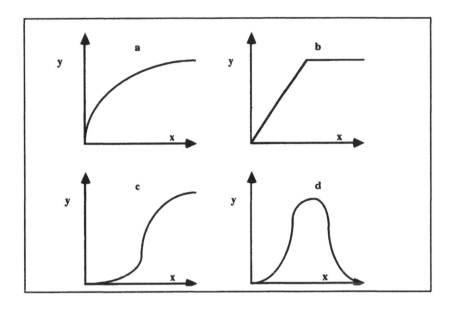

**Figure 7.4:** *Four functional responses (Holling, 1959) - y is number of prey taken per predator per day and x is prey density.*

Holling (1959 and 1966) has developed more elaborate models of prey-predator relationships. He incorporated time lags and hunger levels to attempt to describe the situation in nature. These models are more realistic, but they are also more complex and require knowledge of more parameters. In addition to these complications we have convolution of predators and preys. The prey will develop better and better techniques to escape the predator and the predator will develop better and better techniques to capture the prey. To account for the convolution, it is necessary to have a current change of the parameters according to the current selection that takes place.

**Parasitism** is similar to that of predation, but differs from the latter in that members of the host species affected are seldom killed, but may live for some time after becoming parasitized. This is accounted for by relating the growth and the mortality of the prey, $N_1$, to the density of the parasites, $N_2$. The carrying capacity for the parasites is furthermore dependent on the prey density.

The following equations account for these relations and include a carrying capacity of the prey:

$$\frac{dN_1}{dt} = \frac{r_1}{N_2} \cdot N_1 \left( \frac{K_1 - N_1}{K_1} \right)$$

(7.24)

$$\frac{dN_2}{dt} = r_2 \cdot N_2 \left( \frac{K_2 \cdot N_1 - N_2}{K_2 \cdot N_1} \right)$$

(7.25)

**Symbiotic relationships** are easily modelled with expressions similar to the Lotka-Volterra competition equations simply by changing the signs for the interaction terms:

$$\frac{dN_1}{dt} = r_1 \cdot N_1 \left( \frac{K_1 - N_1 + \alpha_{12} N_2}{K_1} \right)$$

(7.26)

$$\frac{dN_2}{dt} = r_2 \cdot N_2 \left( \frac{K_2 - N_2 + \alpha_{21} N_1}{K_2} \right)$$

(7.27)

In nature, interactions among populations become often intricate. The expressions presented above might be of great help in understanding population reactions in nature, but when it comes to the problem of modeling entire ecosystems, they are in most cases insufficient. Investigations of stability criteria for Lotka-Volterra equations are an interesting mathematical exercise, but can hardly be used to understand the stability properties of real ecosystems or even of populations in nature.

The experience from investigations of population stability in nature shows that it is needed to account for many interactions with the environment to be able to explain observations in real systems, see, for instance, Colwell (1973).

## 7.3 Modeling Age Structure for Management of Natural Resources

It is important in modeling population dynamics to account for the influence of the age structure (age distribution), which establishes the proportion of the population belonging to each age class. Figure 7.4 shows typical age distribution curves.

Two populations with identical $l_x$- and $m_x$-values but with different age distribution will grow differently. If a population has unchanged $l_x$ and $m_x$ schedules, it will eventually reach a stable age distribution. This means that the percentage of organisms in each age class remains the same. Recruitment into every age class is exactly balanced by its loss due to mortality and aging.

The growth equations presented in Section 4.3 and the equations (7.6) and (7.12) all assume that the population has a stable age distribution. The intrinsic rate of increase, r, the generation time, T, and the reproductive value, $v_x$ are all conceptually independent of the age distribution, but might, of course, be different for population of the same species with different age distributions. Therefore, the models presented in the two previous sections do not need to consider age distribution, although the parameters in actual cases reflect the actual age distribution.

A model predicting the future age distribution, was developed by Lewis (1942) and Leslie (1945). The population is divided into n +1 equal age groups - group 0, 1, 2, 3,.....n. The model is then presented by the following matrix equation:

$$
\begin{array}{cccccc}
f_0 & f_1 & f_2 & \cdots & f_{n-1} & f_n \\
P_0 & 0 & 0 & \cdots & 0 & 0 \\
0 & P_1 & \cdots & \cdots & 0 & 0 \\
\cdots & \cdots & \cdots & \cdots & \cdots & \cdots \\
\cdots & \cdots & \cdots & \cdots & \cdots & \cdots \\
0 & 0 & 0 & \cdots & P_{n-1} & 0
\end{array}
\quad
\begin{array}{cc}
n_{t,0} & n_{t+1,0} \\
n_{t,1} & n_{t+1,1} \\
n_{t,2} & n_{t+1,2} \\
\cdot & \cdot \\
\cdot & \cdot \\
n_{t,n} & n_{t+1,n}
\end{array}
\tag{7.28}
$$

The number of organisms in the various age classes at time t + 1 are obtained by multiplying the numbers of animals in these age classes at time t by a matrix, which expresses the fecundity and survival rates for each age class. to, $f_1$, $f_2$.... $f_n$ give the reproduction in the i'th age group and $P_0$, $P_1$, $P_2$, $P_3$, $P_4$....$P_n$ represent the probability that an organism in the i'th age group still will be alive after promotion to the $(i + 1)^{th}$ group. The model can be written in the following form:

$$A * a_t = a_{t+1} \tag{7.29}$$

where A is the matrix, $a_t$ is the column vector representing the population age structure at time t and $a_{t+1}$, is a column vector representing the age structure at time t + 1. This equation can be extended to predict the age distribution after k periods of time:

$$a_{t+k} = A^k * a_t \tag{7.30}$$

The matrix $A$ has $n + 1$ possible eigenvalues and eigenvectors. Both the largest eigenvalues, $l$, and the corresponding eigenvectors are ecologically meaningful. $l$ gives the rate at which the population size is increased:

$$A * v = l * v \qquad (7.31)$$

where $v$ is the stable age structure. In $l$ is the intrinsic rate of natural increase. The corresponding eigenvector indicates as seen the stable structure of the system.

If the harvest exceeds this value the population will decline. Population models of r-strategies might generally cause some more difficulties to develop than models of K-strategies due to the high sensitivity of the fecundity. The number of offspring might be known quite well, but the number of survivors to be included in the first age class, the number of recruits, is difficult to predict. This is the central problem of fish population dynamics, since it represents natures regulation of population size (Beyer, 1981). Beverton and Holt (1957) have suggested the following equation for recruitment:

$$R = \left( \frac{E}{E + yR_{max}} \right) R_{max} \qquad (7.32)$$

where the number of recruits (R) increases toward an asymptotic level $R_{max}$, when the egg production, E, increases.

Ricker (1954) suggests another equation, in which the number of recruits decreases from a maximum level toward zero as the production of eggs decreases:

$$R = R_1 * E * e^{-R_2 \cdot E} \qquad (7.33)$$

Here, $R_1$ and $R_2$ are constants. The decline of recruitment is explained by cannibalism by adults. The graphic representation of the two different approaches is shown in Figure 7.5. The need for models of recruitment is obvious and much effort is devoted to achieve a better quantitative description of this process in fishery management models, see Beyer and Sparre (1983). Often population models are coupled with growth models, which attempt to predict the growth of the individuals. As the population model gives the number and the growth model the weight, the total biomass is easily found for the different age classes. The von Bertalanffy equation is quite widely used to describe the growth of fish species and it constitutes the growth element of the classical Beverton and Holt (1957) one species model:

$$dw / dt = HW^{2/3} - kw \tag{7.34}$$

w is the weight, H and k are constants.

The value 2/3 is based upon the assumption that the area of the intestine is proportional to the surface area and thereby to weight in 2/3.

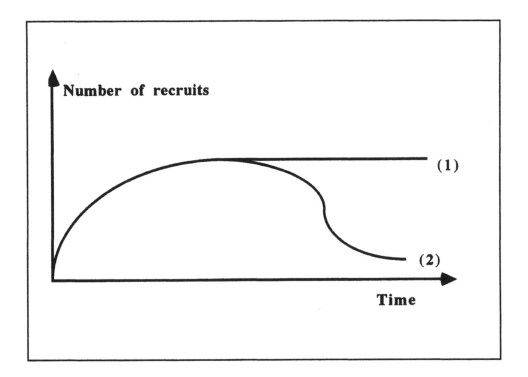

**Figure 7.5**: *Recruitment curves. (1) is Beverton and Holts equation, (2) is Rickers.*

Equation (7.34) can be solved analytically:

$$w(t) = w_\infty (1 - \exp(-K(t - t_0)))^3 \; ; \; w_\infty = (H / k)^3 \text{ and } K = k / 3 \tag{7.35}$$

Growth in weight can be derived into growth in length, l, according to the similar body assumption:

515

$$l(t) = L_\infty(1\text{-}exp(\text{-}K(t\text{-}t_0))); \ L_\infty = \frac{H}{kq^{1/3}}^3 \ and \ K = \frac{k}{3} \tag{7.36}$$

q is the condition factor. For fish of ordinary shape, we find q = 0.01 $g/cm^3$. The von Bertalanffy equation is a special version of the more general growth model developed by Ursin 1967, 1979, and Andersen and Ursin 1977.

## 7.4 Modeling the Primary Production/the Photosynthesis: An Overview

The photosynthetic process may be divided into independent reaction systems, the light absorption-energy producing system (known as the light reaction), and the reductive system of carbon dioxide fixation (known as the dark reaction).

The light reaction transmutes the energy of sunlight into the two biochemical energy sources, ATP and $NADPH_2$, via the two main photochemical pathways. Absorption of photons excite chlorophyll electrons to higher energy levels, which are then utilized via either cyclic photophosphorylation or non-cyclic photophosphorylation.

The dark reaction uses the biochemical energy sources ATP and $NADPH_2$ to reduce carbon dioxide to organic carbon. $CO_2$ is combined with ribulose 1.5-diphosphate and forms two molecules of 3-phosphoglyceric acid, which form one molecule of fructose 1.6-diphosphate.

Obviously, photosynthesis involves two sets of external limiting factors, availability of energy and inorganic elements ($CO_2$), and these two elements govern the rates of the light and dark reactions. Additionally, internal limiting factors are involved since transport mechanisms are involved in providing the elements essential for synthesis of organic matter. Besides this, organisms need time to adapt to fluctuations in environment conditions, for instance, a change in radiant intensity, and so both internal pools of essential elements (C, P, $H_2O$, S, etc.) and the "reaction tools" (enzymes, transport mechanisms, respiration level, leaf index, reproductive stage, etc.) may limit the rate of photosynthesis, see Figure 3.4.

The common mathematical description of photosynthesis normally involves a coupling of light and essential element dependency, and may consequently be categorized as an empiric model. If no changes in adaption occur, then photosynthesis may be quoted as

PHOTO = k · f (Max. requirement of limiting factors) (7.37)

516

where PHOTO is the photosynthesis measured as uptake of $CO_2$, $O_2$ produced, increased organic energy, or similar units, and f(x) represents the optimal yield of the maximal available elements, external as well internal. Figure 7.6 gives some basic experimental results to illustrate different types of limiting factors and adaptation cases.

The use of photosynthetic equations in models must be related to the total system or subsystems. In aquatic models, normal demographic equations are used, e.g., equations treating an average exposed biomass of one or more species. In terrestrial systems, where the number of single species is low, intraspecific adaption plays an important role.

In algal populations, where the photosynthetic description probably is the best-attempted simulation of nature, the models have been extensively developed. Thus, only a few examples will be quoted in this context. Chen et al., 1975, considered four external factors, nitrogen (N), phosphate (P), radiant intensity (I) and temperature (T), given by a product of Michaelis-Menten expressions (q.v.) and an optimal temperature dependency (f(T)),

$$\mu = \mu_{max} \cdot f(T) \cdot I \frac{1}{I + k_i} \cdot N \frac{1}{1 + k_n} \cdot P \frac{1}{1 + k_p} \qquad (7.38)$$

where m is the actual specific photosynthesis and $m_{max}$ is the maximal specific photosynthesis.

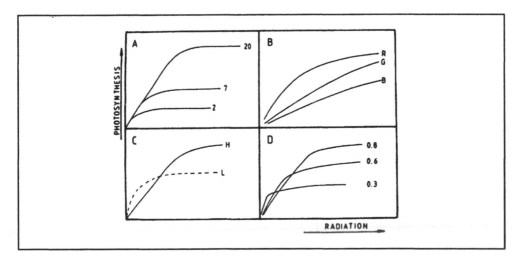

**Figure 7.6**: *Rate of photosynthesis as a function of radiation energy. A: at different temperatures (in °C). B: at different chromatic intervals, R = red, B = blue, G = green. C: of two light-intensity adapted stages, H = high, L = low. D: at different leaf index stage.*

If several limiting factors regulate the growth simultaneously, there are description methods referred to in the literature for the total effect of interacting factors. The most important description methods are summarized as follows:

1) The minimum value of the limiting factors is used. This is in accordance with Liebig's minimum law.
2) The product of the limiting factors is used.
3) It is presumed that the factors work in parallel.
4) The average of the limiting factors is used.

Nyholm (1976) and Jørgensen (1976) also considered internal concentrations as limiting factors. These describe pools of phosphorus and nitrogen, which are readily available during periods of exponential growth. This principle is shown here by phosphate-limited photosynthesis, but may be used for each essential element vital to the process

$$\mu = \mu_{max} \cdot \frac{k + CM - CA}{CM - CA} \cdot \frac{CI - CA}{k + CI - CA} \tag{7.39}$$

where CM is maximal internal concentration of phosphorus, CA is minimal internal concentration of phosphorus and CI is actual internal concentration of phosphorus.

The use of this particular expression involves a separate description of the nutrient uptake from the exterior.

Temperature dependency f(T) has been described in different ways. Chen et al., 1975 gives a simple equation, which does not take temperature optimization into account

$$f(T) = k_{20} \cdot V^{(T-20)} \tag{7.40}$$

where V normally is recorded between 1.0 and 1.2.

Many processes in biological systems have a temperature optimum and some models take this into consideration, as e.g., Jørgensen (1976):

$$(T) = K_{opt} \cdot exp\left(-2.3 \cdot \frac{|T - T_{opt}|}{15}\right) \tag{7.41}$$

where $K_{opt}$ is the optimal process rate and $T_{opt}$ the corresponding temperature. The equation assumes that the temperature dependency is symmetrical around the optimal value.

Different types of light functions have been used including or excluding light inhibition due to photo-oxidation in the chloroplasts. In a homogeneous dispersed biomass (e.g., a phytoplankton community) Vollenweider (1965) has formulated the radiant dependency as

$$f(I) = \frac{I}{k_m\sqrt{1 + (I/k_m)^2}} \cdot \frac{1}{\left(\sqrt{1 + (aI)^2|}\right)^{k_2}}$$

(7.42)

where $k_m$ is the light saturation value, $I$ is the actual light intensity, and $a$ and $k_2$ are constants.

Beside light intensity, the day length influences photosynthesis in different species. Many terrestrial plants contain day-length "receptors", which govern photosynthesis efficiency or more commonly the minimum threshold for starting photosynthesis. Data from studies on phytoplankton indicate that day length may be mainly responsible for population succession. These mechanisms will not be treated further in this chapter. Figure 7.7 gives an example.

Most plants contain mechanisms to adapt to the actual light regime by changing the orientation of the chloroplasts in leaves, or the leaves themselves, or by altering enzymatic concentrations. The processes have not yet been modeled satisfactorily.

In many higher plants there is a well-developed stomata-controlling system, which regulates the $CO_2/O_2$ exchange between leaf and atmosphere. Accordingly, very complex models must be constructed to simulate actual photosynthesis in most terrestrial plants.

The number of external factors controlling photosynthesis and primary production in higher plants, or terrestrial ecosystems, is very high. Some of the most potent environmental factors are the intensity of solar and diffuse radiation; humidity and wind; infrared radiation, and day length. In addition to environmental factors, the infra-organization of the terrestrial subsystem is a most important element. The determination of reflection and transmission in the leaf system, leaf-area index, orientation, and the $CO_2/O_2$ gas-exchange system are the preliminary processes for the development of a mathematical description of production.

Obviously, two basic types of process description can be developed, namely,

I) by the use of single compartments (leaves, galleries, individuals or species) or II) by using total community production, Figure 7.7

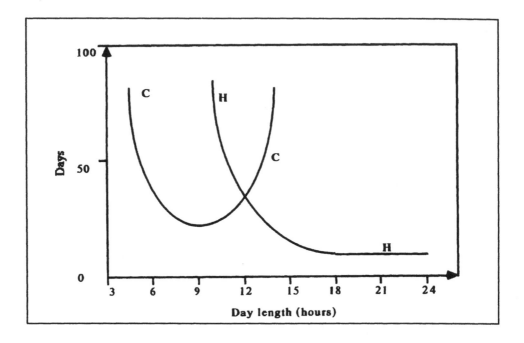

**Figure** **7.7**: *Time of flower development (maturity) as a function of day length. C = Chrysanthemum sp.m "short day plant", H = Hyascomus sp. "long day plant".*

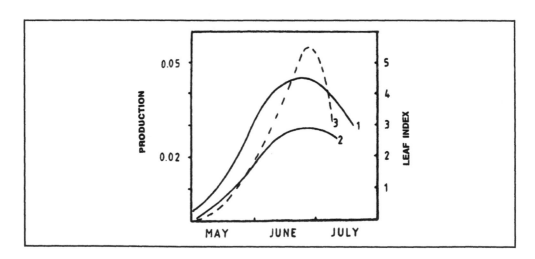

**Figure 7.8:** *Production, respiration and leaf index during growth period for Hordeum sp. (Barley). (1) and (2) are gross primary production, and net primary production, (g dry matter m⁻²) and (3) is leaf index.*

520

Most difficult is the modeling of the effects of soil moisture and temperature profiles in the system. Hence, simple models describing production in terrestrial environments use solar radiation and leaf-area index as variables and let all other factors be constants or stochastic elements.

This section gives in addition to the basic models mentioned above, an overview of the many available models of primary production/photosynthesis. Details of these models are given in the next section, while an overview is provided in this section with a summary in Table 7.3, where the model characteristics and the main references are given for the here presented models and a few older models.

The so-called C4-photosynthesis model simulates the response of C4 leaf photosynthesis and stomatal conductance to atmospheric carbon dioxide, light temperature and humidity. The model is prepared for inclusion into a whole plant growth model for C4 grasses. The model named Clipsim attempts to simulate negative, neutral and positive plant growth response to defoliation, nitrogen and water. The model considers roots, shoots and flowers and the partition of the assimilated carbon in these three compartments.

A model of the macroalgae growth dynamics describes the cycles of the macro-nutrients and the influence of the concentrations of nutrients, the light and the concentration of dissolved oxygen on the competition between macroalgae and phytoplankton. The model has 12 state variables including Ulva, zooplankton, phytoplankton, inorganic nitrogen as ammonium and nitrate, inorganic phosphorus, nitrogen detritus, phosphorus detritus, dissolved oxygen, nitrogen and phosphorus in the upper sediment and intracellular nitrogen in Ulva. The model denoted LeafC3 estimates short-term steady state fluxes of carbon dioxide, water vapor and heat from broad-leaved and needle-leaved coniferous shoots of C3 plant species. It is designed to be used as a fine scale simulation tool as well as a component in process oriented models.

Parker has developed a dynamic model for ammonium inhibition of nitrate uptake by phytoplankton. The inhibition is described as an inverse Michaelis-Menten function.
Rego et al. have developed a model describing the primary production of a shrub communities under different fire regimes. The model has been used as a management tool in prediction of community succession in 6 case studies.

The model Mpk01 simulates the chlorophyll as an expression of the phytoplankton abundance in the sea water during an up-welling event. The main forcing functions are light, nitrate and advection.

Irm is a method to calculate a "plant growth index". It is a number between zero and one that is able to modify the relative growth rate according to the levels of available resources and other factors that affect the growth. The method is prepared to be used to model the growth of any plant entity of interest.

The model denoted Passm simulates the exchange processes and microenvironment of a plant community. It is a steady state, one-dimensional model, that is able to predict the microclimate and the properties of energy exchange processes between plant communities

521

and their ambient environment. The model has three major components: 1) formulation of radiation distribution; 2) formulation of momentum, heat and water exchange processes; and 3) formulation of an energy balance of the plant community.

The model Iyo-Nada is able to simulate the red tide of *Gymnodinium mikimotoi*. It is a three-dimensional numerical assimilation model which includes the important biological processes determining the occurence of red tide.

Sievanen et al. have developed a model for the study of photosynthate partitioning in plants. The model distinguishes between above-ground and below-ground parts. The dry-weight and nitrogen contents of these parts are considered. The photosynthetic rate depends on nitrogen concentration and dry-weight of above-ground parts. The nitrogen uptake rate is proportional to soil nitrogen concentration and dry-weight of below-ground parts.

The model Papran provides a comprehensive assessment of the response of primary production (above ground herbaceous biomass) to environmental factors such as precipitation, nitrogen availability and soil depth in the context of grassland growing in a Mediterranean-type climate. The 12 state variables are dry weight and nitrogen content of leaves, non-leaf vegetative material, seeds, roots, detritus, development stage and LAI, soil moisture content, stable organic material, fresh organic material, nitrogen in stable organic matter, nitrogen in fresh organic matter and inorganic nitrogen.

The model Transit was developed in order to estimate transpiration of single trees, hedgerows or similar plant configurations instead of horizontally homogeneous vegetation canopies. It investigates land surface - atmosphere exchange processes.

Pukkula has developed a model that is able to predict the distribution of photosynthetically active solar radiation in tree crowns and forest canopies of defined structure and to estimate photosynthetic production.

The model Pinogram simulates the growth, the competition and appearance of individual trees in mono-culture influenced by different planting and thinning regimes. The model is originally developed for *Pinus sylvestris*. The program presents images of stems and abstract forms of crowns in a stand of 20 x 50 m$^2$ in three dimensions.

The model denoted Vctm is a steady state thermal radiance model to compute directional thermal energy balance within forest canopies. A forest canopy is described as three vegetation layers. The canopy is characterized by the leaf angle distribution, LAI for each layer, background optical scattering and emission properties and leaf stomatal resistance. The forcing functions are solar radiance, wind speed, air temperature, relative humidity and soil temperature. The model named Tlaloc simulates the dynamics of a vegetation mosaic, depending of the surface water redistribution. There are two groups of state variables: 1) the soil described to a depth of 2 m mainly by its volumetric water content. It changes according to infiltration, soil evaporation and plant transpiration 2) the plants are characterized by their cover, i.e., their transpiring surface for each species and age class. Transpiration determines the photosynthesis and growth, taking into account the light extinction. The seed production, dispersal and germination are also simulated.

Webb has developed a model able to relate the tree growth to increased carbon dioxide concentration, air temperature and vapor pressure, soil water content, short wave radiation and precipitation.

Ramge et al. have developed a model of the uptake of nitrogen dioxide via stomata and its transformation into cellular nitrate and nitrite pools.

The model Specom uses the concept of "single plant ecosystem" tasking into account separate trees growth and its spatial disposition in dependence of soil processes, meteorological conditions and aerial pollution.

The model denoted Spm is used to investigate details of the carbon dynamics and its influence on the stand and to the ecosystem. The state variables are: new foliage, second year foliage, branches, stern, coarse roots, fine roots in the litter layer and fine roots in the mineral soil, needle litter, dead fine roots and soil organic matter. The canopy is divided into 9 layers.

The model Photofact is a simple mechanistic model for photosynthesis and photoinhibition. The model can be used to fit steady production curves and to analyze dynamic incubation experiments. Three states are considered: resting, activated and inhibited. The transition between the states are driven by light and biochemical reactions. The model Alsim simulates alfalfa growth and regrowth as a function of soil and moisture on a daily basis.

Von Stamm has developed a model which links two submodels of the stomata and of the photosynthesis. The model has two hierarchical levels: the basic model which contains the mathematical equations of stomatal reactions, assimilation and respiration as functions of irradiance, temperature and vapor pressure difference between leaf and air, and two submodels of stomatal conductance and carbon dioxide exchange processes.

Hara has proposed a diffusion model for growth and size-structure dynamics of plant population and communities. It is a useful tool for analyzing actual data of growth, mortality and reproduction of individuals. The results are comparative studies of plant species, considering the environmental conditions.

**Table 7.3** Models of the Primary Production/the Photosynthesis

| Model name/ component | Model characteristics | Reference |
|---|---|---|
| Single trees | Growth model: stand, size class | Leary, 1979 |
| Single trees | Diameter increment, death, thinning | Opie, 1972 |
| C4 photosynthesis | Leaf photosynthesis, stomatal cond. | Chen et al., 1994 |
| Clipsim | Growth responses to defoliation, nitrogen and water | Shrub, 1991 |

**Table 7.3 (continued) Models of the Primary Production/The Photosynthesis**

| Model name/ component | Model characteristics | Reference |
|---|---|---|
| Ulva | Detailed growth model incl. intracell. nitrogen | Dejak et al., 1993 |
| LeafC3 | Fluxes of $CO_2$, water and heat C3 plant species | Nikolov et al., 1995 |
| N-uptake | Comp. ammonium, nitrate uptake | Parker, 1993 |
| Shrub communities | Diff. fire regimes | Rego et al., 1993 |
| Algal growth | Site specific coeff. for algal growth | See model |
| Mpk01 | Algal growth in up-welling system | Coutinho |
| Irm, plant growth | Plant gr. index: available resources | Wu, 1995 |
| Passm, plant comm. | Energy exchange processes | Wu, 1990 |
| Iyo-Nada, red tide | 3-D assimilation, incl. biol. processes | Yanagi |
| Nitrogen | Allocation of N, infl. on plant growth | Sievanen et al., 1989 |
| Papran, prim. prod. | Response to rain, N and soil depth model N and C cycles | Seligman et al., 1992 |
| Transit, single trees | Transpiration. land - atmosphere exch. | Janssen |
| Tree crowns | Photosynthesis, active solar radiation | Pukkala |
| Pinogram, pine thinning | Growth , competition, planting and | Leersnijder, 1992 |
| Vctm, canopy | Energy balance of forest canopies | Smith and Goltz, 1994 |
| Tlaloc, vegetation | Mosaic dependence of water | Monta et al., 1990 |
| Tree growth | Infl. of carbon dioxide and climate | Webb |
| Stomata | Uptake of nitrogen dioxide | Ramge et al., 1993 |
| Specom, single trees | Growth/soil dynamics, biomass and N | Chertov, 1990 |
| Spm, carbon | C-dynamics stands and ecosystem level | Cropper, 1993 |
| Photofact | Phytoplankton, photosynthesis and photoinhibition | Eilers and Peeters, 1993 |
| Alsim, alfalfa | Growth/soil and moisture | Fick |
| Corylus avellana | Assimilation, respiration, stomatal reactions/microclimatic factors | Von Stamm 1992 |
| Plant communities | Comparative studies/env. factors | Hara, 1988 |

# 7.5 Models for the Primary Production and the Photosynthesis

**Model identification**
Model name/title: C4 photosynthesis model

**Model type**
Media: terrestrial
Class:
Substance(s)/compounds: carbon
Biological components: ecosystem, grassland

**Model purpose**
To simulate mechanistically the response of C4 leaf photosynthesis and stomatal conductance to atmospheric $CO_2$, light, temperature and humidity; to provide C4 photosynthesis model as a submodel for inclusion in a whole plant growth model for C4 grasses.

**Short description**
The model is based on the scheme for C4 plant photosynthesis which consists of the PEPcase-dependent C4-cycle and the Rubisco-dependent C3-cycle. The C4-cycle consists of the fixation of $CO_2$ by PEPcase in mesophyll cells, the transport and decarboxylation of C4 acids. In the model, the PEPcase-dependent C4-cycle was described in terms of $CO_2$ concentration in the mesophyll space using Michaelis-Menten kinetics, and the activity of PEPcase was related to the incident PAR to take account of the influence of light on the activity of C4-cycle processes. The $CO_2$ refixation by Rubisco in the bundle sheath was described using a widely accepted C3 photosynthesis biochemical model. The model assumes a steady state balance among $CO_2$ diffusion from surrounding atmosphere into the mesophyll space, $CO_2$ transport into the bundle sheath by the C4-cycle, $CO_2$ refixation by the C3-cycle in the bundle sheath, and $CO_2$ leakage from the bundle sheath. The response to temperature of photosynthesis was incorporated via the temperature dependence of model parameters. The photosynthesis model was coupled with a stomatal conductance model in order to predict laef photosynthesis rates at different atmospheric conditions. The input of the model includes incident PAR, atmospheric $CO_2$ concentration, relative humidity, temperature and leaf nitrogen content. The output of the model includes net photosynthesis rate and stomatal conductance.

**Model application(s) and State of development**
Application area(s): Plant eco-physiology, grassland ecosystem and biosphere-atmosphere interaction.
Number of cases studied:
State: validation for C4 grasses

---

**Model software and hardware demands**
Platform: PC-based, IBM
OS: DOS
Software demands: N/A, Fortran
Hardware demands: N/A

---

**Model availability**
Purchase: software available as executable code and sourcecode
Contact person(s) Dr. De-Xing Chen
Natural Resource Ecology Laboratory
Colorado State University
Fort Collins, CO 80523, USA

Phone:
Fax: +1 303-491-1965
E-mail: dexing@NREL.Colostate.EDU

---

**Model documentation and major references**
Chen, D.-X., Coughenour, M.B., Knapp, A.K., and Owensby, C.E., 1993. Mathematical simulation of C4 grass photosynthesis in ambient and elevated $CO_2$. *Ecol. Modelling*.

---

**Other relevant information**
Chen, D.-X., and Coughenour, M.B., 1994. GEMTM: a general model for energy and mass transfer of land surfaces and its application at the FIFE sites. *Agricultural and Forest Meteorology*.

---

**Model identification**
Model name/title: CLIPSIM

---

**Model type**
Media: terrestrial
Class: biogeochemical
Substance(s)/compounds: carbon, nitrogen
Biological components: organism - plant

---

**Model purpose**
The purpose of this model is to mechanistically simulate negative, neutral and positive plant growth responses to defoliation, and to nitrogen and water as parsimoniously as possible. Additional code uses the model to simulate controlled factorial experiments involving regular defoliation, watering and nitrogen fertilization treatments.

---

**Short description**
This model simulates assimilation and allocation of C and N above and below ground and to leaves, stems and flowers. Carbon shoot: root ratio. Assimilated carbon is partitioned between roots and shoots in reponse to shoot:root ratio, water and phenology. The phenological state of the plant is affected by defoliation. Nitrogen uptake is responsive to plant and soil nitrogen and phenology. Tissue senescence is affected by water, nitrogen and

phenology. Nitrogen is conserved during senescence. Attached software reads in data from factorial experiments involving defoliation, watering and N fertilization treatments and executes the model for each of the treatments. Model predictions are compared to output from a "data model", which computes total production and plant N uptake rates from the factorial experimental data. The sum of deviations between observed and predicted C and N in all plant biomass components, including clipped yield, is minimized during the model calibration. As few parameters and processes as necessary are invoked to simulate observed results.

## Model application(s) and State of development
Application area(s):
Number of cases studied:
State:

## Model software and hardware demands
Platform:
OS:
Software demands:
Hardware demands:

## Model availability

Purchase:
Contact person(s)

Phone:
Fax:
E-mail:

## Model documentation and major references
Dwarf Shrub and graminoid responses to clipping, nitrogen and water: simplified simulations of biomass and nitrogen dynamics. *Ecol. Modelling*, 54, pp. 81-110.

## Other relevant information

## Model identification
Model name/title: Macroalgae growth dynamic

## Model type
Media: water, lake, estuary
Class: biogeochemical
Substance(s)/compunds: nutrients, phosphorus, nitrogen, oxygen
Biological components: populations, phytoplankton, zooplankton, macroalgae

## Model purpose
Simulation of the Ulva population dynamic as well as phyto- and zooplanktonic evolution, therefore competition phenomena, by modeling nutrient cycles and dissolved oxygen evolution in a shallow water environment such as the Venice Lagoon. The model can be seen as biological compart of larger water quality.

## Short description
The cycle of macronutrients is described through the major compartments and, at the same time, plantonic activity and macroalgae (ulva) growth is simulated as influenced by nutrient availability and meteorclimatic conditions, namely light irradiance and water temperature, used as forcing functions. Simulated dissolved oxygen evolution, while providing a sound index to verify the quality of the water body, plays an important role inducing feedback autocatalitic effect on Ulva dynamics and nutrient recycles through the mineralization of detritus and sediment respiration. Nitrogen uptake is uncoupled from accresciment in Ulva growth, which follows a two-step kinetic mechanism involving a nutrient internal storage, wheras light and temperatures influences are thought as asymtotic functions. Photo and termal hinibiton are instead considered modeling phytoplantonic dynamics, which is described by Steel formulation for light influnce, Lassiter and Kearns for temperatures, and monod kinetics for nutrients. Intraspecific competition is modeled by assuming nutrient uptake rates and respiration demand, the discriminating factors. The 12 state variables are: Ulva, zoo- and phytoplankton, dissolved inorganic nitrogen (both reduced and oxided), ortophosphate, nitrogen detritus, phosphorous detritus, nitrogen in upper sediment, phosphorus in upper sediment, dissolved oxygen and Ulva intracellular nitrogen. Oxygen reaereation is simulated and nutrients input can be taken into account.

## Model application(s) and State of development
Application area(s): Venice Lagoon
Number of cases studied:
State: conceptualization, verification, calibration

## Model software and hardware demands
Platform: PC-based, IBM
OS: UNIX, DOS
Software demands: FORTRAN
Hardware demands:

## Model availability
Purchase: no, executable code and sourcecode
Contact person(s) Solidoro Cosimo, Pecenik Giovanni, Dejak Camillo
        Universit' di Venezia
        Dip. chimica fisica,
        Dorsoduro 2137
        I-30123 Venezia, Italy
Phone:
Fax:
E-mail: cosimo@pcnk.dcf.unive.it

**Model documentation and major references**
Solidoro, C., 1993, Ph.D. Thesis. University of Venice.
Solidoro, C., Dejak, C., Franco, D., Patres, R., and Pecenik G. 1993. Modeling macroalgae phytoplankton growth and competition in the venice lagoon. International Congress on modeling and simulation. University of Western Australia, December 6-10, 1993. Proceedings (4) 1371-1376.
Pecenik G., Solidoro, C., Franco, D., Dejak C., and Franco, D., 1991. Modeling of the macroealgae growth: an eutrophic-diffusion macromodel of the Venice Lagoon. S.I.T.E. Zara Edf. Parma (Italy) 12: 881-890.
Solidoro, C., Pastres, R., Franco, D., Pecenik, G., and Dejak, C., 1994. Modeling Ulva R in Venice Lagoon. (submitted for publication to *Ecol. Modeling*).

**Other relevant information**

**Model identification**
Model name/title: LEAFC3: An equilibrium photosynthesis model for leaves of C3 plants

**Model type**
Media: forest and agricultural
Class: biogeochemical and hydrology
Substance(s)/compounds: carbon, oxygen, sensible and latent heat
Biological components: ecosystem, leaf level responses

**Model purpose**
The model estimates short-term steady-state fluxes of $CO_2$, water vapor and heat from broad-leaved and needle-leaved coniferous shoots of C3 plant species. It is designed to be used as a fine scale simulation tool as well as a component in process-oriented models of vegetative canopies and ecosystems.

**Short description**
The model explicitly couples all major processes and feedbacks that are known to impact leaf biochemistry and biophysics of C3 plants including biochemical reactions, stomatal responses, and leaf-boundary layer mass and heat transport mechanisms. The implementation of a highly efficient solution technique allows the model to be directly incorporated into plant-canopy and ecosystem models. The following variables and parameters are required for input by the LEAFC3 model. Species-specific input parameter: maximum carboxylation velocity at 25°C, light-saturated potential rate of electron transport at 25°C, kinetic parameters for $CO_2$ and $O_2$ at 25°C, PPFD (photosynthetically photo flux density) efficiency factor, composite stomatal sensitivity factor, leaf width (or needle diameter for conifers), shoot diameter (for coniferous only), two parameters characterizing the response of stomata to xylem water potential. Environmental input parameters include: xylem leaf water potential, atmospheric pressure, ambient concentrations of $CO_2$ and $O_2$, ambient temperature, relative humidity, total bi-directional absorbed radiation by the leaf, incident PPFD, wind speed.

Output variables include: net $CO_2$ assimilation rate, stomatal conductance to water vapor, leaf-boundary layer conductance to water vapor, leaf-temperature, latent heat flux.

---

**Model application(s) and State of development**
Application area(s): terrestrial ecosystem cycles of carbon, water and nitrogen
Number of cases studied: 3
State: validation

---

**Model software and hardware demands**
Platform: PC-based
OS: DOS
Software demands: Turbo Pascal Compiler ver. 4.0 or higher, Turbo Pascal
Hardware demands: IBM-type PC with RAM of minimum 256 Kb

---

**Model availability**
Purchase: yes, free if You send a disk
Contact person(s) Dr. N.T. Nikolov
         USDA FS/ Rocky Mountain Forest and Range Experimental Station
         240 West Prospect Rd.
         Fort Collins, CO 80526 USA

Phone: +1 303-498-1315
Fax: +1 303-498-1010
E-mail:

---

**Model documentation and major references**
Nikolov, N.T., Massman, W.J., and A.W. Shoettle. 1995. Coupling biochemical and biophysical processes at a leaf level: An equilibrium photosynthesis model for leaves of C3 plants. *Ecol. Modeling.*

---

**Other relevant information**

---

**Model identification**
Model name/title: Dynamic models for ammonium inhibition of nitrate uptake by phytoplankton

---

**Model type**
Media: ocean
Class: biogeochemical
Substance(s)/compounds: nutrients, nitrogen
Biological components: population, *Ditylum brightwellii*

## Model purpose
To couple $NO_3^-$ and $NH_4^+$ uptake dynamically at moderate concentrations in such a way that $NO_3^-$ uptake falls in the presence of $NH_4^+$, first without, then with, a nitrate reductase component.

## Short description
Two systems of differential equations are proposed to describe the dynamic effects of $NH_4^+$ on $NO_3^-$ uptake. The inhibitory function is of the form $1/(1+[NH_4^+]/K4)$, where K4 is the Michaelis constant for $NH_4^+$ uptake. In the first system, concentrations of particulate nitrogen, $NO_3^-$ and $NH_4^+$ are modeled, and $NH_4^+$ inhibits $NO_3^-$ uptake directly.
In the second, a pool of cellular reductant is added into which both reduced $NO_3^-$ and $NH_4^+$ flow. Nitrate reductase also is included, and its formation is inhibited by $NH_4^+$. In each case, solutions agree well with published observation on a deep-tank culture of *Ditylum brightwellii*.

## Model application(s) and State of development
Application area(s): Nutrient uptake
Number of cases studied: 1
State: calibration

## Model software and hardware demands

Platform: any
OS:
Software demands: FORTRAN
Hardware demands:

## Model availability
Purchase: no
Contact person(s) Dr. R.A. Parker
        School of Electrical Engineering and Computer Science
        Washington State University
        Pullman, WA 99164-2752 USA

Phone:
Fax:
E-mail:

## Model documentation and major references
Parker, R.A., Dynamic models for ammonium inhibition of nitrate uptake by phytoplankton. *Ecol. Modelling*, 66, pp. 113-120.

## Other relevant information

531

## Model identification
Model name/title: MARKOV CHAIN

## Model type
Media: regional scale, terrestrial, forest
Class:
Substance(s)/compounds:
Biological components: ecosystem, Shrubland, community, *Quercus coccifera garrigue*

## Model purpose
Describes, predicts and simulates dynamics of shrub communities under different fire regimes.

## Short description
Markovian models rely on the fact that plant succession is an orderly process allowing the determination of constant transition probabilities from one vegetation state to another. The number of instances of any one state Si changing into another state Sj during a certain period can be computed and its proportion to the total number og units previously in category Si determined and written as Pij. The resulting matrix of probabilities can be used to generate a model of successional changes.

## Model application(s) and State of development
Application area(s): Vegetation succession from Europe to Japan
Number of cases studied: Half-a-dozen cited in the paper
State: validation

## Model software and hardware demands
Platform: PC-based, MAC, IBM. Mainframe, Digital, IBM. Workstation
OS: Any
Software demands: Any spreadsheet or statistical software with matrix algebra
Hardware demands: Any

## Model availability
Purchase: no
Contact person(s) Dr. F. Rego
        Instituto Superior de Agronomia
        Tapada da Ajuda
        PT-1399 Lisboa
        PORTUGAL
Phone:
Fax:
E-mail:

## Model documentation and major references

Rego, F., Pereira, J., and Trabaud, L., 1993. Modeling community dynamics of a Qurcus coccifera L. garrigue in relation to fire using Markov Chains. *Ecol. Modelling*, 66, pp. 251-260.

Usher, M.B. 1981. Modeling ecological succession with particular reference to Markovian models. *Vegetation*, 46, pp. 11-18.

## Other relevant information

One of the best qualities of the model is in its simplicity, which allows its use with minimal software and hardware demands.

## Model identification

Model name/title: Site-specific coefficients for modeling algal growth

## Model type

Media: water, lake
Class: biogeochemical
Substance(s)/compounds: light and temperature
Biological components: community, natural phytoplankton assemblage

## Model purpose

Mathematical models used to simulate phytoplankton growth require specification of a suite of coefficients. The work described here utilizes laboratory experiments and field measurements to estimate and verify values for those coefficients. The impact of uncertainty in estimates of model coefficients on model performance is also considered.

## Short description

Simple models describing algal growth as a function of light and temperature are examined. Site-specific values for the kinetic coefficients embodied in these models are determined through laboratory experimentation. Rates of algal photosynthesis and community respiration are measured over a range of light and temperature conditions using natural phytoplankton assemblages from two eutrophic systems. Values for the maximum specific growth rate, the half-saturation constant for light, the specific respiration rate, and the temperature response coefficients for photosynthesis and respiration are estimated. The utility of the models and attendant, experimentally derived model coefficients is tested through field measurement of net photosynthesis by the natural phytoplankton assemblage.

## Model application(s) and State of development

Application area(s):
Number of cases studied:
State:

---

**Model software and hardware demands**
Platform:
OS:
Software demands:
Hardware demands:

---

**Model availability**
Purchase:
Contact person(s)

Phone:
Fax:
E-mail:

---

**Model documentation and major references**

---

**Other relevant information**

---

**Model identification**
Model name/title: MPKØ1: Modeling Maximum Chlorophyll in the Cabo Frio (Brazil), Upwelling = A Preliminary Approach

---

**Model type**
Media: water, ocean
Class: hydrology
Substance(s)/compounds: nutrients, nitrogen, chloropyll
Biological components: ecosystem - upwelling

---

**Model purpose**
To simulate a maximum of chlorophyll as an expression of the phytoplankton abundance in the sea water during an upwelling event.

---

**Short description**
State variables: chlorophyll,
Forcing functions: light (function of hour), nitrate (function of temperature), advection (empirically adjusted),
Parameters: Growth rate (= Michaelis-Menten equation), mortality rate, advection rate, maximum growth rate (Vm),
Inputs: initial value of chlorophyll, temperature range, simulation time.

---

**Model application(s) and State of development**
Application area(s): Marine production in upwelling zone.
Number of cases studied: 1
State: conceptualization, calibration

**Model software and hardware demands**
Platform: MAC
OS:
Software demands: STELLA (High Performance Systems, Inc.)
Hardware demands: Macintosh Plus - basic configuration.

**Model availability**
Purchase: yes, free. Executable code, source code
Contact person(s) Dr. Ricardo Coutinho
        IEAPM
        Rua Kioto 253, 21930.000
        Arraial do Docabo
        28910 Rio de Janeiro, Brazil
Phone:
Fax:
E-mail:

**Model documentation and major references**

**Other relevant information**

**Model identification**
Model name/title: An Integrated Rate Methodology (IRM) for Multi-factor Growth Rate Modeling

**Model type**
Media: local scale, general growth model
Class: biogeochemical
Substance(s)/compounds: nutrients
Biological components: community - plants, population - plant, organism - plant

**Model purpose**
IRM is a method to calculate "plant growth index", a number between zero and one, that modifies potential relative growth rate according to the levels of available resources and other factors that affect growth. The methodology can also be used to model the growth of any entity of interest.

**Short description**
The methodology was developed in the context of plant growth modeling. The equation for growth index rn is given by

$$\text{rn} = \text{sigma}\ \overset{n}{\underset{i=1}{Wi}}/\text{sigma}\ \overset{n}{\underset{i=1}{[Wi/(nyi * xi)]}}$$

where n denotes a n-component system, xi are the ratio of available to optimal resources, wi are weightings or sensitivities of resources with respect to one resource chosen as a reference, and nyi are the multiplicative modifiers such as temperature, salinity, pH values, etc.

The actual relative growth rate (ARGR) is the product of rn and the potential relative growth rate (PRGR). The model can be applied when (1) the potential growth can be realistically estimated, and (2) calculation of the weighting factors depends on the availability of suitable data or on theability to infer reasonable values when data are sparse or lacking. The method can be used to simulate growth dynamically. The IRM framework can be used for both qualitative and quantitative modeling where the growth rate of a given entity or quantity is relevant. There are many applications in biological, economic and engineering areas.

---

**Model application(s) and State of development**
Application area(s): plant growth, self-sustaining agricultural systems, integration of multiple factor effecting growth, extensive grassland ecosystems.
Number of cases studied: none
State: conceptualization

---

**Model software and hardware demands**
Platform:
OS:
Software demands:
Hardware demands:

---

**Model availability**
Purchase: no
Contact person(s) Dr. Hsin-i Wu
           Center for Biosystems Modeling
           Department of Industrial Engineering
           Texas A&M University
           College Station, TX 77843-3131 USA
Phone:
Fax: +1 409-847-9005
E-mail: wally@carnivore.tamu.edu

---

**Model documentation and major references**
Citation to *Ecol. Modelling* (in press).

---

**Other relevant information**
IRM is a single equation mathematical model that can also be used as a component of a complex simulation model.

**Model identification**
Model name/title: A Plant-Air-Soil-System Model (PASSM)

**Model type**
Media: local scale, terrestrial
Class:
Substance(s)/compounds:
Biological components: ecosystem - terrestrial

**Model purpose**
To simulate the energy exchange processes and microenvironment of a plant community.

**Short description**
This is a steady-state, one-dimensional model based on the gradient diffusion approach, which is designed to predict the microclimate and properties of energy exchange processes between plant communities and their ambient environment. The architecture of the simulation model consists mainly of three major components:
    (1) formulation of radiation distribution,
    (2) formulation of momentum, heat, and water exchange processes,
    (3) formulation of energy balance in plant communities.
Major variables in the model include air and leaf humidities, air and leaf temperatures, latent and sensible heat flux densities, wind velocity, and radiation. The driving forces are meteorological observations above canopy, and the inputs are several geometrical, optical and ecophysiological properties of the plant community under investigation.

    The model outputs include the vertical profiles of radiation, wind velocity, temperature, humidity, source strength of heat and water vapor, and total sensible and latent heat fluxes, throughout the community height. The results of simulation agree reasonably well with two sets of field measurements reported by other researchers.

**Model application(s) and State of development**
Application area(s): Ecosystem Ecology, Meteorology
Number of cases studied: 2
State: validation, prognosis made

**Model software and hardware demands**
Platform: Mainframe, digital
OS: VMS
Software demands: FORTRAN compiler, FORTRAN
Hardware demands: no peculiar demands

## Model availability
Purchase: no, executable code and source code
Contact person(s) Prof. Jianguo Wu, Ph.D
          Biological Sciences Center
          Desert Research Institute
          P.O. Box 60220
          Reno, NV 89506-0220
          USA
Phone: +1 702-673-7419
Fax: +1 702-673-7397
E-mail: jwu@maxey.unr.edu

## Model documentation and major references
Wu, J., 1990. Modeling the energy exchange processes between plant communities and environment. *Ecol. Modelling*, 51, pp. 233-250.
Wu, J., 1987. A simulation model of energy exchange processes and microenvironment of plant communities. M.S. Thesis. Miami University, Oxford, OH, USA. 120 pp.

## Other relevant information
Wu, J., Paw, K.T. U and J.L. Vankat. 1987. A simulation model of energy exchange processes between plant communities and environment. *Bulletin of Ecological Society of America*, 68: 453.

## Model identification
Model name/title: Numerical Simulation on Red Tide of Gymnodinium in Suo-Nada and Iyo- Nada.

## Model type
Media: water, estuary, local scale
Class: biogeochemical
Substance(s)/compounds:
Biological components: community - *G. mikimotoi*

## Model purpose

## Short description
A numerical simulation on red tide of *Gymnodinium mikimotoi* at Suo-Nada and Iyo-Nada, the Seto Inland Sea, in summer 1985 was tried. The red tide occurrence in Suo-Nada is well reproduced by the 3-dimensional numerical assimilation model, including the biological processes. Gymnodinium, which formed red tide at Iyo-Nada in summer 1985, is considered to be transported from eastern part of Suo-Nada or the southern part of Hiroshima Bay based on the results of numerical experiment.

**Model application(s) and State of development**
Application area(s): Seto Inland Sea, Japan
Number of cases studied:
State: verification

**Model software and hardware demands**
Platform: FACOM
OS:
Software demands: FORTRAN
Hardware demands:

**Model availability**
Purchase:
Contact person(s) Dr. T. Yanagi
            Department of Ocean Engineering
            Ehime University
            Bunkyo 3
            Matsuyama 790
            Japan
Phone:
Fax: 81 899 21 5852
E-mail:

**Model documentation and major references**

**Other relevant information**

**Model identification**
Model name/title: A model for the effect of photosynthesis allocation and soil nitrogen on plant growth.

**Model type**
Media: forest
Class: biogeochemical
Substance(s)/compounds: nitrogen
Biological components: organism, a woody plant

**Model purpose**
A study of photosynthate partitioning in plants.

## Short description

The model consists of above-ground and below-ground parts. The dry-weights and nitrogen contents of these parts are considered. The photosynthetic rate (carbon uptake) depends on nitrogen concentration and dry-weight of above-ground parts. The nitrogen uptake rate is proportional to soil nitrogen concentration and dry-weight of below-ground parts. With the aid of an assumption on division of nitrogen between above-ground and below-ground parts the dynamics (two first order differential equations) of the model can be given in terms of total dry-weight and nitrogen concentration of above-ground part. They are the state variables of the model. The share of photosynthates allocated below-ground is a parameter, the optimal choice of which is studied.

## Model application(s) and State of development

Application area(s):
Number of cases studied: 1 (the one reported in paper)
State:

## Model Software and hardware demands

Platform: Mainframe
OS: VAX/VMS
Software demands: FORTRAN
Hardware demands:

## Model availability

Purchase: no, source code
Contact person(s) Dr. Risto Sievänen
            The Finnish Forest Research Institute
            Department of Mathematics
            PB 18
            01301 Vantaa
            Finland

Phone: + 358-0-85705373
Fax: + 358-0-85705361
E-mail: risto.sievanen@metla.fi

## Model documentation and major references

Sievänen, R., Hari, P., Orava, P.J. and Pelkonen, P., A model for the effect of photosynthate allocation and soil nitrogen on plant growth. *Ecol. Modelling*, 41, pp. 55-65.

## Other relevant information

## Model identification
Model name/title: PAPRAN

## Model type
Media: regional scale, terrestrial, agricultural
Class: biogeochemical
Substance(s)/compounds: nutrients, carbon, nitrogen
Biological components: populations, grassland canopy

## Model purpose
To provide a comprehensive assessment of the response of primary production (above ground herbaceous biomass) to environmental factors (rainfall, nitrogen availability and soil depth) in the context of a grassland growing in a Mediterranean-type climate.

## Short description
The model represents the canopy, its biomass and nitrogen content, its phenological development stage, its leaf area index, and the soil water and nitrogen balance.
State variables:
Canopy: Dry weight of: leaves, non-leaf vegetative material, seeds, roots, dead biomass.
Nitrogen content of: leaves, non-leaf vegetative material, seeds, roots, dead biomass.
Development stage, Leaf area index
Soil: Per soil compartment (in standard version ten): soil moisture content, stable organic material, fresh organic material, nitrogen in stable organic material, nitrogen in fresh organic material, mineral nitrogen, soil temperature.
Inputs:
Environment, weather, daily values of: temperature (minimum and maximum), solar radiation, air humidity, wind speed, precipitation.
Soil: soil moisture content at: field capacity, wilting point, air dryness stable organic matter content fresh organic matter content.
Parameters/Functions (selected); decomposition rate constants for: sugars and proteins, cellulose and hemicellulose, lignin, relation temperature/decomposition rate, relation moisture content/decomposition rate, maximum rate of photosynthesis of individual leaves, initial light use efficiency, maintenance respiration factor for vegetative material, maximum rooting depth, daily root extension rate, maximum nitrogen content (function of development stage) of: leaves, non-leaf material, roots, assimilate partitioning factors (function of development stage) for: leaves, non-leaf material, roots, seeds, relation temperature/development rate.

## Model application(s) and State of development
Application area(s): Range management; regional production estimates; development planning in the Mediterranean region.
Number of cases studied: The model was developed from experiments at one site in the region.
State: validation, prognosis made

**Model software and hardware demands**
Platform: PC-based, Mainframe
OS:
Software demands: CSMP
Hardware demands:

**Model availability**
Purchase: yes, no. Source code
Contact person(s) Dr. H. van Keulen
　　　　　AB-DLO,
　　　　　P.O. Box 14
　　　　　NL-6700 AA Wageningen
　　　　　The Netherlands

Phone: +31 8370 75955
Fax: +31 8370 23110
E-mail: keulen@ab.agro.nl

**Model documentation and major references**
Seligman, N.G. and H. van Keulen, 1981. PAPRAN: A simulation model of annual pasture production limited by rainfall and nitrogen. In: (Frissel, M.J. and J.A. van Veen, Eds.) Simulation of Nitrogen Behaviour of Soil-Plant Systems. Pudoc, Wageningen. pp. 192-221.
Seligman, N.G. and H. van Keulen, 1989. Herbage production of a mediterranean grassland in relation to soil depth, rainfall and nitrogen nutrition: a simualtion study. *Ecol. Modelling* 47: 303-311.
Seligman, N.G., H. van Keulen and C.J.T. Spitters, 1992. Weather, soil conditions and the interannual variability of herbage production and nutrient uptake on annual Mediterranean grasslands. *Agric. Forest Meteorol.*, 57: 265-279.

**Other relevant information**

**Model identification**
Model name/title: TRANSIT: TRANspiration of SIngle Trees.

**Model type**
Media: air, local scale
Class: hydrology
Substance(s)/compounds:
Biological components: organism

**Model purpose**
TRANSIT was developed in order to estimate transpiration of single standing trees, hedgerows or similar plant configurations instead of horizontally homogenous vegetation canopies. Its purpose was to investigate land surface-atmosphere exchange processes.

## Short description
TRANSIT is a grid model, requiring input data such as undisturbed air temperature and moisture profiles as well as the flow configuration in and around the tree have to be provided by other (available) models or data sets. Other input data including stem position, tree shape and leaf area data relate a certain leaf area and leaf angle distribution to each grid volume of the model. The spatial heterogenity allows to study the radiation regime as well as the heat and water vapor exchange within the tree crown. Radiation is only subdivided into shortwave and longwave radiation fluxes, photosynthesis is not considered explicitly. Functions of stomatal resistances are required as input for every new species and are then estimated in dependence of the various environmental factors. Finally, the individual leaf temperatures are calculated by solving an energy balance equation with the help of a Newton-Raphson-iteration. From these leaf temperatures, transpiration rates are calculated using air and leaf moisture, as well as the stomatal and dynamical resistance. In its actual version, the model has a temporal resolution of one hour.

## Model application(s) and State of development
Application area(s): Transpiration of plants, water balances
Number of cases studied: 3
State: verification

## Model software and hardware demands
Platform: CYBER (model is to be transformed for PC/MS-DOS)
OS:
Software demands: FORTRAN
Hardware demands:

## Model availability
Purchase: yes, only for academic exchange
Contact person(s) Dr. Ulrike Janssen
    Ludwigstr. 24
    D-63067 Offenbach
    Germany

Phone: +49 69 8003705
Fax:
E-mail:

## Model documentation and major references

TRANSIT - model documentation (in German) The transpiration rate of a single tree - numerical simulations. *Ecol. Modelling* (in press).

## Other relevant information

## Model identification
Model name/title: Modeling the distribution of photosynthetically active solar radiation in tree crowns and forest canopies of defined structure.

## Model type
Media: forest
Class: biogeochemical
Substance(s)/compounds: solar radiation
Biological components: populations - trees

## Model purpose
To predict the distribution of photosynthetically active solar radiation in tree crowns and forest canopies of defined structure and to estimate photosynthetic production.

## Short description
The model predicts the momentary within-tree between-tree shading (the relative decrease in irradiance from shading) of direct solar radiation at different heights in a tree stand. The stand is composed of identical trees, and the spatial distribution of trees on the ground is Poisson. The crown shape is conical, and the distribution of the foliage elements a crown is assumed to be Poisson, so that the attenuation of direct solar radiation within the crown is described by a negative-exponential function of the projected leaf area density and the length of the path of the solar beam within the crown.

Input parameters are number of trees, crown length and width, projected leaf area density in the crowns, and sun elevation. These can be varied to explore the interactive effects of crown form (height-width ratio), leaf area density, stand density, and sun elevation on the spatial distribution of direct solar irradiance within the tree crowns.

## Model application(s) and State of development
Application area(s):
Number of cases studied:
State: prognosis made

## Model software and hardware demands
Platform: mainframe - vax
OS: VMS
Software demands: FORTRAN 77
Hardware demands:

## Model availability
Purchase: sourcecode
Contact person(s)Professor Timo Pukkala
    University of Joensun
    Faculty of Forestry
    P.O. Box 111
    SF-80101 Joensun,
    FINLAND

Phone: + 358 73 1511
Fax:   + 358 73 1513590
E-mail: pukkala JOYL.JOENSUU.F1

## Model documentation and major references


## Other relevant information


## Model identification
Model name/title: Pinogram: a pine growth area model

## Model type
Media: forest
Class: architecture, growth, development
Substance(s)/compounds: growth, area, (light)
Biological components: population - monoculture of trees, organism - tree

## Model purpose
The PINOGRAM model simulates growth, competition and appearance of individual trees in monoculture influenced by different planting anf thinning regimes. The model was initially developed for Scots pine (*Pinus sylvestris L.*).

## Short description
Growth and dynamics of individual Scots pine trees (*Pinus sylvestris L.*) were studied in several stands of different ages representing different growth stadia. Of each tree height, dbh, growth, area, crown length and crown projection were measured as were of some trees the stem volume and mean branch diameter in the lower quarter of the crown. By means of multiple non-linear regression growth functions for diameter, crown length and crown width are derived with age, height and growth area as predictor variables. With the help of these growth functions a growth simulation model for Scots pine was constructed and a computer program, based on this model, was written. Growth of individual trees may be simulated by entering planting distance, S-value per tree, age and thinning regime. The program presents images of stems and abstract forms of crowns in a stand of 20*50 meters in three dimensions and in this way it gives a good depiction of possible forest structures.
Key-words: computer graphics, crown dimensions, distance dependent model, growing space, growth model, Scots pine, simulation, stem form, thinnings, yield.

## Model application(s) and State of development
Application area(s): Monoculture of trees in a stand
Number of cases studied: 1: Scots pine in the Netherlands
State: validation, prognosis made, validation of prognosis (limited)

**Model software and hardware demands**
Platform: PC-based IBM
OS: DOS
Software demands: PINOGRAM, English
Hardware demands: at least a 286-processor with coprocessor and color screen

**Model availability**
Purchase: yes, $ 30, executable code
Contact person(s) Dr. R.P. Leersnijder
            Tarthoust 923
            NL-6703 JE Wageningen
            The Netherlands
Phone: +31 8370-24109
Fax:
E-mail:

**Model documentation and major references**
See program: a Pine growth area model. 1992. by R.P. Leersnijder in *Ecol. Modelling,* May 1992.

**Other relevant information**
Also Windows 3.1 version is available: price $100.

**Model identification**
Model name/title: Vegetation Canopy Thermal Model (VCTM)

**Model type**
Media: local scale, forest, terrestrial
Class: hydrology, thermal
Substance(s)/compounds: none
Biological components: ecosystem - forest

**Model purpose**
VCTM is a steady-state thermal radiance model to compute directional thermal existance and energy balance within forest canopies. The model treats fully leafed canopies as discrete ensembles of leaves partitioned into slope-angle and height classes.

**Short description**
The model abstracts a forest canopy as consisting of three vegetation layers, the sky and a ground layer, which act as both sources and sinks of thermal energy. Steady-state energy budget equations for each canopy layer are solved for canopy layer temperatures which are the used in view factor matrices to compute directional thermal exitance.

The canopy is parameterized by leaf angle distribution and leaf area index for each layer, leaf and background optical scattering and emission properties, and leaf stomatal resistance. Meteorological driving variables include solar irradiance, wind speed, air temperature, relative humidity and soil temperature.

## Model application(s) and State of development
Application area(s): Northern Experimental Forest, Howland, Maine (International Paper)
Number of cases studied:
State: (mixed) verification, calibration, validation

## Model software and hardware demands
Platform: Workstation SUN
OS: UNIX
Software demands: FORTRAN 77
Hardware demands:

## Model availability
Purchase: the model cannot be purchased, available as source code
Contact person(s) Dr. James A. Smith
        NASA/Goddard Space Flight Center
        Code 920/Biospheric Sciences Branch
        Greenbelt, MD 20771 USA
Phone: +1 301-286-4950
Fax: +1 301-286-1757
E-mail: jasmith@ltpsun.gsfc.gov

## Model documentation and major references
McGuire, M.J., Smith, J.A., Balick, L.K., and Hutchison, B.A., 1989. Modeling directional thermal radiance from a forest canopy. *Remote Sens. Environ.*, 27: 169-186.
Smith, J.A., and Goltz, S.M., 1994. Updated thermal model using simplified short-wave radiosity calculations. *Remote Sens. Environ.* (in press).

## Other relevant information
The model, optimally, requires inputs generated from other geometric and radiative transfer models.

## Model identification
Model name/title: TLALOC: Simulating the dynamics of a vegetation mosiac: a spatialized functional model

## Model type
Media: local scale, terrestrial, soil
Class: hydrology
Substance(s)/compounds: carbon, water
Biological components: ecosystem - Arid zone vegetation, community - Vegetation patches, populations - Shrub and grass species

## Model purpose
In arid areas, vegetation patches depends on surface water redistribution and are characterized by a slow movement in the opposite direction to the water flow. The model was designed to simulate the water relations in a simple functional unit and to analyse the central role of water in the whole systems dynamics.

## Short description
TLALOC is a spatialised model describing a transect folowing the main slope. All state variables correspond to 1 m² plots. There are 2 groups of state variables:
(1) the soil is described to a depth of 2 m mainly by its volumetric water content. It changes according to infiltration, soil evaporation and plant transpiration.
(2) the plants are characterised by their cover, i.e., their transpiring surface for each species and age class. Transpiration determines photosynthesis and growth (PS - maintenance) taking into account a negative effect of light extinction. Transpiration is the link between the plants and the soil. The seed production, dispersal and germination are also simulated. The necessary soil parameters are minimum soil water content and field capacity (that is water available to plants) and the parameter of the Ritchie evaporation model, depending on texture. Plant parameters include water use efficiency, biomass/cover relationships, temperature dependence of photosynthesis (which make the difference between C3 and C4 species) and maintenance costs. Seed dispersal distances and minimum soil water content for germination are needed.
Climatic inputs are daily rainfall and potential evapotranspiration and an initial state of the vegetation cover for each species is required (including in particular the ratio of bare areas to vegetated areas).

## Model application(s) and State of development
Application area(s): All arid or semi-arid systems were an horizontal redistribution of rain water induces spatial heterogeneity.
Number of cases studied: 1
State: conceptualization, verification, calibration

## Model software and hardware demands
Platform: PC-based IBM
OS:
Software demands: language FORTRAN
Hardware demands:

## Model availability
Purchase: no, but freely available as source code
Contact person(s) Dr. A. Mauchamp or Dr. S. Rambal
                Centre d'Ecologie Fonctionelle et Evolutive
                L. Emberger, C.N.R.S
                B.P. 5051
                F-34033 Montpellier, France

Phone: +33 67 61 32 89
Fax: +33 67 41 21 38
E-mail: mauchamkfrmop22.cccnusc.fr

## Model documentation and major references
Monta, A.C., Lopez Portillo, J., and Mauchamp, A., 1990. The response of two woody species to the conditions created by a shifting ecotone in an arid ecosystem. *J. Ecol.*, 78: 789-798.
Mauchamp, A., Monta, A.C., Lepart, J., and Rambal, S., 1993. Ecotone dependent recruitment of a desert shrub (Flourensia cernua) in vegetation stripes. *Oikos*, 68: 107-116.
References for the model:
Fisher, R.A. and Turner, N.C., 1978. Plant productivity in the arid and semi arid zones. *Annu. Rev. Plant. Physiol.*, 29: 277-317.
Molz, F.J., 1981. Models of water transport in the soil plant system: a review. *Water Resour. Res.*, 17: 1245-1260.
Miller, P.C., 1981. *Resource Use by Chapparal and Matorral: A Comparison of Vegetation Function in two Mediterranean Type Ecosystem.* Springer-Verlag, Berlin.
Rambal, S. and Cornet, A., 1982. Simulation de l'utilisation de l'eau et de la production végétale d'une phytocénose sahélienne du Sénégal. *Acta Oecologica, Oecol. Plant.*, 3: 381-397.
Ritchie, J.T., 1972. Model for predicting evaporation from a row crop with incomplete cover. *Water Resour. Res.*, 8: 1204-1212.

## Other relevant information
The different processes considered in TLALOC were simulated with different time steps ranging from the day (transpiration, rainfall and infiltration) to the year (seed procution and dispersal) according to the level of precision we needed and to a necessary compromise between precision and realism.

## Model identification
Model name/title: Modeling the effects of increased atmosphere $CO_2$ and effects of climatic changes on tree growth.

## Model type
Media: local scale, terrestrial, forest
Class:
Substance(s)/compounds: carbon
Biological components: ecosystem - forest

**Model purpose**
Predict effects of increased atmosphere $CO_2$ and effects of climatic changes on tree growth.

**Short description**
The model is driven by atmosphere $CO_2$ concentration, air temperature and vapor pressure, soil water content, shortwave radiation, and precipitation. The output is growth of a Douglas-fir seedling. The model has been partially validated. State variables are seedling biomass components such as neddle and root biomass. This is a process model whose main processes are photosynthesis, respiration, stomatal conductance, and conversion of photosynthate into biomass.

**Model application(s) and State of development**
Application area(s):
Number of cases studied:
State: verification, calibration

**Model software and hardware demands**
Platform: PC-based
OS:
Software demands: language BASIC
Hardware demands:

**Model availability**
Purchase:
Contact person(s) Dr. Warren Webb
                Forest Science Dept.
                Oregon State University
                Concollis, OR 97331 USA
Phone: 503-737-6558
Fax:
E-mail:

**Model documentation and major references**

**Other relevant information**

**Model identification**
Model name/title: Model on uptake of $NO_2$ via stomata

## Model type
Media: air, terrestrial, single leaf level
Class: toxicology
Substance(s)/compounds: nitrogen dioxide
Biological components: leaves on higher plants

## Model purpose
Evaluation of various hypotheses on the fate of $NO_2$ taken up by plants via the stomata.

## Short description
Uptake of $NO_2$ via stomata and its elimination in the mesophyll are investigated by use of mathematical models. Systems of differential equations describe simultaneous diffusion and reaction of the chemical species taken into account. The common hypotheses, postulating the disproportionation reaction of $NO_2$ being the predominant path for the transformation of $NO_2$ into the cellular nitrate and nitrite pools, could be false because of its failure to reproduce the measured sorption characteristics. If, however, the reduction of the pro-oxidative $NO_2$ by apoplastic ascorbate is taken into account, the calculated uptake rates and the effects of changes in immission concentration and vegetational characteristics fit well within corresponding experimental results. The onset of acute injury in case of a fall in apoplastic ascorbate concentration below a critical level is postulated.
State variables: $NO_2$ concentration in several leaf compartments.
Important parameters: volume and surface area of leaf compartments, ascorbate concentration, concentration of unsaturated fatty acids in the plasmalemma, several kinetic constants.

## Model application(s) and State of development
Application area(s):
Number of cases studied:
State: verification performed, application to evaluation of general hypotheses published, including some tests on sensitivity

## Model software and hardware demands
Platform: personal computer
OS: DOS
Software demands: language: TURBO PASCAL
Hardware demands:

## Model availability
Purchase: software available as source code
Contact person(s) Dr. Peter Ramge
        Institut für Physikalische und Theoretische Chemie
        J-W Goethe Universität Frankfurt
        Marie Curie Str. 11
        D-60439 Frankfurt am Main, Germany

Phone: +49 69 5800 9416
Fax: +49 69 5800 9445
E-mail: ramge@chemi.uni-frankfurt.d400.de

## Model documentation and major references

Ramge, P., Badeck, F.-W., Ploechl, M., and Kohlmaier, G.H., 1991. Apoplatische Antioxidantien als massgebliche Eliminierungsfaktoren bei der Aufnahme von NO2 über den Blattpfad. Berichte des Zentrums für Umweltforschung der Johann Wolfgang Goethe-Universität, Frankfurt am Main No. 18.

Ramge, P., Badeck, F.-W., Ploechl, M., and Kohlmaier, G.H., 1993. Apoplatische antioxidants as decisive elimination factors, within the uptake of nitrogen dioxide into leaf tissues. *New Phytologist*, 125: 771-785.

## Other relevant information

## Model identification

Model name/title: SPECOM: a spatial single tree model of Scots pine stand/ raw humus soil ecosystem for middle taiga sub-zone.

## Model type

Media: forest
Class: biogeochemical
Substance(s)/compounds: organic matter (carbon), nitrogen
Biological components: ecosystem - coniferous forest ecosystem

## Model purpose

The purpose of SPECOM is to realize the concept of "single plant ecosystem"- SPE (Chertov, 1983) and to compile simplest simulation of forest ecosystem taking into consideration separate tree growth and its spatial disposition in dependence of soil processes (soil organic matter transformation, nutrient, release and soil moisture), meteorological conditions and aerial pollution.

## Short description

The model consists of submodel of Scots pine (*Pinus sylvestris L.*) growth and raw humus soil dynamics. The follow parameters have been used in Pine models: Bp total plant mass; Bl needles' mass; Br sucking roots' mass; H' tree height; D diameters of tree crown and root system; [Greek alpha] biological productivity of needles; np specific consumption of nitrogen on an unit of increment; kw coefficient of water supply; Ip, Il total tree and needles' increment; Lp litter fall; Ll needles' fall; Lr sucking roots fall; table of Pine morphometrical parameters (D and H') in dependence of Bp and shadowing in community. Soil sub-model: H forest floor mass of SPE; N SPE nitrogen mass in H; Nl nitorgen of litter fall in SPE; Natm input of nitrogen from the atmosphere; Nm available for plant nitrogen in SPE; k1 rate of Lp mineralization; kh rate of H mineralization; Mn rate of nitrogen mineralization in parts of that for organic matter. Full SPECOM model includes also subroutines for determination of separate trees shadowing in a stand and redistribution of SPE area in dependence of tree mass.

Dimension of the values: kg, m, m², kg*m-²*year-1, year-1.

**Model application(s) and State of development**
Application area(s): middle taiga sub-zone of European Russia
Number of cases studied: one
State: validation, prognosis made

**Model software and hardware demands**
Platform: Russian PC Elektronika 0585
OS:
Software demands: language TURBO PASCAL
Hardware demands:

**Model availability**
Purchase: no
Contact person(s) Dr. Oleg G. Chertov
    Dept. of Forestry, Faculty of Forestry
    Sankt-Petersburg Forest Academy
    5, Institutsky per.
    St.Petersburg 194018, Russia

Phone: office +7 812 550 0785, home +7 812 272 7353
Fax: +7 812 550 0756
E-mail: chertov@itec.spb.su

**Model documentation and major references**
Chertov, O.G., 1990. SPECOM - a single tree model of pine stand/raw humus soil ecosystem. *Ecol. Modelling*, 50: 107-132.

**Other relevant informations**
At the moment, the SPECOM model is "under restoration": i. submodels of other tree species have been elaborated; ii. a more comprehensive submodel of soil organic matter dynamics and nutrient supply has been completed; iii. rewriting the programs for IBM in TURBO PASCAL 5.0 has been in progress.

## Model identification
Model name/title: SPM

## Model type
Media: terrestrial
Class: biogeochemical
Substance(s)/compounds: carbon
Biological components: ecosystem - slash pine forest

## Model purpose
The SPM was developed as a tool to investigate a diverse set of physiological and ecological measurements and allow scaling-up of carbon dynamics to the stand and ecosystem level.

## Short description
SPM state variables are new foliage; second-year foliage; branches; stern; coarse roots; fine roots in the litter layer; and find roots in the mineral soil. The canopy is divided into 9 vertical layers with the relative gap. Frequency simulated to provide a more realistic light environment. Additional state variables are: needle litter; dead fine roots; and soil organic matter. Processes include photosynthesis, maintenance respiration, growth respiration, carbon partitioning (growth), soil $CO_2$ evolution, decomposition, and mortality. Physiological processes are simulated with an hourly time step.

## Model application(s) and State of development
Application area(s):
Number of cases studied:
State: verification, calibration

## Model software and hardware demands
Platform: PC-based MAC
OS: MS DOS, SYSTEM 7 (MAC)
Software demands: language BASIC, FORTRAN
Hardware demands:

## Model availability
Purchase: no, available as executable code
Contact person(s) Dr. Wendell Cropper, Jr.
                Department of Forestry
                University of Florida
                Gainesville, FL 32611  USA

Phone: +1 904-392-1850
Fax: +1 904-392-1707
E-mail: wpc@nerum.nerdc.ufl.edu

## Model documentation and major references
*Ecol. Modelling*, 66: 213-249 (1993).
*Climate Research*, 3: 7-12 (1993).
*Ecological Bulletins*, 43, Stockholm (in press).

## Other relevant information


## Model identification
Model name/title: PHOTOFACT: a simple mechanistic model for photosynthesis and photoinhibition.

## Model type
Media: water
Class:
Substance(s)/compounds:
Biological components: organism - phytoplankton

## Model purpose
A simple mechanistic model for dynamic and steady-state behaviour of photosynthesis and photoinhibition in phytoplankton. The model can be used to fit steady production curves, and to analyze dynamic incubation experiments. Special emphasis is laid on the slow dynamics of photoinhibition and resulting hysteresis effects.

## Short description
The photosynthetic system of phytoplankton is modeled as an ensemble of "photosynthetic factories" (PSF). A PSF has three states: resting, activated and inhibited. The transitions between the states are driven by (sun) light and biochemical reactions. The response of the model to (changes in) light intensity is described by a system of two linear differential equations with nonconstant coefficients. The time scales of the two equations differ several orders of magnitude: the photoproduction process reacts within a second, while (recovery from) photoinhibition has a reaction time in the range of thousands of seconds. A very good quasi-steady-state approximation is possible, leading to one linear differential equation with coefficients that depend non-linearly on light intensity.
For stepwise changes in intensity analytical solutions exist. Simulation with changing intensities is easy.
The steady state of the model leads to a three-parameter rational photosynthesis curve that can be fitted relatively easily to steady-state incubation data consisting of intensity and corresponding rates of photosynthesis. The fitting of the dynamic response of the model to dynamic incubations is in progress.

## Model application(s) and State of development
Application area(s): eutrophication studies
Number of cases studied: numerous routine applications
State: calibration, validation. The model is simple and for each new incubation parameters are fitted.

## Model software and hardware demands
Platform: Simple spreadsheets and Pascal programs are sufficient. These can be used on any personal computer or workstation.
OS:
Software demands:
Hardware demands:

## Model availability
Purchase: The spreadsheets and Pascal program are available for free.
Contact person(s) Dr. H.C. Eilers
                DCMR Milieudienst Rijnmond
                's- Gravelandsweg 565
                NL-3119 XT Schiedam
                The Netherlands
Phone: +31 10 4273217
Fax: +31 10 4273283
E-mail: paul@dcmr.nl

## Model documentation and major references
Eilers, P.H.C. and Peeters, J.C.H.. A model for the relationship between light intensity and the rate of photosynthesis in phytoplankton. *Ecol. Modelling*, 42 (1988) 199-215.
Eilers, P.H.C. and Peeters, J.C.H.. Dynamic behaviour of a model for photosynthesis and photoinhibition. *Ecol. Modelling*, 69 (1993), 113-133.

## Other relevant information

## Model identification
Model name/title: ALSIM (level 2), i.e., ALfalfa SIMulation

## Model type
Media: water, local scale, agricultural
Class:
Substance(s)/compounds: nutrients, carbon
Biological components:

## Model purpose
Simulation of alfalfa growth and regrowth as a function of soil and moisture on a daily basis.

## Short description

The model cited in the Parsch reference was developed by another researcher. This paper simply validates the model using statistical procedures. The original model (biophysical simulation) was developed by Gary Fick, Dept. of Agronomy, Cornell University, Ithaca NY USA.

## Model application(s) and State of development

Application area(s):
Number of cases studied:
State: validation

## Model software and hardware demands

Platform:
OS:
Software demands:
Hardware demands:

## Model availability

Purchase:
Contact person(s) Dr. Gary Fick
          Dept. of Agronomy
          Cornell University
          Ithaca NY USA
Phone:
Fax:
E-mail:

## Model documentation and major references

## Other relevant information

## Model identification

Model name/title: Linked Stomata and Photosynthesis Model for *Corylus avellana* (hazel)

## Model type

Media: local scale, terrestrial
Class: biogeochemical
Substance(s)/compounds: carbon, water (vapour)
Biological components: organism - *Corylus avellana* (hazel)

## Model purpose
The model describes the seasonal variation of the processes of assimilation and respiration as well as the stomatal reaction of sun leaves of *Corylus avellana* in dependency of the microclimatic factors irradiance, temperature and vapour pressure difference between leaf and air.

## Short description
The model consists of two hierarchical levels. The "basic model" contains the mathematical equations which describes the dependencies of stomatal reaction, assimilation, and respiration on the external variables irradiance, temperature, and vapour pressure difference between leaf and air (VPD). The "basic model" consists of two submodels "stomatal conductance" and $CO_2$-exchange". The equations of the "basic model" are fitted to data sets measured at different phenological phases of *Corylus avellana* during the vegetation period, leading to the "short time parameters". In the other level "main model", the seasonal, time-dependent alteration of the short-time parameters are formulated mathematically.
Described processes: assimilation (photosynthesis), respiration, transpiration.
State variables (output): sum of assimilates, transpired vapour.
External variables (input for the basic model): irridiance, temperature, vapour pressure difference between leaf and air (VPD).
Auxiliary variables: stomatal conductance.
Parameters: minimal stomatal conductance, quantum yield of assimilation, netto photosynthetic capacity, temperature optimum of assimilation, conductance-dependant increase of assimilation, further empirical parameters.
Input for the main model: temperature sum to compute the starting day of the model.

## Model application(s) and State of development
Application area(s):
Number of cases studied:
State: calibration, validation

## Model software and hardware demands
Platform: PC-based
OS:
Software demands:
Hardware demands:

## Model availability
Purchase: no.
Contact person(s) Dr. Susanne von Stamm
        Hauptstr. 3
        D-24589 Schulp
        Germany
Phone:
Fax:
E-mail:

**Model documentation and major references**
Von Stamm, S., 1992. Untersuchungen zur Primärproduktion von Corylus avellana an einem Knickstandort in Schleswig-Holstein und Erstellung eines Produktionsmodelles. Ecosys, Beiträge zur Ökosystemforschung, Kiel, Suppl. 3, pp. 166.
Von Stamm, S., 1994. Linked stomata and photosynthesis model for Corylus avellana (hazel).- *Ecol. Modelling*, 5 (1976): 345-357.

**Other relevant information**

**Model identification**
Model name/title: Diffusion Model for Growth and Size-structure Dynamics of Plant Populations and Communities

**Model type**
Media: terrestrial, local scale, forest, grassland
Class: ecological
Substance(s)/compounds: nutrients, nitrogen
Biological components: community - plant, populations - plant

**Model purpose**
This model presents a conceptual framework for plant population ecology and is a tool for analyzing actual data (growth, mortality and reproduction of individuals). From these results, comparative study of plant species is possible, considering the environmental conditions each species lives in, to investigate life histories, processes and mechanisms in plant populations and communities (e.g., species diversity, regeneration, succession, etc.).

**Short description**
The diffusion model is a size-structured stochastic model for the growth dynamics of plant populations based on a diffusion equation, where $f(t,x)$ is the size distribution function of size $x$ (plant height, stem diameter or plant mass) at time $t$, $G(t,x)$ is the mean of absolute growth rates of individuals of size $x$ at time $t$, $D(t,x)$ is the variance of absolute growth rates of individuals of size $x$ at time $t$, and $M(t,x)$ is the mortality rate of individuals of size $x$ at time $t$. The $G(t,x)$ and $M(t,x)$ functions represent the averaged size-dependent characteristics of species, whilst $D(t,x)$ function represents variations in the species characteristics caused by environmental heterogeneity, genetic variation, variation in the neighbourhood competition effect due to spatial distribution pattern of individuals, etc. The left boundary condition is given as a recruitment rate function $R(t)$. The $G(t,x)$, $D(t,x)$, $M(t,x)$ and $R(t)$ functions are formulated based on physiological processes of an individual such as photosynthesis, respiration, allocation pattern, etc., and competition between individuals in plant stands. Using these functions, the dynamics of $F(t,x)$ is calculated based on the diffusion equation. Then the dynamics of plant population and communities can be described.

---

**Model application(s) and State of development**
Application area(s): both natural and artificial plant populations and plant communities plant ecology as a basic science forestry and agronomy as applied sciences
Number of cases studied: about 30 (including those not published yet).
State: validation

---

**Model software and hardware demands**
Platform: PC-based - IBM, Mainframe - IBM
OS:
Software demands: MS-DOS, Windows, language: TURBO PASCAL for Windows and Visual BASIC for PC , FORTRAN 77 for Mainframe
Hardware demands: IBM-compatible PC, Mainframe (preferably IBM)

---

**Model availability**
Purchase: no. Software available both as executable and sourcecode.
Contact person(s) Dr. Toshihiko Hara
　　　　　　　　Department of Biology
　　　　　　　　Tokyo Metropolitan University
　　　　　　　　Tokyo 192-03, Japan

Phone: +81 0 426 77 2584
Fax: +81 0 426 77 2559
E-mail: A910818@ibm3090.comp.metro-u.ac.jp

---

**Model documentation and major references**
Hara, T., 1984. A stochastic model and the moment dynamics of the growth and size distribution in plant populations. *Journal of Theoretical Biology*, 109: 173-190.
Hara, T., 1984. Dynamics of stand structure in plant monocultures. Journal of Theoretical Biology 110, 223- 239.
Hara, T., 1985. A model for mortality in a self- thinning plant population. *Annals of Botany*, 55, 667-674.
Hara, T., 1986. Growth of individuals in plant populations. *Annals of Botany*, 57, 55-68.
Hara, T., 1986. Effects of density and extinction coefficient on size variability in plant popualtions. *Annals of Botany*, 57, 855-892.
Hara, T., 1988. Dynamics size structure in plant populations. *Trends in Ecology and Evolution*, 3, 129-133.
Kohyama, T., and Hara, T., 1989. Frequency distribution of tree growth rate in natural forest stands. *Annals of Biology*, 64, 47-57.
Hara, T., Van Rijnberk, H., During, H., Yokozawa, M., and Kikuzawa, K., 1990. Competition process and spatial pattern formation in a Betula ermanii population. In: *Spatial Processes in Plant Communities* (F. Krahulec, A.D.Q. Agnew, S. Agnew and J.H. Willems, Eds.), 127-143. SPB Academy Publishing, The Hague.
Hara, T., Kimura, M., and Kikuzawa, K., 1991. Growth patterns of tree height and stem diameter in populaitons of Abies veitchii, A. mariesii and Betula ermanii. *Journal of Ecology*, 79, 1085-1088.
Hara, T., 1992. Effects of the mode of competition on stationary size distribution in plant populations. *Annals of Botany*, 69, 509-513.

Yokozawa, M. and Hara, T., 1992. A canopy photosynthesis model for the dynamics of size structure and self-thinning in plant populations. *Annals of Botany*, 70, 305-316.

Hara, T., Van der Toorn, J., and Mook, J.H., 1993. Growth dynamics and size structure of shoots of Phragmites australia a global plant. *Journal of Ecology*, 81, 47-60.

Hara, T., 1993. Effects of variation in individual growth on plant species coexistance. *Journal of Vegetation Science*, 4: 409-416.

Hara, T., 1993. Competitive asymmetry and size-structure dynamics in plant populations. *Plant Species Biology*, 8: 75-84.

Hara, T., and Wakahara, M., 1944. Variation in individual growth and the population structure of a woodland perennial herb, Paris tetraphylla. *Journal of Ecology*, 82: 3-12.

Hara, T., and Yokozawa, M., 1994. Effects of physiological and environmental variation on size- structure dynamics in plant populations. *Annals of Botany*, 73: 39-51.

Hara, T. and Wyszomirski, T., 1994. Competitive asymmetry reduces spatial effects on size-structure dynamics in plant populations. *Annals of Botany*, 73: 285-297.

Takada, T., and Hara, T., 1994. The relationship between the transition matrix model and the diffusion equation model. *Journal of Mathematical Biology*, in press.

Hara, T., 1994. Growth and competition in clonal plants - Persistance of shoot populations and species diversity. *Folia Geobotanica et Phytotaxonomica*, 29: 181-201.

Hara, T., Yokoishi, E., Ishiwata, M., and Kimura, M., 1994. Competition and species coexistance in an Abies veitchii and A. mariesii mixed stand. *EcoScience*, 1, in press.

## Other relevant information

### Model identification
Model name/title: Diffusion Model for Growth and Size- structure Dynamics of Plant Populations and Communities.

### Model type
Media: terrestrial, local scale, forest, grassland
Class: ecological
Substance(s)/compounds: nutrients, nitrogen
Biological components: community - plant, population - plant

### Model purpose
This model presents a conceptual framework for plant population ecology and is a tool for analysing actual data (growth, mortality and reproduction of individuals). From these results, comparative study of plant species is possible, considering the environmental conditions each species lives in, to investigate life histories, processes and mechanisms in plant populations and communities (e.g., species diversity, regeneration, succession, etc.).

### Short description
The diffusion model is a size-structured stochastic model for the growth dynamic of plant populations based on a diffusion equation, where $f(t,x)$ is the size distribution function of size x (plant height, stem diameter or plant mass) at time t, $G(t,x)$ is the mean of absolute growth rates of individuals of size x at time t, $D(t,x)$ is the variance of absolute growth rates of individuals of size x at time t, and $M(t,x)$ is the mortality rate of individuals of size

x at time t. The G(t,x) and M(t,x) functions represent the average size-dependent characteristics of species, whilst the D(t,x) function represents variations in the species characteristics caused by environmental heterogeneity, genetic variation, variation in the neighbourhood competition effect due to spatial distribution pattern of individuals, etc. The left boundary condition is given as a recruitment rate function R(t). The G(t,x), D(t,x), M(t,x) and R(t) functions are formulated based on physiological processes of an individual such as photosynthesis, respiration, allocation pattern, etc. and competition between individuals in plant stands. Using these functions, the dynamics of f(t,x) is calculated based on the diffusion equation. Then the dynamics of plant populations and communities can be described.

---

## Model application(s) and State of development
Application area(s): both natural and artificial plant populations and plant communities, plant ecology as a basic science, forestry and agronomy as applied sciences.
Number of cases studied:  about 30 (including those not published yet)
State: validation

---

## Model software and hardware demands
Platform: PC-based - IBM, Mainframe - IBM
OS:
Software demands: MS- DOS, Windows.
Language - TURBO PASCAL for Windows and Visual Basic for PC,
FORTRAN 77 for Mainframe
Hardware demands: IBM - compatible PC, Mainframe (preferably IBM)

---

## Model availability
Purchase: no. Software available both as executable code and sourcecode.
Contact person(s) Dr. Toshihiko Hara
        Department of Biology
        Tokyo Metropolitan University
        Tokyo 192-03
        Japan
Phone: +81 (0)426 77 2584
Fax: +81 (0)426 77 2559
E-mail: A910818@ibm3090.comp.metro-u.ac.jp

---

## Model documentation and major references
Hara, T., 1984. A stochastic model and the moment dynamics of the growth and size and distribution in plant populations. *Journal of Theoretical Biology*, 109, pp. 173-190.
Hara, T., 1984. Dynamics of stand structure in plant monocultures. *Journal of Theoretical Biology*, 110, pp. 223-239.
Hara, T., 1985. A model for mortality in a selfthinning plant population. *Annals of Botany*, 55, pp. 667-674.
Hara, T., 1986. Growth of individuals in plant populations. *Annals of Botany*, 57, pp. 55-68.
Hara, T., 1986. Effects of density and extinction coefficient on size variability in plant populations. *Annals of Botany*, 57, pp. 885-892.
Hara, T., 1988. Dynamics of size structure in plant populations. *Trends in Ecology and Evolution*, 3, pp. 129-133.

Kohyama. T., and Hara, T., 1989. Frequency distribution of tree growth rate in natural forest stands. *Annals af Botany*, 64, pp. 47-57.

Hara, T., Van Rijnberk, H., During, H., Yokozawa, M., and Kikuzawa, K., 1990. Competition process and spatial patter formation in a Betula ermanii population. In: Spatial Processes *in Plant Communities* (F. Krahulec, A.D.Q. Agnew, S. Agnew, and J.H. Willems, Eds.), pp. 127-143. SPB Academic Publishing, The Hague.

Hara, T., Kimura, M., and Kikuzawa, K., 1991. Growth patterns of tree height and stem diameter in populations of Abies veitchii, A. mariesii and Betula ermanii. *Journal of Ecology*, 79, pp. 1085-1098.

Hara, T., 1992. Effects of the mode of competition on stationary size distribution in plant populations. *Annals of Botany*, 69, pp. 509-513.

Yokozawa, M., and Hara, T., 1992. A canopy photosynthesis model for the dynamics of size structure and selfthinning in plant populations. *Annals of Botany*, 70, pp. 305-316.

Hara, T., Van der Toorn, J., and Mook, J.H., 1993. Growth dynamics and size structure of shoots of Phragmites australis, a clonal plant. *Journal of Ecology*, 81, pp. 47-60.

Hara, T., 1993. Effects of variation in individual growth on plant species coexistence. *Journal of Vegetation Science*, 4, pp. 409-416.

Hara, T., 1993. Competitive asymmetry and size-structure dynamics in plant populations. *Plant Species Biology*, 8, pp. 75-84.

Hara, T., and Wakahara, M., 1994. Variation in individual growth and the population structure of a woodland perennial herb, Paris tetraphylla. *Journal of Ecology*, 82, pp. 3-12.

Hara, T., and Yokazawa, M., 1994. Effects of physiological and environmental variation on size- structure dynamics in plant populations. *Annals of Botany*, 73, pp. 39-51.

Hara, T., and Wyszomirski, T., 1994. Competitive asymmetry reduces spatial effects on size-structure dynamics in plant populations. *Annals of Botany*, 73, pp. 285-297.

Takada, T., and Hara, T., 1994. The relationship between the transition matrix model and the diffusion equation model. *Journal of Mathematical Biology*, in press.

Hara, T., 1994. Growth and competition in clonal plants - Persistence of shoot populations and species diversity. *Folia Geobotanica et Phytotaxonomica*, 29, pp. 181-201.

Hara, T., Yokoishi, E., Ishiwata, M., and Kimura, M., 1994. Competition and species coexistence in an Abies veitchii and A. mariesii mixed stand. *EcoScience*, 1, in press.

**Other relevant information**

## 7.6 An Overview of Modeling Population Dynamics

Application of models and system analysis in wildlife and fisheries science is a natural result of our growing need to quantify complex problems. A wide spectrum of models has been developed in this area during the last two decades. They focus on a wide variety of problems, related to management of natural resources.

Models of wildlife management are typically represented in Starfield and Bleloch´s book, *Building Models for Conservation and Wildlife Management*, from 1986. Some of the classic examples in this book are included in Table 7.4. W. Grant's book *System Analysis and Simulation in Wildlife and Fisheries Sciences*, can be used as a basic textbook in this area, and many of the basic equations for fishery models can be found in Pauly, *Fish Population Dynamics in Tropical Waters: A Manual for Use with Programmable Calculators*, 1984.

Models of population dynamics are based upon the equations presented in Section 7.2 with modifications and additional equations to account for the characters of the individual problems. If age-structure should be included, the matrix representation presented in Section 7.3 may be applied.

Table 7.4 gives a summary of the models included in Section 7.7. with the addition of a few classic examples to make a complete overview of the model available in this area.

Many models of aquatic and terrestrial ecosystems encompass population dynamics as an integrated part of the entire modeling system. Such models are included in Chapters 2 and 4, while this chapter will contain the models that can be considered to cover mainly population and food-chain dynamics.

Kumar et al. have developed a model for food chain dynamics in seawater with vertical mixing. The model considers three phytoplankton species, two zooplankton species and as nutrients are only ammonium and nitrate included.

The model developed by Sarkar et al. focuses on the role of herbivore attack patterns on growth of plant populations. Three kinds of distribution patterns of the hervi- vore are taken into account.

Jørgensen et al. has developed a fishery model for lakes considering three fish species and 15 age classes using a matrix representation. The length, weight, and number of fish are state variables for each of the 15 age classes for each of the three fish species, in total 135 variables. The forcing functions are climatic factors, fishery, water quality influencing the parameters (growth, mortality and fecundity) and available food.

Kishi has developed a biomass model for the sand lance in Seto Inland Sea, Japan. It is the scope of the model to study the importance factors for stock fluctuation and the role of young sand lance.

The model Crevet simulates the energy flow through a population of shrimps. The model accounts for respiration, excretion, reproduction and various energy loss processes. The effect of temperature on the biological function is also considered.

Rai uses a simple food chain model to study the dynamics of food chain in ponds or other closed aquatic systems. It is a useful model for aquaculture management.

Sawyer et al. apply a model describing the physiological development in the egg of the gypsy moth during diapause. The response to the temperature gradually changes over the course of the development. The egg-stage is modeled as a series of first-order distributed delays, each with its own temperature-dependent rate function.

The model denoted DeerMgmt is a demographic model able to examine the consequences of hunting, predation and sterilization as methods of deer herd reduction strategies.

Dabrowski simulates larvae and juvenile fish growth and survival depending of density, type of prey, water temperature and duration of feeding by the use of a bioenergetic model for analysis of the ontogenetical aspects of fish growth.

DeAngelis et al. have developed a model for the young-of-year small mouth bass. The model is an individual-based simulation model containing many details of the growth and development of the small mouth bass.

Hommel et al. are using another individual-based model of a zooplankton community. The life history of each individual is simulated in the model which accounts for the following processes: filtration, growth, juvenile development, embryonic development, determination of brood size, survival and algal concentration.

A model developed by Ebenhöh considered the influence of oxygen and suspended matter on the mortality of brown trouts.

French and Reed have developed a model for the Northern fur seal population. The model tracks the movement in time and space of specific groups of seals differentiated by age, sex and reproductive status as they migrate reproduce and feed.

A stochastic model is used to characterize migration of marine shrimps. The model has three spatial compartments: two in an estuary and the third is the Gulf of Mexico.

The reproductive effects of three species are studied by a mathematically simple model developed by Antonelli. The model has been successfully applied on two marine and one terrestrial cases.

A model has been developed to study hypotheses on the regulation mechanisms for whole populations. The model is a predator-prey type model.

The model denoted Patch treats phytoplankton and zooplankton as dependent dynamical variables with nutrient concentrations and planktivorous fish as forcing functions. The model accounts for species diffusion and locomotion as well as their transport by the flowing water as spatial processes.

Karås has developed a bioenergetic model for the Eurasian perch. The model simulates the growth of perch by the use of calculations of consumption.

"Fish-School" is a three-dimensional simulation model for the motion of fish. A new position and swimming direction of a fish is calculated on the basis of the positions, orientations and velocities of a number of neighbouring fish. The influence of several neighbours are averaged. The behaviour of a single fish is based upon four patterns: attraction, repulsion, parallel orientation and averaging the influences of the last four neighbours.

A model for self organization of fish school has also been developed. The model is individual-oriented. A fish is represented by an object having a set of parameters (location, swimming direction and velocity) and a set of behaviour possibilities regarding the interaction with other fish. The new swimming direction of a fish is calculated in five steps: influence of neighbour fish in sight range, turning angle assuming all fish preferred distances to each other, determination of an attraction point by adding the weighted vectors to each visible neighbour, average distance of all visible neighbours and finally the component of the parallel orientation.

The model Perch is a tool for fishery management. It simulates the dynamics of fish populations, exploited by size-selective gears. The model considers an age-and sex-structured exploited population with density-dependent reproduction. User-friendly software allows application of different mathematical methods, from calculation of time-series statistic to bifurcation analysis and comparative examination of different exploitation strategies with graphical display of multicriterion optimization. The main state variables are abundance of age classes and individual body size.

A model for determination of the optimal sex ratio for mass rearing of insects for the purpose of sterile releases, genetic alterations etc. has been proposed by Fitz-Earle and Barclay.

A model named Popcycle simulates the numerical abundance of phenology of a pelagic copepod in the Subarctic Pacific Ocean. It is an individual based model in which growth, development, reproduction and mortality of individual organisms are tracked.

The model denoted MiteSim is able to describe the population dynamics of a mite predator-prey system. The system is directly affected by temperature and humidity within the boundary layer of air on a maize plant leaf.

A nonlinear time series model of measles epidemics is developed by Drepper et al. The model may by applied to describe complex self-generated fluctuations in biological and ecological systems. The unpredictability of the incidence pattern of measles is characterized by a subtle interaction between chaotic nonlinear determinism and stochastic fluctuations.

A model named Competition and Coexistence can be used to investigate the coexistence problem. The state variables are densities of many similar species and a few resource determining their survival and growth. Coexistence in the model is due to variable environmental conditions, instability of steady state or direct interactions between species, for instance, competing predator-prey pairs.

The population dynamics of generic large, medium and small body size birds and mammals were studied by using a network model which considers age-specific birth and death rates. The model may be applied to assess relative sensitivity of population to impact on age-specific mortality and fecundity.

A model developed by Gaedke describes the detailed life cycle of two calanodi copepods that coexist. The model comprises five spatial compartments which differ in scelinity, water exchange rates, temperature, predator abundance and predator size.

Biological control of aquatic vegetation by using grass carp has been modeled by Grant. The model can be applied to evaluate alternative aquatic vegetation management strategies. The model consists of three submodels: 1) biomass vegetation management dynamics, 2) grass carp bioenergetics and growth, and 3) management operations.

Grant has also developed a model for migration of marine shrimps by using a stochastic compartmental analysis. The basic model consists of two spatial compartments and three transfer pathways. It is a Markov process model.

A model studying a complex natural system used to predict population curves under varying environmental conditions has been developed by Nutall. The population dynamics of *Ptinus tectus* is represented by a difference equation model. It has 21 state variables and 7 forcing functions.

Bwmod is a spruce budworm model which is able to depict development, survival, feeding and defoliation by a spruce budworm population living on foliage of balsam fir and white spruce.

Artificial intelligence may be used to model animal movements in a heterogeneous habitat. Object-oriented programming, dynamic linkage, rule-based decision procedures and several other concepts from the filed of artificial intelligence ???. An object-oriented model of a deer that learns about habitat is described and used to simulate effects of patch size on deer movements.

Grant has developed a model which focuses on helicopter disturbance of molting Pacific Black Brant. The model determines the behaviour and energetic response of birds encountered by the aircraft during an overflight. The model provides the degree of weight loss these birds experienced due to helicopter disturbance.

Rose has developed a multi-species phytoplankton-zooplankton model with a stepwise iterative calibration. The model contains eight phytoplankton and seven zooplankton species plus various forms of nitrogen and phosphorus. The dominant zooplankton Daphnia is divided into three size classes.

A model is developed by Rose to improve predictions of striped bass-recruitment and population dynamics. The model consists of a detailed young of the year module optionally coupled with an age-structured matrix adult module. The first mentioned module follows individuals and operates on a daily time step, while the latter module is cohort-based and follows age-classes on an annual time step.

Virtala uses a lichen and reindeer model to find the optimal sequence of the yearly harvest of reindeer. The two main variables are the amount of lichen biomass and the size of the reindeer population.

Wissel has developed a model which provides an estimation of the chance of survival of a small population with overlapping generations. The main state variable is the probability of having $n$ individuals in a population.

Woolhouse has proposed a model which describes dynamics of *Aphis pomi* on apple trees under varying environmental conditions. Abundance of *Aphis pomi* is reproceed as presence or absence of each of the 50-100 patches (trees). Change of the state of patch are described by a 2 x 2 transition matrix. Transition probabilities are functions of *Aphis pomi* abundance, temperature and day-length.

A model able to forecast the abundance of European red mite in managed apple orchards is available. Changes in states are functions of overall mite abundance and temperature.

Yanagi has provided a model which is able to simulate the dispersion of the swimming crab, *Portunus tretuberculatur*. The dispersal pattern is calculated in a three-dimensional numerical model.

Yanagi has also developed a model for numerical simulation of dispersal patterns of red sea bream juveniles. Three groups of juvenile are considered. The model has included the flow filed and the behavioral characteristics of red sea bream juveniles. The model calculates a dispersal pattern.

Woolhouse has developed a model for relative abundance of two pests and two predator mite species in an apple orchard. The state variables are presence/absence of each of the four mite species on leaf samples from a number of apple trees.

The model denoted Aphid-par-pred optimizes the aphid biocontrol. The model focuses on the interaction between the aphid predator, the aphid parasite and two species of aphids.

A model named Crop Water Balance Simulation Model, attempts to estimate the water balance components and its population dynamical implications for crops with the simplest formulation and the minimum set of input data. It is suited to agro-meterological applications for rainfed crop management, yield prediction and sowing data determination.

McLellan and Rowland have developed a model for seasonal changes of drone numbers in a colony of honeybee, *Apisl mellifare*. It is the purpose of the model to predict the number of both immature and adult drones in a honeybee colony during a year. It includes the drone brood destruction and adult drone eviction.

Gretjetzky investigates how different factors, such as temperature, food availability and clutch size, govern the population dynamics of the European robin.

Gayanilo has developed a model which is able to describe the seasonally oscillating growth of fish and invertebrates, including situations where the growth stops entirely for longer periods.

The model named Wcrpop simulates the dynamics of insect pest of corn in the Midwestern United States from the egg development in the spring to the death of adults the following autumn. The state variables are the number of corn rootworms per square meters on day x in life stage i, sex j and cohort k, the number of eggs laid per female corn rootworm beetles by day x, and the corn plant growth stage.

The model Drosophila is used for a critical evaluation of recommendation for the genetic management of endangered species.

Schmitz has proposed a model for a grassland variables: population size of plants, herbivores and carnivores as function of time.

Roese has developed a spatially explicit ruminant foraging model which represents the foraging behaviour of an individual ruminant within a spatially and temporally heterogeneous habitat. The model is object-oriented and event-driven.

Hall is using a model to understand the dynamics of steelhead trout population in the Great Lakes. It simulates the growth and consumption by separate life history forms in lakes Michigan and Ontario. The model is age structured.

The model Lowb is an individual-orineted wading bird nesting colony model. It has been applied in the Florida Everglades. As it considers the impact on changes in hydrology on the bird population, it has not been validated. It would have required changes in the entire hydrology of the Everglades.

An iterated logistic linear model has been developed by Badia to deal with the population dynamics of a wild populations using the data from hunting statitics on the hunting effort and the approximate age of the killed wild animals. It differs from the Seber's catch per unit effort model, as it uses the logistic linear model in order to estimate the capture probability for each hunting.

## 7.7 Population Dynamic Models

**Model identification**
Model name/title: Dynamical System Model of Plankton Growth and Nutrient uptake

**Model type**
Media: ocean
Class: hydrology
Substance(s)/compounds: nutrients, nitrogen, plankton
Biological components: ecosystem, population - animal and plant plankton, nutrients

**Model purpose**
The purpose is to simulate via a dynamical system the detailed plamkton-nutient interactions in seawater with vertical mixing and with components determined by size within the phytoplankton (three), zooplankton (two), and nutrient (two) groupings. The interactions are described by well-known responses. Emphasis is on vertical diffusions and the effects of up-welling events.

**Short description**
Phytoplankton: NP1: picoplankton, with cell diameters less than 2 µm
  NP2: nanoplankton, with cell diameters between 2 and 20 µm
  NP3: microplankton, with cell diameters greater than 200 µm
Zooplankton:  NQ1: microzooplankton, which pass through a 200 mym net
  NQ2: macrozooplankton, retained by a 200 µm net
nutrients:  $NH_4^+$ nitrogen in the form of ammonium
  $NO_3^-$ nitrogen in the form of nitrate
Nutrient limitation follows Michaelis-Menton kinetics.
Light intensity follows a double exponential expression (Platt et al., 1980)
Ammonium inhabiting nitrate uptake is represented as an exponential with an inhibition parameter (Wroblewski, 1977).
Grazing is described by the Iuler equation.
Diffusion coefficients are important parameters and may be variable.

**Model application(s) and State of development**
Application area(s): Food-chain dynamics in seawater with vertical mixing
Number of cases studied: 1
State: conceptualization, validation and prognosis made

**Model software and hardware demands**
Platform: mainframe
OS: Vax
Software demands: Nay routines, FORTRAN
Hardware demands: large mainframe computer

## Model availability
Purchase: yes, US $1,000 (negotiable, depending on support and documentation required).
Contact person(s)

> Professor Graeme C. Wake
> Mathematics Department, Massey University,
> Private Bay 11-222, Palmerstone North,
> NEW ZEALAND

Phone: +64 6 350-5081
Fax:    +64 6 350-5611
E-mail:G.WAKEamassey.ac.nz

## Model documentation and major references
Kumar, S.K., Vincent, W.F., Austin, P.C. and Wake, G.C. 1991. Picoplankton and marine food chain dynamics in a variable mixed layer: a reaction-diffusion model. *Ecol. Modelling*, 57, 193-219.
Hadfield, M., Progress on modeling Plankton Population Dynamics. New Zealand Oceanographic Institute, NIWA-Oceanographic, Biology Section Report, 1993/1.
Platt, T.C., Gallegos, C.L., and Harrison, W.G. 1980. Photoinhibition of photosynthesis in natural assemblages of marine phytoplankton. *J. Mar. Res.*, 38, 687-701.
Wroblewski, J.S. 1977. A model of phytoplankton plume formation during variable Oregon upwelling. *J. Mar. Res.*, 35, 357-394.

## Other relevant information
In the simulations done in Reference 1, one coding error resulted in an underestimate by a factor of ten in the rate of photosynthetic production in the westcoast, South Island, New Zealand application. This is corrected in Reference 2.

## Model identification
Model name/title: Role of herbivore attack pattern in growth of plant populations.

## Model type
Media: terrestrial, forest, agricultural
Class:
Substance(s)/compounds:
Biological components: ecosystem - terrestrial, population - plant and insects

## Model purpose
To study the growth of plant populations under different attack patterns of herbivores by means of mathematical models.

## Short description
The model deals with the growth of plant populations under herbivore attack. The probability distribution of herbivore on plants are assume to follow three different kinds of distributions (e.g., negative binomial, Poisson, binomial). The state variables in the model are the number of plants (V) and the number of herbivores (N) distributed over plants.

The spatial distribution of herbivore reflecting the extent of clumping are the main parameters of the model.

## Model application(s) and State of development
Application area(s): Growth of seasonal crops (e.g., paddy, wheat in India) and the growth of man-made forest (social forests).
Number of cases studied:
State: conceptualization

## Model software and hardware demands
Platform: PC-based
OS:
Software demands: BASIC/ FORTRAN 77
Hardware demands:

## Model availability
Purchase:
Contact person(s) Dr. A.K. Sarkar
        Departement of Mathematics
        Jadavpur University
        Calcutta 700- 032
        India

Phone: 72- 4044
Fax:
E-mail:

## Model documentation and major references
Sarkar, A.K. and Roy, A.B., 1993. Oscillatory behaviour in a resource based plant-herbivore model with random herbivore attack, *Ecol. Modelling*, 68: 213-226.

## Other relevant information
The above model is being extended to cover resource based plant-herbivore systems with nutrient recycling and assuming the probability distribution of herbivore follows Poisson distribution.
From the model study it appears that the supply rate of external resources, herbivore mortality rate and the loss of plant biomass due to grazing and litter-fall play an important role in shaping the dynamics of such system. The model analysis produces the interesting result that increased nutrient input to the system can increase the tendency towards a large amplitude periodic oscillations.

## Model identification
Model name/title: Modeling management of fishery in a lake.

**Model type**
Media: water, lake
Class:
Substance(s)/compounds:
Biological components: population - fish, 3 species

**Model purpose**
Management of fishery in a lake.

**Short description**
State variables: 3 fish species each with 15 age classes indicating length, weight and number of fish. In total 135 st. var.
Forcing functions: fishery, climate, available food, water quality.
Important parameters: growth rate and fertility of fish, influence of water quality on fish mortality.

**Model application(s) and State of development**
Application area(s):
Number of cases studied: 1
State: calibration

**Model software and hardware demands**
Platform: PC-based
OS:
Software demands: language SYSL
Hardware demands:

**Model availability**
Purchase:
Contact person(s) Dr. S.E. Jørgensen
          Sektion for Miljøkemi
          Danmarks Farmaceutiske Højskole
          Universitetsparken 2
          DK-2100 København Ø
          Denmark

Phone: +45 35 37 08 50
Fax: +45 35 37 57 44
E-mail: snors@charon.dfh.dk

**Model documentation and major references**
Jørgensen, S.E., Kamp Nielsen, L., Jørgensen L.A., and Mejer, H., 1982. An environmental management model of the Upper Nile lake system. *ISEM Journal*, 4: 5-72.

**Other relevant information**

**Model identification**
Model name/title: A biomass-based model for the sand lance in Seto Inland Sea, Japan

**Model type**
Media: ocean, regional scale
Class: biogeochemical
Substance(s)/compounds: biomass, organic components
Biological components: ecosystem - biomass-based, prey-predator

**Model purpose**
To study important biological parameters for stock fluctuation and the role of young sand lance, zooplankton, aestivation of sand lance.

**Short description**
State variables: egg, larvae, young, one-year-old adult, more than one-year-old adult of sand lance and copepoda.
Forcing functions: catch data of sand lance.
Important parameters: initial state of zooplankton, grazing rate.
Necessary inputs: selection constants for sand lance.

**Model application(s) and State of development**

Application area(s): Seto Inland Sea, Japan
Number of cases studied: 30
State: verification

**Model software and hardware demands**
Platform: Mainframe IBM, Workstation IBM
OS:
Software demands: GDDM, FORTRAN 77
Hardware demands:

**Model availability**
Purchase: case by case (can be distributed), as source code
Contact person(s) Dr. Michio J. Kishi
        Ocean Research Institute
        University of Tokyo
        Minamidai 1-15-1, Nakano-ku
        Tokyo, 113 Japan

Phone: 81-3-5351-6507
Fax: 81-3-5351-6506
E-mail: kishi@ori.u-tokyo.ac.jp

## Model documentation and major references

*Ecol. Modelling,* 54 (1991): 247-263

## Other relevant information

## Model identification
Model name/title: CREVET

## Model type
Media: ocean, local scale
Class: biogeochemical
Substance(s)/componds: energy (Joules or calories)
Biological components: population

## Model purpose
The purposes of this model are, first to simulate how the structure of population varies, and second to describe how energy flow is channeled through this population.

## Short description
The model uses the individual as the basis of the simulated population. It follows the evolution over three of a group of animals essentially as a function of growth, death and reproduction.
The effect of temperature on the biological function is also taken into account. To simulate the energy fluxes, it takes into account the respiration, excretion, reproduction and energy loss.

## Model application(s) and State of development
Application area(s): One shrimp population of Mediterraenean Sea.
Number of cases studied: 1
State: validation

## Model software and hardware demands
Platform: Workstation HP
OS: UNIX
Software demands: PASCAL
Hardware demands:

## Model availability
Purchase: yes, executable code
Contact person(s) Dr. Jean-Philippe Labat
    Observatoire Océanologique
    BP28-06230 Villefranche-sur-Mer
    France

Phone: 93 76 38 45
Fax: 93 76 38 93
E-mail: labat@ccrv.obs-vlfr.fr

## Model documentation and major references

## Other relevant information

## Model identification
Model name/title: A Simple Food Chain Model

## Model type
Media: lake, pond
Class:
Substance(s)/compounds:
Biological components: predator-prey ecosystems

## Model purpose
It has been designed to a study the temporal dynamics of simple food chains in a pond or in closed aquatic systems.

## Short description

## Model application(s) and State of development
Application area(s): Ecosystem Modeling, Aquaculture Management
Number of cases studied:
State: conceptualization

## Model software and hardware demands
Platform: VAX System operated on VMS
OS:
Software demands: Graphics Works Station: Tecktronix, FORTRAN 77
Hardware demands:

## Model availability
Purchase: Source code and other necessary codes/subroutines are available for a cost which involves only handling charges
Contact person(s) Dr. Vikas Rai
        Centre for Atmospheric Sciences
        Indian Institute of Technology
        New Delhi 110016 India

Phone: +91-11-656197
Fax: +91-11-6862037
E-mail:

## Model documentation and major references
Period-doubling bifurcations leading to chaos in a model food chain; *Ecol. Modelling*, 69, 63-77, Sept. 1993.

## Other relevant information
The model predicts chaotic dynamics for ecosystems with high rate of self-reproduction and high carrying capacity of the environment for prey species. This prediction can be easily verified in the laboratory. Moreover, it can provide valuable guidelines for managers of ponds or other aquatic ecosystems.

## Model identification
Model name/title: Gypsy Moth Egg Diapause (no acronym)

## Model type
Media: forest
Class:
Substance(s)/compounds:
Biological components: organism- Gypsy moth, *Lymantria dispar*

## Model purpose
To evaluate a theoretical model of physiological development in the egg of the gypsy moth during diapause.

## Short description
A simulation model was used to examine the idea that diapause development in the egg of the gypsy moth is a dynamic process in which the response to temperature gradually changes over the course of development. No clear demarcation of diapause and post-diapause development was presumed. Rather, an initial developmental response having low thresholds and optimal temperatures was assumed to give way gradually to a response having higher thresholds and optima. The egg stage was modeled as a series of first-order distributed delays, each with its own temperature-dependent rate function. State variables describes the number of individuals in successive substages, the substages representing physiological age classes.

Parameters describe how the rate functions change as individuals progress through the developmental stages. Any pattern of temperature (the sole driving variable) may be input to simulate field conditions or complex laboratory experiments. The model was used to evaluate and reconcile conflicting experimental data on the development and hatch of gypsy moth eggs as a function of temperature conditions during diapause.

**Model application(s) and State of development**
Application area(s): insect physiology and development (theoretical), forest pest management (insect phenology prediction)
Number of cases studied: 5
State: validated and published

**Model software and hardware demands**
Platform: workstation Symbolics 3650
OS: Genera 6.0, Symbolics Lisp
Software demands:
Hardware demands:

**Model availability**
Purchase: no
Contact person(s) Dr. Alan J. Sawyer
        U.S. Department of Agriculture
        Agricultural Research Service
        U.S. Plant, Soil & Nutrition Laboratory
        Tower Rd., Ithaca, NY 14853 USA
Phone: +1 607-255-2458
Fax: +1 607-255-2459
E-mail: ajs10@cornell.edu

**Model documentation and major references**
Sawyer, A.J., Tauber, M.J., Tauber, C.A, and Ruberson, J.R., Gypsy moth (Lepidotera: Lymantriidae) egg development: a simulation analysis of laboratory and field data. *Ecol. Modelling*, 66, 121-155 (1933).

**Other relevant information**

---

**Model identification**
Model name/title: DeerMgmt

**Model type**
Media: local scale, terrestrial
Class:
Substance(s)/compounds: others - immunosterilants
Biological components: population - white-tail deer (*Odocoileus virginianus*), semi-closed populations, organism - white-tail deer (*Odocoileus virginianus*)

**Model purpose**
Predeict the outcome of various management options (sterilization, hunting, predation).

**Short description**
We developed a closed-population demographic model to examine consequences of hunting, predation, and sterilization as methods of deer herd (*Odocoileus* sp.) reduction strategies. Sterilization appears attractive where legislation or public pressure prevent hunting or other forms of culling. Sterilizing a fixed number of animals annually during the initial phase of a control program, and then reducing the number of annual sterilizations as herd size declines, will probably not be effective, as our model predicts only two outcomes: no reduction in herd size, or extinction. The number of annual sterilizations must be calculated based on the number of fertile females remaining in the herd, not on total herd size, and the number of annual sterilizations must be greatly reduced before total herd size decreases significantly. For Cumberland Island National Seashore, GA, the model predicts that the herd of 1500 deer can be controlled at 750. Over three years, an initial rate of 200 sterilizations per year is reduced to a constant rate of 42 per year. If the current levels of hunting and predation continue, the initial number of sterilizations is reduced from 200 to 81, but the constant annual rate increases to 58.

**Model application(s) and State of development**
Application area(s): deer herd management
Number of cases studied: 1
State: verification, prognosis made

**Model software and hardware demands**
Platform: MAC
OS: Macintosh 68030 or less
Software demands: none-stand alone, language BASIC
Hardware demands: none

**Model availability**
Purchase: no, free. Available as executable code
Contact person(s) Dr. Jim Boone
            Museum of National History
      University of Georgia
            Athens, GA 30602 USA
Phone: +1 706-542-1663
Fax: +1 706-542-3920
E-mail: jboone@zookeeper.zoo.uga.edu

**Model documentation and major references**


**Other relevant information**

## Model identification
Model name/title: Bioenergetic model for the analysis of the ontogenetical aspects of fish growth.

## Model type
Media: water, lake, ocean
Class:
Substance(s)/compounds: nutrients, energy
Biological components: ecosystem - aquatic, organism - fish

## Model purpose
Simulation of larvae and juvenile fish growth and survival depends on density, type of prey, water temperature and duration of feeding.

## Short description
Results from studies on whitefish (*Coregonus* sp.) from many different locations are incorporated into a deterministic model to estimate relationships between physical (temperature, light) and biological (prey type, density, energy value) environmental factors and fish growth. The optimum swimming speed for maximum growth of fish increased with fish body length. Further analysis of the relatonships between growth rate and mortality allows the assessment of young fish recruitment.

## Model application(s) and State of development
Application area(s): fisheries biology
Number of cases studied:
State:

## Model software and hardware demands
Platform: PC-based IBM
OS:
Software demands:
Hardware demands:

## Model availability
Purchase: no
Contact person(s) Dr. K. Dabrowski
         School of Natural Res.
         Ohio State University
         Columbus, OH 43210
         USA
Phone:
Fax:
E-mail:

**Model documentation and major references**
Dabrowski, K., 1989. Formulation of a bioenergetic model for coregorine early life history. *Trans. Am. Fish. Soc.*, 118: 138-150.

**Other relevant information**

---

**Model identification**
Model name/title: Young-of-year smallmouth bass model

**Model type**
Media: water, lake
Class:
Substance(s)/compounds:
Biological components:

**Model purpose**
A cohort of young-of-year smallmouth bass is described by means of an individual- based simulation model. The purpose is to determine the factors influencing recruitment to the yearling population.

**Short description**
The dynamics of a young-of-year cohort of a fish species, smallmouth bass (*Micropterus dolomieui*), are modeled using an individual-based computer simulation model. The young-of-year fish are simulated from the egg stage until recruitment into the yearling age class. The model assumes an initial population of adults. The spawning time is predicted by an empirical regression between the size of the nest-building male smallmouth bass and the number of accumulated degree days. The survival of nests, hatching of eggs, and swimup of larval smallmouth bass depend on temperature. After swimup the juvenile smallmouth bass are modeled as individuals. Several hundred model fish are used to simulate the cohort. Daily foraging is predicted from the encounter rate of the fish with zooplankton and benthic invertebrate prey, whose population dynamics are modeled. The encounter rate depends on swimming speed and reactive distance, which are size dependent, while the probability of capture depends on the relative sizes of the fish and its prey. Growth for each individual is predicted by a bioenergetics model. Size-dependent and size independent sources of mortality are included. The model predicts the changing size distribution of smallmouth bass through time.

**Model application(s) and State of development**
Application area(s): Fish population dynamics in temperate lakes
Number of cases studied:
State: conceptualization, calibration

**Model software and hardware demands**
Platform: PC-based IBM
OS:
Software demands: language FORTRAN
Hardware demands:

**Model availability**
Purchase: no
Contact person(s) Dr. D.L. DeAngelis
         Environmental Science Div.
         Oak Ridge National Laboratory
         P.O. Box 2008
         Oak Ridge, TS 37831 USA
Phone:
Fax:
E-mail:

**Model documentation and major references**
DeAngelis, D.L., Godbout, L., and Shutter, B.J., 1991. An individual-based approach to predicting density- depending dynamics in smallmouth bass populations. *Ecol. Modelling*, 57: 91-115.

**Other relevant information**

---

**Model identification**
Model name/title: Individual Based Model of a Zooplankton community

**Model type**
Media: water
Class: toxicology, ecology
Substance(s)/compounds: 3,4-Dichloroaniline (conc. as forcing function)
Biological components: population - *Scenedesmus subspicatus, Dapnia magna, Ceriodaphnia quadrangula, Keratella quadrata*, community - consisting of the four species

**Model purpose**
Causal analysis of aquatic laboratory systems. Extrapolation of toxic effects on individuals to population level. Interpolation to not tested toxicant concentrations. Risk assessments = 20.

**Short description**
Individual-based model of the zooplankton species: life history of each individual is simulated based on own life. Table data taken from measured probability distributions (i-state model), resulting in stochastic simulations. Simulated processes are filtration, juvenile development, growth, embryonic development, determination of brood size, survival depending on individual state, algae concentration, population density and the forcing functions (toxicant concentration, temperature, light). Functional dependencies are achieved by life-table experiments, model results are validated using population and community experiments. Algae concentration are simulated via differential equation (Lottka-Volterra type).

**Model application(s) and State of development**
Application area(s): ecology, ecotoxicology, laboratory systems
Number of cases studied: 4
State: validated using 3 case studies (each study is consisting of 1 untreated and 3 treated experiments (3,4-dichloroaniline), general risk assessment, not validated yet

**Model software and hardware demands**
Platform: PC-based
OS: DOS 3.3 or higher
Software demands: 286 PC or higher, language TURBO PASCAL (Borland) Version 6.0 or higher
Hardware demands:

**Model availability**
Purchase: model cannot be purchased at the moment
Contact person(s) Dr. Ute Duelmer
        Dept. of Biology V
        Technical University Aachen
        Worringerweg 1
        D-52056 Aachen, Germany

Phone: +49 241 803693
Fax: +49 241 8888 182
E-mail: ratte@rwth.aachen.de

**Model documentation and major references**
Dissertation in prep. (German language).
Hommen, U., Poethke, H.J., Duelmer, U., and Ratte, H.T., 1993. Simulation models to predict ecological risk of toxins in freshwater systems. *ICES Journal of Marine Science*, 50: 337-347.
Hommen, U., Duelmer, U., and Ratte, H.T., (in press). Monte-Carlo simulations in ecological risk assessment. To be published in: *Proceedings in Nonlinear Systems*. John Wiley and Sons.
Fitsch, V., and Kaiser, H., 1987. Population dynamics of Daphnia magna - Simulations using the individuals' approach. In: Moeller D.P.F. (Ed.): *Advances in System Analysis*, Vol. 2, Vieweg Verlag Braunschweig.

## Other relevant information

---

### Model identification
Model name/title: Modeling the time necessary for reaching a certain degree of mortality among a trout sample exposed to a turbid water with low oxygen concentration.

---

### Model type
Media: water
Class: toxicology
Substance(s)/compounds: oxygen, suspended solids
Biological components: organism - brown trout fry

---

### Model purpose
Model the time necessary for reaching a certain degree of mortality among a trout sample exposed to a turbid water with low oxygen concentration.

---

### Short description
The model is based on results from ecotoxicological tests based on a factorial plan with 2 variables ($O_2$ concentration and suspended matters concentration). The mortality is measured over time and the datasets (% mortality versus time) are fitted to a logistic equation with Marquardt algorithm. The lethal time for n% mortality is then fitted to a function of turbidity.

The interpolation of different plots corresponding to different oxygen concentrations give a contour plot for each level of mortality (isochrone lines) and can be used to determine the risk due to a peak of pollution.

---

### Model application(s) and State of development
Application area(s): Effect of dam draining in rivers
Number of cases studied: 1 as research, then transfered to industry
State: verification

---

### Model software and hardware demands
Platform:
OS: VMS
Software demands: language FORTRAN
Hardware demands: VAX

---

### Model availability
Purchase: no, available as executable code. "not maintained".
Contact person(s) Dr. W. Ebenhöh
        Universität of Oldenburg
        Ammerländerstr. 114-118
        D-26129 Oldenburg, Germany

Phone:
Fax:
E-mail:

## Model documentation and major references

## Other relevant information

The model can be applied to different multivariable ecotoxicological studies. We recently used a polynomial regression on a PC rather than complex nonlinear Marquardt on a VAX with good results. A multipurpose statistical software can be used for that as well, and we stopped the development of the FORTRAN model for that reason.

## Model identification

Model name/title: Seal migration model: a simulation model of seasonal migration and daily movements of the Northern fur seal.

## Model type

Media: ocean, regional scale
Class: population and behaviour
Substance(s)/compounds:
Biological components: population - Northern fur seal

## Model purpose

To simulate movements of seals on a daily time scale by sexual status and age; to estimate impacts of interactions with oil spills and other localized pollutants.

## Short description

A simulation model of the movements of the Northern fur seal (*Callorhinus ursinus*) population within the Bering Sea has been developed based on reported observations of fur seal distributions in the field and tagging studies. The model tracks the movements in time and space of specific groups of seals differentiated by age, sex and reproductive status as they migrate, reproduce and feed. The movements are based on probability distributions and include stochastic components in many parameters. This results in a realistic distribution of seals in space and over time which compares favorably observations and hypothesized migration patterns.

The migration model is useful both in understanding the movement patterns of fur seals and in identifying where they are vulnerable to impacts following interaction with the results of man's activities. The model has been used to estimate impacts resulting from hypothetical oil spills in the Bering Sea. The model could also be used to estimate impacts of other localized pollutants or from entanglement in marine debris.

State variables include number, body weight, reproductive condition, location, and oiling status by age and sex in the population. Important population parameters are age- and sex-specific survival rates, age-specific reproductive rates, and density-dependent survival function constants. Necessary inputs include details of migratory patterns such as seasonal migration, pathway, locations of and time spent on rookeries, and time spent on feeding at sea. Forcing functions are oil spill scenario specifics and probability distribution for feeding at sea.

---

**Model application(s) and State of development**
Application area(s): Bering Sea, Alaska
Number of cases studied:
State: calibration

---

**Model software and hardware demands**
Platform: PC-based IBM
OS: MS-DOS
Software demands: language FORTRAN
Hardware demands:

---

**Model availability**
Purchase: no
Contact person(s) Dr. Deborah P. French
              Applied Science Associates, Inc.
              70 Dean Knauss Drive
              Narragansett, RI 02882 USA

Phone: +1 401-789-6224
Fax: +1 401-789-1932
E-mail:

---

**Model documentation and major references**
A simulation model of seasonal migrating and daily movements of the northern fur seal, *Ecol. Modelling*, 48: 193-219.
Simulation modeling of the effects of oil spills on population dynamics of northern fur seals. OCS-MMS 86-0045, Final Report to US DOI, MMS, Alaska.

---

**Other relevant information**
French, D.P., and Reed, M., 1990. Potential impact of entanglement in marine debris on the population dynamics of the nortehrn fur seal, Caalorhinus ursinus. pp. 431-452. In: R.S. Shomura and M.L. Godfrey (Eds.) *Proceedings of the Second International Conference on Marine Debris*, April 2-7, 1989, Honolulu, HI.
Reed, M. et al., 1989. Simulation modeling of the effects of oil spills on population dynamics of northern fur seals. *Ecol. Modelling*, 49: 49-71.

## Model identification
Model name/title: Stochastic (semi-Markov process) Model

## Model type
Media: estuary, local scale
Class:
Substance(s)/compounds:
Biological components: population of shrimp

## Model purpose
Characterize migration of marine shrimps using stochastic (semi-Markov process) models.

## Short description
This model has three spatial compartments - two in an estuary and the third being the Gulf of Mexico. Data were available on the catch over time of marked shrimp released in the estuary. The model parameters defined the migration rates in the model. The parameters were estimated using a new non-linear least squares program. The model fits the date well, improving on a previous Markov model. Occupation probabilities and mean residence times are also estimated.

## Model application(s) and State of development
Application area(s): Animal migration models
Number of cases studied: 1
State: conceptualization, verification, validation

## Model software and hardware demands
Platform: PC-based
OS: DOS
Software demands: KINETICA
Hardware demands:

## Model availability
Purchase: yes, free at present, available as executable code
Contact person(s) Dr. James Matis
              Department of Statistics
              Texas A&M University
              College Station, TX 77843-3143 USA

Phone: +1 409-845-3187
Fax: +1 409-845-3144
E-mail: matis@picard.tamu.edu

## Model documentation and major references

**Other relevant information**
This model demonstrates the statistical analysis of migration data using a semi-Markov model with non-exponential time intervals between events.

**Model identification**

Model name/title: Marine Model of predator reproductive effects of 3 species (3 level) systems.

**Model type**
Media: ocean, regional scale, terrestrial
Class: hydrology (one only)
Substance(s)/compounds: organic compounds (one only) secondary compounds (terpens, tannins etc.)
Biological components: ecosystem - marine or terrestrial population - starfish, sea Urchins, lynx, hares

**Model purpose**
To study the predator reproductive effects of 3 species (3 level) systems.

**Short description**

**Model application(s) and State of development**
Application area(s):
Number of cases studied: 3 separate ones: 2 marine, 1 terrestrial
State: conceptualization, verification, prognosis made, validation of prognosis

**Model software and hardware demands**
Platform: Workstation SUN
OS:
Software demands: 1. AUTO, 2. BIFOR2, 3. BIFPACK, text and paper gives the 3 software packages I used. They must be purchased.
Hardware demands:

**Model availability**
Purchase: yes, cost ?
Contact person(s) Dr. Peter L. Antonelli
    Department of Mathematics
    University of Alberta
    Edmunton, Alta
    T6G 2G1, Canada

Phone:
Fax: +1 403-492-3396
E-mail: mathdept@sirius.math.ualberta.ca

## Model documentation and major references

Potts, D.C., 1981. Crown-of-thorns starfish-man induced pest or natural phenomena. Chap. 4 in: R.L. Kitching and R.E. Jones (Editors), *The Ecology of Pests (Some Australian Case Histories)*, CSIRO, Melbourne.

Mann, K., 1982. Kelp, sea Urchins and predators: a review of strong interactions in rocky subtidal systems of Eastern Canada, 1970-1980. Nethal, *J. of Sea Research*, 16: 414-423.

Rhoadys, D.F., 1985. Offensive-defensive interactions between herbivores and plants: their relevance in herbivore population dynamics and ecological theory. *Amer. Nat.*

## Other relevant information

## Model identification

Model name/title: Predator-prey Model

## Model type

Media: terrestrial, regional scale
Class:
Substance(s)/compounds:
Biological components: Predator-prey populations

## Model purpose

To "test" a hyposynthesis about the vole population regulation. The hypothesis (Sam Erlinge, Görgen Göransson, Göran Jansson, Olof Liberg, Jon Loman, Ingvar Nilsson, Torbjörn van Schantz and Magnus Sylvén. 1984. Can vertebrate predators regulate their prey? Am. Nat., 123: 125-133) proposes that vole populations that have cyclic dynamics in a simple system will have a non-cyclic dynamics if there are abundant alternative prey in the system. A model system that behaves in the predicted way, with reasonable functions and parameter values is presented. This is considered support (but not "proof") of the hypothesis.

## Short description

The vole population is modeled by a differential equation to follow logistic growth up to a carrying capacity. As an alternative, the carrying capacity is assumed to vary seasonally. Another alternative vole model introduces a time lag that produces a cyclic population (this could mimic voles interacting with vegetation). Predators are modeled with difference equations and have a birth rate proportional to feeding rate. Two other alternative predators are modeled. One is assumed to have constant number, irrespective of prey density (i.e., prey included in the system). The second model is based on a predator with territorial behaviour, as density approaches saturation, birth rate approaches death rate, whatever the prey density.

The alternative prey is either modeled like the vole population or assumed to have a constant density, unaffected by predation. Predation is modeled as a functional response where feeding rate asymptotically approaches a fixed upper limit. When feeding on two prey types (voles and alternative) the predators are either assumed to take them in proportion to their abundance or to exhibit switching, feeding disproportionately on both prey. Different combinations of these submodels are simulated. It appears that there are several situations when a model of only voles and predators produces a cyclic behaviour but the same model, with alternative prey is not cyclic.

---

## Model application(s) and State of development

Application area(s): The functions and parameter values used are based on a field study of predators and their prey in the Revinge area in south Sweden. This is the field study that inspired the hypothesis that the model attempts to "test". Let me explain what I mean by testing in this context. I think that, if you propose a hypothesis (like the one proposed by Erlinge et al.), the least you can ask for support is that it is possible to produce a model that, with reasonable functions and parameter values, produces a behaviour like the one hypothesized. If this is not possible, the hypothesis must be looked at skeptically. If it is possible, this is support, but not "proof" for the model. The present model is such an exercise, which shows that the system studied at the Revinge area, and possibly other similar models, may behave like it does because of the mechanism suggested by Erlinge et al.

Number of cases studied:
State:

---

## Model software and hardware demands

Platform: Mainframe UNIVAC, PC
OS:
Software demands: SIMNON program
Hardware demands:

---

## Model availability

Purchase: The SIMNON code is available through me, but being fairly specialized, it really only serves to further document the model as it is presented in the publications listed below.

Contact person(s)  Dr. Jon Loman
                    Dept. of Ecology
                    University of Lund
                    S-22362 Lund
                    Sweden

Phone: +46 46 120782  or  +46 46 109315
Fax: +46 46 104716
E-mail:

**Model documentation and major references**
Loman, J., 1988. Alternative prey that decreases vole population cyclicity: A simulation study based on field data. *Ecol. Modelling*, 40: 265-310.
Erlinge, S., Görasson, G., Högstedt, G., Jansson, G., Liberg, O., Loman, J., Nilsson, I., von Schantz, T., and Sylvén, M., 1988. More thoughts on vertebrate predator regulation of prey. *Am. Nat.*, 132: 148-154.

**Other relevant information**

**Model identification**
Model name/title: PATCH

**Model type**
Media: water, local scale
Class: biogeochemical
Substance(s)/compounds:
Biological components: community - Phytoplankton - Zooplankton

**Model purpose**
Demonstration of multiple stability and spatio-temporal pattern formation in a minimal model of nonlinear phytoplakton-zooplankton interactions far from equilibrium.

**Short description**
Temporal and spatial patterns in natural plankton communities are interpreted as transient and stationary non-equilibrium solutions of dynamical nonlinear interaction-diffusion-advection systems. Phytoplankton and zooplankton biomass are treated as the dependent dynamical variables, whereas the environmental factors of nutrient concentration and planktivorous fish act as the external control parameters, driving the system away from equilibrium. Species diffusion and locomotion as well as their transport by the flowing water are the spatial processes, interfering with the nonlinear species interactions.

**Model application(s) and State of development**
Application area(s): Theoretical ecology, nonlinear dynamics, physical-biological coupling.
Number of cases studied: none
State:

**Model software and hardware demands**
Platform: Workstation SUN SPARCstation 10
OS: UNIX
Software demands: usual, FORTRAN 77
Hardware demands: usual

## Model availability
Purchase: no, available as source code
Contact person(s) Dr. Horst Malchow
  Institute of Physics
  GKSS Research Ctr.
  POB 11 60
  D-21494 Geesthacht
  Germany

Phone: +49 4152 87 1846
Fax: +49 4152 87 1888
E-mail: malchow@gkss.de

## Model documentation and major references
Scheffer, M., 1991. *OIKOS*, 62, 271-282.
Malchow, H. and Shigesada, N., 1994. *Nonlinear Processes in Geophysics*, 1: 3-11.
Malchow, H., 1993. *Proc. R. Soc. Lond.*, B 251: 103-109.

## Other relevant information


## Model identification
Model name/title: An application of a bioenergetics model to Eurasian perch (*Perca fluviatilis L.*).

## Model type
Media: water
Class:
Substance(s)/compounds:
Biological components: organism - perch (*Perca fluviatilis L.*)

## Model purpose
To perform calculations of consumption and simulate growth of perch.

## Short description


## Model application(s) and State of development
Application area(s): Calculations of consumption, growth simulations
Number of cases studied: at least 5 (5-10)
State: verification, calibration, validation, prognosis made, validation of prognosis

## Model software and hardware demands
Platform: No special demands
OS:
Software demands:
Hardware demands:

## Model availability
Purchase: no
Contact person(s) Dr. P. Karås
                Swedish Environmental Protection Agency
                Coastal Laboratory
                P.O. Box 584
                S-74071 Öregrund, Sweden
Phone:
Fax:
E-mail:

## Model documentation and major references
An application of a bioenergetics model to Eurasian perch (*Perca fluviatilis L*).
*J. Fish. Biol.*, Vol. 41 (1992) pp. 217-230.

## Other relevant information

## Model identification
Model name/title: FISH-SCHOOL: simulation model for the movement of fish schools

## Model type
Media: lake, ocean
Class:
Substance(s)/compounds:
Biological components: population - fish schools

## Model purpose
Fish do not need a leader or external stimuli for their school organization. We developed the three-dimensional simulation model FISH-SCHOOL for the movement of fish schools to understand how a school is formed and maintained. The model shows that the group movement of a school can be maintained by interactions in which each fish controls its movement in relation to its neighbours.

## Short description
In the simulation model, the motion of a fish is divided into steps. The new position and swimming direction of a fish is calculated on the basis of the positions, orientations and velocities of a fixed number of neighbour fish. The influence of a neighbour depends on its relative position (attractive influence, parallel influence.). The influences of several neighbours are averaged. The new velocity and the new orientation of each fish is calculated by probability distributions taking into account random influences. The movement of each fish is based upon the same behaviour model.

The simulation model reproduces the typical characteristics of real schools if the behaviour of a single fish is based on four patterns: attraction, repulsion, parallel orientation and averaging the influences of the last four neighbours. The results of our simulations agree with experimental data in many points (polarization, nearest neighbour distance, internal dynamics,etc.). Moreover, we simulated the behaviour of schools in relation to external stimuli, e.g., obstacles, food clouds, other schools and gradient. Also, in these cases our simulated schools show a realistic behaviour (merging of schools, schools in a fish net, etc.).

## Model application(s) and State of development
Application area(s):
Number of cases studied: 10
State: conceptualization, verification, calibration, validation, prognosis made, validation of prognosis

## Model software and hardware demands
Platform: PC-based digital
OS: MS-DOS, VMS (VAX)
Software demands: Pascal
Hardware demands:

## Model availability
Purchase: yes, executable code
Contact person(s) Dr. Andreas Huth
   University of Kassel
   Environmental System Research Group FB17
   Mönchbergstr. 11
   D-34109 Kassel
   Germany

Phone: 0561 8042496
Fax: 0561 8043176
E-mail: andi@usys.informatik.uni-kassel.de

## Model documentation and major references

Huth, A., and Wissel, C., 1990. The movement of fish schools: a simulation model. In: Alt, W., Hoffmann, G. (Eds.). *Biological Motion*. Berlin, (Springer) pp. 578-590.

Huth, A., 1992. Ein Simulationsmodell zur Erklärung der kooperativen Bewegung von polarisierten Fischschwärmen. Ph.D. Thesis, University Marburg, Germany.

Huth, A., and Wissel, C., 1992. The simulation of the movement of fish schools. *J. Theor. Biol.*, 156: 365-385.

Huth, A., and Wissel, C., 1993. Analysis of the behaviour and the structure of fish schools by means of computer simulations. *Comments on theoretical biology*, 3: 169-201.

Huth, A., Wissel, C., 1994. The simulation of fish schools in comparison with experimental data. *Ecol. Modelling* (in press).

## Other relevant information


## Model identification

Model name/title: Self organisation of Fish School

## Model type

Media: ocean, local scale
Class:
Substance(s)/compounds:
Biological components: organism - fish forming school

## Model purpose

The goal of the model is the investigation of specific fish behaviour leading to tight and highly polarized fish schools. The model integrates the biologic knowledge on fish schooling and shows the necessity for an individual fish to take far away neighbours into account when determing its own swimming direction and speed. Without this, fish schools tend to break in heterogenous surroundings or even fail to form.

## Short description

The model is individual oriented. A fish is represented by an object having a set of parameters (location, swimming direction, velocity) and a set of behaviour possibilities regarding the interaction with other fish. These are non-consideration, attraction, parallel orientation and repulsion.

The new swimming direction of a fish is calculated in five steps:

1. Computation of the weighing factor of the influences of each neighbour fish in sight range (reciprocal to distance).

2. Calculation of the turning angle assuming all fish have preferred distances to each other. This is done by adding the weighted differences of the swimming direction between the respective fish and each of his neighbours.

3. Determination of an attraction point by adding the weighted vectors to each visible neighbour.

4. Calculation of the average distance of all visible neighbours and adjusting this measure (ma) according to the number of visible fish. The adjustment is done on this basis on the basis of a hypothetical triangular packing.

5. This measure (ma) determines the component of the parallel orientation (2) which contributes to the new swimming direction. Repulsion and the swimming speed are modeled in a similar way. Any change of velocity direction is overlayed with stochastic aspect.

Results obtained with the model show that it performs the main characteristics of real fish school (parallel alignment and compactness) regardless of how many fish it is started with.

---

**Model application(s) and State of development**
Application area(s): the application of the model is research on fish behaviour
Number of cases studied:
State: the model is calibrated

---

**Model software and hardware demands**
Platform: The model is written in SIMULA and was run with a SIMULA Compiler on a Sparc 19/41 Workstation under Solaris 2.3. With CIM (a Simula to C translator) it runs on DEC-Alpha Workstations under OSF1. Due to the high Standardisation of SIMULA the model should run with all Simula Compiler or CIM Translators.
OS:
Software demands:
Hardware demands:

---

**Model availability**
Purchase: The model cannot be purchased, but within scientific exchange the sourcecode is available from the authors.
Contact person(s)  Dr. Hauke Reuter
                 Projektzentrum Ökosystemforschung
                 Universität Kiel
                 Schauenburgerstr. 112
                 D-24118 Kiel
                 Germany

Phone:  0431 880 4029
Fax: 0431 880 4083
E-mail: hauke@pz.oekosys.uni-kiel.d400.de

---

**Model documentation and major references**
Aoki, I., 1982. A simulation study on the schooling mechanisms in fish, *Bull. Jpn. Soc. Sci. Fish.*, 48: 1081-1088.
Douglas, R. and Djamgoz, M., 1990. *The Visual System of Fish*, Chapman and Hall, London.
Huth, A. and Wissel, Chr., 1992. The simulation of the movement of fish schools, *J. Theor. Biol.*, 156: 365-385.
Patridge, B.L. and Pitcher, T.J., 1980. The sensory basis of fish schools: relative roles of lateral lines and version, *J. Comp. Physiol.*, 135: 315-325.
Reuter, H. and Breckling, B., 1994. Self organisation of fish schools; an object-oriented model. *Ecol. Modelling*, 75/76: 147-157.

Other relevant information

## Model identification
Model name/title: PERCH (Population Effects and Research on Commercial Harvesting)

## Model type
Media: lake, local scale
Class:
Substance(s)/compounds:
Biological components: population - age- and sex-structured exploited population of fish with density-dependent reproduction.

## Model purpose
PERCH is a modeling tool for fishery managers and experts, simulating the dynamics of fish population, exploited by size-selective gears. User-friendly software allows application of different mathematical methods, from calculation of time-series statistics to bifurcation analysis and comparative examination of different exploitation strategies with graphical display of multi-criterion optimization.

## Short description
Main state variables are abundance of age classes. Including individual body size characters and the growth process into the age-structured population model offers a natural description of both size-dependent reproduction and size-selective harvesting. Main processes (forcing functions) are individual growth, reproduction, natural mortality and harvesting.

Growth curves are described by the von Bertalaffy model, with the assumption of a winter delay in growth. It is also assumed that, at any moment, length is normally distributed with an average value given by the growth model. To calculate individual weight and number of eggs produced, empirical allometrics are used.

Density-dependence is essential in the larval stage only and can be modeled with the Ricker function. For other ages, exponential survival is assumed, where contrary to constant natural mortality, fishing mortality depends on quantity and quality (selectivity) of gears. The user can easily simulate combination of gears of different selectivities (size-selective gears, like gillnets, and fishing gears with a lower threshold size, like bownets or sport fishing), including seasonal distribution of fishing effort for each kind of gear.

Parameters are the characteristics of main processes (growth, reproduction, mortality, and harvesting with the use of gears with different forms of selectivities). Most parameters have a natural meaning and their evaluation are provided by usual ichtiological data.

## Model application(s) and State of development
Application area(s): Recommendation on optimal mesh size and fishing effort for the harvested perch population in Lake Constance was given and the nature of a steady 3-year cycle observed in Lake Neuchatel was explained using this software. Currently, PERCH is used by the Fauna Conservancy of Canton Vaud in Switzerland for planning harvesting

in Lake Geneva. The software can be used to investigate a broad class of age-structured populations, including species with artificial stocking.

Number of cases studied: 2, perch (*Perca fluviatilis*) populations in Lake Constance and Lake Neuchatel. State: conceptualization, verification, calibration, validation, prognosis made?

## Model software and hardware demands
Platform: PC-based, IBM
OS: MS-DOS 3.x or higher
Software demands: no, TURBO PASCAL 6.0 (Borland International, Inc.)
Hardware demands: IBM PC compatible computer with VGA or EGA color monitor, 3 Mb space of hard disk, preferably not slower than 33 MHz and mouse.

## Model availability
Purchase: yes, free - $20 for mailing and diskette. Software available as executable code.
Contact person(s) Dr. Bernard Büttiker
        Fauna Conservancy of Canton Vaud
        Marquisat 1
        CH-1025 Saint-Sulpice
        Switzerland

Phone: +41 21 691 63 14
Fax: +41 21 691 63 14
E-mail:

## Model documentation and major references
Tyutyunov, Yu., Arditi, R., Büttiker, B., Dombrovsky, Yu., and Staub, E., 1993. Modeling fluctuations and optimal harvesting in perch populations. *Ecol. Modelling*, 69: 19-24.
Tyutyunov, Yu. and Arditi, R., 1993. Modeling dynamics of harvested fish populations - a computer software. Final report to Conservation de la Faune du Canton de Vaud, IZEA, Université de Lausanne, Switzerland, 26 pp.

## Other relevant information

## Model identification
Model name/title: Modeling the optimal sex ratio in the lab for mass rearing insects (either males or females or both) for the purpose of sterile releases, genetic alterations, etc.

## Model type
Media: terrestrial, agricultural
Class: population, genetic
Substance(s)/compounds:
Biological components: population, insects

## Model purpose
To determine what is the optimal sex ratio in the lab for mass rearing insects (either males or females or both) for the purpose of sterile releases, geneticalterations, etc.

**Short description**
A population genetic approach is used to assess the output of insects of each gendre under conditions of mass rearing for a variety of possible sex-ratios. Under certain conditions it may be possible to manipulate the sex ratio in the lab by either pre-zygotic or post-zygotic mechanisms. When the production of one sex is the goal of mass rearing, the other sex is either unimportant, or may actually be a nuisance if it consumes limited resources. Thus, manipulation of the sex ratio may allow faster and more economical rearing of the desired sex. We found that the production of females can be maximized by having a sex ratio heavily biased towards females. The production of males can be maximized with a parent sex ratio of between 0.1 and 0.5 females. Required information is egg production per female, method of manipulating the sex ratio, cost of raising each sex, total budget (time and cost), desired gender to be produced.

**Model application(s) and State of development**
Application area(s): This is a theoretical model and has not yet been tested to our knowledge. Thus, a prognosis has been made, but validation has not occured.
Number of cases studied:
State: prognosis made

**Model software and hardware demands**
Platform: Anyone is suitable. A small computer and very simple programming would suffice to quantity the theoretical model for any given species once the relevant information is available.
OS:
Software demands:
Hardware demands:

**Model availability**
Purchase: free. The model is in public domain in the publication *Ecol. Modelling*. Either Dr. Fitz-Earle or I would be happy to talk to anyone about the application of the model to a specific case.
Contact person(s) Dr. Hugh Barclay or     Dr. Malcolm Fitz-Earle
      Pacific Forestry Centre                Dept. of Biology
      506 W. Burnside Rd.                   Capilano College
      Victoria, British Columbia          North Vancouver, British Columbia
      Canada V8Z 1M5                    Canada V7J 3H5

Phone: +1 604-363-0604            +1 604-986-1911 ext. 2445
Fax: +1 604-363-0775             +1 604-984-4985
E-mail: hbarclay(tegn)a1.pfc.forestry.ca     mfitzear(tegn)instruct.capcollege.bc.ca

**Model documentation and major references**
M. Fitz-Earle and H. Barclay. 1989. Is there an optimal sex ratio for insect mass rearing? *Ecol. Modelling*, 45: 205-220.

**Other relevant information**

## Model identification
Model name/title: POPCYCLE: acronym for Population Cycle.

## Model type
Media: water, developed for open ocean, but suitable for other pelagic, aquatic populations as well.
Class:
Substance(s)/compounds: used carbon as currency
Biological components: population, *Metridia pacifica* (a pelagic copepod) organism, *Metridia pacifica*

## Model purpose
The model was developed to describe the numerical abundance and phenology (life history timing) of a pelagic copepod, *Metridia pacifica*, in the Subarctic Pacific Ocean.

## Short description
POPCYCLE is an individual based model (IBM) in which growth, development, reproduction and mortality of individual organisms are tracked. The model as developed operates in one dimension (depth in the vertical), but can be readily extended to two- or three-dimensions (by embedding it within a model of three-dimensional transports and adding additional state variables representing position). Equations describing growth, mortality, and reproduction of individuals are formulated and used to describe and generate a population dynamics history of the population for year-long model runs. Time step is one day. Growth is described using carbon-based, input-output model. Mortality is implemented as a constant daily probability of predation, but is allowed to vary with development. Reproductive parameters of clutch size, clutch frequency, and total number are specified based on literature or experimentally-determined values. Growth in the published version of the model is assumed to be food-limited. Subsequently, an influence of temperature has been added to the model. The principal forcing functions are food resources available (each day and by depth) and temperature. In the published version, state variables tracked for each individual are its birth date, current weight, life stage, and number of days as reproducing adult. Subsequently, other state variables, such as degree of recent hunger/satiation have been added to keep track of feeding history. Other state variables can be easily added, as appropiate, to the model. Initial conditions that need to be specified for the one-dimension version are the numbers and weights of individuals present on day one of the model run. Because individual organisms are tracked and potentially the number of organisms can exceed 100 thousand with some parameter selections, the model uses a factor of five scalling to reduce population size whenever it gets so large that run times exceed several hours (on an IBM-PC).

## Model application(s) and State of development
Application area(s):
Number of cases studied: 2. The model was initially developed for the copepod, *Metridia pacifica*, in the Pacific (as published see reference below). Subsequently, the model has been applied to *Metridia lucens* (sibling species) in the North Atlantic with the goal of evaluating the spatio-temporal distribution of zooplankton-derived bioluminescence (unpublished).

## Model software and hardware demands

Platform: The model operates on an IBM-PC with 640 K and at least 5 Mb of free space on a hard disk. A custom written disk paging routine is used to implement "virtual memory" using hard disk space (can also use a ram disk if available).

OS: PC-DOS 4.0 or later. The code is written in Borland Turbo Pascal (vers. 5.0 or later). Results are saved to files which are imported into spreadsheets for display/ graphing/analysis.

Software demands:

Hardware demands:

## Model availability

Purchase: Model is not available commercially. Sourcecode can be made available, but is not very well documented.

Contact person(s) Dr. Harold P. Batchelder
Division of Environmental Studies
University of California
Davis, CA 95616-8576 USA

Phone: +1 916-752-8576
Fax: +1 916-752-3350
E-mail: hpbatchelder(tegn)ucdavis.edu

## Model documentation and major references

Batchelder, H.P. and C.B. Miller. 1989. Life history and population dynamics of *Metridia pacifica*: results from simulation modeling. *Ecol. Modelling*, 48:113-136.

## Other relevant information

## Model identification

Model name/title: MiteSim: A Simulation Model of the Banks Grass Mite (*Acari: tetranychidae*) and the Predatory Mite, *Neoseiulus fallacis* (*Acari: phytoseiidae*) on Maize.

## Model type

Media: agricultural

Class:

Substance(s)/compounds:

Biological components: populations (Spider mites) organism (maize plant)

## Model purpose

MiteSim can simulate the population dynamics of a mite predator/prey system. The system is directly affected by microenvironment (temperature and humidity) within the boundary layer of air on a maize plant leaf.

## Short description
A simulation model (MiteSim) of the mite predator/prey system consisting of Banks grass mite (BGM), *Oligonychus pratensis* (Banks), and the predatory mite (NEO), *Neosiulus fallacis* (Garman), was developed and validated. This model included the effects of temperature, humidity and predation and was coupled to a comprehensive plant-microenvironment model (Cupid). Leaf temperatures, stomatal and boundary layer resistances, and canopy air temperature and humidity are calculated by Cupid.

A submodel (POPVPD) uses this information from Cupid to calculate leaf surface hunidity. Mite developmental rates and fecundity are each calculated as a function of both leaf surface temperature and hunidity in MiteSim. Neo consumption rates are calculated as a function of temperature and prey density. The model predicted the predator/prey population dynamics well for a laboratory microcosm. However, MiteSim consistently overestimated the number of spider mites in field situations, especially toward the end of the simulations.

These results suggest that substantial mortality factors occur in the field that are not accounted for from laboratory data.

## Model application(s) and State of development
Application area(s): ecology, agriculture
Number of cases studied: 1

## Model software and hardware demands
Platform: IBM Mainframe FORTRAN, could be compiled and run on other platforms
OS:
Software demands:
Hardware demands:

## Model availability
Purchase: Source code can be obtained free. Send a formatted MS-DOS diskette
Contact person(s) Jim Berry
        USDA-ARS
        Rangeland Insect Lab
        Bozeman, MT 59717-0366 USA

Phone: +1 406-994-3051
Fax: +1 406-994-3566
E-mail: A03LCBOZEMAN(@)ATTMAIL.COM

## Model documentation and major references
Berry, J.S., T.O. Holtzer and J.M. Norman. 1991. MiteSim: a simulation model of Banks grass mites (*Acari tetranychidae*) and the predatory mite, *Neoseiulus fallacis* (*Acari: phytoseiidae*) on maize: model development and validation. *Ecol. Modelling*, 53: 291-317.

Berry J.S., T.O. Holtzer and J.M. Norman. 1991. Simulation experiments using a simulation model of the Banks grass mite (*Acari: Tetranychidae*) and the predatory mite, *Neoseiulus fallacis* (*Acari: Phytoseiidae*) in a corn plant microenvironment. *Environ. Entomol.*, 20: 1074-1078.

---

**Other relevant information**

---

**Model identification**
Model name/title: Nonlinear time series model of measles epidemics

---

**Model type**
Media: regional scale, terrestrial
Class:
Substance(s)/compounds:
Biological components: ecosystem: human, population: human, organism: Man and Virus

---

**Model purpose**
Phenomenological description of complex behaviour based on empirical data.

---

**Short description**
The incidence pattern of measles epidemics in New York City is used to demonstrate that highly nonlinear autoregressive models can serve as a new semi-phenomenological level of description for complex self-generated fluctuations in biological or ecological systems. The nature of the unpredictability of the incidence pattern of measles is characterized by a subtle interaction between a chaotic nonlinear determinism and stochastic fluctuations. The sensitivity dependence on initial conditions, characteristics for chaos, is not limited to chaotic attractors, but can also occur on so called chaotic transients. The dynamics of recurrent outbreaks of measles in New York City turns out to be close to a so called boundary crisis which converts a stable chaotic attractor to metastable chaotic transients leading finally to a period-3 attractor. However, the demographic noise destabilizes the periodic orbit as well, and creates a situation of intermittent jumps between episodic periodicities and longer chaotic transients. A simple autoregressive model is used to achieve a plausible geometric understanding of the noise induced intermittency switching between episodic periodicity and transient chaos.

---

**Model application(s) and State of development**
Application area(s): Epidemiology, limnology, biotechnology
Number of cases studied:
State: conseptualization, verification, calibration, validation

---

**Model software and hardware demands**
Platform:
OS: e.g.,UNIX workstation, NAG library
Software demands:
Hardware demands:

## Model availability

Purchase: no, sourcecode

Contact person(s) Dr. F. Drepper

        Arbeitsgruppe Modellierung

        fuer Umweltforschung und

        Lebenswissenschaften

        Research Center Juelich

        Postcode 1913

        D-52425 Juelich

        Germany

Phone:

Fax:

E-mail: F.Drepper@kfa-juelich.de

## Model documentation and major references

F.R. Drepper, R. Engbert and N. Stollenwerk. 1995; Nonlinear Time Series Analysis of Emperical Population Dynamics. *Ecol. Modelling.*

## Other relevant information

## Model identification

Model name/title: Competition and Coexistence

## Model type

Media: water

Class: biogeochemical

Substance(s)/compounds: nutrients, carbon

Biological components: community, competing species

## Model purpose

Investigation of the coexistence problem

## Short description

The models are abstract and relatively simple community models. Variables:

        $X1.......Xn$ densities of many similar species

        $S1.......Sk$ few resource variables, relevant for the dynamics of the $Xi$.

Coexistance is enabled not due to many different resources, but due to

        a) variable environmental conditions

        b) instabilities of the steady state

        c) direct interaction between species (e.g., competing predator-prey pairs)

## Model application(s) and State of development
Application area(s): Theoretical studies
Number of cases studied:
State:

## Model software and hardware demands
Platform: Workstation, SUN
OS:
Software demands: FORTRAN
Hardware demands:

## Model availability
Purchase: no
Contact person(s) Prof. Dr. W Ebenhöh
        Universität Oldenburg
        Ammerländerstr. 114- 118
        D-26129 Oldenburg
        Tyskland

Phone: +49-441-7983231
Fax: +49 -441-79833004
E-mail:

## Model documentation and major references
Ebenhöh, W., Competition and coexistance: modeling approaches. 1995, *Ecol.Modeling*.

## Other relevant information


## Model identification
Model name/title: Modeling sensitivity of populations to impact on age-specific mortality and fecundity.

## Model type
Media: terrestrial
Class: toxicology
Substance(s)/compounds: others, any substance. General stress effects on birth, death
Biological components: population, terrestrial vertebrates

## Model purpose
To assess relative sensitivity of populations to impact on age-specific mortality and fecundity. Purpose: to provide guidance to researches on which demographic parameters they might most profitably direct their efforts.

**Short description**
The population dynamics of generic large, medium and small body size birds and mammals, with appropriately corresponding, age-specific birth and death rates, were simulated using a network model. Density dependence was of Richer type, and presumed to affect either birth and/or death rates. Sensitivity of both frequency of pseudoextinction and mean population size to changes in means and variances of birth, death rates were examined.

Small animals, both with respect to pseudoextinction and mean population size, were most sensitive to mean early survival. Therefore, vis a vis analysis, it is most important to obtain information on how stress affects early survival. Large animals proved most sensitive to adult mortality, and medium-sized animals showed roughly equal sensitive to all birth and death impacts.

---

**Model application(s) and State of development**
Application area(s):
Number of cases studied: N.A.
State: N.A.

---

**Model software and hardware demands**
Platform: PC-based
OS: DOS
Software demands: FORTRAN
Hardware demands: PC-IBM386 was used

---

**Model availability**
Purchase: no
Contact person(s)


Phone:
Fax:
E-mail:

---

**Model documentation and major references**


---

**Other relevant information**


---

**Model identification**
Model name/title: Copepod life cycle model

**Model type**
Media: estuary
Class: biogeochemical
Substance(s)/compounds: carbon
Biological components: population

**Model purpose**
To understand mechanisms of coexistance of calanoid copepods which share mostly common resources and predators and largely occur in the same habitat.

**Short description**
The model describes in great detail the life cycle of the 2 calanoid copepods, *Eurytemora affinis* and *Acartia tonsa* which coexist for example in the Eins-Dollart-Estuary. The model comprises 5 spatial compartments which differ in salinity, water exchange rates, temperature predator abundance and size. It includes many details of the life history of the strongly age-structured populations (e.g., preferences for temperatures, salinity, prey items; trade-off of number and mortality of eggs). The numerous predators feed indiscriminantly on both copepod species besides fish which prefers the more common species slightly and migrates to areas with high coepod abundance. Coexistance originates mostly from the spatial pattern in copepod abundance.

**Model application(s) and State of development**
Application area(s):
Number of cases studied:
State: conceptualization, verification, calibration, validation

**Model software and hardware demands**
Platform: PC-based, Hewlett Packard
OS: Hewlett Packard
Software demands: HP basics, HP basics
Hardware demands: at least medium size HP computer

**Model availability**
Purchase: yes, executable code
Contact person(s) Dr. Ursula Gaedke
             Limnologisches Institut
             Mainaustr. 212
             D-78434 Konstanz, Germany

Phone: +49 7531-882969
Fax: +49 7531-883533
E-mail:

## Model documentation and major references
Gaedke, U., 1990. Population dynamics of the calanoid copepods Eurytemora affinis and *Acartis tonsa* in the Eins-Dollart Estuary: Numerical simulation. *Arch. Hydrobiol.*, 1: 185-226

## Other relevant information
The model has not been used since 1988 but is still available.

## Model identification
Model name/title: Biological control of aquatic vegetation using grass carp: Simulation of alternative strategies

## Model type
Media: lake
Class:
Substance(s)/compounds: biomass
Biological components: community, aquatic vegetation/grass carp community

## Model purpose
We describe the development and use of a simulation model to evaluate alternative aquatic vegetation management strategies that involve both stocking and harvesting of grass carp (*Ctenopharyngodon idella* VAL.) under a variety of environmental conditions characteristic of the southern United States. The model consists of three submodels representing (1) biomass dynamics of the vegetation, (2) grass carp bioenergetics and growth, and (3) management operations. The model accurately simulated biomass dynamics of aquatic vegetation under various intensities of grass carp grazing in Texas ponds, and also simulated grass carp growth rates comparable to those reported in the literature. Simulation of alternative management strategies suggested that a major disadvantage of using grass carp as a biological control agent, i.e., the complete elimination of aquatic vegetation and concurrent destruction of habitat for sport fishes, can be avoided through sequential harvesting and restocking.

## Short description
The model consists of three submodels representing (1) biomass dynamics of the vegetation, (2) grass carp bioenergetics and growth, and (3) management operations. The vegetation submodel simulates changes in the biomass of submersed aquatic vegetation (kg wet weight ha$^{-1}$ day$^{-1}$) as the net result of the processes of growth, senescence, and herbivory. All three processes are driven in part by environmental temperature, with plant growth and herbivory also responding to day length and senescence responding to a seasonal effect represented by degree-days. Herbivory is additionally influenced both by vegetation biomass and total biomass of grass carp and, hence, links the vegetation submodel to the grass carp growth submodel. The grass carp growth submodel simulates weight changes of the 'average' fish (kg live weight d$^{-1}$) as the net result of consumption (herbivory) and maintenance costs, both of which are functions of environmental temperature and fish size. The management submodel simulates changes in the number of grass carp in the system as the net result of stocking and harvesting operations.

The model, which runs on a Macintosh computer using STELLA software (STELLA, 1987), simulates the average weight and number of fish in each cohort (each group of fish stocked at one time) separately.

## Model application(s) and State of development
Application area(s):
Number of cases studied:
State: prognosis made

## Model software and hardware demands
Platform: MAC
OS:
Software demands: STELLA
Hardware demands:

## Model availability
Purchase: no
Contact person(s) Dr. W.E. Grant
　　　　　　　Dept. of Wildlife and Fisheries Science
　　　　　　　Texas A&M University
　　　　　　　College Station, TX 77843 USA

Phone: +1 409-845-5702
Fax: +1 409-845-3786
E-mail: wegrant@orca.tamu.edu

## Model documentation and major references


## Other relevant information


## Model identification
Model name/title: A stochastic compartmental model for migration of marine shrimp

## Model type
Media: estuary
Class:
Substance(s)/compounds: animals
Biological components: population, marine shrimps

---

## Model purpose
This paper demonstrates the application of stochastic compartmental analysis to model migration of a marked population of white shrimp (*Penaues setiferus*) from an estuarine bay along the Texas coast. We first develop a stochastic continuous time Markov process model, and fit the model to time series data from a mark-recapture field experiment. We then illustrate the use of mean residence times and other related moments derived from the model in providing additional insight into the kinetics of shrimp migration. Finally, we suggest extensions to this approach and discuss the general applicability of the approach to other types of ecological questions dealing with animal movement and mortality.

---

## Short description
The data analyzed in this paper come from a study of white shrimp migration conducted by the U.S. National Marine Fisheries Service (NMFS) in Moses Lake, an estuary in Galveston Bay off the Gulf of Mexico. Shrimp can migrate from any given area into another area of the lake, however they can leave the lake only from area II. Shrimp were released on several different days with tags that identified the site and date of release.

Commercial shrimpers were given incentives to return the tags of recaptured shrimp to NMFS. Since data from the area III were sparse, they were merged with data from area I. The basic model consists of two spatial compartments and three transfer pathways. Obviously, there is no physical barrier between the two compartments that would divide the lake into two distinct, homogeneous areas. However, there are a number of continuous gradients of such characteristics as salinity and turbulence that fisheries scientists have used to define the two discrete compartments. We assume that the likelihood of reporting a tag remained constant throughout the study and that shrimpers expended equal effort during each day of the study. (Actually, small differences in effort are inconquential because of the aggregation of data over several release dates.) Under these assumptions we computed the dependent variables, $Y_{ij}(t)$ $(i, j = 1,2)$, which represent the number of tagged shrimp released in area j and caught in area i per unit effort after elapsed time t. Recent reviews of the general theory of compartmental modeling are given in Jacquez (1985) and Godfrey (1983). Most of the theory and virtually all the applications have assumed a deterministic model to describe the kinetics of substances in the compartmental system. This paper instead develops a stochastic model based on the Markov process (MP) model and fits this model to the shrimp data (see e.g. Seber and Wild, 1989, section 8). The use of mean residence times and related moments for describing system kinetics is outlined, and such moments are used to analyze the movements of shrimp under the assumed stochastic model.

---

## Model application(s) and State of development
Application area(s):
Number of cases studied:
State: prognosis made

---

## Model software and hardware demands
Platform: Mainframe
OS:
Software demands:
Hardware demands:

**Model availability**
Purchase: no
Contact person(s) Dr. W.E. Grant
         Dept. of Wildlife and Fisheries Science
         Texas A&M University
         College Station, TX 77843 USA
Phone: +1 409-845-5702
Fax: +1 409-845-3786
E-mail: wegrant@orca.tamu.edu

**Model documentation and major references**

**Other relevant information**

**Model identification**
Model name/title: Simulated population dynamics of *Ptinus tectus*

**Model type**
Media: local scale, terrestrial, agricultural, food storage
Class: biogeochemical
Substance(s)/compounds: nutrients, stored foods
Biological components: organism, *Ptinus tectus*

**Model purpose**
A case study of a complex natural system used to predict population curves under varying environmental conditions in teaching autecology.

**Short description**
The closed system population dynamics of *Ptinus tectus* are represented by a difference equation model involving 7 forcing functions and 21 system variables. Simulation indicate that in favourable conditions density-dependent mortality and fecundity induce a basic periodicity in the population trajectories. Such periodicity reflects generation time. Environmental factors dictate the position and shape of the trajectory. Time delays and environmental oscillations induce increasing amounts of chaotic perturbation.

**Model application(s) and State of development**
Application area(s): Food storage pest survey
Number of cases studied: 3
State: validation of prognosis

**Model software and hardware demands**
Platform:
OS:
Software demands:
Hardware demands:

**Model availability**
Purchase: ues, listing only. Documentation price £20. Sourcecode, listing only.
Contact person(s) Dr. R. M. Nuttall
        Div. of Biology
        Thames Valley University
        Slough Berkshire
        England SL1 1YG
Phone:
Fax:
E-mail:

**Model documentation and major references**


**Other relevant information**


**Model identification**
Model name/title: BWMOD - Spruce budworm seasonal biology model

**Model type**
Media: local scale, forest
Class: physiology
Substance(s)/compounds:
Biological components: population, spruce budworm

**Model purpose**
Simulation of seasonal biology including phenology, feeding, outposition, mortality, migration for prediction.

**Short description**
Model depicts development, survival, feeding, and defoliation by a spruce budworm population living on foliage of balsam fir and white spruce. It also simulates moth flight, attraction to pheromone traps and migration behaviour. Input is time series of daily minimum and maximum air temperature, initial population density, shoot density, state-specific survival rates. Output includes life-stage frequencies, egg deposition, and pheromone trap catches on daily basis.

**Model application(s) and State of development**
Application area(s): Forest protection planning
Number of cases studied: 3
State: validation

**Model software and hardware demands**
Platform: PC-based, Digital
OS: VAX/VMS DOS
Software demands: FORTRAN 77
Hardware demands:

**Model availability**
Purchase: no,
Contact person(s) Dr. Jacques Régnière
              Natural Resources Canada
              Quebec Region
              P.O. BOX 3800, STE-FOY
              CANADA G1V 4C7
Phone: 418-648-5257
Fax: 418-648-5849
E-mail:

**Model documentation and major references**
Régnière, J. and M. You. 1990. A simulation model of spruce budworm (*Lepidoptera tortricidae*) feeding on balsam fir and white spruce. *Ecol. Modelling*, 54: 227-2

**Other relevant information**

**Model identification**
Model name/title: AI modeling of animal movements in a heterogeneous habitat

**Model type**
Media: terrestrial
Class:
Substance(s)/compounds: animal
Biological components: organism, white-tailed deer

## Model purpose

We demonstrate use of object-oriented programming, dynamic linkages, rule-based decision procedures, and several other concepts from the field of artificial intelligence (AI) for other modeling animal movements in a heterogeneous habitat. An object-oriented model of a deer that learns about habitat structure, plans movements, and accomodates to changes in a patchy brushland habitat is described and used to simulate effects of patch size on deer movements. Innovative features of this model include: (a) representation of habitat as a network of heterogeneous patches, (2) representation of an individual's knowledge of the environment (memory network) as different from but related to the habitat (habitat network), (c) individual's use of knowledge of the environment to plan paths to goal, and (d) ability for an individual to change its knowledge base when it encounters changes in the environment. Decision rules in the model are hypothetical, but the current application suggests that object-oriented modeling provides a concise yet detailed technology for modeling animal movements.

## Short description

Recent thinking about the use of artificial intelligence (AI) in ecology and natural resource management suggests that AI techniques may provide new ways to model a variety of ecological systems, including problems concerning animal behaviour and animal-habitat interactions (Coulson et al., 1987). Such AI techniques as object-oriented programming and rule-based decision procedures seem particularly well-suited to modeling animal movements in spatially heterogeneous habitats (Graham, 1986; Saarenmaa et al., 1988).

In this paper we further demonstrate the use of several AI concepts for modeling animal movements in a spatially heterogeneous habitat. More specifically, we (a) discuss events and dynamic-linkages as concepts with great potential for ecological modeling; (b) describe an object-oriented model of a deer that learns about habitat structure, plans movements, and accommodates to dynamic habitat changes in a patchy brushland habitat in south Texas; (c) simulate the effect of habitat modifications on movements of the model deer; and (d) evaluate the general utility of an AI approach for modeling animal movements in a spatially and temporally heterogeneous environment.

## Model application(s) and State of development

Application area(s):
Number of cases studied:
State: prognosis made

## Model software and hardware demands

Platform: Workstation, UNIX
OS:
Software demands: C++
Hardware demands:

**Model availability**
Purchase: no
Contact person(s) Dr. W.E. Grant
        Dept. of Wildlife and Fisheries Science
        Texas A&M University
        College Station, TX 77843 USA

Phone: +1 409-845-5702
Fax: +1 409-845-3786
E-mail: wegrant@orca.tamu.edu

**Model documentation and major references**

**Other relevant information**

**Model identification**
Model name/title: A simulation model of helicopter disturbance of molting Pacific Black Brant

**Model type**
Media: regional scale, terrestrial
Class:
Substance(s)/compounds: animals
Biological components: population, Pacific Black Brant or North Hope of Alaska

**Model purpose**
We describe a simulation model designed to study the effects of helicopter disturbance on molting Pacific black brant near Teshekpuk Lake, Alaska. Locations of 18.118 brant were digitized into the model based on 10 years of population survey data. Bell 206 and Bell 412 helicopters were simulated flying across the molting grounds along 2 routes between 2 airfields. The model determined the behavioral and energetic response of birds encountered by the aircraft during an overflight. Altitude and frequency of overflights were held constant during a simulated 28-day molting period, but were varied among simulations. The model provided the degree of weight loss these birds experienced due to helicopter disturbance. The effects of overflights on brant were classified into 5 risk categories based on weight. For both routes, the number of flocks and birds in each category was determined for each altitude, aircraft type, and overflight frequency. Simulation results indicated that the model can be used to identify flight-line modifications that result in significantly decreased disturbance to the birds.

**Short description**
The disturbance model consists of 2 submodels, a Flight-Line (FL) submodel and a Behavior and Energetics (BE) submodel. The FL submodel contains the distribution of molting brant in the TLSA. The program can simulate a helicopter flying over the molting ground in any direction while recording the number, size and location of all flocks encountered along individual flight lines. The BE submodel determine proximity of the aircraft to each to each flock recorded by the FL submodel, and describes the reaction of these flocks to overflights of various altitudes. The magnitude of disturbance generated is a function of aircraft type and proximity, as well as flock site. We parameterized this program for both the Bell 206 and the Bell 412 helicopter. The BE submodel also determines the amount of energy brant expend in response to overflights of various frequencies, then calculates their weights at the end of molt. To do so, the daily energy budget for brant first is calculated by estimating the number of calories the birds ingest while foraging, then allocating that energy to 3 major functions: (1) production, (2) behaviour, and (3) maintenance. Productive energy is that used to grow new flight feathers. Behavioral energy is that used to perform all behaviors (alert, walk, run, swim, feed, rest, preen, bathe, and aggression). Maintenance energy primarily is comprised of basal metabolic rate (BMR) and thermoregulation. If the daily energy budget is positive, the birds gain weight, if negative, they lose weight. Helicopter disturbance results in weight loss by (1) causing an increase in the amount of energetically-expensive disturbance behaviors that brant exhibit, and (2) decreasing the amount of foraging time available to the birds. Both the number of flocks and birds found along the flight lines and the departure weight of those birds at the end of molt are compared to assess relative impact of aircraft flying alternative routes. Models were written using the software package STELLA and a Macintosh II TM Microcomputer.

**Model application(s) and State of development**
Application area(s):
Number of cases studieds:
State: prognosis made

**Model software and hardware demands**
Platform: MAC
OS:
Software demands: STELLA
Hardware demands:

**Model availability**
Purchase: no
Contact person(s) Dr. W.E. Grant
Dept. of Wildlife and Fisheries Science
Texas A&M University
College Station, TX 77843 USA

Phone: +1 409-845-5702
Fax: +1 409-845-3786
E-mail: wegrant@orca.tamu.edu

Model documentation and major references

Other relevant information

## Model identification
Model name/title: Multi-species Phytoplankton-Zooplankton Model [Stepwise iterative calibration of a multi-species phytoplankton-zooplankton simulation model using laboratory data]

## Model type
Media: lake
Class: biogeochemical
Substance(s)/compounds: nutrients
Biological components: community, phytoplankton- zooplankton food web

## Model purpose
Simulate the food web and nutrient cycling dynamics of a multi-species phytoplankton and zooplankton community in a controlled laboratory environment.

## Short description
The model is a system of ordinary differential equations describing the daily dynamics of eight phytoplankton species, seven zooplankton groups, and nitrogen and phosphorus forms.

Processes represented in the model include nutrient uptake, photosynthesis (dependent on internal stores of nutrients and light), respiration, excretion, egestion, mortality, reproduction, and nutrient recycling. Biomass is tracked for phytoplankton and zooplankton, with the biomass of the dominant *Daphnia zooplanctor* further separated into numbers and average weight per individual in each of 3, size classes. Calibration was based on single and paired phytoplankton species experiments under low and high initial nutrients conditions, an experiment with all phytoplankton species but no zooplankton (i.e., ungrazed), and a full 63-day experiment involving all phytoplankton and zooplankton. The final calibrated set of parameter values for each species had to generate satisfactory model predictions for all experiments involving that species.

## Model application(s) and State of development
Application area(s):
Number of cases studied: 1
State: validation

**Model software and hardware demands**
Platform:PC-based, IBM
OS:
Software demands: FORTRAN
Hardware demands:

**Model availability**
Purchase: no, executable code
Contact person(s) Dr. Kenneth Rose
          P.O. Box 2008
          Oak Ridge National Laboratory
          Oak Ridge, TN 37831-6036 USA

Phone: +1 615-574-4815
Fax: +1 615-576-8646
E-mail: ikr@stc1ø.ornl.gov

**Model documentation and major references**
Rose, K. et al. 1988. *Ecol. Modelling*, 42: 1-32.
Swartzman et al. 1989. Modeling the direct and indirect effects of streptomycin in aquatic microcosms. *Aquatic Toxicology*, 14: 109-130.

**Other relevant information**

**Model identification**
Model name/title: Individual-based model of striped bass population dynamics

**Model type**
Media: estuary
Class:
Substance(s)/compounds:
Biological components: population, striped bass

**Model purpose**
To improve predictions of striped bass recruitment and population dynamics. By following individuals, we can better simulate the dynamics of populations with high mortality (i.e., rare individuals are survivors).

**Short description**
The striped bass (*Morone saxatilis*) model consists of a detailed young of the year (yoy) module optionally coupled with an age-structured matrix adult module. The yoy module follows individuals and operates on a daily time step. The adult module is cohort-based and follows age-classes on an annual time step. The yoy model is used to predict the number and mean length of age-1 survivors from eggs. Yoy model predictions can be used alone to better understand factors affecting first year growth and survival, or the yoy model can be coupled with the matrix model for long-term simulations.

The yoy and adult models are coupled by using the predictions of age-1 survivors from the yoy model as input into the adult model, which the projects these yoy through the rest of their lives. Each year the adult model is used to compute the number and size distribution of spawners as input to the yoy model for the next year. The environment consists of daily temperature, fraction of the day that is daylight, and time-dependent maximum densities of zooplankton and benthic prey. In the yoy module, day of spawning and development rates of eggs and yolk-sac larvae depend on temperature. Daily growth of feeding individuals is represented by a bioenergetics equation, for which consumption is based on random encounters by individuals with different types of prey. Larvae feed on 4 zooplankton types and juveniles feed exclusively on size classes of 4 benthic types. Egg and yolk-sac larva mortality has both temperature-dependent and constant terms; larva and juvenile mortality depends on an individual's weight and length. Most of the computations in the yoy module involve determining the daily number of each prey type eaten by each striped bass. When used, the adult module utilizes age-specific growth, mortality (natural and fishing), and maturity schedules to annually update the number and mean size of each adult age cohort, and predict the number and sizes of spawners as input to the yoy module.

## Model application(s) and State of development
Application area(s): examine biotic (e.g., size of spawners) and abiotic (e.g., temperature, toxics) factors affecting recruitment; also perform cross population comparisons of striped bass responses to disturbances.
Number of cases studied: model has been (or will be) applied to 4 or more sites in the USA; has been used to examine factors (including episodic and chronic toxics effect) affecting recruitment in the Potomac River, Maryland.
State: validation, prognosis made

## Model software and hardware demands
Platform: PC-based, IBM. Workstation, UNIX-based
OS: DOS, Windows or UNIX
Software demands: DOS extender required, FORTRAN
Hardware demands: 486 PC

## Model availability
Purchase: yes, no charge, executable code, sourcecode
Contact person(s) Dr. Kenneth Rose
P.O. Box 2008
Environmental Sciences Division
Oak Ridge National Laboratory
Oak Ridge, TN 37831-6036 USA

Phone: 615-574-4815
Fax: 615-576-8543
E-mail: ikr@stclø.ctd.ornl-gov

**Model documentation and major references**
Rose, K.A., and J.H. Cowan (1993). Individaul-based model of young-of-the-year striped bass population dynamics. I. Model description and baseline simulations. *Transactions of the American Fisheries Society*, 122: 415-438.
Cowan, J.H., K.A. Rose, E.S. Rutherford, and E.D. Houde (1993). Individual-based model of young-of-the-year striped bass population dynamics. II. Factors affecting recruitment in the Potomac River, Maryland. *Transactions of the American Fisheries Society*, 122: 439-458.
Rose, K.A., J.H. Cowan, E.D. Houde, and C.C. Coutant (1993). Individual-based modeling of environmental quality effects on early life stages of fishes: A case study using striped bass. *American Fisheries Society Symoposium* 14.

**Other relevant information**
(1) Distribution of model also requires permission of project sponsor (Electric Power Research Institute).
(2) Similar models of other fish species are also available.

**Model identification**
Model name/title: Lichen and reindeer in northern Finland

**Model type**
Media: terrestrial
Class:
Substance(s)/compounds:
Biological components: population - lichen and reindeer in Finnish Lapland

**Model purpose**
Given initial lichen and reindeer population levels, the purpose of the model is to find the optimal sequence of yearly harvests of reindeer by which the specified target levels for lichen and reindeer are reached.

**Short description**
The model describes a plant-herbivore system consisting of lichen and reindeer in Finnish Lapland. The model is deterministic, has discrete time and contains two variables: the amount of lichen biomass and the size of the reindeer population. Lichen growth is assumed logistic and the growth of the reindeeer population exponential (the age structure is not considered). The model is used to find the optimal harvesting policy to reach specified target levels for lichen and reindeer. The effects of discounting and the length of the planning period on the optimal solution are also studied.

**Model application(s) and State of development**
Application area(s): Management of plant-herbivore systems
Number of cases studied: 1
State: conceptualization

620

**Model software and hardware demands**
Platform: PC-based
OS: MS-DOS
Software demands: FORTH (note: NOT FORTRAN)
Hardware demands: 387- math coprocessor

**Model availability**
Purchase: no
Contact person(s) Dr. Matti Virtala
          Departement of Applied Mathematics and Statistics
          University of Oulu
          Oulu, Finland
Phone:
Fax:
E-mail:

**Model documentation and major references**
Virtala, 1992. Ecol. Modelling, 60: 233-255.

**Other relevant information**
A corresponding stochastic version of the model has also been developed and the optimal
harvest policy has been studied by using stochastic dynamic programming.

**Model identification**
Model name/title: Stochastic birth and death processes describing minimum viable
populations

**Model type**
Media: local scale
Class:
Substance(s)/compounds:
Biological components: ecosystem - general

**Model purpose**
The model provides a simple way of estimating the chance of survival of a small
population with overlapping generations.

**Short description**
State variable: probability of having n individuals in a population. Parameters: Carrying
capacity, birth and death rate per individual, strength of environmental stochasticity.
The mean lifetime and the probability of recovery of a population with n individuals at
present is calculated. The probability of minimum viable population is determined.

## Model application(s) and State of development
Application area(s): Conservation biology: minimum viable populations
Number of cases studied: 3
State: verification, prognosis made

## Model software and hardware demands
Platform: PC-based
OS:
Software demands: True Basic
Hardware demands:

## Model availability
Purchase: no, executable code and sourcecode
Contact person(s) Dr. Christian Wissel
    UFZ-Centre for Environmental Research, PF 2
    D-04301 Leipzig
    Germany

Phone: +49 341 235 3245
Fax: +49 341 235 3500
E-mail: postman@oesa.ufz.de

## Model documentation and major references
Wissel, C., and Zaschke, S.-H., 1995; Stochastic birth and death processes describing minimum viable populations. *Ecol. Modelling*.

## Other relevant information

## Model identification
Model name/title: Transition Matrix Model of *Aphis pomi* population dynamics

## Model type
Media: agricultural
Class:
Substance(s)/compounds:
Biological components: population - *Aphis pomi* (apple aphid)

## Model purpose
To describe the population dynamics of *Aphis pomi* on apple trees under varying environmental conditions.

## Short description
*A. pomi* abundance is reprocessed as presence or absence on each of 50-100 patches (trees). Changes of the state of a patch are described by a 2 x 2 transition matrix. Transition probabilities are functions of *A. pomi* abundance, temperature and daylength. Inputs are initial *A. pomi* abundance and observed temperatures for one season. Outputs are *A. pomi* abundance (as the fraction of trees with aphids present) for one season.

## Model application(s) and State of development
Application area(s):
Number of cases studied: 1
State: calibration

## Model software and hardware demands
Platform: PC-based
OS: MS-DOS
Software demands: Compiler, Pascal
Hardware demands: none

## Model availability
Purchase: no, sourcecode
Contact person(s) Dr. M.E.J. Woolhouse
Department of Zoology
University of Oxford
South Parks Rd.
Oxford OX1 3PS
United Kingdom

Phone: +44 865 271244
Fax: +44 865 310447
E-mail: mejw@vax.ox.ac.uk

## Model documentation and major references
Woolhouse, M.E.J. and Harmsen, R., 1991. Population dynamics of *Aphis pomi*: a transition matrix approach. *Ecol. Modelling*, 55: 103-111.

## Other relevant information
Parameter estimates are obtained by fitting a log-linear model using the SAS package to observed data on *A. pomi* abundance, temperature and 'daylength.

## Model identification
Model name/title: Transition Matrix Model for ERM

## Model type
Media: local scale, agricultural
Class:
Substance(s)/compounds: pesticides
Biological components: populations, *Panonychus ulmi*

## Model purpose
Forecast the abundance of European red mite in a managed apple orchard.

## Short description
Model has transition matrix format, considering a number of patches (trees) ascribed to states according to high/low abundance of motile mites and high/low abundance of mite eggs. Changes in states are functions of overall mite abundance and temperature. Time step is 14 days. Inputs are initial numbers of trees with high/low mite and egg abundances, plus mean air temperatures (observed or forecast) for successive 14-days intervals. Outputs are fraction of trees with high mite and or egg abundance. 90% or higher trees with high abundance represent risk of exceeding economic threshold. Calibration curves allow transation of pesticide-induced mortality to changes in state frequencies.

## Model application(s) and State of development
Application area(s): Apple orchards in southern Ontario
Number of cases studied: 2
State: validation

## Model software and hardware demands
Platform: Mainframe
OS: VAX
Software demands: Pascal compiler, Pascal
Hardware demands:

## Model availability
Purchase: no
Contact person(s) Dr. M.E.J. Woolhouse
        Department of Zoology
        University of Oxford
        South Parks Rd.
        Oxford OX1 3PS
        United Kingdom

Phone: +1 865 271244
Fax: +1 865 310447
E-mail: mejw@vax.ox.ac.uk

## Model documentation and major references
Woolhouse, M.E.J. and Harmsen, R., 1989. *Ecol. Modelling*, 46: 269-282.

624

---

**Other relevant information**

---

**Model identification**
Model name/title: Numerical Simulation on the Dispersion of Swimming Crab Megaropae
in Hiuchi-Nada

---

**Model type**
Media: water, estuary, local scale
Class: biogeochemical
Substance(s)/compunds:
Biological components: community - juvenile of crab

---

**Model purpose**

---

**Short description**
A numerical simulation on the dispersion of swimming crab (*Portunus trituberculatus
MEIRS*) megaropae is carried out in Hiuchi-Nada, the central part of the Seto Inland Sea.
The calculated dispersal pattern of megaropae in the 3-D numerical model well reproduces
the observed one. The vertical migration of swimming crab megaropae plays an important
role in their return from off-shore to the near-shore region.

---

**Model application(s) and State of development**
Application area(s):
Number of cases studied:
State: verification

---

**Model software and hardware demands**
Platform: FACOM
OS:
Software demands: FORTRAN
Hardware demands:

---

**Model availability**
Purchase:
Contact person(s) Dr. T. Yanagi
        Department of Ocean Engineering
        Ehime University
        Bunkyo 3
        Matsuyama 790
        Japan

Phone:
Fax: 81 899 21 5852
E-mail:

---

**Model documentation and major references**
*Fisheries Engineering*, Vol. 30 No. 1 pp. 15-22. 1993.

---

**Other relevant information**

---

**Model identification**
Model name/title: Numerical simulation of disperal patterns of red sea bream juveniles, *Pagrus major*, in Nyuzu Bay, Japan

---

**Model type**
Media: water, estaury, local scale
Class: biogeochemical
Substance(s)/compunds: organic compounds
Biological components: community - red sea bream juvenile

---

**Model purpose**

---

**Short description**
A numerical simulation of the experimental stocking of red sea bream, *Pagrus major*, was tested. Three groups of juveniles of 20-40 mm T.L. were released at two stations in Nyuzu Bay in the Bungo Channel, Japan, on July 5. 1989. At the entrance of the Bay, 78,800 of 40 mm fish (G40M) and 130,900 of 20 mm fish (G20M) were released; at the head of the Bay, 144,200 of 20 mm fish (G20H) were released. Samples were collected for three days after release. The recapture percentage of the G40M group (0.69%) was about 35 times larger than that of G20M (0.024%). Recapture of the G20H group occurred only near the point of release and its recapture rate (0.10%) was about 5 times larger than that of G20M. The numerical simulation incorporated the flow field and behavioral characteristics of red sea bream juveniles. The calculated dispersal patterns of the three fish groups were consistent with the observed patterns.

---

**Model application(s) and State of development**
Application area(s): Seto Inland Sea, Japan
Number of cases studied:
State: verification

---

**Model software and hardware demands**
Platform: FACOM
OS:
Software demands: FORTRAN
Hardware demands:

**Model availability**
Purchase:
Contact person(s) Dr. T. Yanagi
           Department of Ocean Engineering
           Ehime University
           Bunkyo 3
           Matsuyama 790
           Japan
Phone:
Fax: 81 899 21 5852
E-mail:

**Model documentation and major references**
*Journal of Marine Systems*, 3 (1992) 477-487.

**Other relevant information**

**Model identification**
Model name/title: Transition Matrix Model of Acarid prey-predator dynamics

**Model type**
Media: terrestrial, agricutural
Class:
Substance(s)/compounds:
Biological components: ecosystem - agro-ecosystem - apple orchard, community - Acarine subcommunity of the apple phylloplane anthropod community, population - of for mitetaxa: Phytoseiids, Stigmaeids, Tetranychids and Eryophyids.

**Model purpose**
Forecasting relative abundance of two pests and two predator mite species in an apple orchard through one season.

**Short description**
State variables are presence/absence of each of four mite species on leaf samples from a number of apple trees.
Transition matrix specifies transition between states through times as functions of current state, overall abundance, and temperature.
Inputs are initial presence/absence data early in the season and air temperature data, as degree days over 14 day intervals.
Outputs are presence/absence frequencies for each mite species.

---

**Model application(s) and State of development**
Application area(s): Pest outbreak forecasting
Number of cases studied: 1
State: validation, validation of prognosis

---

**Model software and hardware demands**
Platform: Mainframe
OS: VAX
Software demands: Pascal compiler, Pascal
Hardware demands:

---

**Model availability**
Purchase: no, source code
Contact person(s) Dr. M.E.J. Woolhouse
              Deptartment of Zoology
              University of Oxford
              South Parks Rd.
              Oxford OX1 3PS
              United Kingdom

Phone: +44 9 865 271244
Fax: +44 9 865 310447
E-mail: mejw@vax.ox.ac.uk

---

**Model documentation and major references**
Woolohouse, M.E.J. and Harmsen, R., 1987. *Ecol. Modelling*, 39: 307-323.

---

**Other relevant information**

**Model identification**
Model name/title: APHID-PAR-PRED

**Model type**
Media: agricultural
Class:
Substance(s)/compounds:
Biological components: population - aphids, their parasites and predators

**Model purpose**
Optimization of aphid biocontrol.

**Short description**
Empirical data on trophic interactions between the aphid predator *Metasyrphus corollae,* the aphid parasite *Diaeretiella rapae* and the aphids *Brevicoryne brassicae* and *Myzus persicae* were collected. A skewed curve seemed to express the development of parasitism in the absence of the predator. An increase in the percentage of parasitized aphids of both species occured at the time of intense feeding by the predator released into the system. A simple mathematical model for this situation was developed. The simulation of interactions between aphids and the parasite corresponded well to the empirically obtained curve. Application of the model to a range of situations was investigated by computer simulations. It is practical value for further considerations and predictions in biological control is demonstrated.

**Model application(s) and State of development**
Application area(s): biocontrol
Number of cases studied: 2
State: validation, prognosis made

**Model software and hardware demands**
Platform: PC-based
OS: MS-DOS
Software demands: BASIC
Hardware demands: 640 kb RAM

**Model availability**
Purchase: yes, shipping and handling (US $10)
Contact person(s) Dr. P. Kindlmann
        Laboratory of Biomathematics
        Czechoslovak Academy of Science
        Branisovska 31
        37005 Ceske Budejovice
        Tjekiet

**Model documentation and major references**
*Ecol. Modelling*, 61 (1992): 253-265.

**Other relevant information**

**Model identification**
Model name/title: Crop Water BAlance Simulation Model

**Model type**
Media: water, local scale, agricultural
Class: hydrology
Substance(s)/compounds:
Biological components: ecosystem - agricultural crop

**Model purpose**
The model aims at the best estimate possible of the water balance components with the simplest formulation and the minimum set of input data. It is suited to agrometeorological applications such as rainfed crop management, yield prediction or sowing date determination.

**Short description**
The model works on the basis of a current process with a time step of one day. It simulates the temporal evolution of the different components of water balance. The climatic inputs are daily rainfall and potential evapotranspiration. Soil input is maximum available moisture. Crop inputs are crop coefficients and the corresponding duration of each phenological stage.

**Model application(s) and State of development**
Application area(s): agriculture
Number of cases studied:
State: validation completed

**Model software and hardware demands**
Platform:
OS:
Software demands: FORTRAN
Hardware demands:

## Model availability
Purchase: The model can be obtained on request, free of charge. The software is available as executable code.
Contact person(s) Dr. Jean-Paul Lhomme
     B.P. 5045
     34032 Montpellier Cedex
     France

Phone: 67 61 75 23
Fax: 67 41 18 06
E-mail: lhommej@orstom.orstom.fr

## Model documentation and major references


## Other relevant information


## Model identification
Model name/title: Seasonal changes of Drone numbers in a colony of the honeybee, *Apis mellifare*.

## Model type
Media: terrestrial
Class:
Substance(s)/compounds:
Biological components: society - Honeybee, community - Honeybee colony, population - Drone, organism - *Apis mellifare*

## Model purpose
The purpose of the model was to predict the number of both immature and adult drones in a honeybee colony during a year. This allows the effects of drone brood destruction and adult drone eviction to be investigated.

## Short description
The model is based on the seasonal variation in worker honeybee numbers where the rate lay of eggs by the queen was shown to approximate to a skewed normal distribution curve, (McLellan et al., 1980). This is the primary parameter and input to the model. Biological data analysis shows that drone egg production is 24% of that of workers, however, this is suppressed until sufficient workers are available to forage and nurse the immature stages; chosen as 9500 workers. Egg, larvae and pupae drone numbers are predicted by applying mean developmental times (3, 7 and 15 days) and mortality factors (15.7%, 28.8% and 9.1%) to the eggs laid. Adult numbers are calculated by applying adult longevity figures to those leaving the pupal stage.

The longevity switches from 35 days for eggs laid before June to 75 days for those laid later in the year. The results show that the number of drone adult peaks at the time most likely for colony swarming to occur and are only tolerated in order to spread the genes of the colony.

## Model application(s) and State of development

Application area(s): results from the model have been validated against a significant number of biological observations. The comparisons show that the model not only produces drone brood and adults in similar numbers but also at similar times to nature. The model can be used to investigate the effects of drone brood destruction and adult eviction.

Number of cases studied:

State:

## Model software and hardware demands

Platform: The model is small and consists of around 25 lines of code. The demands are therefore quite small. A PC should be more than sufficient to run the model.

OS: RSTS, VMS EMAS

Software demands: language: BASIC, BASIC IMP

Hardware demands:

## Model availability

Purchase: The model is no longer available in coded form, however the model can be constructed simply from the Model Prerequisites section and the 13 equations published.

Contact person(s) Dr. Chris Rowland
        Institute for Animal Health
        Pirbright Laboratory
        Ash Road, Pirbright
        Woking, Gu24 0NF
        United Kingdom

Phone: Worplesdon +44 483 232441

Fax: +44 483 232448

E-mail:

## Model documentation and major references

McLellan et al., 1980. A monogynous eusocial insect worker population model with particular reference to honeybee. *Insect Soc.*, 27: 305-311.

McLellan and Rowland, 1986. A honeybee colony swarming model. *Ecol. Modelling*, 33: 137-148.

## Other relevant information

## Model identification

Model name/title: Simulation of the Life-History and Time-Energy-Budget of the European Robin (*Aves: Erithcus rubecula*) During the Breeding Season

## Model type
Media: forest, local scale
Class:
Substance(s)/compounds:
Biological components: ecosystem - a beech forest ecosystem (*Asperulofagelum*) population -
European robin (*Erithacus rubecula*)

## Model purpose
The purpose of the model is the investigation of the different factors governing the
population dynamics of the European robin (*Erithacus rubecula*). Special emphasis was put
on the influence of temperature, food availability, and energy constants on the start of
breeding and clutch size.

## Short description
The model is individual based with a single robin represented as a data object in the
computer. Each object has a set of individual (changeable) parameters such as location,
weight, energetic condition, ownership of territory, etc., and a set of common rules
('behavioral patterns') determing how these parameters can be changed. The main factors
influencing the choice of a behaviour sequence are
- The vegetation structure of the location. This is realized through a grid map based on GIS.
- The weather. Data from the years of the field study or scenario data are used.
- The internal condition of the modeled individual robin. This comprises energy demand,
phase of breeding cycle and a parameter describing execution urgency for each possible
activity.
 - Availability of food.
 - Interactions with other robins or predators.
Every few minutes (simulation time) individual parameters are actualized and it is decided
which activity pattern is to perform next.
Data and knowledge used to develop the model were obtained in six years of field studies
in the context of the Project 'Ecosystems Research of the Bornhöved Lakes Region'.
With this model an integration of abiotic influences, physiology and ethology is achieved.
The population dynamics as well as the status of the single robin such as weight, energy
demand, preferred locations, territories, time budgets, etc., is represented by the model.
This enables for instance its applicability for predictions in global change scenarios.
The structure is highly modular allowing easy adaption to other avian species or even
small mammals.

## Model application(s) and State of development
Application area(s): the model is applied for verifying hypothesis concerning factors
governing the breeding success of the European robin and is used to make prognosis for
global change scenarios.
Number of cases studied:
State:

## Model software and hardware demands
Platform: The model is written in SIMULA and runs with a SIMULA COmpiler on a Sparc 19/41 workstation under Solaris 2.3. With CIM (a SIMULA to C translator) it runs on DEC-ALPHA under OSF1. Due to the high standardisation of SIMULA the model should run on all workstations for which SIMULA or CIM is available.
OS:
Software demands:
Hardware demands:

## Model availability
Purchase: The model cannot be purchased. Within scientific exchange the source code may be obtained from the authors.
Contact person(s)  Dr. Hauke Reuter
                Projektzentrum Ökosystemforschung
                Universität Kiel
                Schauenburgerstr. 112
                D-24118 Kiel
                Germany

Phone: 0431 880 4029
Fax: 0431 880 4083
E-mail: hauke@pz-oekosys.uni-kiel.d400.de

## Model documentation and major references
Goldstein, D.L., 1988. Estimates of daily expenditures in birds. The time-energy-budget as an integrator of laboratory and field studies, *Am. Zool.*, 28: 829-844.
Gratjetzky, B., 1992. Nahrung und Brulverhalten von Rothehlchenweibchen *Erithacus rubecula* einer schleswig-holsteinischen Knicklandschaft, *Die Vogelwelt*, 113: 282-288.
Gratjetzky, B., 1993. Nahrungsökologie adulter Rotkehlchen *Erthacus rubecule* einer scleswig-holsteinischen Knicklandschaft, *J.Orn.*, 122: 13-22.
Pätzold, R., 1982. *Das Rotkehlchen*, Ziemsen Verlag, Wittenberg.
Reuter, H., 1994. Die Simulation einer Vogelpopulatio: Ein individueorientierles Modell der Rotkehlchen des Belauer Buchenwaldes, Diplomarbeit, Universität Bremen.
Tatner, P., and Bryant, D.M., 1986. Flight cost of a small passerine measured using doubly labeled water: implications for energetics studies, *AUK*, 102: 169-180.

## Other relevant information

## Model identification
Model name/title: Seasonal growth with NGT

## Model type
Media:
Class:
Substance(s)/compounds:
Biological components: population - fish and aquatic invertebrates

## Model purpose
To describe the seasonally oscillating growth of fish and aquatic invertebrates, including cases where growth stops entirely for longer periods.

## Short description
The model used here is a modification of the von Bertalanffy growth function,

Lt = L(1-exp-(K(t-t0) + A))                    (1)

where: Lt is the predicted length at age t; L(uendelig) is the asymptote length; K is the rate at which L(uendelig) is approached; t0 is the predicted "age" at zero length; and A allows for seasonal oscillations.

Two cases can be considered, without and with no-growth times (NGT), i.e., without longer periods for which dL/dt=0. In the former case, we have:

A = S sin 2 pi (t-ts) - S sin 2 pi (t0-ts)                    (2)

where S= CK/ 2 pi, where C expresses the relative amplitude of the seasonal oscillations, which may range from none (C=0) to one time each year when dL/dt= 0 (when C=1). The parameter ts defines the start of the convex segment of a sinusoid oscillation with respect to t=0.

For growth with annual NGT, we have

A= K/Q[sin Q (t'-ts) - sin Q (t0-ts)                    (3)

where Q= 2 pi/ (1-NGT) , and where t' is obtained by substracting from the real age (t) the total no- growth time occuring up to age t.

In (3), the symbol for the parameter C has disappeared, but the curve still behaves as if C=1, thus enabling smooth transitions between growth and no growth periods. Seasonal growth itself is described by a sine wave of period 1-NGT; the units of K is (1-NGT) i minus 1. instead of year$^{-1}$ (in which case one should refer to K' instead of K.

The data to be fitted using the model consist of length-at-age data taken e.g., at monthly intervals over several years, and displaying seasonal growth oscillations.

A software package is available on a 3½' diskette for MSDOS computer; it will fit equations (1-3) to appropriate data and is distributed with a short manual.

## Model application(s) and State of development
Application area(s):
Number of cases studied: approx. 20
State: can be used routinely

**Model software and hardware demands**
Platform: PC-based
OS: MS DOS
Software demands: Basic compiler, language: BASIC
Hardware demands:

**Model availability**
Purchase: yes, price still to be determined, probably about US $15. Software available as executable code.
Contact person(s)  Dr. Felimon Gayanilo, Jr.
                    M.C. P.O. BOX 2631
                    0718 Makati, Metro Manila
                    Philippines

Phone: +63 2 818 0466
Fax: +63 2 816 3183
E-mail:

**Model documentation and major references**
Panly, D., Soriano-Bortz, M., Moreau, J., and Jarre-Teichmann, A., 1992. A new model accounting for seasonal cessation of growth in fishes. *Anstr. J. Mer. Freshwat. Res.*, 43: 1151-1156.

**Other relevant information**
Software also includes a two phase growth model described in Soriano, M., Moreau, J., Hoenig, J.M., and Panly, D., 1992. New functions for the analysis of two-phase growth of juveniles and adult fishes, with applications to Nile Perch. *Trans. Amer. Fish. Soc.*, 121: 486-493.

**Model identification**
Model name/title: WCRPOP: Western corn rootworm population dynamics simulation model.

**Model type**
Media: terrestrial, local scale, soil, agricultural
Class:
Substance(s)/compounds:
Biological components: populations - Western Corn Rootworm

**Model purpose**
The model simulates the dynamics of a univoltine insect pest of corn in the Midwestern United States, *Diabrotica virgifera virgifera* (Coleoptera:Chrysomelidae), from the beginning of postdiapause egg development in spring to death of adults the following autumn.

## Short description

The models state variables are: (1) Nijk(x), number of corn rootworms per m² on day x in life stage i, sex j, and cohort k; (2) E(x), number of eggs per m² laid by female corn rootworm beetles by day x; and (3) P(x), corn plant growth stage. The main forcing function, r ij (t), gives the development (growth) rate of corn rootworms in stage i and sex j as a function of temperature t. Data inputs required to execute the model are (1) minimum and maximum daily air temperatures in the corn field at 1 m height; (2) the number of live corn rootworm eggs per m² at the beginning of post-diapause development; and (3) the date in spring that the field was planted to corn.

## Model application(s) and State of development

Application area(s): Agricultural pest management.
Number of cases studied: 8
State: validation

## Model software and hardware demands

Platform: PC-based
OS: MS-DOS
Software demands: Fortran compiler, language: FORTRAN
Hardware demands:

## Model availability

Purchase: no, software available both as executable and source code
Contact person(s) Dr. Norman C. Elliott
USDA, ARS, 1301 N. Western St.
Stillwater, OK 74075 USA

Phone: +1 405-624-4244
Fax: +1 405-372-1398
E-mail:

## Model documentation and major references

Elliott, N.C. and Hein, G.L., 1991. Population dynamics of the western corn rootworm: formulation, validation, and analysis of a simulation model. *Ecol. Modelling*, 59:93-122.

## Other relevant information

## Model identification

Model name/title: "Drosophila" as a model in conservation genetics.

## Model type
Media:
Class:
Substance(s)/compounds:
Biological components: populations -  all outbreeding species

## Model purpose
To critically evaluate theory underlying recommendations for the genetic management of endangered species, and to investigate problems in the field using a convenient laboratory species.

## Short description
We use the convenient laboratory organism, *Drosophila melanogaster*, to model problems in conservation genetics because of its short generation interval, low cost, the detailed knowledge of its genetics, availability of innumerable stocks to aid genetic analysis, and its extensive use as a model for related problems in animal breeding, population genetics and quantitative genetics. DROSOPHILA has proven to be a reliable model for naturally outbreeding diploid eukaryotes. I am aware of no case where DROSOPHILA experiments have yielded results at variance with equivalent experiments with other such species.
There is considerable theory that has been used in the genetic management of rare and endangered species, but much had not been experimentally evaluated. However, this theory is typically simple single locus neutralist theory that ignores natural selection, mutation and linkage. Endangered species are unsuitable for evaluating such a theory as they are typically slow breeders, expensive to maintain, and present in low numbers. We have completed studies on the effects of variance in family size, harems, fluctuations in population size and equalisation of founder representation on inbreeding, reproductive fitness and genetic variation. Further, beneficial effects of immigration on the reproductive fitness of small partially inbreed populations has been documented, and rapid genetic adaptation to captivity reported.

## Model application(s) and State of development
Application area(s): Conservation genetics
Number of cases studied:
State:

## Model software and hardware demands
Platform:
OS:
Software demands:
Hardware demands:

## Model availability
Purchase: no.
Contact person(s) Dr. R. Frankham
          School of Biological Sciences
          Macquarie University
          Sydney, New South Wales 2109
          Australia

Phone: 612 850 8186
Fax: 612 850 8245
E-mail: frankham@laurel.ocs.mq.edu.au

## Model documentation and major references
Borlase, S.C., Loebel, D.A., Frankham, R., Nurthen, R.K., Briscoe, D.A., and Daggard, G.E., 1993. Modeling problems in conservation genetics using captive "Drosophila" populations: Consequences of equalizing family sizes. *Conservation Biology*, 7: 122-131.

Briton, J., Nurthen, R.K., Briscoe, D.A., and Frankham, R., 1994. Modeling problems in conservation genetics using captive "Drosophila" populations: Consequences of harems. *Biological Conservation*, 69: 267-275.

Frankham, R., and Loebel, D.A., 1992. Modeling problems in conservation genetics using captive "Drosophila" populations: Rapid genetic adaptation to captivity. *Zoo Biology*, 11: 333-342.

Loebel, D.A., Nurthen, R.K., Frankham, R., Briscoe, D.A., and Craven, D., 1992 Modeling problems in conservation genetics using captive "Drosophila" populations: Consequences of equalizing founder representation. *Zoo Biology*, 11: 319-332.

Spielman, D., and Frankham, R., 1992. Modeling problems in conservation genetics using captive "Drosophila" populations: improvement of reproductive fitness due to immigration of one individual into small partially inbreed populations. *Zoo Biology*, 11: 343-351.

Woodworth, L.M., Montgomery, M.E., Nurthen, R.K., Briscoe, D.A., and Frankham, R., 1994. Modeling problems in conservation genetics using "Drosophila": Consequences of fluctuating population sizes. *Molecular Ecology*, 3: 393-399.

## Other relevant information
This work has been praised by prominent workers involved in the captive breeding of endangered species (see Ralls and Meadows, 1993, *Nature*, 361: 689-690.).

## Model identification
Model name/title: Trophic exploitation model for grassland food webs

## Model type
Media: local scale, terrestrial
Class:
Substance(s)/compounds: nutrients, nitrogen, biomass of vegetation and animal matter
Biological components: ecosystem - grassland, community - insects, arachnids, herbs and grasses

## Model purpose
To understand how herbivores exploit plant populations and how plant- herbivore dynamics are mediated by carnivores and nutrient inputs.

## Short description
Model describes trophic dynamics using differential equations that describe population sizes of plants, herbivores and carnivores as a function of time and parameters describing rates of resource uptake and allocation to survival and reproduction. The predictions are wholy qualitative, and allow general analytical output to describe trophic dynamics of grassland systems.

## Model application(s) and State of development
Application area(s): Understanding dynamics of terrestrial food webs
Number of cases studied: 1
State: validation, prognosis made - model explains dynamics

## Model software and hardware demands
Platform: PC-based
OS: DOS
Software demands: Mathematica
Hardware demands:

## Model availability
Purchase: no, in public domain
Contact person(s) Dr. Oswald Schmitz
        Yale Forestry and Environmental Studies
        370 Prospect St.
        New Haven, CT 06511 USA

Phone: +1 203-432-5110
Fax: +1 203-432-3929
E-mail: schmitz@minerva.cis.yale.edu

## Model documentation and major references

## Other relevant information

## Model identification
Model name/title: A Spatially Explicit Ruminant Foraging Model

**Model type**
Media: terrestrial
Class:
Substance(s)/compounds: nutrients
Biological components: organism - *Alces alces*

**Model purpose**
To demonstrate the influence of structure and variability of forage resources in determing the effiency of an individual forager.

**Short description**
The model represents the foraging behaviour of an individual ruminant (in our case *Alces alces*) within a spatially and temporally heterogenous habitat. The input required includes the physical (i.e., size, stride length etc.), physiological (i.e., passage rate, rumination time, etc.), and cognitive ability (i.e., memory, learning, perceptual distance, etc.) of the forager. Also required are nutritional and distributional properties of the forage resources (i.e., cell-wall, lignin, energy, protein, density, etc.). The model is object-oriented and event- driven.

**Model application(s) and State of development**
Application area(s): Boral forests
Number of cases studied:
State: validation

**Model software and hardware demands**
Platform: PC-based - IBM
OS: DOS
Software demands: language - modula- 2
Hardware demands: 286 w/hard drive

**Model availability**
Purchase: no.
Contact person(s) Dr. John H. Roese
        Dept. of Biology
        Lake Superior State University
   Sault Ste. Marie, MI 49783 USA

Phone: +1 906-635-2648
Fax: +1 906-635-2111
E-mail:

**Model documentation and major references**
*Ecol. Modelling*, 57: 133-143.

**Other relevant information**
Currently being re-written in Turbo-Pascal for Windows. This rewrite will include a user-interface, and a group of tools to analyse the outputs.

---

**Model identification**
Model name/title: Steelhead Trout Population -Energetics Model

---

**Model type**
Media: lake
Class: ecological
Substance(s)/compounds:
Biological components: population

---

**Model purpose**
To formalize our understanding of steelhead trout populations in the Great Lakes (Laurentian). We applied the model to estimate lake-wide rates of prey consumption by steelhead on different prey items. The more general goal was to estimate carrying capacity of Great Lakes for salmonines.

---

**Short description**
We developed a bioenergetics model for steelhead *Oncorhynchus mykiss* that simulates growth and consumption by separate life history forms in lakes Michigan and Ontario. We estimated abundances of hatchery and wild smolts during 1975-90 for both lakes based on stocking rates, survival schedules, and discrimination of the proportions of the lake populations that were naturally recruited. We developed an age-structured population model with separate accounting for s run timing (summer, fall, and spring) using proportions of life history stages observed at weirs and estimated adult survival rates. We summarized data on growth, diet, water temperature, and energy contents of predator and prey to model lakewide prey consumption in both lakes during 1975-90. Sensitivity analysis of the population submodel indicated that most sensitive parameters were lake survival and size at stocking.

---

**Model application(s) and State of development**
Application area(s): Lakes Michigan and Ontario (Laurentian Great Lakes)
Number of cases studied:
State: calibration

---

**Model software and hardware demands**
Platform: PC-based
OS: MS-DOS
Software demands: Lotus 1-2-3 and Fish Bioenergetic Model 2, Communications Office, University of Wisconsin, Sea Grant Institute, 1800 University Avenue, Madison, Wisconsin 53706, USA
Hardware demands:

**Model availability**
Purchase: yes, see above (program written in Pascal, sourcecode available).
Contact person(s) Dr. Peter S. Rand
    207 Jarvis Hall
    University of Buffalo
    Buffalo, NY 14260-440 USA

Phone: +1 716-645-3855 or +1 716-645-2088
Fax: +1 716-645-3667
E-mail: psrand@acsu.buffalo.edu

**Model documentation and major references**
Available with Fish Bioenergetic Model 2.
Steelhead Trout Model: Rand, P.S., Stewart, D.J., Seelbach, P.W., Jones. M.L., Wedge, L.R., 1993. Modeling steelhead population energetics in Lakes Michigan and Ontario. *Trans. Am. Fish. Soc.*, 122: 977-1001.

**Other relevant information**

# REFERENCES

Alcamo, J., Shaw, R. and Leen Hordijk (eds.), 1990. The Rains Model of Acidification. Klüwer Academic Publ. and IIASA.

Allen, T.F.H. and Starr, T.B., 1982. Hierarchy: Perspectives for Ecological Complexity. University of Chicago Press, 310 pp.

Andersen, K.P. and Ursin, E., 1977. A multispecies extension to the Beverton and Holt theory of fishing, with account of phosphrus circulation and primary production. Meddr. Danm. Fisk,- og Havunders. N.S., 7: 319-435.

Armstrong, N.E., 1977. Development and Documentation of Mathematical Model for the Paraiba River Basin Study, Vol 2 - DOSAGM: Simulation of Water Quality in Streams and Estuaries. Technical Report CRWR-145. Center for Research in Water Resources, The University of Texas at Austin.

Arp, P.A., 1983. Modelling the effects of acid precipitation on soil leachates: A simple approach. Ecol. Modelling, 19: 105-117.

Beverton, R.J.H. and Holt, S.J., 1957. On the dynamics of exploited fish populations. Fishery Invest., Ser. 2, 19: 1-533.

Beyer, J.E., 1981. Aquatic Ecosystems - An Operational Research Approach. Univ. Wash. Press, Seattle and London, 315 pp.

Beyer, J. and Sparre, P., 1983. Modelling expolited marine fish stocks. In S.E. Jørgensen (ed.), Application of Ecological Modelling in Environmental Management, Part A. Elsevier Scientific Publishing Company, Amsterdam.

Dobbins, W.E., 1964. BOD and oxygen relationship in streams. J. San. Eng. Div., Proc. ASCE 90, SA 3,: 53.

Eckenfelder, W.W., Jr., 1970. Walter Quality Engineering for Practicing Engineering. Barnes and Noble, Inc., New York. Eckenfelder, W.W., Jr., and O'Connor, D.J., 1961. Biological Waste Treatment. Pergamon Press, New York.

Edwards, R.W. and Rolley, H.L.J., 1965. Oxygen Consumption of River Muds. J. Ecol., 53:1.

EMEP/CCC, 1984. Summary Report 2/84, Norwegian Institute for Air Research, Lillestrøm, Norway.

Fair, G.M., Moore, E.W. and Thomas, H.A., Jr., 1941. The natural purification of river muds and pollutional sediments. Sewage Works Journal, 13: 270.

Hansen, H.H. and Frankel, R.J., 1965. Economic Evaluation of Water Quality. A Mathematical Model of Dissolved Oxygen Concentration in Freshwater Streams. Second Annual Report, Sanitary Engineering Laboratory Report No. 65-11. Sanitary Engineering Research Laboratory, University of California, Berkeley.

Hansen, S. and Aslyng, H.C., 1984. Nitrogen balance in crop production simulation model NITCROS. Hydrotechnical Laboratory, RVAU, Copenhagen, 113 pp.

Hansen, S., Jensen, H.E., Nielsen, N.E. and Svendsen, H., 1990. DAISY - Soil Plant Atmosphere System Model, NPO-forskning fra Miljøstyrelsen Nr. A10. Miljøstyrelsen, Copenhagen.

Henriksen, A., 1980. Acidification of freshwaters - large scale titration. In D. Drabløs and A. Tollan (eds.), Ecological Impact of Acid Precipitation, SNSF-project, pp. 68-74.

Holling, C.S., 1959. Some characteristics of simple types of predation and parasitism. Canad. Entomol., 91: 385-398.

Holling, C.S., 1966. The functional response of invertebrate predators to prey density. Mem. Entomol. Soc. Canada, 48: 1-87.

Jørgensen, S.E., 1976. A eutrophication model for a lake. Ecol. Modelling, 2: 147-165.

Jørgensen, S.E., Kamp-Nielsen, L., Christensen, T., Windolf-Nielsen, J. and Westergaard, B., 1986. Validation of a prognosis based upon a eutrophication model. Ecol. Modelling, 72: 165-182.

Jørgensen, S.E., Hoffmann, C.C. and Mitsch, W.J., 1988. Modelling nutrient retention by a reed-swamp and wet meadow in Denmark. In W.J. Mitsch., M. Straskraba and S.E. Jørgensen, (eds.), Wetland Modelling. Elsevier, Amsterdam, pp. 133-151.

Jørgensen, S.E., 1991. Modelling in Environmental Chemistry, Developments in Environmental Modelling, 17. Elsevier, Amsterdam, 505 pp.

Jørgensen, S.E., 1994. Fundamentals of Ecological Modelling (2nd Edition). Elsevier, Amsterdam, London, New York, Tokyo.

Mesarovic, M.D., Macko, D. and Takahara, Y., 1970. Theory of Hierarchical Multilevel Systems. Academic Press, New York.

Mogensen, B., and Jørgensen, S.E., 1979. Modelling the distribution of chromium in a Danish firth. In S.E. Jørgensen (ed.), Proceedings of 1st International Conference on Stated of the Art in Ecological Modelling, Copenhagen, 1978. International Society for Ecological Modelling, Copenhagen, pp. 367-377.

Nyholm, N., 1976. Kinetics studies of phosphate-limited algae growth. Thesis, Technical University of Copenhagen.

O'Connell, R.L. and Thomas, N.A., 1965. Effects of benthic algae on stream dissolved oxygen. Journal of Sanitary Engineering Division, Proceedings ASCE 91, SA 3:1

O'Connor, D.J., 1962. The Effect of Stream Flow on Waste Assimilation Capacity. Proceedings of 17th Purdue Industrial Waste Conference, Lafayette.

O'Connor, D.J. and DiTorro, D.J., 1970. Photosynthesis of oxygen balance in streams. Journal of Sanitary Engineering Division, Proceedings ASCE 96, SA 2: 547.

O'Neill, R.V., DeAngelis, D.L., Waide, J.B. and Allen, T.F.H., 1986. A Hierarchical Concept of Ecosystems. Princeton University Press.

Overton, W.S., 1972. Toward a general model structure for forest ecosystems. In J.F. Franklin, Proc. Symp. on Research on Coniferous Forest Ecosystems Ecosystems. Northwest Forest Range Station, Portland.

Overton, W.S., 1974. Decomposability: a unifying concept? In S.A. Levin (ed.), Ecosystem Analysis and Prediction. Society for Industrial and Applied Mathematics, Philadelphia, pp. 297-298.

Pattee, H.H., 1969. Physical conditions for primitive functional hierarchies. In L.L. Whyte, A.G. Wilson and D. Wilson (eds.), Hiearchical Structures. Elsevier, New York, pp. 161-177.

Pattee, H.H., 1972. The evolution of self-simplifying systems. In E. Lazlo (ed.), the Relevance of General Systems Theory. Braziller, New York, pp. 31-42.

Patten, B.C., 1978. Systems approach to the concept of the environment. Ohio J. Sci., 78: 206-222.

Patten, B.C., 1982a. Indirect causality in ecosystem: its significance for environmental protection. In W.T. Mason and S. Iker, Research on Fish and Wildlife Habitat, Commemorative monograph honoring the first decade of the US Environmental Protection Agency, EPA-600/8-82-022. Office of Research and Development, US Env. Prot. Agency, Washington, D.C.

Patten, B.C., 1982b. Environs: relativistic elementary particles for ecology. Amer. Nat., 119: 179-219.

Patten, B.C., 1985. Energy cycling in the ecosystem. Ecol. Modelling, 28: 7-71.

Pielou, E.C., 1977. An introduction to mathematical ecology. Wiley-Interscience, New York. 385 pp.

Raytheon Company, Oceanographic Environmental Services, 1973. REBAM - A Mathematical Model of Water Quality for the Beaver River Basin. U.S. Environmental Protection Agency, Washington, D.C.

References

Reuss, J.O., 1983. Implications on the Ca-Al exchange system for the effect of acid precipitation on soils. J. Env. Quality, 12: 15-38.

Ricker, W.E., 1954. Stock and recruitment. J. Fish. Res. Board Canada, 11: 559-623.

Simon, H.A., 1962. The architecture of complexity. Proc. Amer. Phil. Soc., 106: 467-482.

Simon, H.A., 1969. The Sciences of the Artificial . MIT Press, Cambridge, Massachusetts.

Simon, H.A., 1973. The organization of complex systems. In H.H. Pattee, Hierarchy Theory. Braziller, New York, pp. 3-27.

Smith, F.E., 1963. Population dynamics in Daphina magna and a new model for population growth. Ecology, 44: 651-663.

Texas Water Development Board, 1970. DOSAG-I Simulation of Water Quality in Streams and Canals. Program Documentation and User's Manual. Prepared by systems engineering Division.

Thomann, R.V., Di Toro, D.M. and O'Connor, D.J., 1974. Preliminary model Potomac estuary phytoplankton. ASCE, J. Environ. Eng. Div., 100, EE1: 699-715.

Thomas, A.H., Jr., 1961. The dissolved Oxygen Balance in Streams. Proceedings, Seminar on Waste Water Treatment and Disposal, Boston Society of Civil Engineering.

Turner, D.B., 1970. Workbook of Atmospheric Dispersion Estimates. U.S. Public health Service Publication 999-AP-26, revised 1970 ed.

Ursin, E., 1967. A mathematical model of some aspects of fish growth, respiration and mortality. J. Fish. Res. Bd. Can. 13: 2355-2453.

Ursin, E., 1979. On multispecies fish stock and yield assessment in ICES. A Workshop on multispecies approaches to fisheries management advice. St. John's, November 1979.

Vollenweider, R.A., 1965. Calculation models of photosynthesis-depth curves and some implications regarding day rate estimates in primary production. Memorie dell Istituto Italiano di Idrobiologia, 18 Suppl.: 425-457.

Water Resources Engineers Inc., 1973. Computer Program Documentation for the Stream Quality Model QUAL-II. Prepared for U.S. Environmental Protection Agency, Systems Analysis Branch, Washington, D.C.

# INDEX

667